Material Science for Engineers

Material Science for Engineers

Dr. Aparna Gupta Ph.D.

Professor of Physics,
Birla Institute of Technology & Science,
Pilani - 333031 (Rajasthan), India

Dr. Santosh Kumar Ph.D.

Professor of Physics,
Head, Applied Sciences and Humanities,
NC College of Engineering,
Israna, Panipat (Haryana), India

CBS

CBS Publishers & Distributors Pvt. Ltd.

New Delhi • Bengaluru • Chennai • Kochi • Kolkata • Mumbai
Hyderabad • Nagpur • Patna • Pune • Vijayawada

Dedicated To Our Parents

ISBN: 81-239-0992-6

First Edition: 2004
Reprint: 2008, 2011, 2012, 2016

Published by:
Satish Kumar Jain for CBS Publishers & Distributors Pvt. Ltd.,
4819/XI Prahlad Street, 24 Ansari Road, Daryaganj, New Delhi - 110002
delhi@cbspd.com, cbspubs@airtelmail.in • www.cbspd.com
Ph.: 23289259, 23266861, 23266867 • Fax: 011-23243014

Corporate Office: 204 FIE, Industrial Area, Patparganj, Delhi - 110 092
Ph: 49344934 • Fax: 011-49344935
E-mail: publishing@cbspd.com • publicity@cbspd.com

Branches:
• *Bengaluru:* 2975, 17th Cross, K.R. Road, Bansankari 2nd Stage,
  Bengaluru - 70 • Ph: +91-80-26771678/79 • Fax: +91-80-26771680
  E-mail: cbsbng@gmail.com, bangalore@cbspd.com
• *Chennai:* No. 7, Subbaraya Street, Shenoy Nagar, Chennai - 600030
  Ph: +91-44-26681266, 26680620 • Fax: +91-44-42032115
  E-mail: chennai@cbspd.com
• *Kochi:* Ashana House, 39/1904, A.M. Thomas Road, Valanjambalam,
  Ernakulum, Kochi • Ph: +91-484-4059061-65
  Fax: +91-484-4059065 • E-mail: cochin@cbspd.com
• *Kolkata:* 6-B, Ground Floor, Rameshwar Shaw Road, Kolkata - 700014
  Ph: +91-33-22891126/7/8 • E-mail: kolkata@cbspd.com
• *Mumbai:* 83-C, Dr. E. Moses Road, Worli, Mumbai - 400018
  Ph: +91-9833017933, 022-24902340/41 • E-mail: mumbai@cbspd.com

Representatives:
• Hyderabad: 0-9885175004   • Nagpur: 0-9021734563
• Patna: 0-9334159340   • Pune: 0-9623451994
• Vijayawada: 0-9000660880

Printed at:
J.S. Offset Printers, Delhi (India)

Foreword

Rapid developments have been witnessed in the field of materials science and technology during the last few decades. The advances in the technology of semiconductors, ceramics, composites and other materials have brought in revolutions in integrated circuits, silicon sensors, and nanomaterials. Newer applications are a direct result of the various advances in materials science and technology, and materials characterization techniques.

The book 'Material Science for Engineers' by Dr. Aparna Gupta and Dr. Santosh Kumar is an effort in bringing out the basics along with the advances in the subject for the benefit of undergraduate students pursuing the engineering degree programmes. The treatment is exhaustive with a number of solved and unsolved examples spread throughout the chapters. The background and long experience of the authors in teaching the related subjects at this institute makes the presentation lucid and suitable even for self-study.

I am sure that the science and engineering students will greatly benefit from this book.

L.K. Maheshwari
Deputy Director and Professor of
Electrical & Electronics Engineering,
Birla Institute of Technology and Science,
Pilani (Rajasthan)

Foreword

Dr. Aparna Gupta and Dr. Santosh Kumar are famous physicists, who have worked at BITS, Pilani for almost 30 years. Their contributions in the area of material science are well documented. Advances in material science, particularly in the last half century, have affected all branches of engineering. Earlier the engineering designs were conservative due to the approximate models used and the uncertainties in the qualities of raw materials available. Now more accurate models are available for predicting the performance of subsystems and systems which permit optimum designs to be made. Advances in material science have ensured that vastly improved materials are now available. Advances in manufacturing of these materials have permitted safety factors to be reduced. The result is that designs have become much more economical. In addition, many products and systems, which are now possible, are due to the availability of new materials.

The science of materials has become important for practically all engineers encompassing all branches of engineering. This is the reason why most of the universities have included material science in the core programmes of their bachelor of engineering curricula. At that stage students enter only with a basic physics background. Therefore, physicists have an intimate knowledge of the level of understanding of students.

The authors have not stopped at the basics alone. They have covered Natural and Synthetic Polymers, Ceramics, Composites, Conducting Materials, Semiconductors, Magnetic Materials and Dielectric Materials also. Thus, this book will serve chemical engineers, mechanical engineers, metallurgists and electrical engineers in their professional years also. Keeping in mind the needs of students entering the bachelor's degree programme, a large number of solved examples and numerical problems have been included in the book. This is a book which will eminently serve as a textbook and must be part of the collection of all libraries.

I hope the authors have a successful launch of this project and wish them a bright future.

Prof. S. Gupta
Principal, N.C. College of Engineering,
Israna - 132107 (Haryana), India
(Formerly, Professor, IIT, Kanpur)

Preface

We are living in an era of advancement of material technology in which knowledge of materials has become essential for all the students majoring in basic as well as applied sciences and in various branches of engineering. There are many books that cover various aspects of material science and engineering. However, we have always felt the need of a textbook which covers all the basic principles in lucid and simple enough manner, with a large number of illustrative examples covering entire course material. These feelings have culminated into writing this book entitled "Material Science for Engineers". Due to long experience of teaching the course 'Structure and Properties of Materials' at the Birla Institute of Science and Technology (BITS), Pilani (Raj.), many new ideas have come up to make the course more illustrative and interesting. All these ideas are incorporated in this book.

The purpose of writing this textbook is to cater to the needs of a vast majority of students of various branches of science and engineering. This is intended for use as a textbook for those students who have had science up to the school stage and are going to get their first exposure of this subject. Materials are discussed in such a manner that the students who have an elementary knowledge of physics and chemistry should find themselves adequately prepared to go through this course. Additional concepts of thermodynamics, physical chemistry, and atomic structure are explained in the places wherever these are required. The examples of varieties of material used in daily life are included under relevant section. A large number of problems and questions are given at the end of each chapter.

The material presented here is more than what can be covered in a one-semester course. The appropriate topics can be selected with an intended coverage suitable for the first-degree level course for the science and engineering discipline. While writing this book, basic objectives have been the following:

- To project overall view of the field of materials science within the framework of science and engineering disciplines.
- To provide a smooth link between the basic knowledge of science pertaining to first/second level and the engineering courses.
- To better prepare those who will have to tackle day-to-day materials problems in professional engineering careers.

A large portion of this book has actually been used for classroom teaching and has been received quite well by the students. This book covers basics, and applications of material science. Topics in this book include crystal structure, crystal imperfections, polymers, diffusion in solids, phase diagrams, kinetics of phase transformation, and elastic behaviour of materials, viscoelasticity, magnetic materials and dielectric materials. This also includes advanced topics like ceramics, composites, semiconductors, fracture, strengthening and deterioration of materials etc.

SI units are used throughout this book. The appendix on SI units describes the basic units, derived units and also conversion factors connecting CGS, MKS and FPS systems. Many illustrations are introduced for understanding of the basic concepts of material science. More than 180 illustrative examples are fully worked out in the text. In addition there are about 335 problems and questions dealing with materials properties, analysis and design. There are 310 figures and illustrations, 70 tables are presented relating the structure, properties and applications of materials.

The authors are very much interested in getting feedback. Comments and suggestions may be sent to

Dr. Aparna Gupta,
Professor of Physics,
Birla Institute of Technology & Science,
Pilani - 333031 (Rajasthan), India

Dr. Santosh Kumar,
Professor of Physics,
Head, Applied Sciences and Humanities,
NC College of Engineering,
Israna, Panipat (Haryana), India

Acknowledgements

The endeavour to produce this book was helped by a large number of people. We have made a humble attempt to convey our gratitude for the range of help received to bring this book in the present form.

Though it is not practical to mention them all, we thankfully acknowledge the individuals, the sources and organizations, which have been consulted during the process of writing this book. We thankfully acknowledge our indebtedness to Prof. S. Venkateswaran, Director, BITS for his encouragement. We also thank Deputy Directors Prof. L.K. Maheshwari, Prof. K.E. Raman and Prof. V.S. Rao, for their help and encouragement. We are indebted to Prof. I.J. Nagrath who has always been a source of inspiration and encouragement for us. We are also thankful to Prof. R.K. Patnaik, Dean, Instruction Division, and Prof. G.P. Srivastava, Dean, Educational Development Division, for their interest in this work. We express our sincere thanks to our colleagues Prof. R. Dutta, Prof. S.K. Sharma, Prof. P.C. Pande, Prof. S.P. Gupta and Dr. (Mrs.) T. Mandal for their valued discussions. We specially acknowledge the help rendered by many of our students—S. Giridharan, Suresh Kumar, Amrita Gupta and Savitri—to name a few, with whom we had useful discussions and this helped us in understanding the students' viewpoint.

We are especially thankful to Mr. Suresh Kumar Saini for his help to prepare the manuscript well. We deeply acknowledge the help rendered by our daughter Amrita Gupta, who has drawn most of the figures in this book. We are thankful to our colleagues Mr. I.V. Singh and Mr. D.M. Kulkarni, of Mechanical Engineering Department for going through the manuscript and giving their comments and suggestions. We are grateful to our children Ambarish, Amrita and Sumeet who have tolerated us for stealing away the time we could have been with them.

Dr. Aparna Gupta,
Professor of Physics,
Birla Institute of Technology & Science,
Pilani - 333031 (Rajasthan), India

Dr. Santosh Kumar,
Professor of Physics,
Head, Applied Sciences and Humanities,
NC College of Engineering,
Israna, Panipat (Haryana), India

Contents

Introduction to Materials

1.1 INTRODUCTION

Materials have played a very important role in the development of science and technology. Starting from a limited number of materials in the early dawn of civilization, human endeavour has created a revolutionary era of material science where thousands of materials are being used for scientific and engineering applications. Numerous materials have been developed with the advent of new technologies. Many of the materials like ceramics, high temperature superconductors have been developed through experimentation while others are tailor-made for specific applications e.g. duralumin alloy 2024-T6 having tensile strength higher than that of aluminium, is found to be more suitable for aircraft bodies than pure aluminium.

Many engineering products have become better with the discovery of new materials. Each engineering product requires specific types of materials which can be processed into final form satisfactorily such that the desired product

  i) has the required structural strength ;
 ii) is economically viable ;
iii) performs as per desired specifications.

Choice of the material also decides the life of the final engineering product.

Material science has developed as a new discipline encompassing basic sciences, engineering sciences, metallurgy, polymer sciences etc. This is the discipline, which deals with structure, properties and applications of materials. Knowledge of structure related properties helps engineers to choose appropriate materials for development of specific engineering products. Various metallurgical processes like heat treatment, strain hardening etc., change the properties of the materials suitably such that the ultimate engineering product becomes more robust and durable. Remarkable progress in material technology has been achieved through the discovery of new materials and their production techniques.

1.2 BROAD CLASSIFICATION OF MATERIALS

The foundation of material science is based on the fundamental principles of physical sciences and empirical laws derived from experimental observations. While discussing various aspects of material science and engineering, we shall particularly refer to materials, which are important for applications in physical sciences and engineering. Also, we shall restrict ourselves mainly to the study of solid materials. This branch of science has been developed to cater to the needs of both engineers and scientists belonging to the fields of metallurgy, ceramics, polymer science etc. Materials are used in engineering products like

i) Devices e.g. transistors, lasers, photoelectric cells.
ii) Goods of everyday use e.g. utensils, clothes, helmets, bicycle tyre etc.
iii) Structures e.g. buildings, dams, bridges, chemical plants etc.
iv) Machines e.g. boilers, lathes, electric motors, generators etc.

Material science is becoming richer, as more and more information is gathered about materials from experiences of scientists and engineers. Engineering materials can be broadly classified into five groups (Fig. 1.1):

i) Metals and alloys;
ii) Semiconductor & device materials;
iii) Ceramics and glasses;
iv) Polymers;
v) Composites;
and   vi) Biomaterials

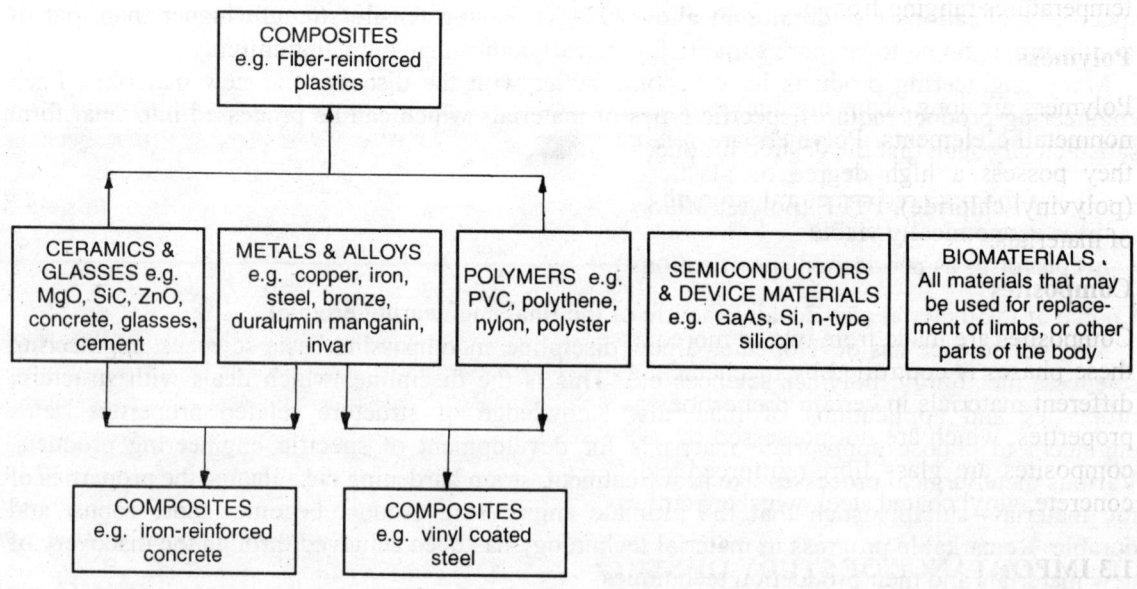

Fig. 1.1 Broad classification of materials

Metals and alloys

Metals have luster and good electrical and thermal conductivity. These properties can be directly attributed to the presence of a large number of nonlocalized electrons. Metals are usually malleable and ductile so that they can be given any shape e.g. aluminium sheets, iron filings, copper vessels, silver foils etc. Alloys are formed from combinations of metals in specific ratios to achieve certain specific properties e.g. stainless steel, which is an alloy of iron, carbon, chromium, phosphorous and manganese has better corrosion resistance than pure iron. Similarly, metallic oxides like $YBa_2Cu_3O_{7-x}$ are superconductors at temperature $T < 100$ °K and have applications in electronic devices. These metallic oxides become insulators at room temperature.

Semiconductors & device materials

Semiconductors have low thermal and electrical conductivity e.g. germanium, silicon, cadmium sulphide, gallium arsenide. Semiconductors have energy gap of the order of eV between the valence band and the conduction band. Depending upon the energy gap and the temperature, a material may behave as an insulator or as a semiconductor or as a conductor. Valence electrons can contribute to conductivity if they occupy partially filled valence band or they are excited to the conduction band. If the energy gap is small, valence electrons can be thermally excited at moderate temperatures to the conduction band, giving rise to conductivity. Semiconductors are widely used in devices like transistors, rectifiers, photoelectric cells etc.

Ceramics and glasses

Ceramics are metallic or nonmetallic oxides, carbides, nitrides or combinations of them. These are normally insulators to the passage of heat and electricity and are usually brittle. Some of these materials are resistant to high temperatures and are used in the fabrication of furnaces. Recent investigations have revealed that a large variety of ceramics behave as superconductors at low temperatures ranging from near 0 °K to 110 °K.

Polymers

Polymers are long chain organic compounds that are composed of carbon, hydrogen and other nonmetallic elements. Polymers are relatively light, inert and can be given any shape because they possess a high degree of plasticity. They are thermal and electrical insulators. PVC (polyvinyl chloride), PTFE (polytetrafluoroethylene) and polyethylene are examples of this class of materials.

Composites

Composites are made from two or more phases of same or different materials. The distribution of these phases is controlled by mechanical as well as by heat treatment processes. By combining different materials in certain proportions and ways, it is possible to obtain new materials having properties, which are not possessed by any of the individual constituent materials. Examples of composites are glass fibre-reinforced plastics, carbon fibre-reinforced rubber, steel-reinforced concrete, vinyl coated steel, whisker-reinforced plastics etc.

1.3 IMPORTANCE OF STUDY OF STRUCTURE OF MATERIALS

Study of the structure of materials reveals how the properties of materials depend on the internal structure. Internal structure refers to the positions of atoms and their constituents i.e. electrons and nucleus, and also to how the atoms coordinate to form phases. Various phases in turn coexist in materials.

Properties of a material depend on the structure of the material at various levels. The study of the structure of materials can be classified as follows:

(a) Macroscopic structure;
(b) Microscopic structure;
(c) Substructure;

(d) Crystal structure;
(e) Atomic structure;
(f) Nuclear structure;

Macroscopic structure

The macroscopic structure of a material can be examined with naked eyes. The external shape of a crystal sometimes depends on the symmetry of arrangement of the constituent atoms e.g. quartz crystal. The limit of resolution of human eye is of the order of 10^{-4}m i.e. human eye can distinguish between two fine lines separated by a distance greater than or equal to 0.1 mm. For finer details, a microscope has to be used.

Microscopic structure

The details of the microstructure of a material can be resolved up to a limit of about 10^{-7}m with the help of an optical microscope. A photomicrograph can achieve linear magnification of 5000 times. The study of microstructure reveals the clustering of crystal grains and phases, the network of boundaries etc. The properties of materials change due to a change in the composition, size or shape of these crystal grains, which form the components of a microstructure.

Substructure

We can obtain a wealth of additional information by studying the substructure with the help of microscopes of higher magnification e.g. a linear magnification of 10^6 can be obtained in the electron microscope. The resolving power of this microscope is more compared to the optical microscope because an electron beam is used in place of visible light. The high resolving power can be explained using the fact that the ultimate limit of resolution depends on the wavelength of the probing beam (wavelength of the visible light is of the order of 5×10^{-7}m whereas that of electron beam is 10^{-12}m). With the help of an electron microscope, the finely distributed phases in grain, dislocations in crystals etc., can be studied in detail. The field ion microscope is another advanced microscope with the help of which location of individual atoms can be observed. Magnification at the centre of the micrograph can be of the order of 10^9. A latest microscope, Scanning Tunneling Microscope (STM), is being used nowadays for studying the position of an atom on the surface of materials.

Crystal structure

Details of the atomic arrangement in a crystal can be studied with the help of x-ray diffraction pattern photograph. It is usually sufficient to describe the arrangement of the atoms in the unit cell of the crystal since crystal structure is a regularly repeating pattern of unit cells in three dimensions.

Atomic structure

If one is interested in even more detailed structure of the solid, one has to investigate the electronic structure of atoms and molecules. Spectroscopic study of atoms and molecules can reveal the electronic structure of the outermost electronic orbits.

Nuclear structure

The study of nuclear structure shows even finer details of atoms than the electronic structure. The nuclear structure can be investigated by using nuclear magnetic resonance technique and also by studying the Mossbauer effect.

1.4 VARIOUS TYPES OF STRUCTURE RELATED PROPERTIES

Structures that are of importance in material science and engineering are microstructure, substructure and crystal structure. These structures are related to the properties and service behavior of the materials. Various properties of materials are briefly discussed in this section. The main purpose of this discussion is to compare certain common properties of different materials and to know how the properties change with changes in the internal structure. Such information will be helpful in selecting materials for a given application. For example, silicon, germanium and compounds of IIIrd and V^{th} group elements or IInd and VIth group elements of the periodic table have the characteristics of semiconductors and are found to be suitable for fabrication of solid state devices like transistors, diodes, radiation detectors etc. The primary criterion used for material selection is proper judgment of properties depending upon the field of application. The properties of interest, in selecting materials may be mechanical properties (e.g. strength, toughness, ductility, hardness), electrical properties (e.g. electrical conductivity), thermal properties (e.g. thermal conductivity, specific heat, free energy), magnetic properties (e.g. magnetic susceptibility, magnetization) and also other properties (e.g. corrosion resistance, specific gravity). Depending upon the application, emphasis is given on the specific properties of the material to be investigated. The dependence of these properties on internal structure is shown schematically in Table 1.1.

Mechanical properties

Mechanical properties are briefly discussed in this section and a detailed discussion will follow in Chapters 14 and 15. In service condition, the mechanical properties of engineering materials play a very important role. While fabricating engineering materials suitable mechanical properties can be attributed to the material by cold working, hot working, heat treatment etc.

Modulus of elasticity is one of the basic mechanical properties of materials. All solid materials undergo changes in shape and size i.e. strain is developed in materials when they are subjected to external force. If the strain is fully recovered i.e. it is reversible when the external force is removed, the material exhibits elastic behavior. According to Hooke's law, stress is directly proportional to strain within the elastic limit. The constant of proportionality between stress and reversible strain is called the modulus of elasticity. The deformation of the material is called elastic deformation, if the stress is within elastic limit and the resulting strain is called the elastic strain (Fig. 1.2a). The modulus of elasticity reveals characteristics of atomic bonding.

When the external force is large, there is increase in stress and as a result the material may undergo an irreversible yielding or permanent deformation. The permanent deformation is also called plastic deformation. In general, plastic deformation causes changes in the cross-section of the material. The actual stress and deformation can be calculated using the instantaneous or true dimensions of the material. Yield strength (σ_y) is the measure of stress at which plastic deformation starts. The maximum strength is the stress at failure and is calculated using the original dimensions. This is called tensile strength (Fig. 1.2b).

Example 1.1

A brass wire 2.5m long and of cross-sectional area 1.0×10^{-3} cm^2 is stretched 1.0 mm by a load of 0.40 kg. Calculate Young's modulus for brass.

Solution:

Using Hooke's law we get

$$\text{Stress} = \frac{F}{A} = \frac{0.40 \times 9.81 \text{ N}}{1.0 \times 10^{-3} \times 10^{-4} \text{ m}^2} = 3.924 \times 10^7 \text{ Nm}^{-2}.$$

$$\text{Strain} = \frac{\text{Change in length}}{\text{Original length}} = \frac{1 \times 10^{-3} \text{ m}}{2.5 \text{ m}} = 4 \times 10^{-4}$$

$$Y = \frac{\text{Stress}}{\text{Strain}} = \frac{3.924 \times 10^7 \text{ Nm}^{-2}}{4 \times 10^{-4}} = 9.81 \times 10^{10} \text{ Nm}^{-2}$$

Table 1.1. Dependence of various properties on the internal structure.

Properties	Dependence on internal structure
Mechanical properties e.g. strength, fracture, ductility etc.	<ul><li>Bonds between atoms</li><li>Coordination of atoms in phases</li><li>Microstructure of phases</li><li>Macrostructure e.g. describing preferred orientation, domain structure etc.</li></ul>
Electrical properties e.g. conductivity, polarization etc.	<ul><li>Types of atoms.</li><li>Bonds between atoms.</li><li>Coordination of atoms in phases.</li><li>Microstructure of phases.</li></ul>
Thermal properties e.g. thermal expansion, thermal conductivity etc.	<ul><li>Bonds between atoms.</li><li>Coordination of atoms in phases.</li><li>Macrostructure.</li></ul>
Chemical properties e.g. corrosion, reactivity, resistance etc.	<ul><li>Types of atoms.</li><li>Bonds between atoms.</li><li>Microstructure of phases.</li><li>Macrostructure.</li></ul>
Magnetic properties e.g. permeability, magnetic susceptibility etc.	<ul><li>Types of atoms.</li><li>Bonds between atoms.</li><li>Coordination of atoms in phases.</li></ul>

The magnitude of plastic strain at fracture is a measure of ductility of the material and the energy required to produce fracture is a measure of its toughness. Toughness is calculated as the integrated product of stress and strain per unit volume. Integration is of importance in the fracture process, since the strain is not uniformly distributed during fracture, especially where a crack has already started.

A material in service conditions may be required to undergo certain mechanical processes. One has to acquire information on the capability of the material to undergo the processes e.g. how much static load a material can withstand. This can be tested when the material is in tension or under compression. This type of testing of materials is called hardness testing and is carried out to get information about its resistance to permanent deformation. During these tests, complex stress patterns are developed. Hardness is the resistance of a material to plastic indentation. Hardness measurement gives a measure of strength and structural coherence.

Impact tests are performed to measure the toughness of a material under shock loading conditions. The lifetime of a material under a cyclic load is measured by fatigue test. Behaviour of a material is also tested when it is subjected to a load at high temperature. This is done by creep or stress rupture test. An engineering material continues to deform for indefinite period of time when under constant stress. This time-dependent deformation is called creep. Results of these tests give quite valuable data on the mechanical properties of a material, which are useful to designers, fabricators and research workers.

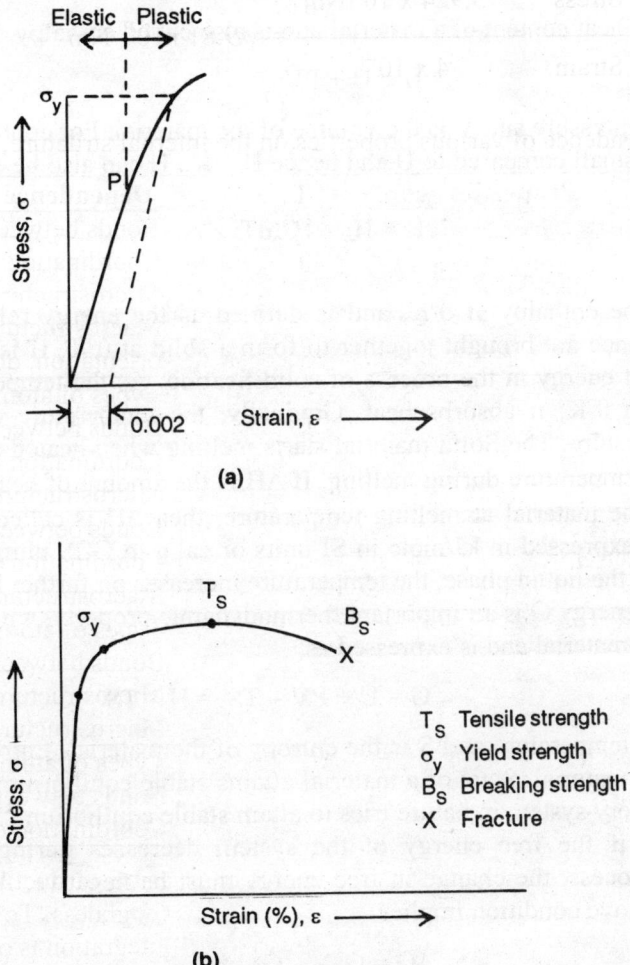

Fig. 1.2. (a) A stress vs. strain curve showing yield strength. (b) A stress vs. strain curve showing tensile strength (T_s) and breaking stress (B_s).

Thermal properties

Thermal properties are closely related to the thermodynamic variables of state e.g..pressure, temperature, internal energy, specific heat etc. Internal energy (U) is a disorganized form of energy stored in matter and is microscopic in nature. It is manifested in the form of movement of electrons and that of atoms. Specific heat of a material is defined as the heat energy required to raise the temperature of one kg of the substance through 1°C. In case of liquid or gaseous substances specific heat can be defined at constant pressure (C_p) or at constant volume (C_v). The relationship between the specific heat at constant volume and internal energy at a temperature T is given as:

$$C_v = (dU/dT)_v \qquad (1.1)$$

Similarly, specific heat at constant pressure is defined by

$$C_p = (dH/dT)_p \qquad (1.2)$$

where H is the heat content of a material and is also called enthalpy. It is defined as,

$$H = U + PV \qquad (1.3)$$

where P is the pressure and V is the volume of the material. For condensed matter (liquid or solid phases) PV is small compared to U and hence $H \approx U$. H can also be expressed as:

$$H = H_o + \int_0^T C_p dT \qquad (1.4)$$

where H_o is the enthalpy at 0°K, and is defined as the energy released when the atoms of a gaseous substance are brought together to form a solid at 0°K. H is taken as negative since the system has lost energy in the process of solidification. As the temperature of a solid material is increased from 0°K, it absorbs heat. Gradually, the temperature of the material reaches the melting temperature. The Solid material starts melting when heated at this temperature. There is no change in temperature during melting. If ΔH is the amount of heat energy needed just to melt unit mass of the material at melting temperature, then ΔH is called the latent heat of melting. Latent heat is expressed in kJ/mole in SI units or cal/g in CGS units. When the entire solid has melted to form the liquid phase, the temperature increases on further heating.

Gibb's free energy G is an important thermodynamic property, which describes the stability of the state of the material and is expressed as,

$$G = U + PV - TS = H - TS \qquad (1.5)$$

where T is the temperature and S is the entropy of the material. Entropy is a measure of internal disorder of the material. State of a material attains stable equilibrium when it acquires minimum free energy. Every system in nature tries to attain stable equilibrium. So a process is said to occur spontaneously if the free energy of the system decreases during the process. Thus, for a spontaneous process, the change in free energy must be negative. At constant temperature and pressure, the above condition implies

$$\Delta G = G_{final} - G_{initial} \qquad (1.6)$$

i.e.
$$\Delta G = (\Delta H - T\Delta S) < 0 \qquad (1.7)$$

During a physical or a chemical process, the change in entropy and enthalpy govern the nature of the process. In a spontaneous process the entropy of the system always increases. If in such a process, change in entropy is such that $T\Delta S > \Delta H$ and ΔH is positive, then the spontaneous process is endothermic i.e. ΔH would be absorbed by the system. If ΔH is negative i.e. heat energy is given away by the system in a spontaneous process, we say that the process is exothermic.

When some impurity is added to a sample or temperature is increased, the entropy of the material increases. When the temperature of a material is increased from 0 °K to T °K at constant pressure, the entropy of the system is given by,

$$\Delta S = \int_0^T (C_p/T') \, dT' \tag{1.8}$$

Entropy of a material is often chosen as zero at 0°K whereas H_o as well as the internal energy at absolute zero is nonzero. Change in entropy can occur due to configurational disorder even in the absence of an increase in thermal energy. Change in configurational entropy can be expressed as

$$\Delta S = k \ln W \tag{1.9}$$

where k is Boltzmann's constant and W is the number of configurations of the system having equal potential energy. From Eq. (1.5) it is quite clear that the free energy of a material is equal to the enthalpy at absolute zero. If the enthalpy H of a system does not change with the change of temperature, then the free energy decreases with an increase of temperature and is given by

$$(dG/dT) = -\{S + (T \, dS/dT)\} \tag{1.10}$$

Hence, the free energy minimizes as the entropy maximizes.

Thermal conductivity κ is an important thermal property of a material and is defined as the rate of flow of heat energy through unit area per unit temperature gradient along the direction of flow of energy. It is expressed as

$$\kappa = \frac{dQ/dt}{A \, (dT/dl)} \tag{1.11}$$

where dQ/dt is the rate of conduction of heat energy through the cross-sectional area A of a bar. Conduction of heat energy takes place along the length of the bar due to temperature gradient dT/dl in the direction of flow. It has been concluded from the experimental observations that although κ can be a function of temperature, it is independent of the temperature gradient and cross-sectional area A. Thermal conductivity depends very much on the electronic structure of the material e.g. metals have much higher thermal conductivity than insulators or semiconductors due to the presence of electrons in the conduction band. Thermal expansion of a material is estimated from the coefficient of thermal expansion, which can be linear, surface or volume in nature. The coefficient of linear expansion is defined as fractional change in linear dimension per unit degree rise in temperature. Similarly, one can define the coefficients of surface or volume expansions. These coefficients depend on the properties of materials like chemical bonding, electronic structure etc., and are also functions of temperature.

Electrical properties

Electrical properties of material, which are of interest in engineering applications, are resistivity, dielectric constant, electrical conductivity etc. Resistance R and resistivity ρ of a material are related as follows:

$$\rho = (\text{Resistance/length}). \, (\text{Area}) = (R/l) \, A \tag{1.12}$$

Unit of resistivity is ohm-meter (Ω-m). Resistivity of a material depends very much on the structure and is also a function of temperature. Thus, the total resistance of a material is temperature dependent.

The electrical conductivity σ is the reciprocal of resistivity ρ and can also be defined as the electrical charge flux j per unit electric field

$$\sigma = \frac{1}{\rho} = \frac{j \ C/m^2 s}{E \ V/m} = \frac{j}{V/d} \ \text{Siemens m}^{-1} \tag{1.13}$$

where E is the applied electric field, which is responsible for conduction of electric current and equals V/d, V being the voltage drop across a length d of the conductor. Dependence of electrical conductivity on density and nature of charge carriers, mobility of the carriers, temperature, strain in the material and composition of the material are discussed in Chapter 17.

Polarization is another important concept, for understanding the structure of the material, which has its origin in the static electric field. Materials, in which polarization occurs, are commonly referred to as dielectrics. Polarization is developed within a material when the negative and positive charge centres of the atoms, ions or molecules get separated due to application of electric field. Polarization can be defined as dipole, quadrupole or higher pole moments per unit volume. Commonly dipole moment being most prominent contributor, polarization will be referred as dipole moment per unit volume. The dipole moment is defined as Qd (Fig. 1.3)

$$\longleftarrow \quad d \quad \longrightarrow$$

$+Q \bullet \qquad\qquad\qquad \bullet -Q$

Fig. 1.3 Dipole moment Qd.

Although the magnitude of atomic, ionic or molecular dipole moment is small, the number of polarized units per unit volume of a dielectric material will be quite large thus resulting in a significant polarization.

Insulators are materials with high resistivity. Resistivity of insulators is nearly 10^{24} times that of metals and nearly 10^8 to 10^{10} times that of semiconductors. Insulators have certain electrical characteristics like dielectric constant, loss angle and dielectric breakdown strength, which are very important in engineering applications. If **E** is the applied electric field in a dielectric medium, the electric flux density **D** is proportional to the electric field and is given by

$$\mathbf{D} = \varepsilon \mathbf{E} = \varepsilon_0 \varepsilon_r \ \mathbf{E} \tag{1.14}$$

where **D** is also called displacement vector (C/m^2). The constant of proportionality between **D** and **E** is ε, the permittivity of the medium. The relative permittivity of the medium ε_r is a useful property for comparing insulators. This is also called as the dielectric constant of the medium and is denoted by K. For example, a capacitor with dielectric spacer of dielectric constant K will be able to hold K times more charge as compared to a capacitor with no spacer. A more detailed discussion on the electrical characteristics of dielectric materials is presented in Chapter 20.

Example 1.2

A rectangular copper block has dimensions 50cm x 1cm x 1cm.

(a) What is the resistance measured between the two

i) square ends,

ii) opposing rectangular faces?

(b) Also calculate the current density when a potential difference of 5V is applied between two square ends. (Resistivity of copper at 20°C is 1.7 x 10^{-8} ohm-m.)

Solution:

(a)   The area of a square end is 1.0 cm^2 = 1.0 x 10^{-4} m^2

$$R = \frac{\rho L}{A} = \frac{(1.7 \times 10^{-8} \text{ ohm-m}) (0.5 \text{ m})}{1.0 \times 10^{-4} \text{ m}^2} = 8.5 \times 10^{-5} \text{ ohm}$$

For the resistance between opposing rectangular faces having 5.0x10^{-3} m^2 areas we have

$$R = \frac{\rho l}{A} = \frac{(1.7 \times 10^{-8} \text{ ohm m}) (10^{-2} \text{ m})}{5 \times 10^{-3} \text{ m}^2} = 3.4 \times 10^{-8} \text{ ohm}$$

(b)   The conductivity of copper at 20°C is

$$\sigma = 1/\rho = 5.88 \times 10^7 \text{ mho m}^{-1} \text{ and } \sigma = j/(V/d)$$

Since potential difference is applied across two square ends, d = 50 cm,

$$j = \frac{\sigma V}{d} = \frac{5.88 \times 10^7 \text{ mho m}^{-1} \times 5 \text{ volts}}{50 \times 10^{-2} \text{ m}} = 5.88 \times 10^8 \text{ A/m}^2$$

Note: The whole block will melt due to intense Joule heating.

Magnetic properties

Magnetic materials are widely used for various engineering applications and can be classified as

i) Diamagnetic

ii) Paramagnetic

iii) Ferromagnetic

Electrons in atoms have magnetic moments since they possess orbital as well as spin angular momenta. Magnetic moment of an atom is the vector sum of these magnetic moments. The resultant magnetic moment of an atom may be zero depending upon whether or not there are unpaired electrons in the incomplete shell. In the presence of an external magnetic field **H**, atomic magnets tend to align in the direction of **H**. Thus, the magnetic moment developed per unit volume of the material is called Magnetization (**M**), which is proportional to **H**,

$$\mathbf{M} = \chi \mathbf{H} \tag{1.15}$$

where χ, the constant of proportionality is called the susceptibility of magnetic material. This parameter is positive for paramagnetic and ferromagnetic materials and is negative for diamagnetic materials.

Diamagnetism results from the changes in the angular velocity of the orbital electrons of the atom, when an external magnetic field **H** is applied. The applied magnetic field induces magnetic flux, which opposes the applied field. As a result, a pure diamagnetic substance expels the applied magnetic lines of force and gives rise to negative magnetic susceptibility.

In case of paramagnetism, the magnetic moments of constituent atoms get aligned in the direction of the applied field (e.g. substances with unpaired electrons). The material attains magnetization only below a critical temperature. When temperature T increases, the alignment of magnetic moments gets disturbed and this tends to decrease **M**.

In an unmagnetized ferromagnetic substance, there exist separate magnetic domains having completely aligned atomic magnetic moments, which line up in such a way that magnetization persists in domains after the removal of applied field **H**. These domains are oriented randomly when there is no external field i.e. net magnetization is zero. In the presence of magnetic field the magnetic moments of these domains orient along **H** and produce strong magnetization. Above a certain temperature ferromagnetic materials become paramagnetic. This is called the Curie temperature. The magnitude of the magnetic induction **B** within the long solenoid can be expressed as,

$$\mathbf{B} = \mu ni \qquad (1.16)$$

where μ is the magnetic permeability of the substance, n is the number of turns per unit length and i is the current through each turn of the coil. When a current flows through a coil it stores magnetic energy. The expression for stored magnetic energy density u is

$$u = \tfrac{1}{2}\,\mu n^2 i^2 = \tfrac{1}{2}\, B^2/\mu \qquad (1.17)$$

The unit of energy density is J/m^3. As in the case of the relation between the displacement vector and the electric field, we can write a similar expression relating magnetic field **H** and magnetic induction **B** as

$$\mathbf{B} = \mu\, \mathbf{H} \qquad (1.18)$$

Hence

$$u = (HB/2) \qquad (1.19)$$

Another important expression relating **B** and magnetization vector **M** is given by

$$\mathbf{B} = \mu_o(\mathbf{H} + \mathbf{M}) \qquad (1.20)$$

where μ_o is called the permeability of free space (vacuum). In SI units, μ_o has a value of $4\pi \times 10^{-7}$ Hm^{-1}. Diamagnetic substances have negative susceptibility whereas other types of magnetic materials have positive magnetic susceptibilities.

Ferromagnetic solids are extremely useful for various industrial applications especially for making permanent magnets. Magnetic materials and their properties are discussed in detail in Chapter 19.

While choosing materials for industrial applications, the corrosion resistance properties of the material should also be considered. Many engineering materials deteriorate and corrode due to the chemical or electrochemical reaction of the material with the environment. These reactions result in the formation of oxides, salts or some other compounds of the material. Scientists and engineers have developed materials such that degradations of materials can be decreased by

(a) avoiding severely corrosive environments,
(b) providing protection against corrosion, using appropriate corrosion resistant material or coatings.

A detailed study on corrosive properties of materials is discussed in Chapter 16.

1.5 NEED FOR TECHNOLOGY ORIENTED MATERIALS

Modern technology needs more and more materials to feed the various developmental engineering projects. The need has also been felt for developing and synthesizing new materials because many of the conventional materials are being obtained from nonrenewable sources. Most of the metals are obtained from ores and polymers are obtained from raw materials like oil, paraffin, metals etc. These nonrenewable resources are getting depleted. Therefore, either new resource has to be discovered or new materials have to be developed.

Energy crisis is another problem being faced by mankind. This problem has to be tackled by developing new technologies. Some of the nonconventional forms of energy are the following:
a) Solar energy
b) Nuclear energy
c) Wind and tidal energy

To produce and supply energy available from the above forms at economically viable prices, we may require developing new materials. Thus, human endeavour will strive to open up the material wealth of nature to meet the ever-increasing demand for energy.

EXERCISES

1.1   What are the main classes of engineering materials? Give examples of each class.

1.2   When a 2V flash battery sends a current of 0.2 A through a flash light bulb, calculate the number of electrons flowing through the bulb each second.
Ans: 1.25×10^{18} electrons/s

1.3 A certain capacitor has a capacitance of 50 pF with a layer of air in between its plates and 350 pF with a plastic sheet of same thickness as that of air in between its plates. Find the dielectric constant of the plastic?
Ans: 7

1.4 Copper has resistivity 1.54×10^{-8} Ω-m. When the temperature of a copper wire is increased by 100°C, the resistivity is increased by 44%. What is the value of dR/dT per meter of the wire if its diameter is 0.5 mm?
Ans: 3.45×10^{-4} Ω m^{-1}°K^{-1}

1.5   A 1.08m long metallic rod (Young's Modulus = 110.2 GN m^{-2}) is to be elongated by 1.016 mm. What should be the diameter of the rod, so that it can carry a tensile load of 2000 kg without undergoing plastic deformation?
Ans: 4.96 mm

1.6   An aluminium bar having cross-sectional area 0.02m x 0.02m shows reduction in area by 5%, when an external stress of 0.193 GNm^{-2} based on original cross-section, is applied. Calculate true stress and true strain of the material.
Ans: 0.203 GNm^{-2}, 0.0526

1.7   Calculate the magnetic energy density and magnetic induction in a solenoid if there are 1000 turns/m. Given that the current flowing through the wire is 5A and the permeability of the medium of the solenoid core is 1.2×10^{-6}.

Ans: 15 J/m^3, 6×10^{-3} Wb/m^2

1.8   Two capacitor plates (2 cm x 2 cm) are parallel and are 0.22cm apart with air in between the plate. Calculate the voltage, which produces a charge of 0.44×10^{-10}C on the plates.

Ans: 27.34V

1.9   Starting from the expression $G = H - TS$, show that $dG/dT = -S$ at constant pressure using $(dH/dT)_p = C_p$ and $(dS/dT)_p = C_p/T$.

1.10 Show that $C_p = (5/2) R$ for an ideal monoatomic gas.

2

Atomic Structure and Related Properties

2.1 INTRODUCTION

Atomic structure plays a very important role in deciding the properties of a material. The chemical bonding between atoms depends upon the atomic structure, which in turn dictates quite a few physical and chemical properties of the material. Thus, the starting point in the study of the structure of a material is its atomic structure. Since the atomic structure is first introduced in the preliminary level physics and chemistry courses, only a brief review of the atomic structure and the periodic table will be presented in the next two sections for the sake of completeness.

2.2 ATOMIC STRUCTURE

The familiar model of an atom is that of a tiny nucleus composed of protons and neutrons with surrounding electronic orbitals, as discovered by the stalwarts of physics like Rutherford, Niels Bohr. Neutron does not carry any charge, whereas proton carries 1.602×10^{-19}C positive charge and the electron carries an equal amount of negative charge. The size of a nucleus is of the order of 10^{-15}m and that of the atom is of the order of 10^{-10}m. Therefore, the fraction of atomic volume occupied by the nucleus is of the order of 10^{-15} only. Thus, most of the space in the atom is unoccupied. Nucleus has an extremely dense structure. Protons, in spite of facing repulsive electrostatic force due to their like charges, are held together along with the neutrons because of the presence of strong nuclear forces. The mass of an atom mostly depends on the mass of protons and neutrons since the mass of electron m_e (9.11×10^{-31} kg) is very small (1/1840th approximately) compared to that of proton or of neutron.

Ions are formed when atoms lose or gain electrons. During the process of ionization as many electrons an atom loses it becomes a positive ion with that many positive charges. Similarly, as many electrons an atom gains, it becomes a negative ion with that many negative charges. Thus, the charges on the ions are integral multiples of the electronic charge. The properties of an ionic material e.g. conductivity, polarization, the strength of the ionic bonds etc., depend on the value of the electronic charge on the ion.

The number of protons in atoms ranges from one for hydrogen to nearly 100 for the transuranic elements, Table 2.1. This is also called the atomic number. Each element has a unique atomic number. The atomic weight of an element is the relative weight of its atom measured on a standard scale in which the weight of the most common isotope, $^{12}C_6$, of carbon atom is taken to be 12 atomic mass units (amu). Different isotopes of the same element have the same atomic number but different atomic weight. This difference in atomic weight is due to the presence of different number of neutrons in their nuclei. The atomic weight of a sample of element is given by the summation of atomic weight times the fraction of various isotopes present in the sample.

Molecular weight is also an important concept. It is the relative weight of a molecule of the substance as measured using the above scale. The molecular weight, expressed in gram is called gram mole and that in kilogram is called kilogram mole or kmole. For example, 1 kmole of $^{12}C_6$ means 12 kg of carbon. One kilogram mole of any substance contains 6.02×10^{26} atoms or molecules, which is referred to as the Avogadro number. Thus the atomic weight of sodium is 22.99 amu /atom or 22.99 kg/kmole.

We often consider an atom as a very small hard sphere. Although this model of atom can explain certain experimental observations while studying the properties and the structure of a crystal, it gives a very superficial view of the atom. The electrons in an atom are attracted towards the nucleus but repel each other. The electrons move in specific paths called electronic orbits. The average radius r of the path of the electron can be calculated from the equation, which states that the centrifugal force experienced by an orbiting electron is equal to the attractive force due to nucleus,

$$\frac{m v^2}{r} = \frac{Ze^2}{4\pi\varepsilon_0 r^2} \tag{2.1}$$

where m is the mass, v is the velocity of the orbiting electron and Z is the atomic number. The kinetic energy of the electron is

$$T = \tfrac{1}{2}\, mv^2 = Ze^2/8\pi\varepsilon_0 r \tag{2.2}$$

i.e.
$$r = Ze^2 / (4\pi\varepsilon_0 mv^2) = Ze^2/(8\pi\varepsilon_0 T) \tag{2.3}$$

Thus, the size of an electronic orbit is different for electrons with different energies.

Uncertainty principle

Electronic orbit with a specific energy cannot be visualized as a discrete path because, according to the Heisenberg's uncertainly principle, the momentum and the position of the electron cannot be determined simultaneously with high order of accuracy. Either of the two quantities can be measured with desired accuracy, but the experimental techniques to do so are such that if the accuracy in the measurement of momentum is increased, the measurement of position becomes less accurate or vice versa. If Δp is defined as the uncertainly in momentum and Δx is defined as the uncertainty in position, then according to Heisenberg's uncertainty principle, we have

$$\Delta p \Delta x > \tfrac{1}{2}\,\hbar \tag{2.4}$$

where h is Planck constant and $\hbar$ is $h/2\pi$. This principle is extremely important in case of microscopic entities e.g. nucleons, electrons etc., whereas it gives trivial results in case of macroscopic objects. So it is quite clear, why the electronic orbit cannot be visualized as a definite path. Instead, this should be visualized as an electron probability density cloud surrounding the nucleus. According to the wave mechanics, an electron is represented by a wave function. The square of the amplitude of this wave function at a given point gives the probability of density for finding the electron at that point. Electrons, in various shells, constitute electron cloud around the nucleus and form the outer region of the atom. Electron cloud of an atom repels that of another atom strongly when the later approaches the former, giving the approximate character of a tiny sphere to an atom.

Energy states and quantum numbers

An electronic orbit with characteristic energy is defined to be the energy state of an electron which is designated with a definite set of four quantum numbers n, l, m_l, m_s. Out of these, n is the principal quantum number, l is the orbital angular momentum quantum number, m_l is the magnetic quantum number and m_s is the spin quantum number. The principal quantum number n takes integer values 1,2,3,4...etc. Higher the value of n, higher is the energy of the electron and such electrons are less bound to the nucleus as compared to those electrons with lower energy. The energy state of an electron with a specific value of n is called an energy shell. Energy shells are designated by the letters K, L, M, N, O and so on corresponding to n = 1,2,3,4, and 5 respectively, Table 2.2.

Table 2.2 Arrangement of electrons in shells and subshells

Principal quantum number, n	Shell designation	Subshells	Number of electrons in	
			subshell	shell
1	K	s	2	2
2	L	s	2	8
		p	6	
3	M	s	2	18
		p	6	
		d	10	
4	N	s	2	32
		p	6	
		d	10	
		f	14	
5	O	s	2	50
		p	6	
		d	10	
		f	14	
		g	18	

An electronic energy state with principal quantum number n has orbital angular momentum expressed as $\sqrt{[l(l+1)]}\hbar$. The quantum number l can have values 0,1,2,3...(n–1) corresponding to the energy state n. The electronic orbit with l = 0 does not have any angular momentum and is circular in shape. Shape of the electronic orbit is elliptic for l > 0. Energy states having specific n, l values form subshells of the main shell. For example, the subshells corresponding to n = 6 are designated by s, p, d, f, g and h with the quantum number l = 0, 1, 2, 3, 4, 5, respectively. The energy state of an electron is designated by the particular values of the two quantum numbers n and l and the number of electrons having same n, l values written in the superscript. For example, if an energy state (n = 3 and l = 2) has 10 electrons, we designate this by $3d^{10}$, Table 2.3.

Magnetic quantum number m_l is also called the projection quantum number. This quantum number defines one of the possible orientations of the electronic orbit or electron probability density cloud in space. The quantum number m_l can take (2l+1) values given as –l, –l+1, –l+2,...0,1,2,..l corresponding to each value of l. In the presence of an external magnetic field the electronic orbit is subjected to a torque, which tries to orient angular momentum vector parallel to the field. As a result, the orbital angular momentum vector precesses about the field. Thus, the spin and orbital angular momentum vectors take up definite positions in space. The projection of orbital angular momentum along the direction of the magnetic field, $m_l h$ is associated magnetic orbital quantum number, whereas the projection of the spin angular momentum along the direction of magnetic field $m_s h$ is associated with spin quantum number m_s. The spin momentum has numerical value $\frac{1}{2}h$ and it has been found experimentally that it must be oriented either up or down in the presence of the magnetic field i.e. parallel or antiparallel to the direction of magnetic field. Thus the fourth quantum number m_s has possible values $+\frac{1}{2}$ or $-\frac{1}{2}$, one for each of the spin orientations. In the absence of a magnetic field there are no unique values of m_l or m_s. The set of four quantum numbers define the state of the electron in an atom in terms of its energy, angular momentum, orientation of electronic orbit and spin with respect to the direction of the external magnetic field.

Table 2.3(a) Electronic configuration for elements of first and second period

Element	Symbol	Atomic number	Electronic configuration	Spin alignment of electrons in outer shells 1s 2s 2p
Hydrogen	H	1	$1s^1$	↑
Helium	He	2	$1s^2$	↑↓
Lithium	Li	3	$He2s^1$	↑↓ ↑
Beryllium	Be	4	$He2s^2$	↑↓ ↑↓
Boron	B	5	$He2s^22p^1$	↑↓ ↑↓ ↑
Carbon	C	6	$He2s^22p^2$	↑↓ ↑↓ ↑ ↑
Nitrogen	N	7	$He2s^22p^3$	↑↓ ↑↓ ↑ ↑ ↑
Oxygen	O	8	$He2s^22p^4$	↑↓ ↑↓ ↑↓ ↑ ↑
Fluorine	F	9	$He2s^22p^5$	↑↓ ↑↓ ↑↓ ↑↓ ↑
Neon	Ne	10	$He2s^22p^6$	↑↓ ↑↓ ↑↓ ↑↓ ↑↓

Table 2.3(b) Electronic configuration for elements of third period

Element	Symbol	Atomic number	Electronic configuration	Spin alignment of electrons in outer shells 3s 3p
Sodium	Na	11	$Ne3s^1$	↑
Magnesium	Mg	12	$Ne3s^2$	↑↓
Aluminum	Al	13	$Ne3s^23p^1$	↑↓ ↑
Silicon	Si	14	$Ne3s^23p^2$	↑↓ ↑ ↑
Phosphorus	P	15	$Ne3s^23p^3$	↑↓ ↑ ↑ ↑
Sulphur	S	16	$Ne3s^23p^4$	↑↓ ↑↓ ↑ ↑
Chlorine	Cl	17	$Ne3s^23p^5$	↑↓ ↑↓ ↑↓ ↑
Argon	Ar	18	$Ne3s^23p^6$	↑↓ ↑↓ ↑↓ ↑↓

Table 2.3(c) Electronic configurations for elements of fourth period

Element	Symbol	Atomic number	Electronic configuration	Spin alignment of electrons in outer shells (4s, 3d, 4p)
Potassium	K	19	$Ar4s^1$	↑
Calcium	Ca	20	$Ar4s^2$	↑↓
Scandium	Sc	21	$Ar3d^14s^2$	↑↓ ↑
Titanium	Ti	22	$Ar3d^24s^2$	↑↓ ↑ ↑
Vanadium	V	23	$Ar3d^34s^2$	↑↓ ↑ ↑ ↑
Chromium	Cr	24	$Ar3d^54s^1$	↑ ↑ ↑ ↑ ↑ ↑
Manganese	Mn	25	$Ar3d^54s^2$	↑↓ ↑ ↑ ↑ ↑ ↑
Iron	Fe	26	$Ar3d^64s^2$	↑↓ ↑↓ ↑ ↑ ↑ ↑
Cobalt	Co	27	$Ar3d^74s^2$	↑↓ ↑↓ ↑↓ ↑ ↑ ↑
Nickel	Ni	28	$Ar3d^84s^2$	↑↓ ↑↓ ↑↓ ↑↓ ↑ ↑
Copper	Cu	29	$Ar3d^{10}4s^1$	↑ ↑↓ ↑↓ ↑↓ ↑↓ ↑↓
Zinc	Zn	30	$Ar3d^{10}4s^2$	↑↓ ↑↓ ↑↓ ↑↓ ↑↓ ↑↓
Gallium	Ga	31	$Ar3d^{10}4s^24p^1$	↑↓ ↑↓ ↑↓ ↑↓ ↑↓ ↑↓ ↑
Germanium	Ge	32	$Ar3d^{10}4s^24p^2$	↑↓ ↑↓ ↑↓ ↑↓ ↑↓ ↑↓ ↑ ↑
Arsenic	As	33	$Ar3d^{10}4s^24p^3$	↑↓ ↑↓ ↑↓ ↑↓ ↑↓ ↑↓ ↑ ↑ ↑
Selenium	Se	34	$Ar3d^{10}4s^24p^4$	↑↓ ↑↓ ↑↓ ↑↓ ↑↓ ↑↓ ↑↓ ↑ ↑
Bromine	Br	35	$Ar3d^{10}4s^24p^5$	↑↓ ↑↓ ↑↓ ↑↓ ↑↓ ↑↓ ↑↓ ↑↓ ↑
Krypton	Kr	36	$Ar3d^{10}4s^24p^6$	↑↓ ↑↓ ↑↓ ↑↓ ↑↓ ↑↓ ↑↓ ↑↓ ↑↓

Energy level diagram for the various shells and subshells using the wave mechanical model is given in Fig. 2.1. We observe here the following important features:

(i) The smaller the value of n, the lower is the energy level for same l value e.g. 1s has lower energy as compared to 2s state.

(ii) Within each shell the energy of a subshell level will be lower as the value of l is lower e.g. the energy of 3p state will be lower than that of 3d state.

(iii) There may be overlap of energy states of one shell with those in an adjacent shell. This is especially true for d, and f states. For example, the energy of a 3d state is greater than that of a 4s state and the energy of 4d state is greater than that of 5s state etc. These energy states also overlap.

Bohr atomic model

According to the Bohr atomic model, each electron has a specific energy. The energy of an atom, as a whole, depends on the way electrons are arranged in the various possible electronic states.

Each arrangement of electrons gives rise to a state of the atom. The arrangement of electrons in the energy state, which gives minimum energy, is called the ground state of the atom. The energy states of atom with energy greater than the ground state are called excited states. Electrons in the lower energy states can be excited to the vacant higher energy states, if the atom absorbs a required amount of energy. This energy can be in the form of electromagnetic radiation. An electron in an excited state (E_f) jumps back to a vacant lower energy state (E_i) giving rise to the emission of radiation whose energy is equal to the energy difference of the two states of the electron and is given by

$$E_f - E_i = h\nu, \tag{2.5}$$

where ν is the frequency of electromagnetic radiation.

Pauli exclusion principle

Arrangement of electrons in the energy states follows Pauli exclusion principle. According to this principle no two electrons can occupy the same energy state which is defined by the four quantum numbers n, l, m_l, m_s. We know that a specific energy shell is designated by the principal quantum number n, and the subshells are characterized by the orbital quantum number l. Each subshell can accommodate $2(2l+1)$ electrons, as m_s can take only two values i.e. $\pm 1/2$ corresponding to each value of m_l. This implies that the maximum possible number of electrons in a shell is given by

$$\sum_{l=0}^{n-1} 2(2l + 1) = 2[1 + 3 + 5 + ... + (2n - 1)] = 2n^2 \tag{2.6}$$

In an atom with Z electrons, there are Z discrete energy states each of which is occupied by an electron starting from the lowest state. The lowest energy level is occupied first and then the next energy level and so on. Since the energy of a state depends both on n and l, the order in which the electrons occupy the energy states are as follows: 1s, 2s, 2p, 3s, 3p, (4s, 3d), 4p, (5s, 4d), 5p, (6s, 4f, 5d), 6p, (7s, 5f, 6d), 7p as indicated in the diagram shown in Fig. 2.2. The states lying on a particular arrow have almost the same energy, especially when they are the outermost states of the atom. Thus, electrons may partially fill up one subshell and then occupy the other or vice versa.

Electronic configuration of an atom represents the arrangement of electrons in it. For example, the electronic configurations of hydrogen, helium and sodium are $1s^1$, $1s^2$ and $1s^2 2s^2 2p^6 3s^1$, respectively. The superscript in the electronic configuration denotes the number of electrons in the particular subshell. The electrons, which occupy the outermost shell of the atom, are called valence electrons. The outermost shell is referred to as the valence shell. The valence electrons play a very important role in the formation of chemical bonds. Many important physical and chemical properties are based on the type and strength of chemical bonding. When the outermost shell of an atom is completely filled with electrons, the atom is said to have a stable configuration. Usually this configuration corresponds to the occupation of just s and p electrons, which is normally 8, in the outermost shell e.g. neon, argon, and krypton, one exception being helium which contains only two 1s electrons. These elements do not interact with other elements. They are called inert gases or noble gases. The elements with partially filled valence shells are involved in chemical reactions to form compounds. These elements gain, lose or share valence electrons while forming chemical bonds such that eight electrons occupy the outermost shell. Atoms form molecules or molecular aggregates through chemical bond formation with the help of valence electrons.

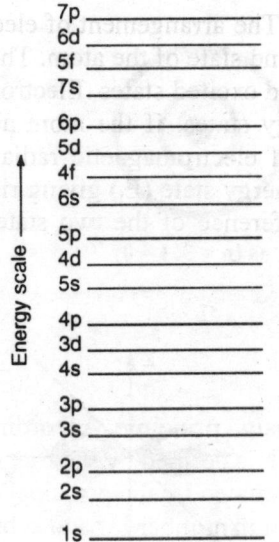

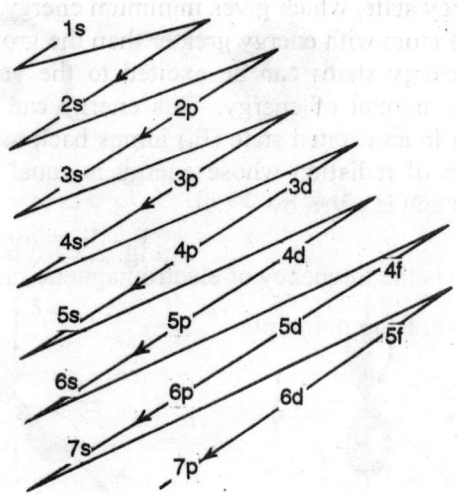

Fig. 2.1 Schematic representation of
electronic energy shells and subshells

Fig. 2.2 Sequence of filling up of electronic
levels following the Afbau principle

Electronic orbitals

It is known that the exact location of an electron in electronic orbital of an atom cannot be determined due to the uncertainty principle. Any attempt to determine the exact orbit of an electron using suitable probe would modify the wave characteristics of the electron. We can only talk of probability of finding the electron at a given distance. Probability of finding an electron, which is measured from the calculation of radial density $4\pi r^2 \rho$, increases to a maximum value at distance 'a' from the nucleus and decreases with further increase of radial distance from the nucleus. 'a' is defined as the radius of the electronic orbit and ρ is probability density of the electronic wave function. Using the wave mechanical concept, electron clouds associated with the orbitals, define the most probable volume in space where an electron can be found. The size and shape of the orbital depends on the energy and angular momentum of the electron i.e. the characteristics of the orbital depend on the set of quantum numbers defining the electronic state.

All s-type orbitals, Fig. 2.3, have spherical surface boundary with centre at the centre of the nucleus. The centre of the nucleus of an atom is normally chosen to be the origin for describing the electronic orbitals. The larger the value of n, the larger is the size of the orbital. The electron density is not constant everywhere within these spheres. Within the boundary surface of 1s orbital there is a single spherical shell of high electron density, there are two dense concentric shells within the 2s boundary surface. All the p orbitals have directional properties in contrast to the s orbitals. In the p orbital (Fig. 2.4), an electron is somewhere in a dumb bell shaped space, each lobe being equally probable. Both lobes together constitute one orbital. The orbitals have a plane of zero electron density. This plane is a nodal plane, which separates the two lobes. The d-orbitals, Fig. 2.5 are shaped like a cloverleaf with four lobes and two nodal planes each excepting the d_z^2 orbital, which is concentrated around the Z-axis with the larger part of the volume being shaped somewhat like a p orbital.

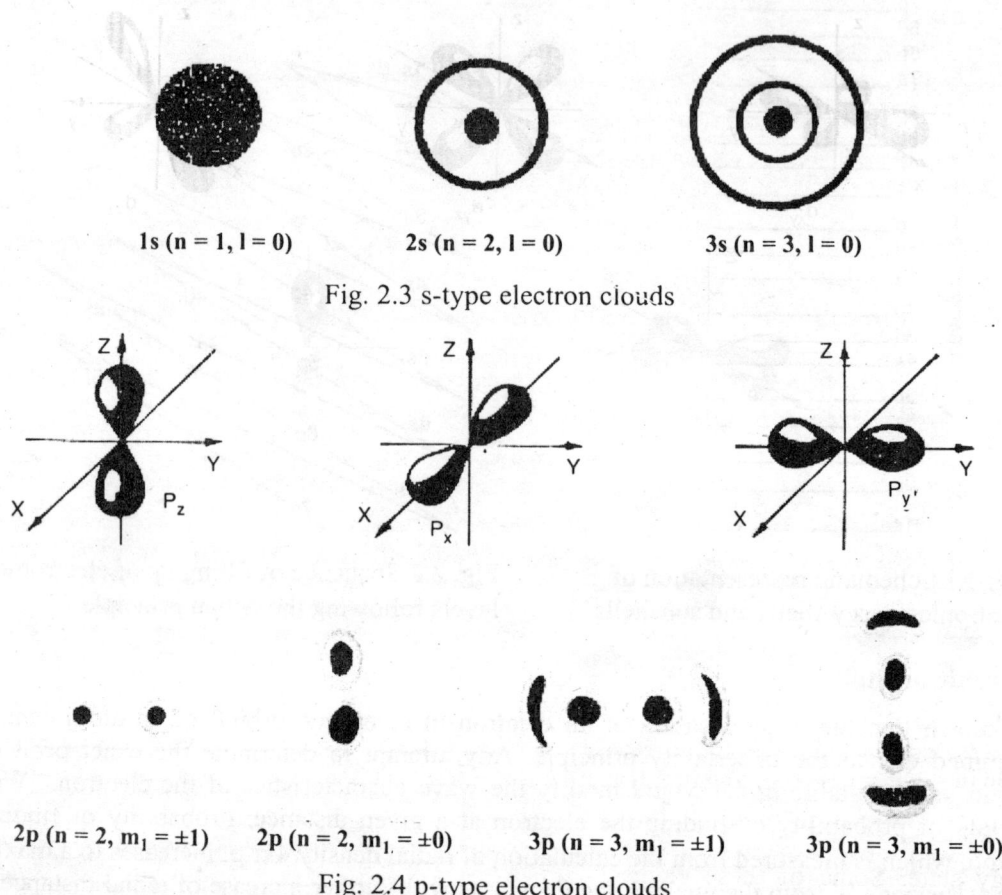

1s (n = 1, l = 0)   2s (n = 2, l = 0)   3s (n = 3, l = 0)

Fig. 2.3 s-type electron clouds

2p (n = 2, m_l = ±1)   2p (n = 2, m_l = ±0)   3p (n = 3, m_l = ±1)   3p (n = 3, m_l = ±0)

Fig. 2.4 p-type electron clouds

2.3 PERIODIC TABLE

The periodicity of chemical properties of elements was found in the nineteenth century. In 1870, the Russian chemist Mendeleev devised a methodology to arrange all the elements systematically in the order of increasing atomic weight such that the elements of similar chemical and physical properties are arranged along a vertical column. Arranging the elements in the order of increasing atomic number, from left to right, modified the idea of Mendeleev and showed marked periodicity in chemical and physical properties. Such an arrangement of elements forms the periodic table, Table 2.1. Success of the modern periodic table is that all the elements are arranged in the increasing order of atomic number in seven horizontal rows called periods. Along the horizontal rows the number of outer electrons increases stepwise by one from left to right. A new row starts whenever a p subshell gets filled up and each row in the periodic table finishes with an inert gas. Elements in a vertical column in the periodic table form a group. They have similar chemical properties and crystal structure since they have the same number of valence electrons. Thus Li, Na, K etc. form group IA and are referred to as the alkali metals. The next column is occupied by group IIA elements i.e. Be, Mg, Ca etc., and are called alkaline earth metals. Alkali and alkaline earth metals have one and two electrons in excess of stable structures, respectively. Elements in group IIIB, IVB, VB, VIB, VIIB, VIII, IB, and IIB are transition metals having partially or fully filled d orbitals and one or two electrons in the next higher

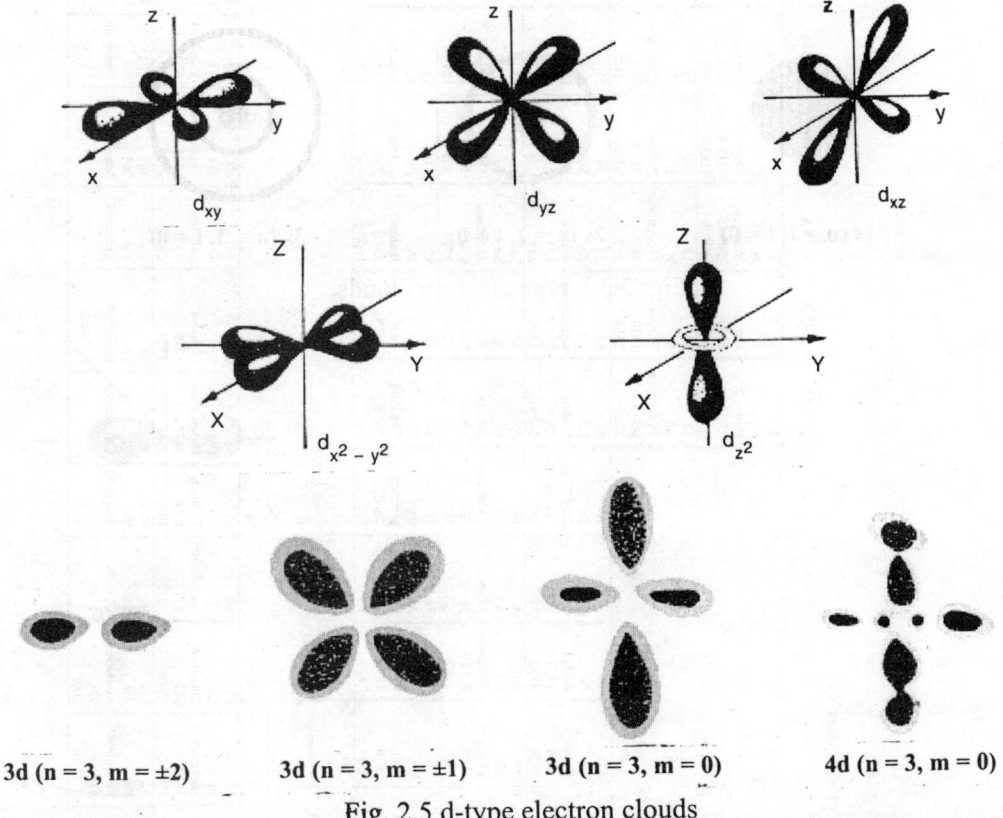

| 3d (n = 3, m = ±2) | 3d (n = 3, m = ±1) | 3d (n = 3, m = 0) | 4d (n = 3, m = 0) |

Fig. 2.5 d-type electron clouds

energy state. Some of these elements in groups IIIA, IVA and VA are metals; some are nonmetals while others have properties intermediate between metals and nonmetals e.g. Bi, As etc. These intermediate characteristics arise due to nearly half-filled valence shell. Elements in group VIIA, (F, Cl, Br, I and At), referred to as halogens, have one electron deficient from stable structures. Elements of group VIA, and VIIA are nonmetals and mostly electronegative. Atoms of these groups share or exchange electrons to form molecules. The elements of the last column are inert gases having completely filled electronic states and are in group zero. Removal of an electron from a stable inert gas configuration requires more energy as compared to that from elements positioned near the inert gas in the periodic table.

The first element of the table is hydrogen which has one electron occupying one of the two 1s states (K shell). The next element is helium which has two electrons in the 1s state with opposite spins. The next element is lithium, which has three electrons. Two of its electrons occupy K shell and the third occupies the lowest subshell of L shell. Electrons get filled up in the L shell in the elements of this row. The filling up of electrons in the energy states follows Hund's rule which states that in order to reduce the electron-electron repulsive energy, the number of electrons of same spin in various orbitals like p, d or f should be maximum (Table 2.3). When a new shell begins to fill up, the element in question is an alkali metal.

A new situation is observed in the fourth period, which starts with potassium with configuration $1s^2 2s^2 2p^6 3s^2 3p^6 4s^1$. Elements in the fourth period have 4s energy level slightly

Period/Group →

Legend:
- Atomic number → 80
- Symbol → Hg
- Name → Mercury
- Melting point (°C) → -38.87
- Boiling point (°C) → 357

Period	IA	IIA	IIIB	IVB	VB	VIB	VIIB	VIII	VIII	VIII	IB	IIB	IIIA	IVA	VA	VIA	VIIA	VIIIA Or Zero
1	1 H Hydrogen -259.14 -252.8																	2 He Helium -272.2 -268.9
2	3 Li Lithium 180.54 1347	4 Be Beryllium 1278 2970											5 B Boron 2300 2550	6 C Carbon 3500 4827	7 N Nitrogen -209.86 -195.8	8 O Oxygen -218.4 -183.0	9 F Fluorine -219.62 -188.2	10 Ne Neon -248.67 -246.1
3	11 Na Sodium 97.5 552.9	12 Mg Magnesium 650 1107											13 Al Aluminium 660.3 2467	14 Si Silicon 1410 2355	15 P Phosphorus 44.1 280	16 S Sulfur 112.8 444.6	17 Cl Chlorine -101 -34.7	18 Ar Argon -189.4 -185.8
4	19 K Potassium 63.6 774	20 Ca Calcium 839 1484	21 Sc Scandium 1539 2832	22 Ti Titanium 1660 3287	23 V Vanadium 1890 3380	24 Cr Chromium 1857 2672	25 Mn Manganese 1245 1962	26 Fe Iron 1535 2750	27 Co Cobalt 1495 2870	28 Ni Nickel 1453 2732	29 Cu Copper 1083 2567	30 Zn Zinc 419.58 907	31 Ga Gallium 29.78 2403	32 Ge Germanium 937.4 2830	33 As Arsenic 817 613	34 Se Selenium 217. 684.9	35 Br Bromine -7.2 58.78	36 Kr Krypton -157 -153.4
5	37 Rb Rubidium 39 688	38 Sr Strontium 770 1384	39 Y Yttrium 1523 3337	40 Zr Zirconium 1850 4377	41 Nb Niobium 2468 4927	42 Mo Molybdenum 2617 4612	43 Tc Technetium 2200 4877	44 Ru Ruthenium 2250 3900	45 Rh Rhodium 1966 3727	46 Pd Palladium 1552 2927	47 Ag Silver 961.9 2212	48 Cd Cadmium 320.9 765	49 In Indium 156.61 2000	50 Sn Tin 231.9 2270	51 Sb Antimony 630.5 1750	52 Te Tellurium 450 989.8	53 I Iodine 113.5 184	54 Xe Xenon -112 -108
6	55 Cs Cesium 28.5 678.4	56 Ba Barium 725 1140	57 La Lanthanum 920 3469	72 Hf Hafnium 2150 5400	73 Ta Tantalum 2996 5425	74 W Tungsten 3410 5660	75 Re Rhenium 3180 5627	76 Os Osmium 3045 5027	77 Ir Iridium 2410 4527	78 Pt Platinum 1772 3827	79 Au Gold 1064.4 2807	80 Hg Mercury -38.87 357	81 Tl Thallium 303.5 1457	82 Pb Lead 327.5 1740	83 Bi Bismuth 271.3 1560	84 Po Polonium 254 962	85 At Astatine 302 337	86 Rn Radon -71 -61.8
7	87 Fr Francium 27 677	88 Ra Radium 700 1737	89 Ac Actinium 1050 3200	104 Rf Rutherfordium	105 Db Dubnium	106 Sg Seaborgium	107 Bh Bohrium	108 Hs Hassium	109 Mt Meitnerium									

Lanthanide series:

58 Ce Cerium 795 3257	59 Pr Praseodymium 935 3127	60 Nd Neodymium 1010 3127	61 Pm Promethium 1027	62 Sa Samarium 1072 1900	63 Eu Europium 822 1597	64 Gd Gadolinium 1311 3233	65 Tb Terbium 1360 3041	66 Dy Dysprosium 1412 2562	67 Ho Holmium 1470 2720	68 Er Erbium 1522 2510	69 Tm Thulium 1545 1727	70 Yb Ytterbium 824 1466	71 Lu Lutetium 1656 3315

Actinide series:

90 Th Thorium 1750 4790	91 Pa Protactinium 1600	92 U Uranium 1132 3818	93 Np Neptunium 640 3902	94 Pu Plutonium 639.5 3235	95 Am Americium 994 2607	96 Cm Curium 1340	97 Bk Berkelium	98 Cf Californium	99 Es Einsteinium	100 Fm Fermium	101 Md Mendelevium	102 No Nobelium	103 Lr Lawrencium

Period/Group →

Legend:

20	Atomic number
Ca	Symbol
Calcium	Name
fcc	Crystal structure
5.58	Lattice parameter (Å)
1550	Density (kg / m³)
25.86	Molar vol (c c)

Periodic table — element data (Atomic number, Symbol, Name, Crystal structure, Lattice parameter(s), Density, Molar volume):

Z	Symbol	Name	Structure	Lattice param (Å)	Density	Molar vol
1	H	Hydrogen	Hexagonal		89.8	11.21
2	He	Helium	hcp	3.57, 5.83	178.5	22.41
3	Li	Lithium	bcc	3.49	530	12.99
4	Be	Beryllium	hcp	2.27, 3.59	1850	4.87
5	B	Boron	Rhom		2340	4.62
6	C	Carbon	Dc (hex)	3.57	2620	4.58
7	N	Nitrogen			1250.6	11.2
8	O	Oxygen			1429	11.2
9	F	Fluorine	Cubic		1696	11.2
10	Ne	Neon	fcc	4.46	901	22.40
11	Na	Sodium	bcc	4.22	970	23.67
12	Mg	Magnesium	hcp	3.21, 5.21	1740	13.98
13	Al	Aluminium	fcc	4.05	2702	9.98
14	Si	Silicon	Dc	5.43	2329	12.06
15	P	Phosphorus	Complex		1820	17.02
16	S	Sulfur	Complex		2070	15.5
17	Cl	Chlorine	Orthorhom		3214	11.03
18	Ar	Argon	fcc	5.31	1784	22.39
19	K	Potassium	bcc	5.22	860	45.36
20	Ca	Calcium	fcc	5.58	1550	25.86
21	Sc	Scandium	hcp	3.31, 5.27	2990	15.04
22	Ti	Titanium	hcp	2.95, 4.68	4540	10.54
23	V	Vanadium	bcc	3.03	6100	8.35
24	Cr	Chromium	bcc	2.88	7190	7.23
25	Mn	Manganese	cubic		7430	7.39
26	Fe	Iron	bcc	2.87	7870	7.1
27	Co	Cobalt	hcp	2.51, 4.07	8850	6.6
28	Ni	Nickel	fcc	3.52	8900	6.59
29	Cu	Copper	fcc	3.61	8960	7.09
30	Zn	Zinc	hcp	2.66, 4.95	7133	9.17
31	Ga	Gallium	Complex		5907	11.8
32	Ge	Germanium	Dc	5.66	5323	13.64
33	As	Arsenic	Rhom		5720	13.10
34	Se	Selenium	Hex		4790	16.48
35	Br	Bromine	Orthorhom		3119	25.62
36	Kr	Krypton	fcc	5.64	3740	22.40
37	Rb	Rubidium	bcc	5.58	1530	55.79
38	Sr	Strontium	fcc	6.08	2600	34.49
39	Y	Yttrium	hcp	3.65, 5.73	4470	19.89
40	Zr	Zirconium	hcp	3.23, 5.51	6490	14.06
41	Nb	Niobium	bcc	3.30	8600	10.8
42	Mo	Molybdenum	bcc	3.15	10220	9.39
43	Tc	Technetium	hcp	2.74, 4.40	11500	8.52
44	Ru	Ruthenium	hcp	2.71, 4.28	12200	8.28
45	Rh	Rhodium	fcc	3.80	12410	8.29
46	Pd	Palladium	fcc	3.89	12020	8.85
47	Ag	Silver	fcc	4.09	10490	10.28
48	Cd	Cadmium	hcp	2.99, 5.62	8650	13.00
49	In	Indium	Tetr	3.25, 4.90	7310	15.71
50	Sn	Tin	bct (dc)		7310	16.24
51	Sb	Antimony	Rhom		6684	18.22
52	Te	Tellurium	Hex		6240	20.45
53	I	Iodine	Orthorhom		4930	25.74
54	Xe	Xenon	fcc	6.13	5897	22.26
55	Cs	Cesium	bcc	6.04	1900	70.95
56	Ba	Barium	bcc	5.02	3500	39.12
57	La	Lanthanum	Hex		6700	20.73
72	Hf	Hafnium	hcp	3.19, 5.05	13200	13.52
73	Ta	Tantalum	bcc	3.30	16654	10.86
74	W	Tungsten	bcc	3.16	19300	9.53
75	Re	Rhenium	hcp	2.76, 4.46	21020	8.86
76	Os	Osmium	hcp	2.74, 4.32	22400	8.49
77	Ir	Iridium	fcc	3.84	22500	8.54
78	Pt	Platinum	fcc	3.92	21450	9.09
79	Au	Gold	fcc	4.08	19320	10.19
80	Hg	Mercury			13456	14.907
81	Tl	Thallium	hcp	3.46, 5.52	11850	17.25
82	Pb	Lead	fcc	4.95	11340	18.27
83	Bi	Bismuth	Rhom		9800	21.32
84	Po	Polonium	monoclinic		9400	22.23
85	At	Astatine				
86	Rn	Radon	Cubic		9730	22.82
87	Fr	Francium				
88	Ra	Radium	Cubic		5000	45.00
89	Ac	Actinium	fcc	5.31	10070	22.54
104	Rf	Rutherfordium				
105	Db	Dubnium				
106	Sg	Seaborgium				
107	Bh	Bohrium				
108	Hs	Hassium				
109	Mt	Meitnerium				

Lanthanides:

Z	Symbol	Name	Structure	Lattice param (Å)	Density	Molar vol
58	Ce	Cerium	Cubic	5.16 / fcc 5.08	6773	20.69
59	Pr	Praseodymium	Hex		6770	20.81
60	Nd	Neodymium	Hex		7000	20.61
61	Pm	Promethium	Hex		6475	20.39
62	Sa	Samarium	Rhom		7540	19.94
63	Eu	Europium	bcc	4.58	5259	28.89
64	Gd	Gadolinium	hcp	3.63, 5.78	7895	20.01
65	Tb	Terbium	hcp	3.60, 5.70	8270	19.22
66	Dy	Dysprosium	hcp	3.59, 5.65	8536	19.04
67	Ho	Holmium	hcp	3.58, 5.62	8540	19.31
68	Er	Erbium	hcp	3.56, 5.59	8795	19.02
69	Tm	Thulium	hcp	3.54, 5.56	9321	18.12
70	Yb	Ytterbium	fcc	5.48	6980	24.79
71	Lu	Lutetium	hcp	3.50, 5.55	9850	17.76

Actinides:

Z	Symbol	Name	Structure	Lattice param (Å)	Density	Molar vol
90	Th	Thorium	fcc	5.08	11720	19.80
91	Pa	Protactinium	Orthorhom		15400	15.0
92	U	Uranium	Orthorhom		18950	12.56
93	Np	Neptunium	Orthorhom		20450	11.59
94	Pu	Plutonium	Monoclinic		19840	12.29
95	Am	Americium	Hex		13600	17.87
96	Cm	Curium			13511	18.28
97	Bk	Beclium				
98	Cf	Californium				
99	Es	Einsteinium				
100	Fm	Fermium				
101	Md	Mendelevium				
102	No	Nobelium				
103	Lr	Lawrencium				

Group headings: IA, IIA, IIIB, IVB, VB, VIB, VIIB, VIII, IB, IIB, IIIA, IVA, VA, VIA, VIIA, VIIIA Or Zero

Periodic Table — Atomic Structure and Related Properties

Legend (key):

11	Atomic number
22.990	Atomic weight
Na	Symbol
Sodium	Name
1.86	Atomic radius
(1+)0.98	Ionic charge, Ionic radius

Period / Group →

Period	IA	IIA	IIIB	IVB	VB	VIB	VIIB	VIII	VIII	VIII	IB	IIB	IIIA	IVA	VA	VIA	VIIA	VIIIA Or Zero
1	1 1.008 H Hydrogen 0.46 (1+)0																	2 4.003 He Helium 1.76
2	3 6.940 Li Lithium 1.52 (1+)0.68	4 9.012 Be Beryllium 1.12 (2+)0.34											5 10.811 B Boron 0.46 (3+)0.25	6 12.011 C Carbon DC 0.77 Gr 0.71 (4+)0.2	7 14.007 N Nitrogen 0.71	8 15.999 O Oxygen 0.68 (2-)1.32	9 18.998 F Fluorine 0.64 (1-)1.33	10 20.183 Ne Neon 1.58
3	11 22.990 Na Sodium 1.86 (1+)0.98	12 24.312 Mg Magnesium 1.59 (2+)0.78											13 26.982 Al Aluminium 1.43 (3+)0.57	14 28.086 Si Silicon 1.18 (4+)0.38	15 30.974 P Phosphorus 1.10 (5+)0.30	16 32.064 S Sulfur 1.04 (2-)1.74 (6+)0.34	17 35.453 Cl Chlorine 0.90 (1-)1.81	18 39.948 Ar Argon 1.88
4	19 39.102 K Potassium 2.31 (1+)1.33	20 40.080 Ca Calcium 1.97 (2+)0.94	21 44.956 Sc Scandium 1.61 (3+)0.68	22 47.900 Ti Titanium 1.46 (4+)0.60	23 50.942 V Vanadium 1.32 (3+)0.65 (5+)0.40	24 51.996 Cr Chromium 1.25 (3+)0.64	25 54.938 Mn Manganese 1.12 (2+)0.91	26 55.847 Fe Iron 1.24 (2+)0.83 (3+)0.67	27 58.93 Co Cobalt 1.25 (2+)0.82	28 58.710 Ni Nickel 1.25 (2+)0.78	29 63.540 Cu Copper 1.28 (1+)0.96	30 65.370 Zn Zinc 1.31 (2+)0.83	31 69.720 Ga Gallium 1.22 (3+)0.62	32 72.590 Ge Germanium 1.22 (4+)0.44	33 74.922 As Arsenic 1.25 (3+)0.69 (5+)0.40	34 78.960 Se Selenium 1.14 (2-)1.91	35 79.909 Br Bromine 1.13 (1-)1.96	36 83.800 Kr Krypton 2.00
5	37 85.470 Rb Rubidium 2.44 (1+)1.48	38 87.620 Sr Strontium 2.15 (2+)1.27	39 88.905 Y Yttrium 1.79 (3+)1.06	40 91.220 Zr Zirconium 1.58 (4+)0.87	41 92.906 Nb Niobium 1.43 (5+)0.69	42 95.940 Mo Molybdenum 1.36 (4+)0.68	43 99 Tc Technetium	44 101.070 Ru Ruthenium 1.35 (4+)0.65	45 102.905 Rh Rhodium 1.34 (3+)0.68	46 106.400 Pd Palladium 1.38	47 107.870 Ag Silver 1.44 (1+)1.13	48 112.400 Cd Cadmium 1.49 (2+)1.03	49 114.820 In Indium 1.63 (3+)0.92	50 118.690 Sn Tin 1.51 (4+)0.74	51 121.750 Sb Antimony 1.45 (5+)0.90	52 127.600 Te Tellurium 1.43 (2-)2.11	53 126.904 I Iodine 1.35 (1-)2.20	54 131.300 Xe Xenon 2.17
6	55 132.905 Cs Cesium 2.62 (1+)1.65	56 137.340 Ba Barium 2.17 (2+)1.43	57 138.92 La Lanthanum (3+)1.22	72 178.490 Hf Hafnium 1.59 (4+)0.84	73 180.948 Ta Tantalum 1.43 (5+)0.68	74 183.85 W Tungsten 1.37 (4+)0.68	75 186.20 Re Rhenium 1.37	76 190.02 Os Osmium 1.37 (4+)0.67	77 192.2 Ir Iridium 1.36 (4+)0.66	78 195.09 Pt Platinum 1.39	79 196.967 Au Gold 1.44 (1+)1.37	80 200.590 Hg Mercury 1.55 (2+)1.12	81 204.370 Tl Thallium 1.70 (3+)1.05	82 207.190 Pb Lead 1.75 (2+)1.32 (4+)0.84	83 208.980 Bi Bismuth 1.56	84 210 Po Polonium 1.7	85 211 At Astatine	86 222 Rn Radon
7	87 223 Fr Francium (1+)1.75	88 226 Ra Radium (2+)1.5	89 227 Ac Actinium (3+)1.3	104 Rf Rutherfordium	105 Db Dubnium	106 Sg Seaborgium	107 Bh Bohrium	108 Hs Hassium	109 Mt Meitnerium									

Lanthanide series:

58 140.13 Ce Cerium (3+)1.11	59 140.92 Pr Praseodymium	60 144.24 Nd Neodymium (3+)1.08	61 145 Pm Promethium	62 150.35 Sm Samarium (3+)1.04	63 152 Eu Europium	64 157.25 Gd Gadolinium (3+)1.02	65 158.92 Tb Terbium	66 162.50 Dy Dysprosium (3+)0.99	67 164.93 Ho Holmium	68 167.3 Er Erbium (3+)0.96	69 169 Tm Thulium	70 173.04 Yb Ytterbium (3+)0.94	71 174.98 Lu Lutetium

Actinide series:

90 232.038 Th Thorium 1.8 (4+)1.10	91 231 Pa Protactinium	92 238.30 U Uranium 1.38 (4+)1.05	93 237 Np Neptunium	94 239 Pu Plutonium	95 241 Am Americium	96 242 Cm Curium	97 249 Bk Berkelium	98 252 Cf Californium	99 254 Es Einsteinium	100 253 Fm Fermium	101 256 Md Mendelevium	102 254 No Nobelium	103 257 Lw Lawrencium

lower in energy as compared to that of the 3d energy level. In this period, 4s states get filled up first, and then starting from scandium to copper, the 3d states get increasingly occupied. These are transition metals, as referred earlier. Similarly, the fifth period contains the second series of transition elements in which 5s states are occupied before the 4d states. We also find the gradual filling up of 4f and 5f orbitals when we scan the elements of the sixth and seventh periods. The elements with partially filled f subshell are called rare earth elements.

We know that the structure of the valence shell determines the chemical properties, establishes the nature of interatomic bonding (which in turn decides the strength and ductility), controls the size of the atom and also decides the electrical and optical properties. Valence electrons play an important role in the ionization of atoms. When an atom absorbs energy, the valence electrons get energized and make transitions to the excited energy states. If sufficient energy is supplied, then the electrons in the valence shell can break away completely from the atom and become free. The atom loses electrons and thus becomes a positively charged ion. The energy required to remove an electron from its orbit such that the electron becomes just free with total energy zero, is called the ionization energy. The ionization energy expressed in eV, gives the ionization potential of that energy state.

The first ionization potential corresponds to the ionization potential for the outermost electronic state. The first ionization potential of elements, Fig. 2.6, is found to increase as we move from left to right in a period. The first element of a period always has a lone s electron in the valence shell. This electron can be removed with relative ease. The inert gases on the other hand, have fully occupied s and p orbitals in their outermost shell. Removal of an electron from the valence shell of these elements requires a relatively large amount of energy.

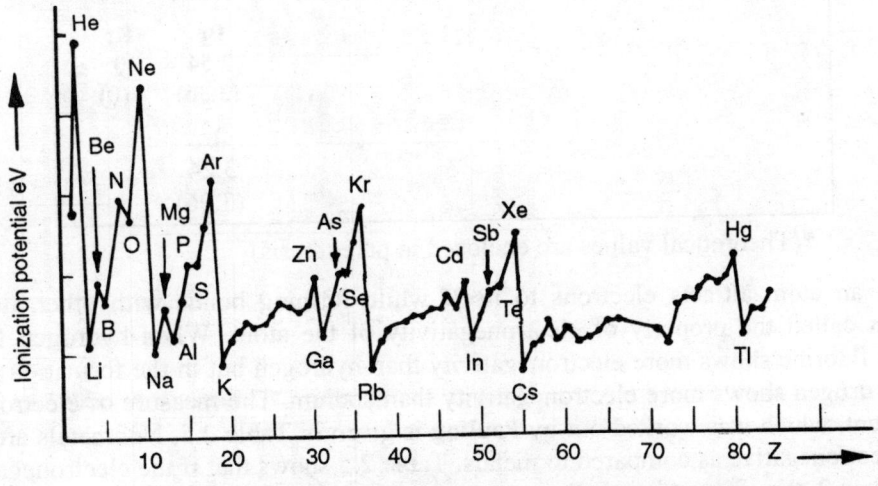

Fig. 2.6 The first ionization potential for elements

Thus, in a period, the ionization potential increases gradually from the minimum value for an alkali metal to the maximum value for an inert gas. Also there is a decrease in the ionization potential as we move from the top to the bottom of a group. For example, we observe that the first ionization potential for Li is highest among all alkali metals, and that for Cs is the lowest. This can easily be explained from the fact that in the atoms with higher atomic number, the outermost electrons will be less and less bound to the nucleus. When an electron is removed from the outer shell of the atom there is a net decrease in mutual repulsive energy and as a result the electronic orbitals shrink and the size of the ion becomes smaller. The remaining electrons in the ion come closer to the nucleus. So removal of the remaining electrons from the electronic orbital of the positive ion requires more energy.

Electrons can join atoms of an electronegative element and make them negative ions. It happens because these atoms have electron affinity. Electron affinity is the work done by the atom to attract an electron from infinity to its outer orbital. The atoms with stable structure like inert gases have no affinity for electrons, which is obvious from their electronic configurations whereas the halogens have high electron affinity, Table 2.4. The alkali metals have very low electron affinity. When an extra electron is added to a neutral atom, there is a weakening of attraction of electrons and as a result the size of the electronic orbitals increases. So the size of the negative ion is larger than that of the atom.

Table 2.4 Electron affinities (eV per atom)

H 0.77 (0.75)							He 0 (0)
Li 0.54 (0.58)*	Be -- (--)	B 0.50 (0.30)	C 1.13 (1.17)	N 0.2 (−0.27)	O 1.48 (1.22)	F 3.62 (3.37)	Ne 0 (0)
Na 0.74 (0.78)	Mg -- (--)	Al 0.40 (0.49)	Si 1.90 (1.39)	P 0.80 (0.78)	S 2.07 (2.12)	Cl 3.82 (3.56)	Ar 0 (0)
						Br 3.54 (3.36)	Kr 0 (0)
						I 3.24 (3.06)	

*(Theoretical values are enclosed in parenthesis)

Sometimes an atom attracts electrons to itself while forming bonds with other atoms. This tendency is called the property of electronegativity of the atom. When hydrogen fluoride is formed the fluorine shows more electronegativity than hydrogen but in the formation of sodium hydride, hydrogen shows more electronegativity than sodium. The measure of electronegativity of an element, which was worked out by Pauling is given in Table 2.5. Nonmetals are found to be more electronegative as compared to metals. Table 2.5 shows that if the electronegativity is 2 or greater than 2, the element is usually a nonmetal.

2.4 BOND ENERGY , BOND LENGTH AND RELATED PROPERTIES

Materials can exist in three possible states – solid, liquid and gaseous. Atoms in the gaseous state are well separated from one another and have random to and fro motion. Atoms can be brought together by applying pressure or by lowering the temperature. Atoms in a liquid have less mobility as compared to those in gaseous state. When the liquid is cooled below the freezing temperature, the mobility of atoms decreases further and they have only vibrational motion about their positions. This is the solid state.

Table 2.5   Pauling's electronegativity values

Group / Period	IA	IIA	IIIB	IVB	VB	VIB	VIIB	VIII	VIII	VIII	IB	IIB	IIIA	IVA	VA	VIA	VIIA	0
1	H 2.1																	He
2	Li 1.0	Be 1.5											B 2.0	C 2.5	N 3.0	O 3.5	F 4.0	Ne --
3	Na 0.9	Mg 1.2											Al 1.5	Si 1.8	P 2.1	S 2.5	Cl 3.0	Ar --
4	K 0.8	Ca 1.0	Sc 1.3	Ti 1.5	V 1.6	Cr 1.6	Mn 1.5	Fe 1.8	Co 1.8	Ni 1.8	Cu 1.9	Zn 1.6	Ga 1.6	Ge 1.8	As 2.0	Se 2.4	Br 2.8	Kr --
5	Rb 0.8	Sr 1.0	Y 1.2	Zr 1.4	Nb 1.6	Mo 1.8	Tc 1.9	Ru 2.2	Rh 2.2	Pd 2.2	Ag 1.9	Cd 1.7	In 1.7	Sn 1.8	Sb 1.9	Te 2.1	I 2.5	Xe --
6	Cs 0.7	Ba 0.9	La 1.1	Hf 1.3	Ta 1.5	W 1.7	Re 1.9	Os 2.2	Ir 2.2	Pt 2.2	Au 2.4	Hg 1.9	Tl 1.8	Pb 1.8	Bi 1.9	Po 2.0	At 2.2	Rn --
7	Fr 0.7	Ra 0.9	Ac 1.1															

Solid materials are large aggregates of atoms. The properties of materials depend on the bonding between the atoms. When two atoms are at a large distance apart, no force acts between them. As they are brought nearer to form bonds, two types of forces – attractive and repulsive – come into play. Initially the resultant force is attractive and it increases as the distance between them decreases and it becomes zero when the atoms occupy equilibrium positions after forming the bond. The distance between the centres of the atoms forming the chemical bond is called equilibrium distance r_o and is also called bond length. Large repulsive force acts between the two atoms when the interatomic distance is less than the equilibrium distance, which forbids them to come nearer. The energy of this system is minimum at equilibrium, Fig. 2.8. The energy required to separate these atoms infinite distance apart is called the bond energy i.e. by supplying energy equivalent to bond energy the bond between the two atoms can be broken.

The potential energy of a pair of atoms is assumed to be zero when the distance between the atoms is infinitely large. The attractive potential energy E_A is negative and E_A increases as the atoms approach nearer to each other due to the increase in attractive force. The work done by the atoms is at the cost of potential energy of the system. As the interatomic distance decreases, the repulsive force begins to dominate and work is done by the repulsive force on the atoms. This contributes repulsive potential energy E_R to the system. The variation of net potential energy $E_N = E_R + E_A$ is plotted as a function of the distance of separation between two atoms, Fig. 2.8. The plot of net potential energy versus interatomic distance has the shape of a trough around minimum potential energy E_o, which is the bond energy. The interatomic distance decreases with the increase of pressure and it increases when the temperature increases.

Bond energy E_o can be expressed either in kJ per mole of bonds or in kcal per mole of bonds or in eV per bond. For example, the bond energy of copper is 56.4 kJ/mole, which means the bond energy for 6.023×10^{23} bonds of copper is 56.4×10^3 Joules i.e. 3.525×10^{23} eV. So,

$$\text{Bond energy per bond} = \frac{3.525 \times 10^{23}}{6.023 \times 10^{23}} \text{ eV} = 0.585 \text{ eV} \tag{2.7}$$

Therefore, the bond energy of copper can also be written as 0.585 eV/ bond.

We can alternatively define the bond energy as the energy required breaking a bond, or the energy required to break one mole (N_{av}) of bonds, where N_{av} is Avogadro number. We also define

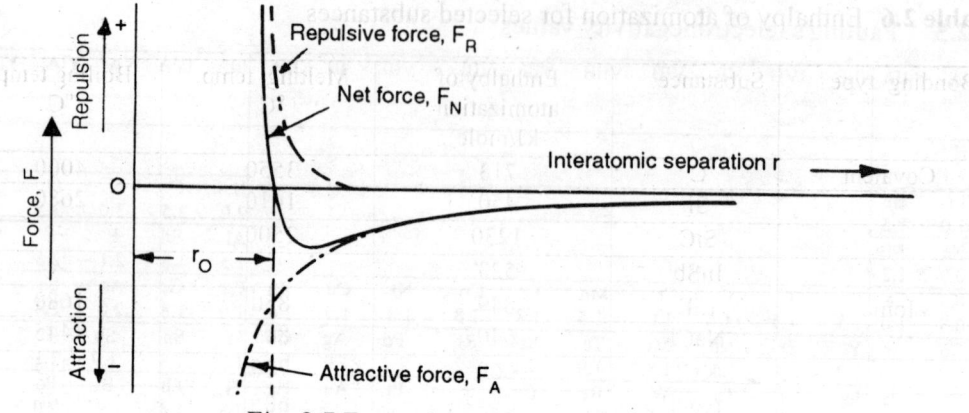

Fig. 2.7 Forces between bonding atoms.

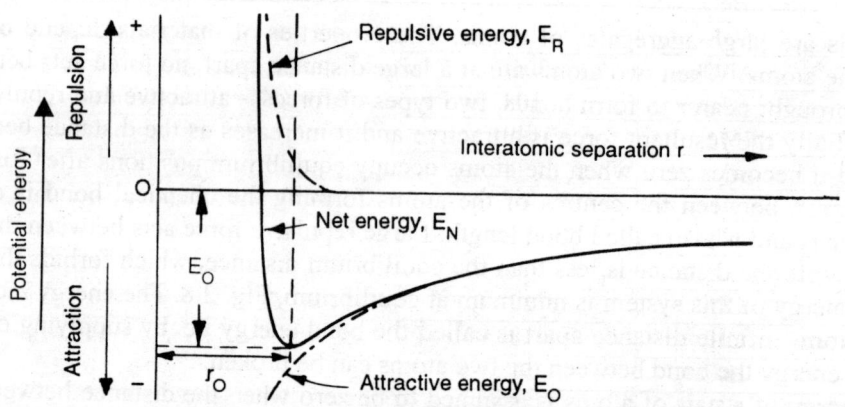

Fig. 2.8 Energy of the bonding atoms.

another quantity called enthalpy of atomization to characterize the physical properties of a material. The enthalpy of atomization is the energy required to break all the bonds of one mole of a substance. The enthalpy of atomization is different from the bond energy of the element. This is because one mole of substance contains more than one mole of bonds. Each atom in a solid forms bonds with a number of neighbours and all the bonds are broken during atomization. Let us consider one mole of an element whose atom has coordination number CN (i.e. the number of nearest neighbours). Two atoms share each bond, so the number of bonds per atom is CN/2. Thus, in one mole of substance there will be N_{av} x CN/2 bonds and hence to vapourize one mole of this substance, N_{av} x CN/2 bonds are to be broken. Thus we obtain,

$$\text{Enthalpy of atomization} = (CN/2) \times (\text{Bond energy/mole}) \quad (2.8)$$

The enthalpy of atomization of certain selected substances are given in Table 2.6. This table shows that the enthalpy of atomization depends on the type of bonding. It is higher for covalent and ionic bonding and lower for metallic bonding. The enthalpy of atomization is also called enthalpy of sublimation for low boiling point solids.

The magnitude of bond energy and the shape of potential energy curve are both dependent on the type of bonding. In case of gaseous substances, the bond energy is low, whereas in case of solid substance bond energy is much higher. Deeper the trough of the energy curve, larger is the

Table 2.6 Enthalpy of atomization for selected substances

Bonding type	Substance	Enthalpy of atomization kJ/mole	Melting temp. °C	Boiling temp. °C
Covalent	C	713	3550	4000
	Si	450	1410	2680
	SiC	1230	2500	--
	InSb	523	--	--
Ionic	LiF	849	850	1680
	NaCl	640	801	1445
	KCl	669	770	1415
	CsCl	649	605	1300
	MgO	1000	2850	3260
	BeO	1170	2550	--
	CaF_2	1548	1420	2510
	Al_2O_3	3060	2050	---
Metallic	Na	108	98	890
	Mg	149	650	1120
	Al	324	660	2447
van der Waals	Ar	75	−189.4	−185.80
	Cl_2	31	−101	−34.7
	O	75	−218.4	−183.0
	CO_2	25	--	--
	CH_4	18	−183	−161
Hydrogen	H_2O	51	0	100
	NH_3	35	--	--

bond energy and shorter is the bond length. Bond energy and bond length are different for different types of bonding in materials as shown in Fig. 2.9. The bonding arises due to exchange or sharing of electrons between the atoms in a molecule. Among the primary bonds, covalent and ionic bonds are generally stronger than the metallic bond. However, there are very few materials, which have bonds of only one type. Occurrence of mixed bonds in materials is most common. We usually classify materials according to the bond type, which is dominant in the material. This classification helps in identifying the properties of the material. Bond energy and bond length also vary from material to material. These variations give qualitative idea about the properties

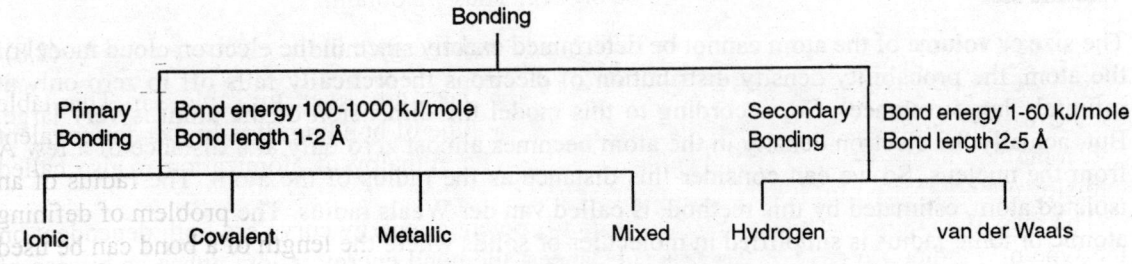

Fig. 2.9 Classification of bonding.

of material like thermal expansion, melting point, boiling point, modulus of elasticity etc. The details of various types of bonding are presented in the next few sections.

Let us consider a substance having bond length r_0 at 0°K and let the corresponding potential energy be E_0. Now if the substance is heated to a temperature T_1, the atoms start vibrating about their equilibrium positions, hence the interatomic distance varies between r_1' and r_2', Fig. 2.10. The average interatomic distance $r_0' = (r_1' + r_2')/2$ becomes more than r_0 at this temperature. Similarly, the average interatomic distance r_0'' at temperature T_2 $(T_2 > T_1)$ becomes greater than r_0'. Thus, the mean bond length increases on heating i.e. the material shows thermal expansion. This is mainly due to the fact that the potential energy curve is asymmetrical. The repulsive energy increases more rapidly than the attractive Coulomb energy as the distance between the atoms decreases. So, the energy curve to the left of r_0 is steeper than that to the right of r_0. The more asymmetrical the energy curve, higher is the coefficient of thermal expansion, Fig. 2.10(a). Hence the thermal expansion coefficient is an inverse function of the slope of the curve for mean vibrational position versus potential energy. It can also be said that higher the bond energy, higher are the melting and boiling points. Roughly, one can say those three-dimensional structures, which have high bonding energy also have deep and more symmetrical energy troughs than those of the weakly bonded materials.

Since most metallic and ionic crystals have high packing factors, the atomic disorder, which accompanies melting, also introduces the expansion of volume. The added volume is due to loose packing in the liquid state and this is known as 'free volume'. This also leads to increase in the amplitude of atomic and molecular vibrations.

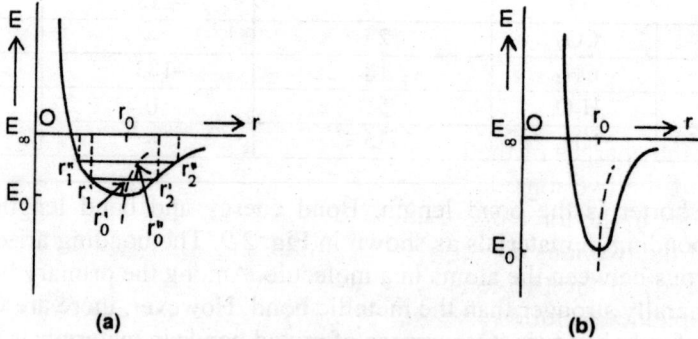

Fig. 2.10 Expansion versus interatomic spacing
(a) Asymmetrical energy curve;
(b) Symmetrical energy curve with strong binding.

Atomic size

The size or volume of the atom cannot be determined exactly since in the electron cloud model of the atom, the probability density distribution of electrons theoretically falls off to zero only at infinitely large distances. So, according to this model the dimension of the atom is very large. But, actually the electron density in the atom becomes almost zero only at a distance of a few Å from the nucleus. So we can consider this distance as the radius of the atom. The radius of an isolated atom, estimated by this method, is called van der Waals radius. The problem of defining atomic or ionic radius is simplified in molecules or solids where the length of a bond can be used to determine atomic or ionic radii. In case of covalent bonding of two similar atoms, the bond length is equal to the atomic diameter. For example, the bond length in a copper crystal is 2.56 Å, which is also the diameter of the copper atom. When a material exists in different crystalline forms at different temperatures, then some ambiguity may arise in this type of definition e.g. atomic diameter of bcc iron is 2.48 Å and that of fcc iron is 2.54 Å. This ambiguity can be explained by the fact that in case of bcc iron the number of nearest neighbours is eight, whereas in fcc iron the nearest neighbours are twelve. The effective distance between the two atoms is more in case the number of nearest neighbours is more. In case of heteropolar bonding, the bond length is the sum of the radii of the two ions. Depending upon the type of bonding, the radius of the atom or ion, as determined from bond length, is called covalent, ionic, or metallic radius.

2.5 PRIMARY BONDS

Relatively stronger bonds e.g. ionic, covalent and metallic are called primary bonds. These types of bonds arise due to sharing or exchange of valence electrons in the s and p orbitals. We shall now discuss various types of primary bonds in more detail.

Ionic bonding

Ionic bonding is the simplest example of interatomic heteropolar bonding. Mostly, ionic bonding forms between a metal and a nonmetal. Metal atom gets ionized positively by donating its outermost electrons and nonmetal atom accepts these electrons to become a negative ion. The resulting ions of opposite nature are held together by the electrostatic attraction. For simplicity, we can assume the ions to be point charges. When the two oppositely charged ions are brought together, the potential energy of the system is given by

$$E_A = \frac{- Z_1 Z_2 e^2}{4\pi \varepsilon_o r} \tag{2.9}$$

where $Z_1 e$, $-Z_2 e$ are the charges of the two ions, e is the electronic charge and r is the distance between the centres of the two ions. E_A is negative since unlike charges release energy as they approach each other due to the force of attraction. The point charge model of the ions is valid till the electron clouds of the two ions do not overlap. If the ions approach still nearer to each other such that electron clouds of the two ions start overlapping then the repulsive force comes into play, which can be explained as follows:

i) The electrons of the overlapping orbitals of the two ions start repelling each other.
ii) When there is an appreciable overlap of the electronic charge cloud of the two ions, then according to the Pauli exclusion principle the force of repulsion avoids the electronic states of the two electrons being same.

The repulsive force is a short-range force and it increases as the interatomic distance r decreases. The energy (E_R) arising due to this repulsive force can be represented as B/r^m, where B and m are empirically determined constants. These constants are different for different materials. Thus, the total energy E is given by

$$E = -\frac{Z_1 Z_2 e^2}{4\pi\varepsilon_o r} + \frac{B}{r^m} + \Delta\varepsilon \tag{2.10}$$

When two atoms form an ionic bond, one of the atoms becomes ionized by losing electrons from its outer orbital and as a result the energy of the ion is increased. The other atom accepts electrons due to its electron affinity and, as a result, the energy of the ion decreases. Thus, the electron transfer at the time of bond formation results in the net change in energy of amount $\Delta\varepsilon$. The coulomb term is dependent on the charges on the ions forming the bond. The ionic bond is stronger for the multivalent ions, since the potential energy will be more negative if Z_1, Z_2 or both have values greater than unity. Substituting the value of r_o, the interatomic separation, for r in Eq(2.10), we obtain bond energy if we know the values of $\Delta\varepsilon$, B and m.

Consider the ionic bonding in KCl. Both ions K^+ and Cl^- are spherically symmetric, having the stable octet electronic structure. All available 3s and 3p states are filled up in these two ions. If the ions approach so close that the n = 3 shells start overlapping then some of the electrons in 3p or 3s shells will be excited to higher energy states such as 3d or 4s. The stable equilibrium of this system is an ion pair bound together, with the nuclei of the two ions at a most probable distance of separation r, Fig. 2.11. If two ion pairs join to form an ion square, Fig. 2.12, the configuration becomes more stable. If a large number of ions cluster together by lining up as the ion squares in all directions, a regular three-dimensional array is generated.

The melting and boiling points are higher for crystals with multivalent ions than those for crystals with monovalent ions. For example, Al_2O_3, MgO etc., are refractory oxides with melting points more than 2000°C. The bond energy for ionic crystals generally lies between 600 kJ/mole to 1500 kJ/mole, Table 2.7. Ionic compounds are usually hard and brittle. Another characteristic of ionic bonds is that, they are nondirectional in nature i.e. a cation in an ionic crystal is usually surrounded by as many anions as possible and vice versa. The bond energy of a crystal depends on the mutual attraction and repulsion between the ions.

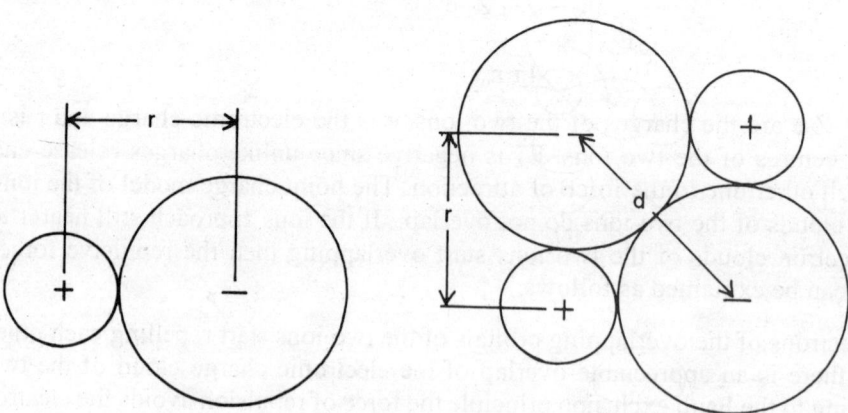

Fig. 2.11 Ion pair   Fig. 2.12 Ion square

To summarize, ionic compounds have the following characteristics:

    i) They are generally found in rigid crystalline solid form.
    ii) They have high melting and boiling points due to strong electrostatic binding forces.
    iii) They are soluble in water but insoluble in organic solvents.
    iv) The bonding is nondirectional in nature.
    v) These compounds are nonconductors of electricity.

Table 2.7 Bond energy in ionic compounds

Compound	Bond energy (kJ/mole)	Bond length (nm)
LiF	1012.8	0.2014
LiCl	831.4	0.2570
LiBr	793.8	0.2751
NaF	896.2	0.2317
NaCl	763.27	0.2820
KCl	693.04	0.3147
KBr	662.53	0.3298

Covalent bonding

Covalent bonding is formed due to sharing of electrons between the bonding atoms. Sharing of electrons results in stable octet electronic structure of atoms. The shared electrons may be considered to belong to both the atoms to achieve the inert gas configuration. Covalent bonding is explained schematically in Fig. 2.13. Carbon atom has four electrons in its valence shell. In order to have stable configuration carbon atom must have four more electrons in its outermost shell. In methane, the carbon atom is covalently bonded to four hydrogen atoms.

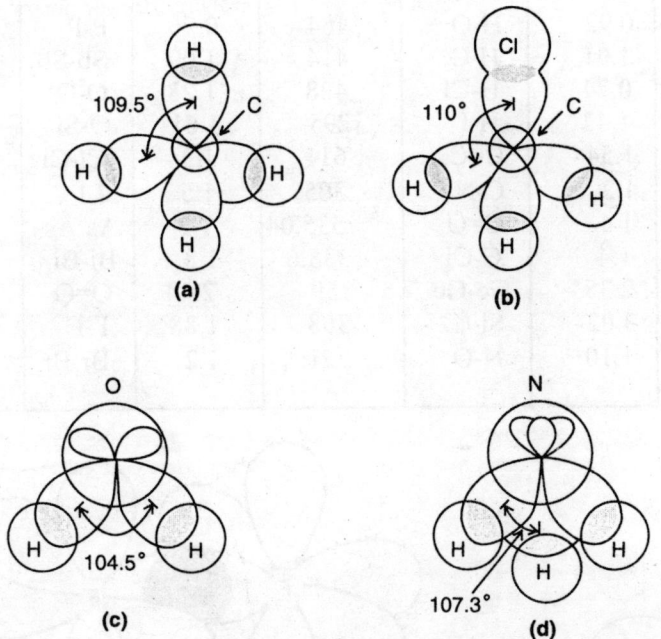

Fig. 2.13 Covalent bonding: (a) In methane, each bond angle (C–H–C bond) is equal to 109.5° (b) Methyl chloride has C–H–C bond angle equal to 110° (c) Water has H–O–H bond angle equal to 104.5° (d) Ammonia has H–N–H bond angle equal to 107.3°.

Hydrogen atom has one electron in the valence shell and it can acquire an inert gas configuration by sharing a pair of electrons with the carbon atom, one each being contributed by carbon and hydrogen, respectively. The carbon now has four additional shared electrons one from each of the four hydrogen atoms and thus attains the stable configuration of neon. Formation of covalent bonding in methane, methyl chloride (CH_3Cl), water (H_2O) and ammonia (NH_3) is shown in Fig. 2.13. Table 2.8 gives the bond energies and bond lengths of specific bonds occurring commonly in covalent compounds. A covalent bond is stereospecific i.e. it has a specific direction in space. Stereospecificity of covalent bonds arises due to strong attraction that exists only between the first neighbours and the directional nature of p and d orbitals. The stereospecific characteristic of bonds gives rise to specific bond angles. As a consequence, carbon has a coordination of only four in methane, although the surrounding space of carbon atom can contain many more of the small hydrogen atoms. Nonmetallic atoms having unfilled p orbitals predominantly form covalent bonds because these orbitals are directional in nature and therefore permit efficient overlapping of orbitals in the direction of maximum electron probability density, Fig. 2.14.

Table 2.8 Covalent bond lengths and associated bond energies

Nature of bond	Bond energy kJ/mole	Bond length(Å)	Nature of bond	Bond energy kJ/mole	Bond length (Å)	Nature of bond	Bond energy kJ/mole	Bond length(Å)
H-F	561	0.92	H-O	464	0.96	P-P	214	1.10
H-N	389	1.01	H-C	414	1.09	Sb-Sb	126	1.45
H-H	436	0.74	H-Cl	428	1.28	O-O	217.36	1.50
H-Br	362	1.42	H-I	295	1.61	O-Si	376.2	1.8
C-C	347	1.54	C=C	611	1.34	Cl-Cl	239	2.02
C≡C	837	1.2	C-N	305	1.5	I-I	149	2.67
C-O	359.5	1.4	C=O	535.04	1.2	As-As	134	1.25
C-F	451.44	1.4	C-Cl	338.6	1.8	Bi-Bi	105	1.56
Si-Si	177.65	2.35	Ge-Ge	159	2.44	O=O	494	1.21
Sn-Sn	146	3.02	Si-C	308	1.88	F-F	154	1.44
N≡N	942	1.10	N-O	251	1.2	Br-Br	190	2.28

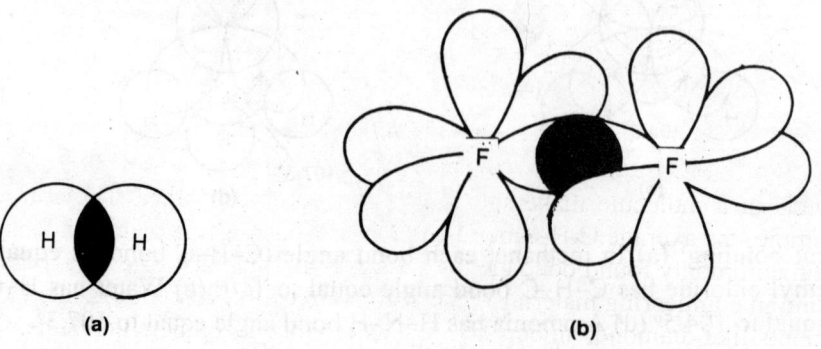

(a)

(b)

Fig. 2.14 Overlapping of orbitals, (a) s-s in H_2, (b) p-p in F_2

The electronic structure of fluorine is $1s^2 2s^2 2p^5$, having a vacant p_z orbital. Two fluorine atoms form a fluorine molecule when they come close to each other such that the half-filled p_z orbitals overlap. In oxygen, the unpaired electrons in p_y and p_z are shared with another atom. In the water molecule, the two unpaired 2p electrons of oxygen atom are shared with the lone electrons of two hydrogen atoms. Formation of this bond is more appropriately visualized to occur between hybridized orbitals (produced by the interaction between 2s and 2p orbitals) of oxygen atom and 1s orbitals of hydrogen atoms. These hybridized orbitals are called sp^3 orbitals and can hold eight electrons. Their orientations are such that they can be thought of parallel to the lines joining the centre of a regular tetrahedron to its four corners, Fig. 2.15. In water molecule, four electrons occupy two of the sp^3- hybridized orbitals of oxygen and other two orbitals contain one electron each. These electrons, along with electrons of hydrogen atoms, are shared between oxygen and hydrogen atoms and thus covalent bonds are formed.

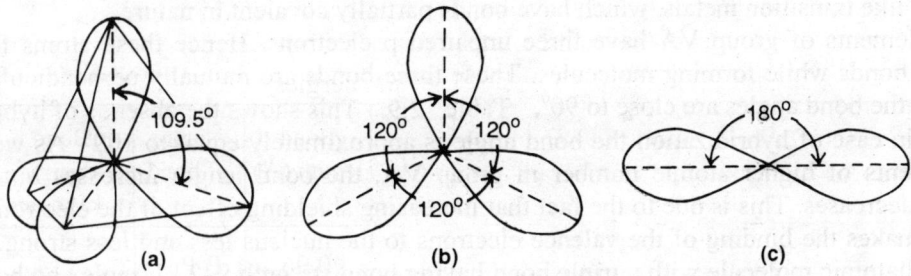

Fig. 2.15 Orientation of (a) sp^3,(b) sp^2 and (c) sp orbitals.

By similar mechanisms other elements of group VIA also form covalent bonds. When there is end to end overlap of p orbitals, σ bond is formed, Fig. 2.16a. π bond is formed due to lateral overlap of p orbitals, Fig. 2.16b.

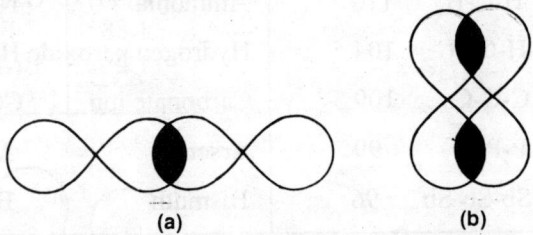

Fig. 2.16 (a) σ bond, (b) π bond

Covalent bonds in a molecule make specific angles with each other. These are called bond angles. For example, the average C-H-C bond angle in various organic compounds is 109.5°. Data regarding certain commonly found bond angles are given in Table 2.9. The tetrahedral bond angle of 109.5° is found in engineering materials which are nonmetallic compounds like CCl_4, SiF_4 etc., as well as elements like diamond, silicon, SiC etc. Many other nonmetallic clemental molecules have covalent bonding e.g. H_2, N_2, O_2, Cl_2. Substances with multiple bonds have shorter bond length and higher bond energy as compared to substances having a single bond, Table 2.8.

The number of covalent bonds that are possible for a particular atom is given by 8-N', where N' is the number of electrons in the valence shell. In diamond, each carbon atom is covalently bonded with four other carbon atoms giving rise to a three-dimensional interconnected structure. Covalent bonds are very strong, as found in diamond; which has high melting temperature 3550°C. In some cases it is weak as found in bismuth, which has melting point of 270°C. The covalent bond forms the skeleton of the polymeric long chain molecules, which are discussed in Chapter 5.

Strength of a covalent bond depends on factors like degree of ionic character, order of the bond (i.e. single bond, double bond etc.) and resonance of bonds (e.g. in H_2) between alternate positions. Most stable covalent bonds are formed between atoms of nonmetals like nitrogen, oxygen, fluorine, chlorine etc. Certain elements of the fourth and fifth groups like germanium, arsenic and selenium form bonds that are partially covalent and partially metallic. There are elements like transition metals, which have bonds partially covalent in nature.

The elements of group VA have three unpaired p electrons. Hence these atoms form three covalent bonds while forming molecules. These three bonds are mutually perpendicular to each other i.e. the bond angles are close to 90°, Table 2.9. This shows the absence of hybridization, because in case of hybridization the bond angle is approximately equal to 109°. As we examine the elements of higher atomic number in group VA, the bond length increases but the bond strength decreases. This is due to the fact that increasing shielding effect of the electrons of inner orbitals makes the binding of the valence electrons to the nucleus less and less strong. Nitrogen forms a diatomic molecule with a triple bond having bond strength 942 kJ/ mole and bond length 1.10 Å.

Table 2.9 Bond angles in covalently bonded compounds

Compound	Chemical formula	Bond	Bond angle in °	Compound	Chemical formula	Bond	Bond angle in °
Methane	CH_4	H-C-H	109.5	Ethane	C_2H_6	H-C-H	109.3
Chloromethane	CH_3Cl	H-C-H	110	Ammonia	NH_3	H-N-H	107.3
Water	H_2O	H-O-H	104.5	Hydrogen peroxide	H_2O_2	O-O-H	100
Diamond	C	C-C-C	109.5	Carbonate ion	CO_3^{-2}	O-C-O	120
Phosphorus	P	P-P-P	99	Arsenic	As	As-As-As	97
Antimony	Sb	Sb-Sb-Sb	96	Bismuth	Bi	Bi-Bi-Bi	94

The diamond form of carbon and other elements of group IVA share all the four hybridized orbitals with four neighbours. These elements form crystals with a three-dimensional network of covalent bonds. Characteristics of compounds having covalent bonds are:

a) Bonds in these compounds are directional in nature.

b) They can exist in all states of matter.

c) They are insoluble in water and soluble in nonpolar solvents like alcohol, benzene, chloroform and paraffin etc.

d) Some of these covalent compounds are soft, rubbery elastomers, while others form plastics.

e) The boiling and melting temperatures of these compounds are low except in case of diamond.

f) The covalent bonding is strong in case of diamond; hence it has a very high melting point.

g) These compounds are homopolar. The bonding valence electrons are usually bound to an individual or pair of atoms in a covalent compound, hence these electrons have less mobility. These compounds are usually insulators or at most semiconductors.

Metallic bonding

Metals can be visualized as a three-dimensional network of positive metal ions embedded in the charge cloud of free electrons, called free electron gas contributed by the metal atoms. In a metal, valence electrons are loosely bound to the nucleus. These electrons resonate between different atoms and thus form metallic bonding. They have freedom to move throughout the crystal and ultimately act as a medium to hold the ions in the crystal lattice. Therefore the origin of metallic bonding lies in the attractive nature of force between positive metallic ions and the electron gas. Metallic bonding is spherical in nature i.e. nondirectional, unlike covalent bonding. The valence electrons forming covalent bond are localized, and stay in the region between the atoms forming covalent bond. Metal atoms have one, two or at most three valence electrons. Electrons other than valence ones and atomic nucleus form ionic core, which possesses a net positive charge equal to the number of valence electrons. Free electron charge cloud surrounding the ion core prevents the positively charged ions from repelling each other. In case of metallic bonding, a metal ion is surrounded by as many numbers of ions as can be accommodated in the space around the ion. Each metal ion is surrounded by 12 metal ions for close-packing. The model of metallic state was first proposed successfully in the classical free electron theory of Drude and Lorentz. The metallic bonding may be weak or strong e.g. bond energy for mercury is 68 kJ/mole whereas bond energy for niobium is 180 kJ/mole. Similarly, melting point also varies a lot from one metal to another, Table 2.10. Metallic bonding is found in elements of groups IA, IIA, and IB to VIIB and VIII of the periodic table. These elements are good conductor of both heat and electricity because of the presence of free electrons. Metals have the following properties:

a) They are crystalline in nature.

b) Since free electrons are present, they are good conductors of heat and electricity.

c) They are opaque to light, since the free electrons in a metal absorb light energy.

d) Metallic bonding are strong and hence metals have high melting points.

e) Metals also have high reflectivity and luster.

2.6 SECONDARY BONDS

Secondary bonding arises due to van der Waals type weak attractive forces acting between the internal dipoles of a material. Internal dipoles are present due to dispersion effect i.e. statistical irregularities of electron distribution in atoms or molecules or due to the existence of molecular asymmetry. Normally the spatial distribution of electrons is symmetrical with respect to the positively charged nucleus. Due to proximity of other atoms or due to constant vibrations of atoms, distortions may be created in the electronic charge distribution of atoms or molecules. These short lived distortions can give rise to small electric dipoles. These induced dipoles can create a displacement of the charge distribution of adjacent molecules or atoms and can induce dipoles.

Table 2.10 Enthalpies of atomization H (kJ/mole) of metals

Metal	H kJ/mole	Melting point °C	Boiling point °C	Metal	H kJ/mole	Melting point °C	Boiling point °C
Li	162	181	1331	B	565	2030	3927
Na	108	98	890	Al	326	660	2447
K	90	64	766	Ga	272	30	2237
Rb	82	39	701				
Cs	78	29	685	Sc	376	1539	2480
				Ti	469	1668	3280
Be	324	1277	2477	V	562	1900	3300
Mg	146	650	1120	Cr	397	1875	2642
Ca	178	838	1492	Mn	285	1245	2041
Sr	163	768	1370	Fe	415	1537	2887
Ba	178	714	1638	Co	428	1495	2887
				Ni	430	1453	2837
				Cu	339	1083	2582
				Zn	130	420	908

The electric dipoles attract each other weakly and form secondary bonding. This type of bonding is also called van der Waals bonding since van der Waals forces come into play when there are electrostatic interactions between adjacent electric dipoles. Other stronger bonding often masks the effect of secondary bonding. The other type of secondary bonding is due to interaction among the permanent dipoles found in asymmetric molecules or polar molecules where the negative and positive charge centres do not coincide. For example, HCl molecule is a polar molecule shown schematically in Fig. 2.17.

$$H^+ \quad :\ddot{C}\!\ddot{l}: \qquad H:\ddot{C}\!\ddot{l}:$$

(a)   (b)

Fig. 2.17 HCl molecule

Liquefaction and solidification of certain inert gases, and gaseous elements of group VA, VIA and VIII A are possible due to existence of van der Waals type of bonding. The melting and boiling points are, in general, quite low in case of materials with this type of bonding. The bond energy corresponding to secondary bonding ranges from a few kJ /mole to about 40 kJ/mole, Table 2.11.

Hydrogen bonding

Hydrogen bonding is a special case of polar molecular bonding in which the bonding occurs between the molecules containing hydrogen. Hydrogen forms covalent bonding with oxygen (in H_2O), nitrogen (in NH_3), fluorine (in HF) etc. In water molecule, two hydrogen atoms form covalent bonding with one oxygen atom. In these molecules, the centres of the positive and negative charges do not coincide since the electronegativity of the oxygen atoms is more than that of the hydrogen atoms. Therefore the bonding electrons have more probability to stay near the oxygen atom than near the hydrogen atoms. Thus the water molecule will have polar character

Table 2.11 Hydrogen bonding energies in some molecules

Molecule	Bond energy (kJ mole^{-1})
Water (H$_2$0)	20.5
Ammonia (NH$_3$)	7.8
Hydrogen fluoride(HF)	31.5

with a net negative charge at the oxygen end and positive charge at the hydrogen end where there is a bare proton. These polar molecules possess permanent dipole moments and hence secondary bonding arises between the water molecules due to attractive force acting between the positively charged end of one molecule and the negatively charged end of the adjacent molecule, Figs. 2.18 and 2.19. Thus we can see that a single proton forms a bridge between two negatively charged atoms. This is called hydrogen bonding. The bond energy of the hydrogen bond is as high as 52kJ/mole. The melting point and boiling point of materials possessing hydrogen bonding are normally higher than those having van der Waals type of bonding. Hydrogen bonding is directional and strong enough to keep the water in liquid state. Water molecules form a relatively open network of hydrogen bonds in the solid state (ice) but a more closely packed structure in the liquid state. This is the reason why the density of water at 4°C is higher than that at 0°C. The strength of the secondary bond decides the structure and properties of many nonmetallic materials e.g. plastics, paraffin, graphite etc. Thus, in conclusion, commonly found secondary bonding are (i) dispersion bonding, (ii) dipolar bonding, and (iii) hydrogen bonding. The compounds with secondary bonding have the following characteristics:

a) Transparent to light and good insulator with the exception of water.

b) Low melting and boiling points.

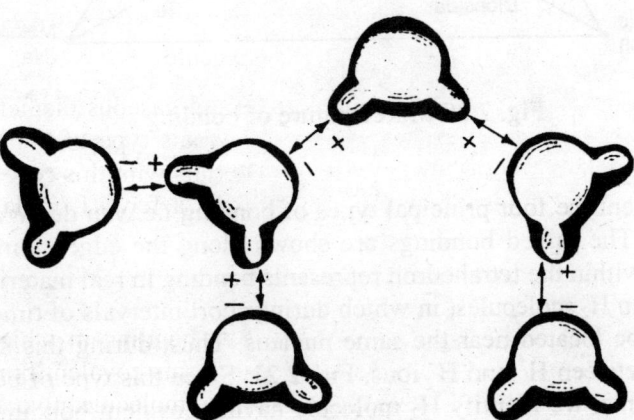

Fig. 2.18 Hydrogen bonding between water molecules.

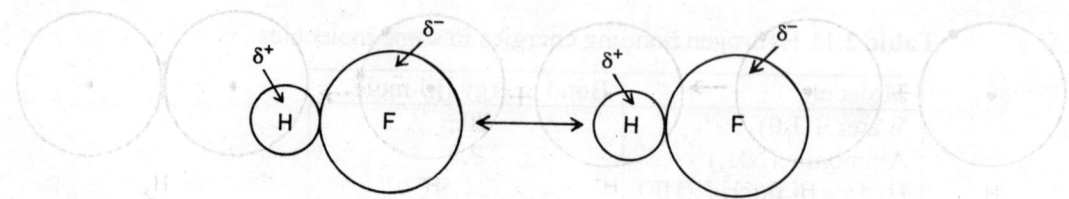

Fig. 2.19 Hydrogen bond between HF molecules.

2.7 MIXED NATURE OF BONDING

The bonding between the atoms in many of the materials cannot be classified purely as one of the ideal bonding types described above. The bonding may have mixed character as indicated in Fig. 2.20. In this figure, a tetrahedron showing the distribution of types of bonding in materials

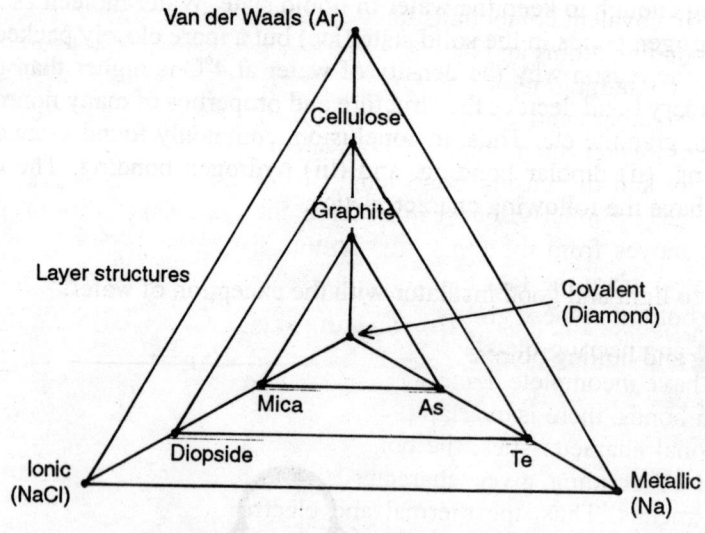

Fig. 2.20 Mixed nature of bonding.

where the apices represent the four principal types of bonding i.e. van der Waals, ionic, metallic and covalent bonding. The mixed bondings are shown along the edges joining the corners. A point on the surface or within the tetrahedron represents bonding in real materials. A specific type of bond mixing occurs in H_2 molecules, in which during short intervals of time ($\sim 10^{-17}$s) both the sharing electrons may be located near the same nucleus. Thus, during this short interval, there will be ionic bonding between H^+ and H^- ions, Fig. 2.21. Since this type of bonding occurs for a very short interval of time, we identify H_2 molecule having covalent bonding. In HCl molecule bonding has more ionic and less covalent character. In this case, ionic character predominates in aqueous solution and covalent character predominates in gaseous form.

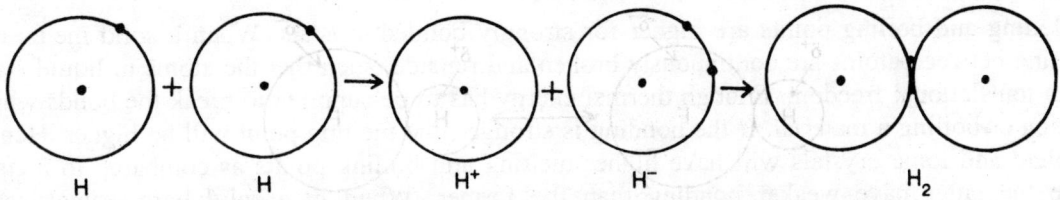

Fig. 2.21 Representation of bond mixing.

2.8 PROPERTIES OF MATERIALS RELATED TO BONDING

Properties of material are related to the types of bonding. As we investigate materials from left to right in the periodic table, we find the character of bonding changes from metallic to covalent. Hence the related properties also change. In case of metallic bonding, the valence electrons are free and mobile as compared to those in covalent and ionic compounds. In case of ionic compounds, the ions attain the stable configuration of inert gases by donating or accepting electrons, so the valence electrons are localized to the region near the specific ions. Hence there are no free electrons. In covalent compounds likewise, since covalent bonding results in inert gas configuration of the bonding atoms and hence no free electrons are here either. This explains the fact that compounds containing ionic and covalent bonds are good thermal and electrical insulators. Metals on the other hand, are good conductors of heat and electricity.

Metallic bonds are weaker than ionic and covalent bonds. This is explained using the increasing bond energy and decreasing bond length from left to right, in the third row of the periodic table, Table 2.1. Similarly, one can observe the transition from covalent to metallic characteristic as one moves from the top to the bottom in groups IIIA to VIA of the periodic table. This is again clear from the fact that there is a decrease in bond energy as one compares elements from top to bottom in these groups.

A large number of engineering metals and component of alloys belong to the three transition series. These metals have incomplete d-shell leaving apart valence electrons in outermost s shell. As these metals form bonds, there is overlap of s orbitals as well as overlap of d orbitals. Since d orbitals have directional characteristics, the bonds in transitional metals have a partial covalent character. This type of bonding gives characteristics, which fall between those of covalent crystals and typical metals. Thus, the thermal and electrical conductivities of the transitional elements are lower than those of typical metals such as copper, sodium, aluminium etc.

Mechanical properties like hardness, strength, ductility and malleability also depend on the type of bonding. Solids with directional covalent bonding are brittle e.g. diamond is brittle and hard since it has strong covalent bonding. The hardness of the material is decided by the strength and type of bonding. Since metallic bonds are relatively weak and nondirectional, metals are soft, ductile and malleable. Ionic solids also show a limited extent of ductility since these bonding are nondirectional. As the incident radiant energy is easily absorbed by the valence electrons in metals, metals are opaque and lustrous.

When we consider properties like the thermal expansion of materials, we find that the symmetry of the potential energy versus interatomic distance curve plays a very important role to decide this type of properties, Fig. 2.10. Deep potential wells are more symmetrical about equilibrium position than shallow potential wells. So thermal expansion and its coefficient is less for covalent and ionic crystals as compared to that for metals.

Melting and boiling points are higher for strongly bonded crystals. When a solid melts, the bonding between atoms are continuously broken and remade, such that the atoms in liquid state attain translational freedom. Enough thermal energy has to be supplied to break the bonds while melting or boiling a material. If the bonding is stronger, the melting point will be higher. Hence covalent and ionic crystals will have higher melting and boiling points as compared to metals since the latter have weaker bonding than the former. When in a solid both primary and secondary bonding are present, the solid starts melting if only secondary bonds between the molecules are broken on heating. The melting point of such materials only reflects the strength of secondary bonding between the molecules, although the covalent bonds in the molecules are not broken during melting. To understand this point, let us consider diamond, which has a three-dimensional network of covalent bonding (i.e. primary bonding) but does not have any secondary bonding. The melting point of diamond is 3500°C. This can be explained if we examine the bonding of methane. Although methane molecules have covalent bonding, they are held together by secondary bonding (van der Waals type) in the solid state. They melt easily at a temperature as low as –182°C. Transition metals have high melting points compared to other metals since they have partial covalent bonding for e.g. Fe (1535°C), Ni (1453°C) Co (1495°C) etc.

EXERCISES

2.1 How many electrons are there in the 3d subshell of Mn, Fe and Co and how are their spins aligned? How are electron spins aligned in 4s subshell of these atoms?
Ans: Mn: $3d^5$; Fe: $3d^6$; Co: $3d^7$; for spin alignment see Table 2.3(c)

2.2 Write the electronic configuration for Be, H^+, Mn^{2+}, Cu, Zn, and Cl^-.
Ans: Be: $1s^2 2s^2$; H^+: no electrons; Mn^{2+}: $1s^2 2s^2 2p^6 3s^2 3p^6 3d^3 4s^2$; Cu: $1s^2 2s^2 2p^6 3s^2 3p^6 3d^{10} 4s^1$; Zn: $1s^2 2s^2 2p^6 3s^2 3p^6 3d^{10} 4s^2$; Cl^-: $1s^2 2s^2 2p^6 3s^2 3p^6$

2.3 If the uncertainty in momentum of a cricket ball of mass 0.25 kg is $\Delta p = 10^{-30}$ kg ms^{-1}, find the minimum uncertainty in determining the position (Δx) of the ball. Repeat the calculation for an electron ($m_0 = 9 \times 10^{-31}$ kg) having same Δp.
Ans: 5.273×10^{-5} m for both.

2.4 (a) Calculate the coulomb attractive force between Na^+ and Cl^- ions in NaCl crystal where Na^+ and Cl^- ions touch each other. Given $r_{Na^+} = 0.098$ nm and $r_{Cl^-} = 0.181$ nm.
     (b) Calculate the net energy (Eq 2.10) of the system if $m = 9$, $B = 4.92 \times 10^{-105}$ Jm^9
Ans: (a) 2.96×10^{-9} N; (b) -3.46×10^{-19} J

2.5 Compare the first ionization energies of elements that are (a) good electrical conductors (e.g. Li, Na, Al, Cu) and (b) poor electrical conductors (e.g. P, S, Cl).

2.6 The potential energy for the formation of a bond between two univalent ions is given by
$$W = -(Ae^2/r) + (B/r^9)$$
where $A = 1.20 \times 10^{11}$ Jm, $B = 4.47 \times 10^{-105}$ Jm^9, determine the equilibrium spacing r_0.
Ans: 2.46 Å

2.7   The bond energy of diamond is 347 kJ mole⁻¹ and the enthalpy of atomization is 713 kJ mole⁻¹. Explain.

Ans: Enthalpy = 347 × (4/2) = 694 kJ/mole theoretically which is very close to the experimental value.

2.8   Explain how the increase in bond energy is related to the increase in melting point and why such trend is absent from Si to Cl.

2.9   Given the following potential energy diagram, Fig. 2.22 for a monoatomic bcc calculate the enthalpy of atomization.

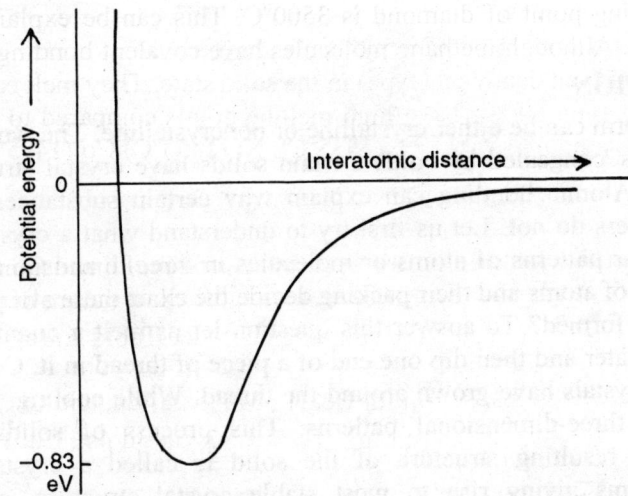

Fig. 2.22 Potential energy diagram for monoatomic bcc crystal.

Ans: 320 kJ

2.10 Explain why the density of the water decreases when it cools below 4°C and ultimately converts to ice.

2.11 We require 849.2 kJ of heat energy to atomize 1 mole of tungsten (bcc). The density of tungsten is 19.3 × 10³ kg/m³ at 0°K. Atomic weight is 183.85 amu. Calculate the
a) bond length
b) bond energy
c) B/A ratio,
where the expression for potential energy of the system is given by

$$W = -A/r + B/r^9$$

Ans: (a) 2.744 × 10⁻¹⁰ m; (b) 212.3 kJ/mole; (c) 3.57 × 10⁻⁷⁸

is coordinated with six Na^+ ions in a similar manner. The distance between Na^+ and nearest Cl^- ion is fixed (a/2=2.8Å being lattice parameter) and is constant throughout the crystal. This occurs for the sodium and/or Cl^- ions and as a result a long-range order exists in which there is order not only at the nearest neighbour but also in more distant neighbours. In case of Cl^- ions, each has twelve Cl^- ions as second neighbours, positioned at a distance of √2 times.

3.1 INTRODUCTION

Materials in solid form can be either crystalline or noncrystalline. The word crystal has its origin in Greek and means 'congealed by cold'. Certain solids have crystal structure while others are glassy substances. Atomic bonding can explain why certain substances form specific crystal structures while others do not. Let us first try to understand what a crystal structure is. Crystal structures are regular patterns of atoms or molecules in three-dimensional space. The nature of bonding, dimension of atoms and their packing decide the exact nature of pattern.

How crystals are formed? To answer this question let us melt a quantity of sugar in half its volume of boiling water and then dip one end of a piece of thread in it. Cool it for 8 to 10 hours. It is seen that the crystals have grown around the thread. While cooling, the atoms in the liquid may form regular three-dimensional patterns. This process of solidification is known as crystallization. The resulting structure of the solid is called a crystal structure. An ideal arrangement of atoms giving rise to most stable crystal structure, satisfies the following conditions:

1. The crystals are electrically neutral.
2. The ion-ion repulsion is minimized.
3. The ions or atoms in crystals are packed as closely as possible under the constraints of specific bonding.
4. The free energy of the system becomes minimum.

In some cases the atomic arrangement in solids may not be regular. Such solids are noncrystalline e.g. glass.

3.2 RANGE, ORDERING AND CRYSTALLINITY

The long-range order is closely related to crystallinity. A solid can become crystalline if long-range order exists in the arrangement of atoms or ions. When in a given solid each similar atom or ion is in the same position relative to its neighbours, (except for the atoms or ions on the surface or on the grain boundary of the crystal) we say that long-range order exists in the solid. As an example, consider NaCl crystal, Fig. 3.1, where each Na^+ ion is surrounded by six Cl^- ions arranged in a specific manner i.e. if we join the nearest Cl^- ions we generate an octahedron, each vertex of which is occupied by Cl^- and Na^+ is located at the centre of the octahedron. Each Cl^- ion is coordinated with six Na$^+$ ions in a similar manner. The distance between Na$^+$ and nearest Cl$^-$ ion is fixed (a/2, 'a' being lattice parameter) and is constant throughout the crystal. This occurs for all sodium and chloride ions and as a result, a long-range order exists in which there is order not only in the first nearest neighbour but also in more distant neighbours. In case of Cl$^-$ ions, each has twelve Cl$^-$ ions as second neighbours, positioned at a distance of $\sqrt{2}$ times the first

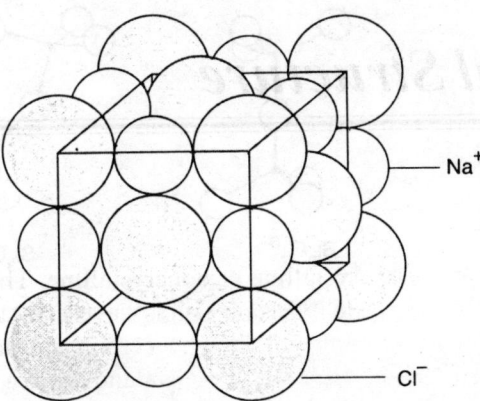

Fig. 3.1 Sodium chloride crystal.

nearest neighbour distance and eight Na$^+$ ions as third neighbours at $\sqrt{3}$ times the first nearest neighbour distance. One can easily discover such regular patterns in more distant neighbours, both for chloride as well as for sodium ions. Nearly all metals, ceramics and certain polymers have crystalline structures, if cooled in a specific manner, which is ideal for crystal formation.

When regular pattern is observed in a solid only in the first nearest neighbour, only short-range order exists in the solid. For example, in silica glass each silicon atom at the centre of the tetrahedron is having covalent bonding with four oxygen atoms situated at the corners of the tetrahedron, Fig.3.2 and each oxygen atom is bonded with two silicon atoms. This geometry can be explained by considering the overlap of the four sp^3 hybrid orbitals of Si with a hybridized orbital of each oxygen atom. Each corner of the tetrahedron is common to two tetrahedra and the tetrahedra are arranged in random manner. Thus, the resulting network structure in silica glass does not have any long range order as shown in Fig. 3.3.

3.3 SPACE LATTICE

In a crystalline solid, identical atoms or groups of atoms are arranged in a periodic and regular three-dimensional array. Now if these identical sets of atoms are considered as points then the crystalline solid can be viewed as a regular three-dimensional array of points extended infinitely in all directions, in which every point has identical surrounding in the array. These points are lattice points and the regular array of points in three-dimensions is called a space lattice. Atomic arrangement in the crystal lattice can be described completely by specifying the positions, arrangements and orientation of molecules or atoms in some repeating unit of space lattice. If this repeating unit is given lattice translations in three-dimensions without leaving any gap, it generates space lattice. The lattice translations are lattice vectors, which join two lattice points.

This repeating unit is defined as a unit cell. In other words, we can define the unit cell as any convenient unit, which when repeated in three-dimensions through lattice translations, generates the crystal structure. The unit cell has the symmetry of the crystal structure. The edges of unit cell are lattice translations.

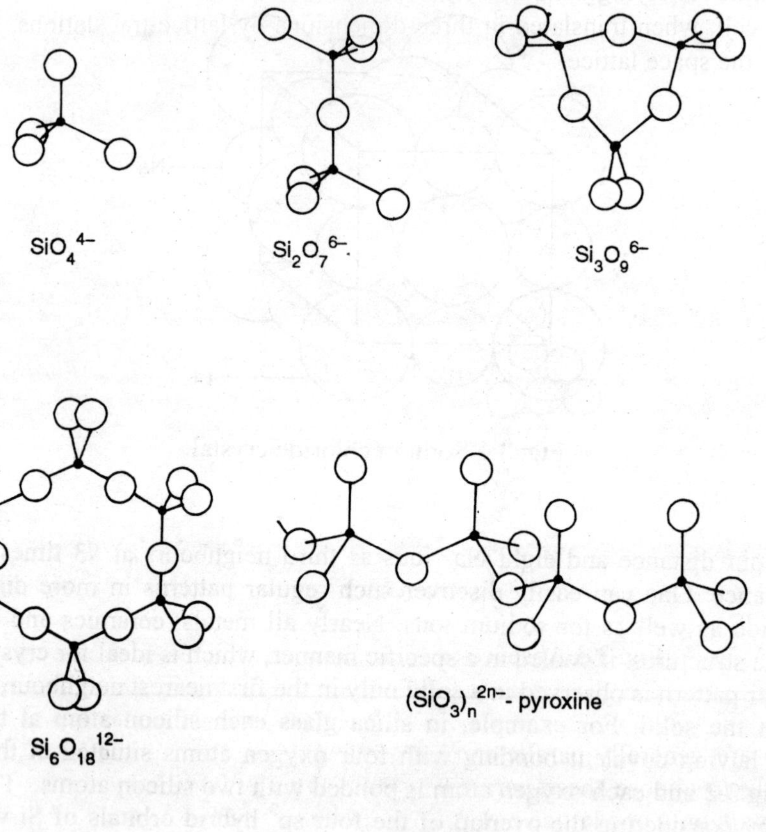

Fig. 3.2 Structure of silica glass.

A space lattice can have a number of unit cells that satisfy the above criteria. The conventional way of choosing a unit cell for a space lattice is the following:

i) The unit cell should have the same geometry as that of the crystal

ii) The unit cell should possess minimum area in case of a two-dimensional lattice and minimum volume in case of three- dimensional lattice i.e. it should contain minimum number of lattice points, while satisfying condition (i)

iii) The unit cell, when translated in three-dimensions by lattice translations, should be able to generate the space lattice.

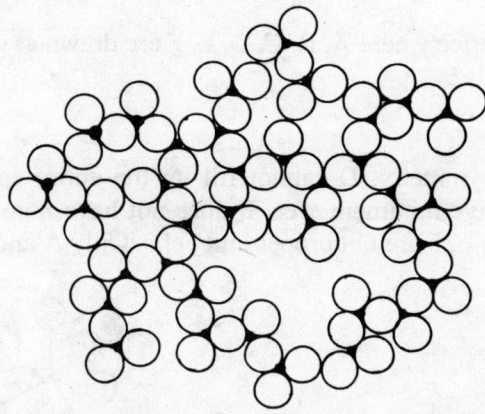

Fig. 3.3 Absence of long-range order in silica glass.

The unit cells are generally parallelepipeds or prisms having three sets of parallel faces. It is always possible to choose these parallelepipeds defined by primitive lattice vectors such that these cells contain only one lattice point per unit cell and are of minimum volume. Such units are called primitive cells.

3.4 BRAVAIS LATTICE AND UNIT CELL

The crystal structures can be classified into seven lattice systems according to the symmetry of the crystal. The symmetry of the crystal depends upon the relative magnitude of the fundamental lattice vectors along crystal axes and also the angles between the crystal axes. The unit cell of the crystal possesses this symmetry. The crystal axes are chosen along the three nonparallel edges of the unit cell. The fundamental lattice vectors **a**, **b**, **c** along the three crystal axes are the vectors joining the nearest lattice points.

The crystal angles α, β, γ, Fig. 3.4, define the relative orientations of the crystal axes where α is the angle between b and c axes; β is the angle between c and a axes; γ is the angle between a and b axes. The fundamental lattice vectors **a**, **b**, **c** and the crystal angles α, β, γ are called the lattice parameters. These parameters completely define the unit cell.

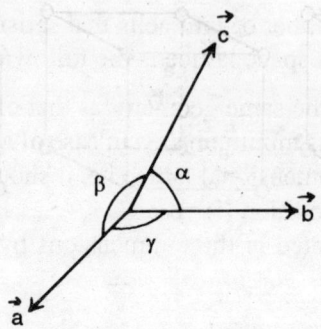

Fig. 3.4 Crystal axes.

Example 3.1

Consider a square space lattice where A, B, C, D, E, F are drawn as unit cells, Fig. 3.5. Which out of these are proper unit cells?

Solution:

B does not have square symmetry. D cannot fill up the entire space by giving simple lattice translations. E does not have minimum area. F does not have minimum area and does not have square symmetry. So B, D, E, F are not proper unit cells. Only A and C are proper unit cells.

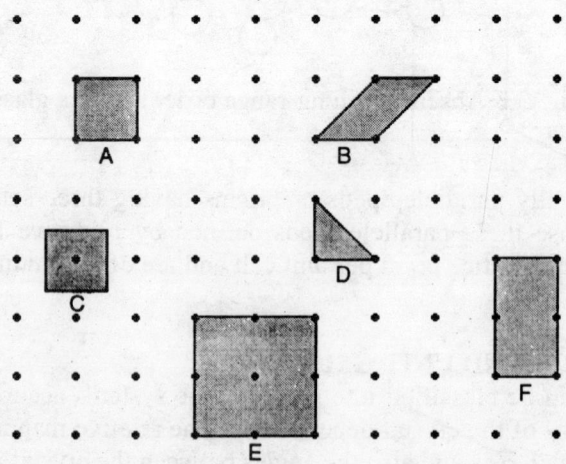

Fig. 3.5 Two-dimensional space lattice.

The seven classes of crystal systems have lattice parameters, which satisfy specific relationships depending upon the geometry. According to the geometry and the arrangement of lattice points in the unit cell there are a total number of fourteen distinct Bravais lattices, Fig. 3.6. The details of crystal structure are presented in Table 3.1.

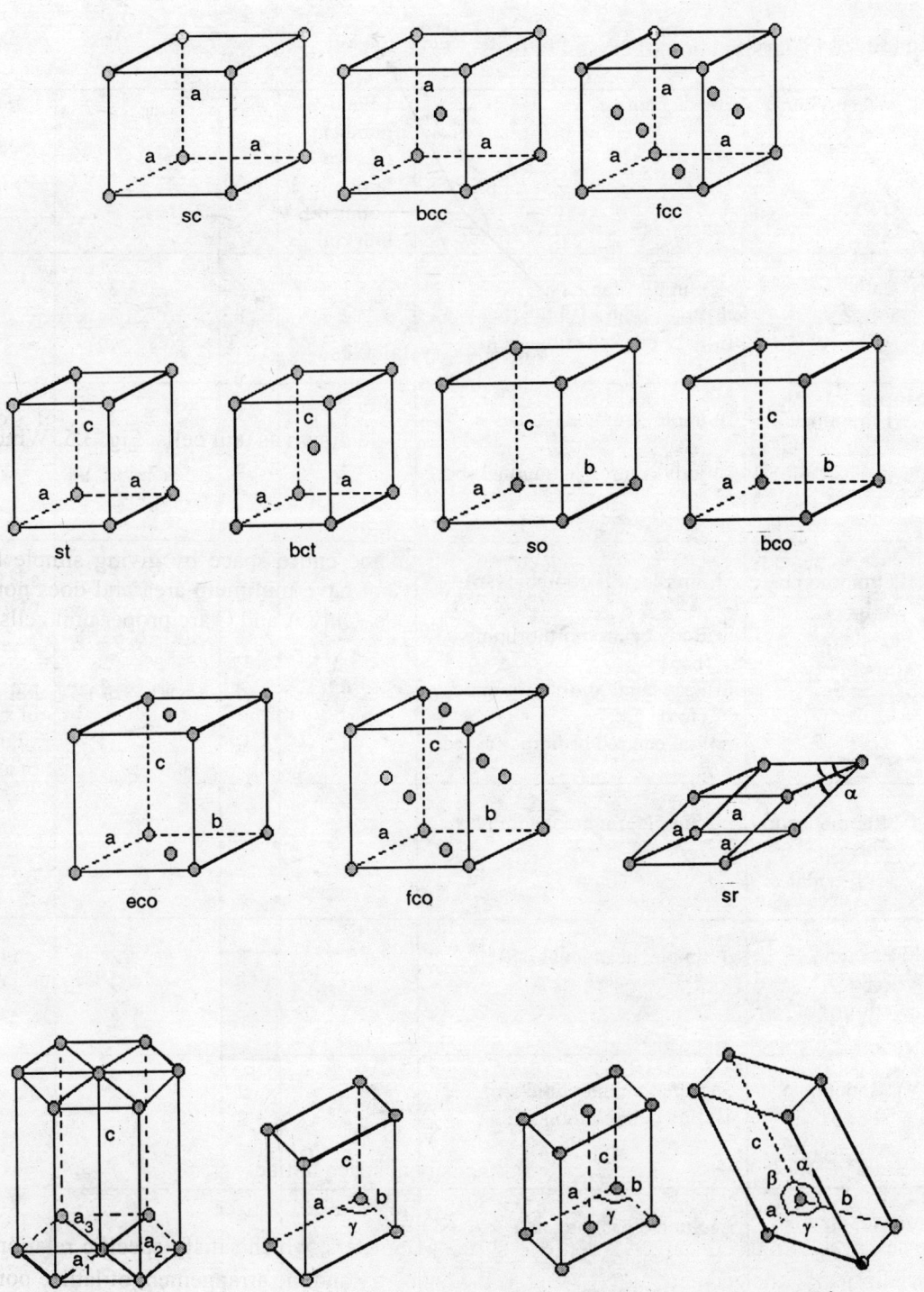

Fig. 3.6. Fourteen Bravais lattices.

Table 3.1 Classification of crystal lattices.

Crystal system	Bravais lattice	Effective number (n_{eff})of lattice points per unit cell	CN	r_{atom}	Remarks
I. Cubic a=b=c $\alpha=\beta=\gamma=90^0$	i)Simple cubic (sc) ii)Body centred cubic (bcc) iii)Face centred cubic (fcc)	1 2 4	6 8 12	a/2 $\sqrt{3}a/4$ $a/(2\sqrt{2})$	
II.Tetragonal a=b≠c $\alpha=\beta=\gamma=90^0$	i)Simple tetragonal (st) ii)Body centred tetragonal (bct)	1 2	2 4 8	c/2 a/2 $\sqrt{(2a^2+c^2)}/4$	for c<a for c>a
III.Orthorhombic a≠b≠c $\alpha=\beta=\gamma=90^0$	i)Simple orthorhombic (so) ii)Body centred orthorhombic (bco) iii)Face centred orthorhombic (fco) iv)End centred orthorhomb (eco)	1 2 4 2	2 8 4 --	s/2, $\sqrt{(a^2+b^2+c^2)}/4$ $\sqrt{(p^2+q^2)}/4$	s is smallest of a,b,c p,q are two of smaller values out of a,b,c
IV. Rhombohedral a=b=c $\alpha=\beta=\gamma\neq90^0$	i) Simple rhombohedral(sr)	1	--	--	
V. Hexagonal a=b≠c $\alpha=\beta=90^0,\gamma=120^0$	i) Simple hexagonal (sh)	3*	8 6 2	a/2 a/2 c/2	For a=c For a<c For a>c
VI.Monoclinic a≠b≠c $\alpha=\beta=90^0\neq\gamma$	i) Simple monoclinic(sm) ii) End centred monoclinic (ecm)	1 2	2 2	-- --	
VII.Triclinic a≠b≠c $\alpha\neq\beta\neq\gamma$	i) Simple triclinic(stri)	1	2	--	

Note: '3*' indicates that the effective number of lattice points per unit cell is 3 if the unit cell is hexagonal prism otherwise it is 1.

The geometry of two different Bravais unit cells may be same but they may contain different number of lattice points. Depending upon the arrangement of the lattice points in the unit cell the lattice may be called simple, body centred, face centred or end centred. In case of simple space lattices, the lattice points are located only at the eight corners of the unit cell. In body centred lattice, the lattice points are located not only at the eight corners but also at the centre of the unit cell. The face centred lattices are those in which the lattice points are located at the eight corners as well as at the centres of six faces of the unit cell. When the lattice points are located at the eight corners and also at the centres of one pair of parallel faces, it is called end centred lattice.

The existence of symmetry in the crystal structure introduces constraints in the system and reduces the number of independent lattice parameters to describe unit cell. In cubic lattice $a = b = c$ and $\alpha = \beta = \gamma = 90°$. Hence to define the unit cell we need to specify only one parameter i.e. a, the side of the cube. The complete information regarding the crystal parameters for various crystal systems is given in Tables 3.1 and 3.2. From Table 3.2 it follows that more symmetric the crystal structure, lesser is the number of parameters required to describe it.

It is of importance to find out the effective number of lattice points in a unit cell. The corner of a cubic unit cell is common to eight adjacent unit cells. So the contribution of a corner lattice point is 1/8th to each unit cell. Thus, the total contribution of all corner lattice points to a unit cell is $8 \times \frac{1}{8} = 1$. In case of unit cells having the crystal angles not equal to 90°, the contributions of the corner points are unequal but the total contribution of all corner lattice points still adds up to one. The lattice point in the body centre position of a unit cell belongs to that particular unit cell only and hence its contribution to a unit cell is one. The lattice point at the centre of a face of a unit cell contributes ½ to each of the two unit cells to which it is common. Thus, in case of face centred unit cell, the six face centre lattice points contribute $6 \times \frac{1}{2} = 3$. For the end centred lattice the two face centres lattice points contribute $2 \times (\frac{1}{2}) = 1$. The effective number of lattice points in Bravais unit cells is also given in Table 3.1. The effective number of lattice points per unit cell is equal to the number of independent lattice point locations, Fig. 3.6. The locations of other lattice points within the unit cell can be generated through fundamental lattice translations of these independent lattice points. Consider the example of a face centred cubic unit cell. The effective number of lattice points per unit cell in fcc lattice is 4. The independent lattice point locations are A(0,0,0), B(½,0,½), C(0, ½,½) and D(½,½, 0). Other lattice points can be obtained through

Table 3.2 Crystal systems and independent lattice parameters.

Crystal system	Independent lattice parameters		
	Lattice vectors	Angles	Total parameters
Cubic	a	--	1
Tetragonal	a, c	--	2
Orthorhombic	a, b, c	--	3
Rhombohedral	a	α	2
Hexagonal	a, c	--	2
Monoclinic	a, b, c	γ	4
Triclinic	a, b, c	α, β, γ	6

fundamental lattice translations of these four lattice points A, B, C and D. These independent lattice point locations from which all other lattice points can be generated through lattice translations are also called equivalent points, Fig. 3.7. The equivalent points for various crystal structures are given in the Table 3.3.

Example 3.2

Consider that one of the atoms in the basis of a fcc unit cell is located at $(0, 0, \frac{1}{2})$. Find the coordinates of the atoms located at other equivalent points.

Solution:

The equivalent points for fcc are at $(0,0,0)$, $(0,\frac{1}{2},\frac{1}{2})$, $(\frac{1}{2},0,\frac{1}{2})$, $(\frac{1}{2},\frac{1}{2},0)$. Consider $(0,0,\frac{1}{2})$ as origin, and then the coordinates of the atoms at other equivalent sites can be obtained by adding $(0,0,\frac{1}{2})$ to the coordinates of the equivalent points i.e. the coordinates are

$$(0, \tfrac{1}{2},1),(\tfrac{1}{2},0,1) \text{ and} (\tfrac{1}{2},\tfrac{1}{2},\tfrac{1}{2})$$

or equivalently at $\qquad (0, \tfrac{1}{2},0),(\tfrac{1}{2},0,0) \text{ and} (\tfrac{1}{2},\tfrac{1}{2},\tfrac{1}{2})$.

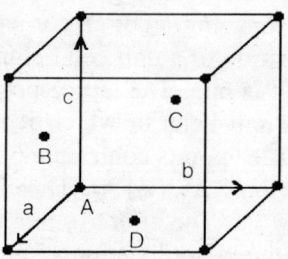

Fig. 3.7 fcc unit cell showing independent lattice point locations.

Table 3.3 Equivalent points in a unit cell

Bravais lattice	Coordinates of equivalents points (in units of a,b,c)
sc,st,so,sr,sh,sm,stri	$(0,0,0)$
bcc,bct,bco	$(0,0,0)$, $(\frac{1}{2},\frac{1}{2},\frac{1}{2})$
eco,ecm	$(0,0,0)$, $(\frac{1}{2},\frac{1}{2},0)$
fcc,fco	$(0,0,0)$, $(0,\frac{1}{2},\frac{1}{2})$, $(\frac{1}{2},0,\frac{1}{2})$, $(\frac{1}{2},\frac{1}{2},0)$

3.5 CRYSTAL STRUCTURE AND BASIS

In nature we find thousands of varieties of crystal structures, which can be classified into one or the other of the fourteen Bravais lattices. This can be understood from the fact that the lattice point of a Bravais lattice may represent a single atom (e.g. Cu) or a group of atoms or molecules (e.g. CH_4 molecules are located at the lattice points of fcc in methane crystal). The lattice points need not coincide with any atom. A group of atoms, molecules or single atom is placed at every lattice point exactly in an identical manner so that the specific crystal structure is generated. This group of atoms, molecules or single atom is called a basis. The crystal structure is generated by proper combination of space lattice and basis. Thus one can express,

$$\text{Crystal structure} = \text{Space lattice} + \text{Basis} \qquad (3.1)$$

In a monoatomic crystal, basis consists of a single atom, which is located at each lattice point whereas in a molecular crystal, the atoms (ions) of molecular basis are arranged in a specific manner around each lattice point. The basis in a crystal structure consists of specific number and types of atoms. The interatomic distances between the atoms and the orientation of the molecules in a basis are same at each lattice point. A large number of crystals may belong to the same space lattice but all of them have different bases. For example, consider the bcc crystals of iron, chromium and manganese. All these crystals have space lattice bcc with different bases. In case of iron crystal, the basis is a single iron atom. Chromium has also a single chromium atom as basis. However, in case of manganese 29 atoms are grouped together to form the basis. In NaCl crystal, one Na^+ and one Cl^- ions form the basis. In certain organic polymers, the basis may consist of even as large as 10,000 atoms or so. In Fig. 3.8, the unit cell of cubic CsCl crystal is shown, in which the Cl^- ions are at the corners and Cs^+ ions are at the body centre positions of the unit cells or vice versa. Now the question is whether CsCl crystal is simple cubic or body centred cubic? The answer to this question is that the CsCl crystal is simple cubic with basis consisting of two ions Cl^- and Cs^+. Cl^- ions are shifted along the body diagonal and occupy body centre positions and Cs^+ ions occupy corner positions. The CsCl should not be confused with bcc crystal, because in case of bcc, the corners as well as body centre positions of the unit cell are occupied by same basis.

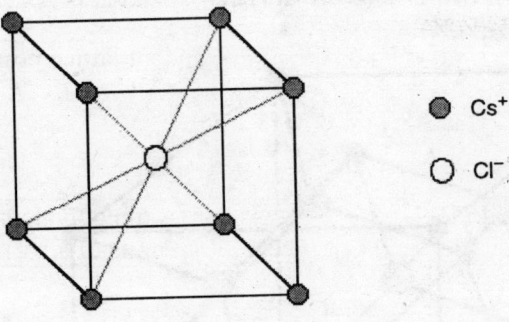

Fig. 3.8. CsCl unit cell.

Example 3.3

Why there is no fct Bravais lattice?

Solution:

Let us consider that there is a fct Bravais lattice. Draw two adjacent unit cells of fct. A unit cell of bct can be drawn as shown in Fig. 3.9. Since the volume of bct unit cell is smaller than that of fct and since the number of lattice points per unit cell is 2 in case of bct whereas it is 4 for fct, so bct is only considered as Bravais lattice.

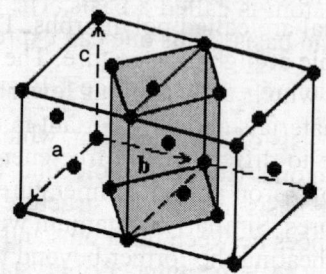

Fig. 3.9 fct is equivalent to bct.

Example 3.4

(a) Determine the Bravais lattice of NaCl (see Fig. 3.1). Find its basis.

(b) Determine the Bravais lattice of CaF_2.

Solution:

(a) Bravais lattice of NaCl is fcc where Na^+ (or Cl^-) ions by themselves form fcc structure such that fcc unit cell of Na^+ ions is displaced by a/2 along the edge with respect to fcc unit cell of Cl^- ions. The basis of NaCl crystal consists of a Cl^- ion at the corner (face centre) of the unit cell and a Na^+ ion at the adjacent edge centre (body centre) or vice versa.

(b) CaF_2 has fcc lattice with each Ca^{2+} ion coordinated with eight F^- ions, but every F^- ion is coordinated with only four Ca^{2+} ions, Fig. 3.10. Basis consists of one Ca^{2+} and two F^- ions. If Ca^{2+} is located at (0,0,0), two F^- ions are at (¼,¼,¼) and at (−¼,−¼,−¼) (equivalent position within unit cell being (¾,¾,¾).

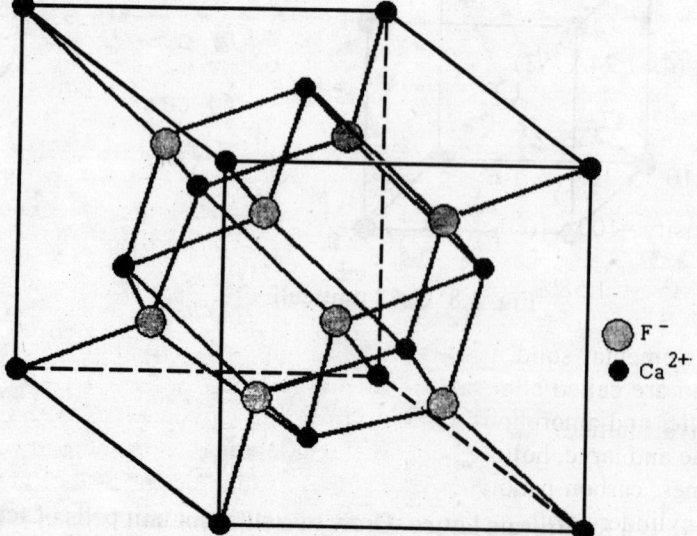

F^-
Ca^{2+}

Fig. 3.10 CaF_2 crystal structure.

Polymorphism and allotropy

Certain materials exhibit diverse forms of crystal structure. This is called polymorphism. The different crystalline forms of a material are called polymorphs. The simplest polymorph of SiO_2 is cristobalite, Fig. 4.22, which is stable at high temperature. The other two polymorphs of silica are tridymite, Fig. 4.23, and quartz, which are stable at lower temperature. Polymorphism is commonly observed in engineering materials and is important in their processing and behaviour. Polymorphic phase change arises due to differences in free energy. The heat treatment of steel utilizes the stability of fcc structure of γ-iron at high temperature and that of bcc structure of α-iron after cooling to service temperatures. Similarly α-titanium with hcp structure transforms into β-titanium having bcc structure after heating the former beyond the temperature 880°C. There is change in volume (density) as a result of this polymorphic transformation since the packing efficiency of hcp is greater than that of bcc.

Example 3.5

Calculate the percentage change in density of iron due to polymorphic transformation from bcc to fcc. Radius of iron in bcc structure is 1.24 Å and that in fcc structure is 1.27 Å.

Solution:

Volume of fcc unit cell $=(4x\ 1.27\text{Å}\ /\sqrt{2})^3$

$$= 46.20\ \text{Å}^3$$

ρ_{fcc} = Number of atoms per unit cell x Atomic weight/(Avogadro number x Volume of unit cell)

$= 4 \times 55.85/\ (6.023\ x10^{26} \times 46.20 \times 10^{-30})$

$= 8028.3876\ \text{kg/m}^3$

Volume of bcc unit cell $=(4x\ 1.24\text{Å}\ /\sqrt{3})^3$

$$= 23.47\ \text{Å}^3$$

ρ_{bcc} = 2x 55.85/ $(6.023\ x10^{26} \times 23.47 \times 10^{-30})$ = 7901.8216 kg/m^3

Percentage change in density =100 x [(8028.3876 –7901.8216)/ 8028.3876]

$$= 1.58\%$$

If the material is an elemental solid, polymorphism is often called allotropy. The different phases of elemental solid are called allotropes. An example of allotropy is carbon, which can exist as diamond, graphite, and amorphous carbon. Pure solid carbon exists in three crystalline forms– diamond, graphite and large, hollow Fullerenes. Fullerenes can be there in various forms, e.g. buckministerfullerenes, carbon nanotube etc. Carbon nanotubes consists of graphite sheets seamlessly wrapped to cylinders with a few nanometer diameter and upto a millimeter long. Their electronic structures depend on their geometry i.e. diameter and chirality, which in turn decides whether it will be metallic or semiconducting. The energy gap varies from zero to 1 eV.

3.6 SYMMETRY IN CRYSTAL STRUCTURE

Crystal structure possesses symmetry. The symmetry group of a crystal consists of a set of symmetry operations which when applied on a crystal carry the crystal onto itself. These include the lattice translations as well as point group operations e.g. rotations, reflections or combination of these operations. The point group operations are applied about lattice points or about certain specific points in the unit cell. There may be compound operations made up of combined translations and point group operations. The rotation operations are applied about certain symmetry axes. For example, a cubic crystal has a total of 13 axes of symmetry passing through the centre of the unit cell: three 4-fold axes (repeat the structure after every 90° rotation) parallel to three edges and perpendicular to each other, four 3-fold axes (repeat the structure after every 120° rotation) along the body diagonals of the unit cell and six 2-fold axes (repeat the structure after every 180° rotation) passing through the centres of the opposite edges i.e. parallel to face diagonals. Also there are nine planes of symmetry and a centre of symmetry, the centre of the cube, in a cubic crystal, Fig. 3.11. The number of symmetry operations is less for tetragonal and orthorhombic crystals.

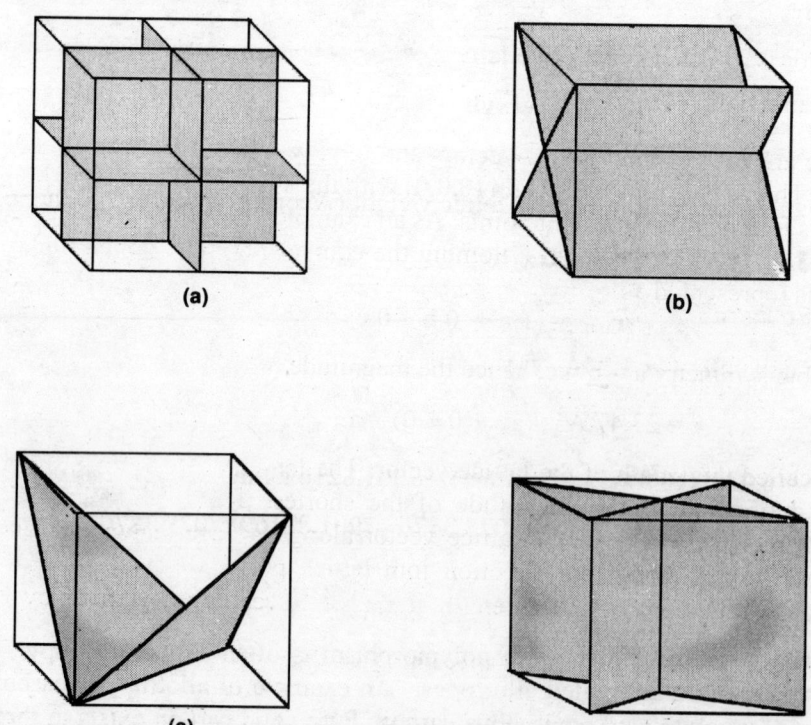

(a)

(b)

(c)

(d)

Fig. 3.11(a) Symmetry operations of cubic crystals: (a), (b), (c) and (d) are showing planes of symmetry

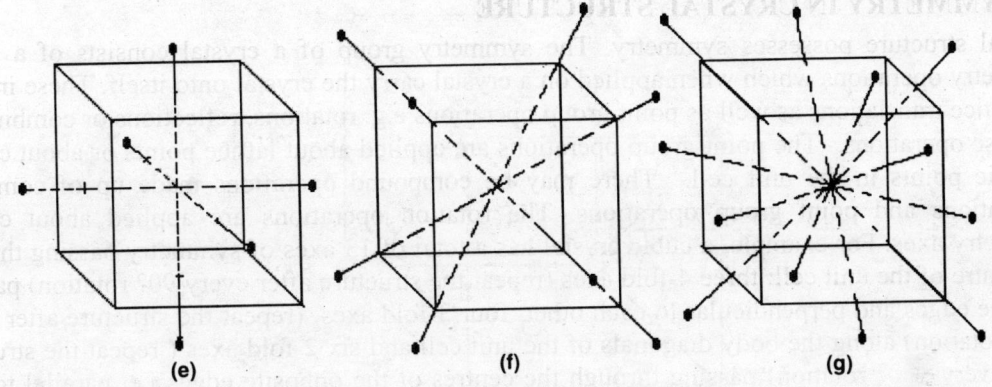

Fig. 3.11(b) Symmetry operations of cubic crystals: (e), (f) and (g) are showing symmetry axes.

3.7 LATTICE VECTORS, DIRECTION INDICES AND LINEAR DENSITY

Lattice vectors

A lattice vector is a vector joining two lattice points. A general lattice vector is represented as:

$$\mathbf{r} = u\mathbf{a} + v\mathbf{b} + w\mathbf{c} \tag{3.2}$$

where **a**, **b**, **c** are fundamental lattice vectors and u, v, w are coefficients along a, b and c axes, respectively. This vector joins the origin (0,0,0) with the lattice point (u, v, w). One can generate lattice vectors joining the equivalent points. As an example, consider the lattice vectors of fcc unit cell, Fig. 3.12 The lattice vector $\mathbf{r}_{100}$, joining the equivalent lattice points (0,0,0) and (1,0,0) of the unit cell is represented as

$$\mathbf{r}_{100} = 1.\mathbf{a} + 0.\mathbf{b} + 0.\mathbf{c} \tag{3.3}$$

In case of cubic symmetry a = b = c, hence the magnitude of $\mathbf{r}_{100}$ is given by

$$r_{100} = \sqrt{(a^2 + 0 + 0)} = a \tag{3.4}$$

This is also called the length of the lattice vector. The length of the lattice vector in a particular direction is defined to be the magnitude of the shortest lattice vector in that direction. For example, consider the length of the lattice vector along the face diagonal of fcc unit cell. The shortest lattice vector along this direction join lattice points (0,0,0) and (½, ½, 0) and it is expressed as $\mathbf{r}_{110}$ and not $\mathbf{r}_{½½0}$. The length of the lattice vector $\mathbf{r}_{110}$ is given by

$$r_{110} = \sqrt{(a^2/4 + a^2/4 + 0)} = a/\sqrt{2} \tag{3.5}$$

Similarly $$r_{111} = \sqrt{(a^2 + a^2 + a^2)} = \sqrt{3}a \qquad \text{for fcc},\tag{3.6}$$

whereas $$r_{111} = \sqrt{(a^2/4 + a^2/4 + a^2/4)} = \sqrt{3}a/2 \quad \text{for bcc}\tag{3.7}$$

Now we define the directions in space lattice and correlate these directions with the length of the lattice vector in that direction for various types of space lattices.

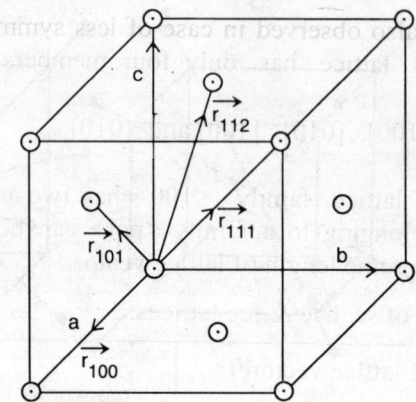

Fig. 3.12 Lattice vectors of fcc unit cell.

Direction indices

The direction indices of a lattice vector **r** (Eq. 3.2) is given by [hkl] which is obtained from u,v,w, by reducing them to the smallest appropriate integers. [hkl] is also called Miller Indices (MI) of direction. For example $r_{\frac{1}{2}\frac{1}{2}\frac{1}{2}}$, r_{333} and r_{111} have same direction indices [111]. In case one or more of the indices are negative, a bar is put over such indices. For example if h=1, k= –3 and l=2, we write it as

$$[1\bar{3}2].$$

In case some of the indices have more than 1 digit, we separate the indices by commas. Thus for the case h=11, k=3, l=5 we write [11,3,5]. The body diagonal of bcc and fcc will have same direction indices [111] but the length of the lattice vector r_{111} for bcc will be $\sqrt{3}a/2$, whereas that of fcc is $\sqrt{3}a$ since there is a lattice point in the body centre position of bcc. Table 3.4 shows certain lattice vectors, direction indices and the length of lattice vectors for sc, fcc and bcc structures.

Family of directions

There are a number of lattice vectors, within a cubic crystal, which have same length but different directions. These lattice vectors form a family of directions and are denoted by < hkl>. All the six directions along the edge of cubic unit cell are

$$[100] \quad [\bar{1}00] \quad [010] \quad [0\bar{1}0] \quad [001] \quad [00\bar{1}].$$

These form the family of directions <100>. The parallel edges will have same direction indices.

The direction [010] is antiparallel to [0$\bar{1}$0]. Crystals have similar properties along all directions of the same family but these properties may be different for different family of directions. This is particularly observed in certain anisotropic crystals. In aluminium, elastic modulus is 9.2 x 10^6 psi along <100> but it is 10.5 x 10^6 psi along <110> and 11.0x10^6 psi along <111>. But there are crystals in which measured properties are of same value in all lattice directions i.e. these properties are independent of crystal directions. Such crystals are called isotropic.

The family of directions is also observed in case of less symmetric Bravais lattices. Family <100> for tetragonal Bravais lattice has only four members

$$[100], [010], [\bar{1}00] \text{ and } [0\bar{1}0]$$

and for orthorhombic Bravais lattice, family <100> has two members directions [$\bar{1}$00] and [100]. Thus, the directions belonging to a family <pqr> can be generated by permuting the indices such that these have the same length of lattice vector.

Table 3.4 Lattice vectors of sc, bcc & fcc lattices.

Lattice vector	Length of lattice vector			Family of Directions
	sc	bcc	fcc	
$\mathbf{r}_{100}$	a	a	a	[100] [010] [001]
				[$\bar{1}$00] [0$\bar{1}$0] [00$\bar{1}$]
$\mathbf{r}_{110}$	√2a	√2a	a/√2	[110] [$\bar{1}$10] [1$\bar{1}$0] [$\bar{1}\bar{1}$0]
				[101] [$\bar{1}$01] [10$\bar{1}$] [$\bar{1}$0$\bar{1}$]
				[011] [0$\bar{1}$1] [01$\bar{1}$] [0$\bar{1}\bar{1}$]
$\mathbf{r}_{111}$	√3a	√3a/2	√3a	[111] [$\bar{1}$11] [1$\bar{1}$1] [11$\bar{1}$]
				[$\bar{1}\bar{1}\bar{1}$] [1$\bar{1}\bar{1}$] [$\bar{1}$1$\bar{1}$] [$\bar{1}\bar{1}$1]

For example, if a=b≠c then the possible directions are

[pqr],[qpr], [p$\bar{q}$r], [q$\bar{p}$r], [$\bar{p}$qr], [$\bar{q}$pr], [$\bar{p}\bar{q}$r], [$\bar{q}\bar{p}$r], [pq$\bar{r}$], [qp$\bar{r}$], [p$\bar{q}\bar{r}$], [q$\bar{p}\bar{r}$], [$\bar{p}$q$\bar{r}$], [$\bar{q}$p$\bar{r}$], [$\bar{p}\bar{q}\bar{r}$], [$\bar{q}\bar{p}\bar{r}$]

belonging to family <pqr>. As we examine the lower symmetry crystals like tetragonal or orthorhombic etc. we find that a family of directions in cubic splits up as the symmetry of the crystal reduces. For example, we examine the family of directions <100> and <110> for cubic, tetragonal and orthorhombic crystals in Table 3.5.

Drawing directions

To draw a direction in space lattice choose the origin suitably such that the direction is drawn within the unit cell. In Fig. 3.13, O is chosen as origin and OA, OB and OC are the crystal axes. The direction [hkl] can be drawn within unit cell as follows:

(i) If h, k, l are positive and not greater than 1 then one should simply choose a lattice point P with coordinates (h, k, l) and join this lattice point with the origin. The direction should be drawn with an arrow pointing to the lattice point P. For example, the direction [110] is obtained by joining the origin with P.

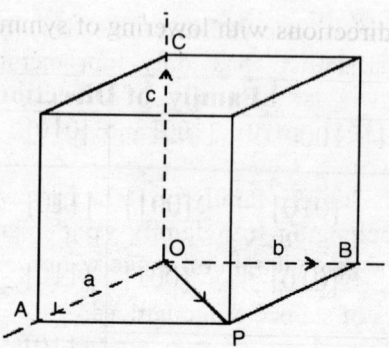

Fig. 3.13 Direction of lattice vectors in unit cell

(ii) If the indices h, k, l are positive but some of them are greater than 1, then one should divide h, k, l by largest of the three indices to generate the coordinates u, v, w which will be within unit cell. Joining (0, 0, 0) and (u, v, w) one can draw the direction [hkl]. For example the direction [230] can be drawn by joining (0, 0, 0) and (2/3,1,0).

(iii) If some of the indices h, k, l are negative one can draw the direction by using the methods suggested above. The direction in that case is drawn outside the unit cell. The direction can be drawn within unit cell by shifting the origin from O. For example, if l index is negative one can choose the origin at C, Fig. 3.14, where [101] is drawn within unit cell. Similarly, if k (h) index is negative, origin should be chosen at B (A). If all the indices are negative the origin should be taken at P. For example, [111] direction is drawn from P to O, Fig. 3.14.

(iv) Directions can always be drawn when two points are given. For example, the direction joining the points (½,½,0) and (½,0, ½) in simple cubic is calculated as follows:

$$s_x = (½ - ½) = 0$$
$$s_y = (½ - 0) = ½$$
$$s_z = (0 - ½) = -½$$

Direction indices joining the two points are $[01\bar{1}]$

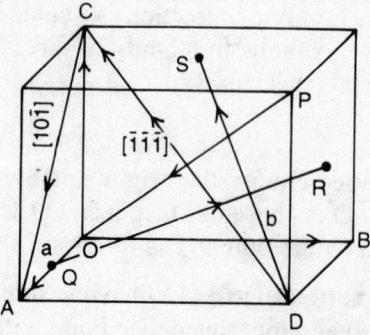

Fig. 3.14 Direction of lattice vector

Table 3.5. Splitting of family of directions with lowering of symmetry

Crystal class	Family of Directions					
	<100>			<110>		
Cubic	[100]	[010]	[001]	[110]	[101]	[011]
	[1̄00]	[01̄0]	[001̄]	[1̄10]	[1̄01]	[01̄1]
				[11̄0]	[101̄]	[011̄]
				[1̄1̄0]	[1̄01̄]	[01̄1̄]
				<110>	<101>	
				[1̄10]	[1̄01]	[01̄1]
Tetragonal				[11̄0]	[101̄]	[011̄]
	<100>		<001>	[1̄1̄0]	[1̄01̄]	[01̄1̄]
	[100]	[010]	[001]	[110]	[101]	[011]
	[1̄00]	[01̄0]	[001̄]			
	<100>	<010>	<001>	<110>	<101>	<011>
Orthorhombic	[100]	[010]	[001]	[110]	[101]	[011]
	[1̄00]	[01̄0]	[001̄]	[1̄10]	[1̄01]	[01̄1]
				[11̄0]	[101̄]	[011̄]
				[1̄1̄0]	[1̄01̄]	[01̄1̄]

Example 3.6

A lattice vector is drawn by joining O (0,0,0) to P (4Å, 4Å, 4Å) in a tetragonal unit cell. If a=8Å and c=4 Å, find the direction indices of the lattice vector.

Solution:

The coordinates of P in the units of a, b, c are (½, ½, 1) i.e.
$x_p - x_o = ½$, $y_p - y_o = ½$, $z_p - z_o = 1$,
Taking ½ common, direction indices of OP are [112].

Example 3.7

Find the direction indices of the lattice vector joining the lattice points p(½, ½, 0) and q(1, 1, ½). Also find the length of the lattice vector.

Solution:

$x_q - x_p = ½$,
$y_q - y_p = ½$,
$z_q - z_p = ½$,
Taking ½ common, MI of the direction are [111].
The length of the lattice vector is $½ \sqrt{(a^2 + b^2 + c^2)}$; a, b and c are the lattice parameters.

Example 3.8

Find the direction indices of lattice vectors QR, DC and DS as shown in Fig. 3.14. Draw these directions taking (½, ½, 0) as origin.

Solution: QR => [021], **DC** => [$\bar{1}11$] and **DS** => [$\bar{1}12$]

Example 3.9

A lattice direction [121] passes through (1/5, 1/5, 0) as it enters unit cell. Find the coordinates of the point through which it comes out of the unit cell.

Solution:

Let us consider the point through which the direction comes out of the unit cell is (x_1, y_1, z_1). The lattice direction [121] joins the origin (0,0,0) with the point (½, 1, ½). Thus, one can express

$$x_1 - (1/5) = c(½)$$
$$y_1 - (1/5) = c(1)$$
$$z_1 - \quad 0 = c(½)$$

Choice of constant 'c' should be made such that none of the coordinates is greater than one and at least one of the coordinates should be one or zero. If we choose c=4/5, it satisfies all the above conditions and we get:

$$x_1 = (4/10)+(1/5) = 3/5$$
$$y_1 = 1$$
$$z_1 = 4/10 = 2/5$$

Hence the line comes out at the point (3/5,1,2/5).

Linear density

Linear density is defined as the number of lattice points per unit length along a given crystal direction. Linear density L_d is also expressed as

$$L_d = 1/r_{hkl} \qquad (3.8)$$

where r_{hkl} is the length of the lattice vector along [hkl] direction. One can also define linear density of atoms in case of molecular crystals, in which one should take into account the number of atoms per unit length. While considering linear density the atoms or lattice points are considered to be circular. The contribution of the point in taken ½ or 1 depending upon the projection of the atom on the line, Fig. 3.15.

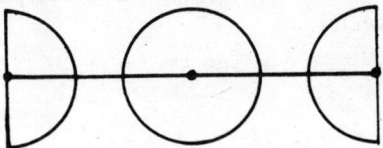

Fig. 3.15 Contribution of lattice points to the linear density

Example 3.10

a) Consider a simple tetragonal unit cell whose lattice parameters are a=b=1.5 Å and c/a=2. Find the Miller indices of the direction obtained by joining lattice points at (0,0,0) and (3Å, 3Å, 3Å).

b) Find the length of a lattice vector having direction indices as [321] of a simple cubic unit cell with lattice parameter a Å and hence find the linear density (L_d). What is the value of L_d in the case of fcc unit cell?

Solution:

a) To find out the values of coordinate in terms of units of a, b, c parameters, we divide the corresponding coordinates by a, b and c values respectively. Thus, (3Å, 3Å, 3Å) can be expressed as (3/1.5, 3/1.5, 3/3) i.e. (2,2,1) and Miller indices of direction of the lattice vector are [221].

b) The length of the lattice vector is

$$r_{321} = [(3a)^2 + (2a)^2 + (a)^2]^{1/2} == \sqrt{14}a \text{ Å}$$
$$L_d = 1/(\sqrt{14}a) \text{ Å}^{-1}$$

In case of fcc unit cell, the shortest length of the lattice vector [321] is obtained by joining (0,0,0) and face centre lattice point (3/2,1,1/2). Hence

$$r_{321} = [(3a/2)^2 + (a)^2 + (a/2)^2]^{1/2} = \sqrt{14}a/2 \text{ Å}$$
$$L_d = 2/(\sqrt{14}a) \text{ Å}^{-1}$$

Angle between crystal directions

The angle between the lattice vectors r_1 and r_2 for cubic crystal can be expressed as

$$\theta = \cos^{-1}\{(r_1.r_2)/(|r_1||r_2|\} \qquad (3.9)$$

$$\theta = \cos^{-1} (h_1h_2 + k_1k_2 + l_1l_2) /[\sqrt{ (h_1^2 + k_1^2 + l_1^2)}\sqrt{(h_2^2 + k_2^2 + l_2^2)}] \qquad (3.10)$$

where $$r_1 = h_1a + k_1b + l_1c \qquad (3.11a)$$

and $$r_2 = h_2a + k_2b + l_2c \qquad (3.11b)$$

Example 3.11

(a) Calculate the tetrahedral angle of CH_4 molecule.

(b) A force of 1500 dynes is applied in [221] direction of a cubic crystal. What is the resolved force in [201] direction?

Solution:

(a) Angle between the direction $[1\bar{1}\bar{1}]$ and $[111]$ is given by

$$\theta = \cos^{-1} \frac{1 - 1 - 1}{\sqrt{3}\sqrt{3}} = \cos^{-1}(-\tfrac{1}{3}) = 109.5°$$

This procedure is valid since CH_4 molecule has cubic symmetry.

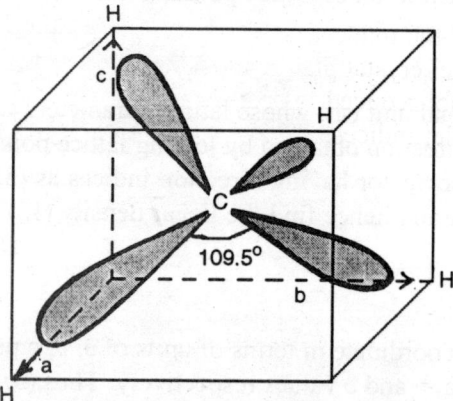

Fig. 3.16 Tetrahedral angle of CH_4 molecule.

b) Let θ be the angle between [221] and [201] directions. Then

$$\cos\theta = (2\times2 + 2\times0 + 1\times1)/(\sqrt{9} \times \sqrt{5}) = \sqrt{5}/3$$

Resolved force in [201] direction is

$$F\cos\theta = 1500 \times \sqrt{5}/3 = 1118 \text{ dynes}$$

Example 3.12

Find the angle between [111] and $[11\bar{1}]$ directions in case of orthorhombic unit cell with a= 2Å, b= 5Å and c= 3Å.

Solution:

Let the angle be θ, then

$$\cos\theta = \frac{a^2hh' + b^2kk' + c^2ll'}{\sqrt{[(ah)^2 + (bk)^2 + (cl)^2]}\sqrt{[(ah')^2 + (bk')^2 + (cl')^2]}} = \frac{a^2+b^2-c^2}{a^2+b^2+c^2} = 10/19$$

or $\theta = \cos^{-1}(10/19) = 58.24°$

3.8 MILLER INDICES OF CRYSTAL PLANES AND PLANAR DENSITY

Crystal plane

A large number of crystal planes can be visualized within the crystal passing through the lattice points. There are crystal planes parallel to each other with a fixed distance of separation between two consecutive planes. The distance of separation is called the interplanar spacing. The crystal planes are specified by a set of indices called Miller indices. The Miller indices can be defined as follows:

Step 1. Find the intercepts of the plane along the crystal axes a, b, c and express the intercepts in units of crystal parameters a, b, c respectively.

Step 2. Find the reciprocal of the intercepts.

Step 3. Reduce the reciprocals to the three smallest integers h,k,l keeping the ratio same.

Step 4. Enclose these three integers into parenthesis (hkl).

The indices (hkl) are called the reduced Miller indices or simply the Miller indices of the plane. The Miller indices of parallel crystal planes are same. If the reciprocals of the intercepts are only expressed as integers and not reduced to smallest integers (see Step 3), the indices obtained are called as the unreduced Miller indices.

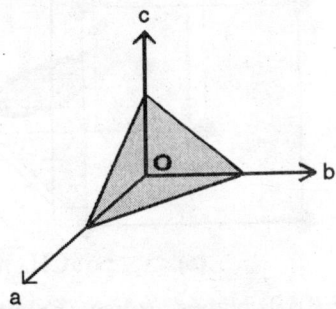

Fig. 3.17 Determination of Miller indices of a crystal plane.

Example 3.13

Find the Miller indices of a plane that makes intercepts (in the units of lattice parameters) of 1 and 2 on a- and b-axes, respectively. The plane is parallel to c-axis.

Solution:

	a	b	c
Intercepts	1	2	∞
Take reciprocal	1	$\frac{1}{2}$	0
Reduce to simple integers	2	1	0

Hence Miller indices are: (210).

The interplanar spacing of the parallel crystal planes is equal to the distance of the crystal plane nearest to the origin, from the origin. The set of parallel planes for cubic crystals are shown in Figs. 3.18, 3.19 and 3.20.

Directions in a plane

There are a large number of directions parallel to a crystal plane. If a lattice vector $[h_1k_1l_1]$ is parallel to the crystal plane $(h_2k_2l_2)$ the direction indices of the lattice vector $[h_1k_1l_1]$ will satisfy the condition

$$h_1h_2 + k_1k_2 + l_1l_2 = 0 \qquad (3.12)$$

For example, in a cubic crystal (110) plane will contain following directions, among many others:

$$[1\bar{1}2], \ [1\bar{1}0], \ [001] \text{ etc.}$$

The above relation can be proved very easily for cubic crystals; in this case, the direction of the normal to the plane $(h_2k_2l_2)$ is $[h_2k_2l_2]$. As $[h_1k_1l_1]$ direction lies in the crystal plane, it is perpendicular to $[h_2k_2l_2]$ and hence the dot product of $[h_1k_1l_1]$ and $[h_2k_2l_2]$ will be equal to zero as given by Eq. (3.12).

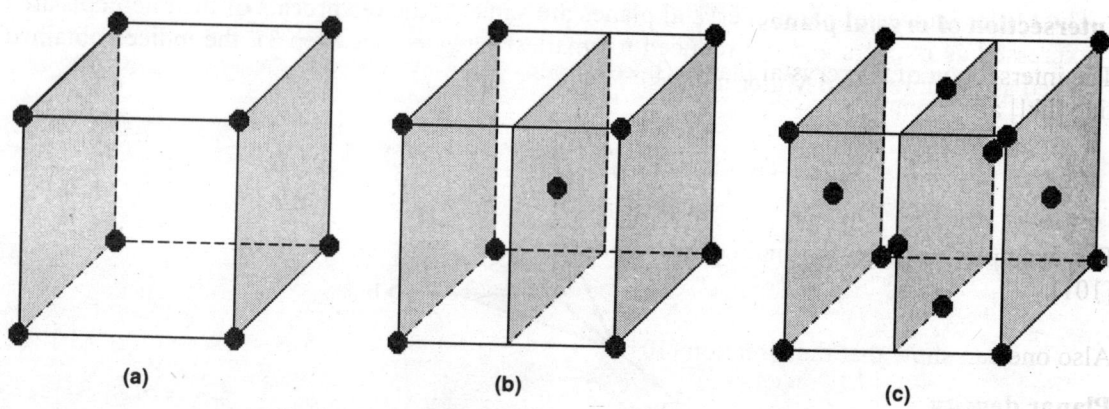

(a)   (b)   (c)

Fig. 3.18 (010) planes : (a) sc ,(b) bcc, (c) fcc.

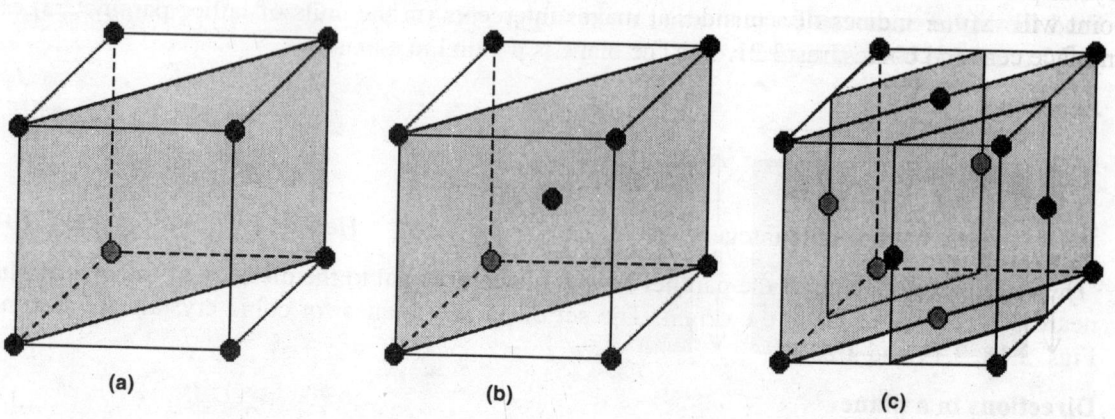

(a)   (b)   (c)

Fig. 3.19 (110) planes : (a) sc ,(b) bcc, (c) fcc.

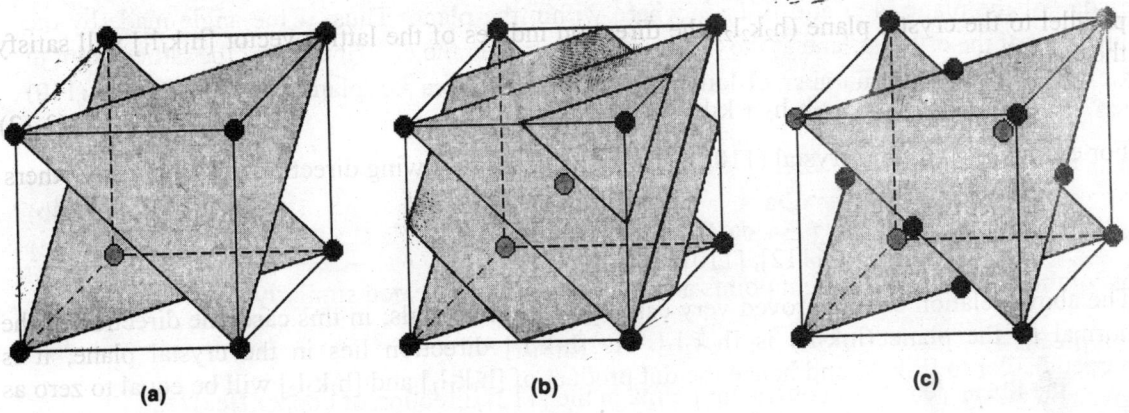

Fig. 3.20 (111) planes : (a) sc ,(b) bcc, (c) fcc.

Intersection of crystal planes

The intersection of two crystal planes $(h_1k_1l_1)$ and $(h_2k_2l_2)$ is a line. The direction indices of the line [hkl] are given by

$$h = k_1l_2 - l_1k_2$$
$$k = l_1h_2 - h_1l_2 \qquad\qquad (3.13)$$
$$l = h_1k_2 - k_1h_2$$

For example, the direction indices of the line of intersection of the planes $(10\bar{1})$ and $(11\bar{1})$ is [101].

Also one can show that the direction [101] lies in both $(11\bar{1})$ and $(10\bar{1})$ planes using Eq.(3.12).

Planar density

The number of lattice points per unit area is defined as the planar density. The number of lattice points on a crystal plane within the unit cell has to be calculated properly. The lattice points on crystal plane are considered as circles. Depending upon the location, the contribution of lattice point will vary. For example, consider (110) plane in the unit cell of simple cubic, body centred and face centred cubic, Fig. 3.21. The contribution of the lattice point at the corner is ¼,   that at

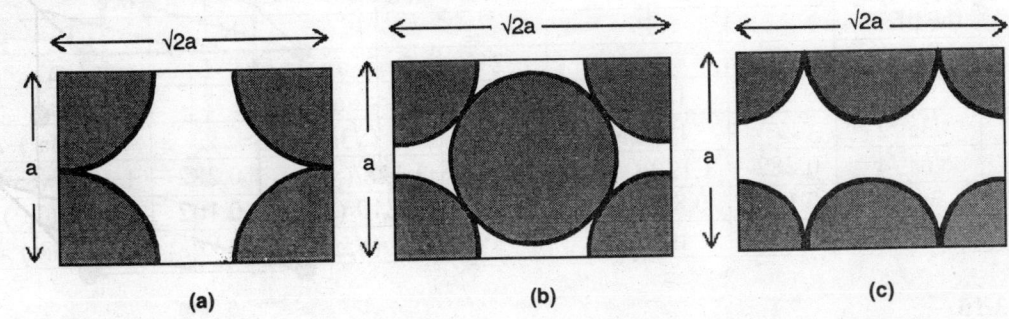

Fig. 3.21 Lattice points in (110) plane (a) sc (b) bcc and (c) fcc

the edge of the plane is ½ and is 1 anywhere within the plane. Thus, if the angle made by the boundaries of the crystal plane at the lattice point is $\theta°$ then the contribution of the lattice point to the plane is $\theta/360$. The number of lattice points per unit area i.e. planar density (P_d) for (110) plane is

$$P_d = (4 \times ¼)/\sqrt{2}a^2 = 1/\sqrt{2}a^2 \qquad \text{for sc} \qquad (3.14a)$$
$$= (4 \times ¼ + 1)/\sqrt{2}a^2 = \sqrt{2}/a^2 \qquad \text{for bcc} \qquad (3.14b)$$
$$= (4 \times ¼ + 2 \times ½)/\sqrt{2}a^2 = \sqrt{2}/a^2 \qquad \text{for fcc} \qquad (3.14c)$$

Most of the times, the equivalent points and lattice points are treated similarly.

Example 3.14

What is the linear density of equivalent points in the [112] direction of copper (fcc)?

Solution:

Since it is fcc structure and $r_{Cu} = 1.278$ Å, $a = 2\sqrt{2}r_{Cu} = 3.614$ Å
The length of the lattice vector in [112] direction $= \sqrt{[(a/2)^2 + (a/2)^2 + a^2]} = a\sqrt{(3/2)} = 4.426$ Å
Hence the linear density $L_d = 1/[a\sqrt{(3/2)}] = 0.226/$ Å $= 2.26 \times 10^9$ atoms/m

Table 3.6(a) Linear density of cubic lattices

Crystal direction	sc		fcc		bcc	
	Length	L_d	Length	L_d	Length	L_d
[100]	a	1/a	a	1/a	a	1/a
[110]	$\sqrt{2}a$	$1/(\sqrt{2}a)$	$a/\sqrt{2}$	$\sqrt{2}/a$	$\sqrt{2}a$	$1/(\sqrt{2}a)$
[111]	$\sqrt{3}a$	$1/(\sqrt{3}a)$	$\sqrt{3}a$	$1/(\sqrt{3}a)$	$\sqrt{3}a/2$	$2/(\sqrt{3}a)$
[112]	$\sqrt{6}a$	$1/(\sqrt{6}\,a)$	$\sqrt{(3/2)}a$	$\sqrt{2}/(\sqrt{3}a)$	$\sqrt{6}\,a$	$1/(\sqrt{6}\,a)$
[122]	3a	1/(3a)	3a	1/(3a)	3a	1/(3a)
[120]	$\sqrt{5}a$	$1/(\sqrt{5}a)$	$\sqrt{5}a$	$1/(\sqrt{5}a)$	$\sqrt{5}a$	$1/(\sqrt{5}a)$
[123]	$\sqrt{14}a$	$1/(\sqrt{14}a)$	$\sqrt{(7/2)}a$	$\sqrt{2}/(\sqrt{7}a)$	$\sqrt{14}a$	$1/(\sqrt{14}a)$

Table 3.6(b) Planar desity of cubic lattices

Crystal plane	Area of the plane	sc		fcc		bcc	
		$N_{lattice}$	P_d	$N_{lattice}$	P_d	$N_{lattice}$	P_d
(100)	a^2	1	$1/a^2$	2	$2/a^2$	1	$1/a^2$
(110)	$\sqrt{2}a^2$	1	$1/(\sqrt{2}a^2)$	2	$\sqrt{2}/a^2$	2	$\sqrt{2}/a^2$
(111)	$\sqrt{3}a^2/2$	½	$1/(\sqrt{3}a^2)$	2	$4/(\sqrt{3}a^2)$	½	$1/(\sqrt{3}a^2)$
(112)	$\sqrt{6}a^2/4$	0.282	$1.128/(\sqrt{6}a^2)$	0.782	$3.128/(\sqrt{6}a^2)$	0.282	$1.128/(\sqrt{6}a^2)$
(122)	$3a^2/8$	0.102	$0.819/(3a^2)$	0.102	$0.819/(3a^2)$	0.102	$0.819/(3a^2)$
(120)	$\sqrt{5}a^2/2$	½	$1/(\sqrt{5}a^2)$	1	$2/(\sqrt{5}a^2)$	½	$1/(\sqrt{5}a^2)$

Example 3.15

Calculate the planar density of equivalent points on the (021) plane of lead (fcc) ?

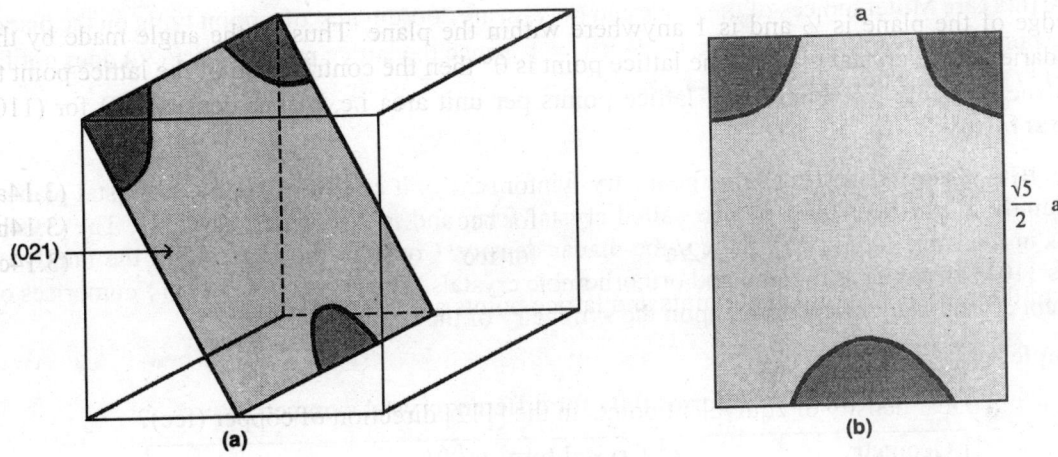

Fig. 3.22(a) (021) plane in lead (fcc); (b) lattice points in(021) plane.

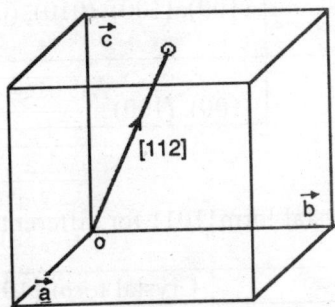

Fig. 3.23 [112] direction of fcc unit cell.

Solution:

In fcc lead each lattice point has a basis of only one atom. The lattice parameter is given by
$$a_{Pb} = 4 (1.750)/1.414 = 4.95 \text{ Å} \quad \text{since } r_{Pb} = 1.750 \text{ Å}.$$

Fig. 3.22(b) shows that the (021) plane contains ¼ + ¼ + ½ = 1 atom.
The area A of (021) plane = $(4.95)^2 \sqrt{5/2}$ Å^2 = 27.4 Å^2
Planar density P_d = 0.0365 atoms/ Å^2 = 3.65 x 10^{18} atoms/m^2

Example 3.16

Verify whether the lattice point at (½, ½, ½) lies on (111) plane of a bcc crystal.

Solution:

In case of cubic crystal, lattice planes satisfy the equation
$$hx + ky + lz = 1$$

where (hkl) are Miller indices of the plane and (x, y, z) are coordinates of a point lying on the plane. In the present problem, $\frac{1}{2} \times 1 + \frac{1}{2} \times 1 + \frac{1}{2} \times 1 \neq 1$, hence the lattice point at $(\frac{1}{2}, \frac{1}{2}, \frac{1}{2})$ does not lie on (111) plane of a bcc crystal, Fig. 3.20.

Crystal forms

The crystal planes possessing same geometry lying in the unit cell form family of crystal planes. The family of crystal planes is also called crystal form and is designated as {hkl}. The crystal planes in the same family will have same planar density. Consider the example of the family of planes {101} in cubic, tetragonal and orthorhombic crystals. The crystal form {101} comprises of different crystal planes depending upon the symmetry of the unit cell, Table 3.7.

Table 3.7 (a) Crystal form{100} for different crystal symmetries.

Geometry	Crystal form {100}
Cubic	(100), ($\bar{1}$00), (010), (0$\bar{1}$0), (001), (00$\bar{1}$)
Tetragonal	(100), ($\bar{1}$00), (010), (0$\bar{1}$0)
Orthorhombic	(100), ($\bar{1}$00)

Table 3.7 (b) Crystal form{101} for different crystal symmetries.

Geometry	Crystal form {101}
Cubic	(101), ($\bar{1}$01), (10$\bar{1}$), ($\bar{1}$0$\bar{1}$), (110), ($\bar{1}$10), (1$\bar{1}$0), ($\bar{1}$$\bar{1}$0), (011), (0$\bar{1}$1), (01$\bar{1}$), (0$\bar{1}$$\bar{1}$)
Tetragonal	(101), ($\bar{1}$01), (10$\bar{1}$), ($\bar{1}$0$\bar{1}$), (011), (0$\bar{1}$1), (01$\bar{1}$), (0$\bar{1}$$\bar{1}$)
Orthorhombic	(101), ($\bar{1}$01), (10$\bar{1}$), ($\bar{1}$0$\bar{1}$)

Example 3.17

Draw the crystal planes (110), (010), (111) in simple cubic unit cell. In these figures, the origin is chosen at O to draw the crystal planes within the unit cell.

Solution:

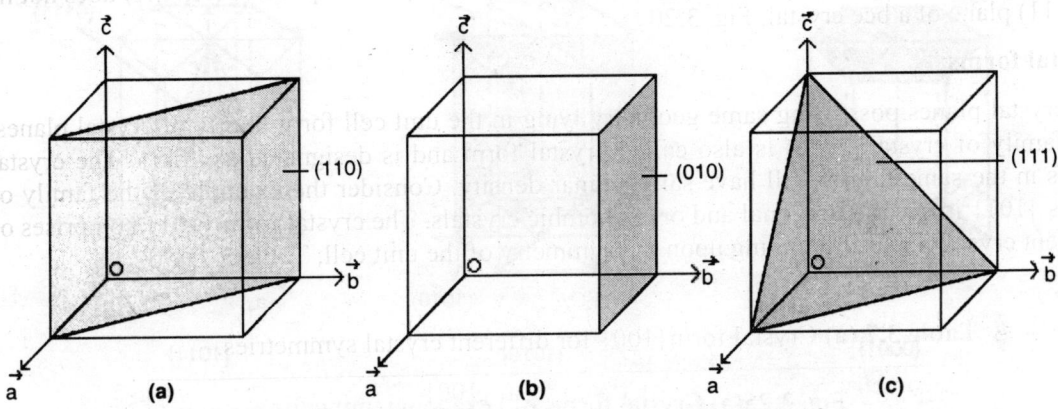

Fig. 3.24 Crystal planes (a) (110), (b) (010), (c) (111).

If any of the Miller indices is zero the plane is parallel to the corresponding axis. If two of the Miller indices are zeroes, the plane is parallel to both the corresponding axes. For example, (110) plane is parallel to c-axis and (010) plane is parallel to both a- and c-axes.

Miller-Bravais indices

Any crystal plane can be completely specified by three Miller indices. In hexagonal crystal the unit cell can be taken as parallelepiped of crystal parameters $a = b \neq c$ and $\alpha = \beta = 90°$ but $\gamma = 120°$. The symmetry of the hexagonal system can be described more appropriately if the hexagonal prism is chosen as unit cell. There are altogether four axes of the hexagonal unit cell. The axes a_1, a_2, a_3, are in basal plane of the unit cell and are at 120° to one another. The fourth axis c is perpendicular to the basal plane. To define a plane in hexagonal unit cell, a set of indices (hkil) is used where i is the additional index. These are called Miller-Bravais indices (MBI). These indices are used to reveal the hexagonal symmetry and are related by the equation

$$h + k + i = 0 \tag{3.15}$$

As shown in Fig. 3.25, the crystal form $\{11\bar{2}0\}$ contains 6 members

$$(11\bar{2}0), (1\bar{2}10), (\bar{2}110), (\bar{1}\bar{1}20), (\bar{1}2\bar{1}0), (2\bar{1}\bar{1}0).$$

The crystal form $\{10\bar{1}1\}$ contains 12 planes, out of these six planes form hexagonal pyramid on the basal plane as shown above and six other planes form inverted hexagonal pyramid below the basal plane; $\{0001\}$ includes 2 basal planes. The six faces of hexagonal unit cell are represented by the family.

$$\{10\bar{1}0\}$$

Example 3.18

Find the planar density of (0001), $(11\bar{2}0)$, $(10\bar{1}0)$ and $(10\bar{1}1)$ planes of simple hexagonal cell.

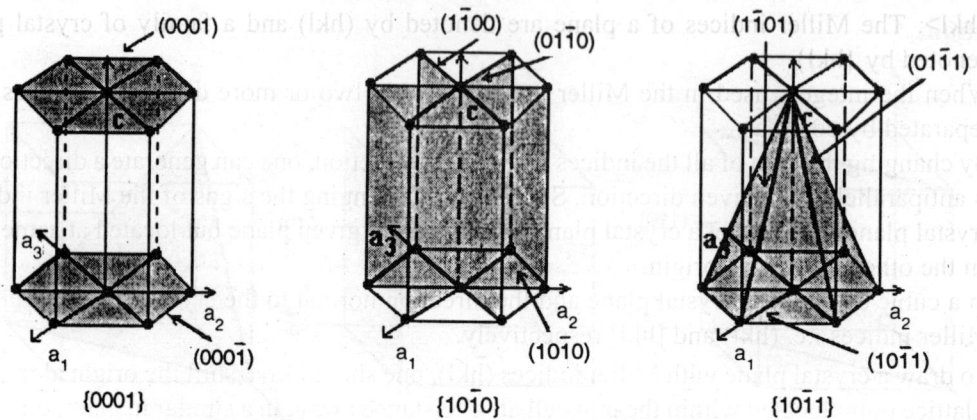

Fig. 3.25(a) Crystal forms of hexagonal unit cell.

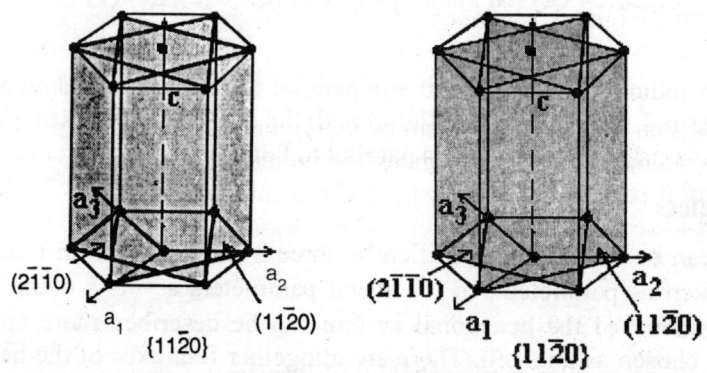

Fig. 3.25(b) Crystal forms of hexagonal unit cell.

Solution:

a) Area of (0001) plane = 6 x ($\sqrt{3}a^2/4$) = $3\sqrt{3}a^2/2$

Contribution of each corner lattice point is (120°/360°)= ⅓ to (0001) lattice plane and there are six such corner lattice points. Contribution of the lattice point at the centre of the basal plane is one. Thus, the effective number of lattice points in the plane is 6 x (⅓) + 1 = 3.

Thus, the planar density = $3x2/(3x\sqrt{3}a^2)=2/(\sqrt{3}a^2)$.

b) Area of (11$\bar{2}$0) plane = $\sqrt{3}ac$

Effective number of lattice points on this plane = 4x(1/4) = 1. Planar density = $1/(\sqrt{3}ac)$

c) Planar density of (10$\bar{1}$0) plane = $1/(ac)$

d) Planar density of (10$\bar{1}$1) plane = $(½)/(½ a\sqrt{((3a^2/4)+c^2)}) = 1/[a\sqrt{((3a^2/4)+c^2)}]$

Summary on notation of Miller indices

i) Miller indices of a direction are denoted by [hkl] whereas family of directions is denoted by

<hkl>. The Miller indices of a plane are denoted by (hkl) and a family of crystal planes is denoted by {hkl}.

ii) When the integers used in the Miller indices contain two or more digits, the indices must be separated by commas.

iii) By changing the sign of all the indices of a crystal direction, one can generate a direction, which is antiparallel to the given direction. Similarly, by changing the signs of the Miller indices of a crystal plane, we can get a crystal plane parallel to the given plane but located at same distance on the other side of the origin.

iv) In a cubic crystal, the crystal plane and the direction normal to the crystal plane have the same Miller indices i.e. (hkl) and [hkl] respectively.

v) To draw a crystal plane with Miller indices ($\bar{h}$kl), one should first shift the origin along a axis to a lattice point located within the unit cell at 'a' distance away. In a similar manner, one can draw other crystal planes for the cases where other Miller indices are negative.

vi) To draw a plane or a direction in hexagonal unit cell one should remember that out of the Miller-Bravais indices (hkil), i index is given by,

$$i = - (h + k)$$

as per Eq. (3.15). Miller-Bravais indices can be determined exactly in the way Miller indices are determined. In Fig. 3.26, the direction indices are indicated in basal plane of the unit cell. For example, the direction parallel to a_2 axes is resolved into components along $-a_3$ and $-a_1$ axes. Thus indices h and i are -1, -1 respectively. The index $l = 0$ and $k = -(h + i) = 2$. Thus the direction indices of a_2 are

$$[1\bar{2}10].$$

Similarly other directions are also represented by Miller-Bravais indices in Fig. 3.26.

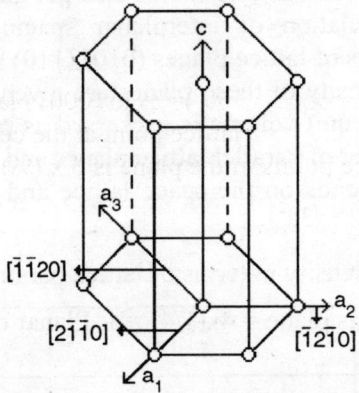

Fig. 3.26. Miller- Bravais indices of direction in the basal plane of hexagonal unit cell.

3.9 INTERPLANAR SPACING

The interplanar spacing is the perpendicular distance between two adjacent planes of a set of parallel crystal·planes in the space lattice. The interplanar spacing d_{hkl} is measured between the first such plane and a parallel plane passing through the origin. The interplanar spacing for cubic lattice can be expressed as,

$$d_{hkl} = \frac{a}{\sqrt{(h^2 + k^2 + l^2)}} \tag{3.16}$$

Table 3.8 shows the interplanar spacing for various crystal geometries.

Table 3.8 Interplanar spacing

Crystal System	Spacing (d_{hkl})
Cubic	$[(h^2 + k^2 + l^2)/a^2]^{-1/2}$
Tetragonal	$[(h^2 + k^2)/a^2 + (l^2/c^2)]^{-1/2}$
Orthorhombic	$[(h^2/a^2)+(k^2/b^2) +(l^2/c^2]^{-1/2}$
Hexagonal	$[2(h^2+k^2+i^2)/3a^2 + (l^2/c^2)]^{-1/2}$

While determining the Miller indices, we normally reduce the indices to the lowest integer set. While calculating interplanar spacing one should not reduce the indices. In order to calculate the interplanar spacing for a specific set of parallel lattice planes, all possible lattice planes having same MI are to be drawn within the unit cell. One has to locate the plane from this set, which is nearest to the origin. Interplanar spacing can be calculated using Eq(3.16) in case of cubic crystal, where h,k,l are unreduced Miller indices of the plane nearest to the origin. To explain this, let us consider the case of bcc crystal as compared to sc. In case of simple cubic lattice interplanar separation d_{100} is given by

$$d_{100} = a$$

In bcc unit cell there is a lattice plane in the midway between two parallel faces (100), which can be drawn within the unit cell. One has to locate the lattice plane from this set, which is nearest to the origin. In this case it is the plane passing through the lattice point at the body centre, Fig. 3.18(b). This lattice plane has unreduced Miller indices (200) and is identical to the (100) plane. This plane is at distance a/2 from the origin. We also get the same result considering unreduced Miller indices for the calculation of interplanar spacing d_{200} using Eq.(3.16). Figs. 3.18, 3.19 and 3.20 show parallel sets of lattice planes (010),(110) and (111), respectively.

The interplanar distance and planar density of these planes are given in Table 3.9 for sc, bcc and fcc. The density of lattice points per unit volume is expressed as planar density divided by interplanar spacing. For sc, consider any set of parallel lattice planes and find the value of P_d/d. It is always equal to $1/a^3$; so this ratio depends on the space lattice and does not depend on the lattice planes.

Table 3.9 Interplanar spacing and planar density in terms of lattice parameter a for cubic crystals

Lattice→	sc		bcc		fcc	
Lattice plane↓	d	P_d	d	P_d	d	P_d
(100)	a	$1/a^2$	a/2	$1/a^2$	a/2	$2/a^2$
(110)	$a/\sqrt{2}$	$1/(\sqrt{2}a^2)$	$A/\sqrt{2}$	$\sqrt{2}/a^2$	$a/(2\sqrt{2})$	$\sqrt{2}/a^2$
(111)	$a/\sqrt{3}$	$1/\sqrt{3}a^2$	$a/(2\sqrt{3})$	$1/(\sqrt{3}a^2)$	$a/\sqrt{3}$	$4/\sqrt{3}a^2$
P_d/d	$1/a^3$		$2/a^3$		$4/a^3$	

Similarly, this ratio has values $2/a^3$ and $4/a^3$ for bcc and fcc crystals, respectively.

3.10 ORIGIN OF X-RAYS

When fast moving electrons are incident on a metal target, the electron beam loses its energy due to collision with the target. This results in the emission of x-rays. The generation of x-rays is described below:

(i) An electron may lose whole of its energy in one collision or may do so while undergoing several collisions. This gives rise to the generation of white radiation or continuous spectrum of x-ray. This type of x-ray beam will consist of radiations of continuous range of frequencies limited by a minimum wavelength. The minimum wavelength (λ_{min}) in the continuous spectrum will correspond to the x-ray beam generated when the electron in x-ray tube loses whole of its energy in one collision i.e. the energy of x-ray will be equal to the energy eV of the electron. Thus,

$$\lambda_{min} = (hc/eV) = (1.24 \times 10^{-6}/V)m \qquad (3.17)$$

where V is the applied voltage which accelerates the electrons towards the target and e is the electronic charge. If the electron loses its energy in more than one collision, it gives rise to white x radiation i.e. x-ray consisting of continuous spectrum.

ii) Electrons can acquire high energy while getting excited through a very high voltage. These electrons can collide with the core electrons of the target atoms and cause inelastic excitation of core electrons. In this process the electrons from the inner shell of the target atoms are excited to higher energy states and the electrons from other higher energy states may jump into these vacant energy states giving rise to x-radiation of specific wavelengths. This process generates sharp x-rays of specific wavelengths, λ called characteristic x-radiation. The wavelength of the characteristic x-ray beam will depend upon the target material e.g. when copper target is bombarded by electrons a strong line Cu-Kα (= 1.542 Å) is generated. Both types of x-radiations emitted by molybdenum target are shown in Fig. 3.27.

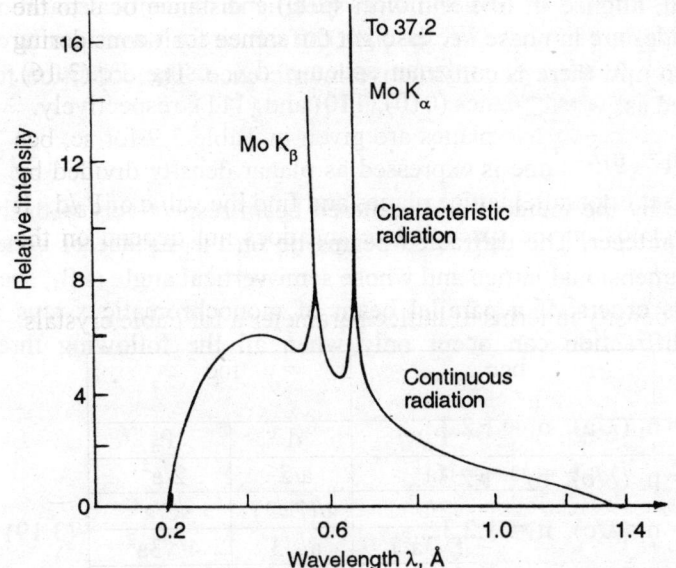

Fig. 3.27 The x-ray spectrum using molybdenum target at 35 KV.

3.11 X-RAY DIFFRACTION AND STRUCTURE DETERMINATION

The phenomenon of diffraction is commonly observed when the electromagnetic radiation, in the visible range of the spectrum, is incident on an optical grating. The diffraction occurs only when wavelength of the incident radiation is of the order of grating spacing. When the x-ray beam is incident on a crystal, it gets diffracted, if the wavelength of x-ray is of the order of the interplanar spacing. In fact, the crystal acts as a three-dimensional grating. In other words the study of monochromatic x-ray diffraction reveals the details of crystal structure and is the most commonly used technique for studying the crystal structure.

When an x-ray beam is incident on a crystal, the x-ray photons interact with atoms, which start oscillating and become the secondary source of electromagnetic radiation. The electromagnetic radiation, emitted by the oscillating electrons propagates in all directions, which have same or different frequencies as that of the incident x-ray. In case of elastic collision between x-ray photons and electrons, the emitted radiation has same frequency as that of the incident x-ray. The emitted secondary radiations from various crystal planes are in phase only in certain specific directions, and reinforce to give constructive interference. The directions in which constructive interference takes place depend on

(a) the angle of incidence of x-ray;
(b) the interplanar spacing;
(c) the wavelength of the incident x-ray.

The constructive interference in certain specific directions, arising out of scattering of x-ray beam by crystal lattice, culminates into the diffraction pattern of x-ray. Von Laue as well as Bragg explained the origin of x-ray diffraction patterns:

(i) Von Laue Method

Laue explained the x-ray diffraction by considering the interference of the scattered x-radiation from the lattice points of the material, aligned in line and with specific distance of separation between them. If the scattered amplitudes are in phase i.e. the path difference for waves scattered from adjacent lattice points is equal to $n_1\lambda$, there is constructive interference. The condition for constructive inference can be expressed as:

$$a \cos \alpha_1 - a \cos \beta_1 = n_1\lambda \qquad (3.18)$$

where α_1 and β_1 are the angles made by the incident and scattered beam respectively with the a-axis, Fig. 3.28 and n_1 is a positive integer. The diffracted beams lie on the surface of cones whose common axis will be the one-dimensional lattice and whose semi-vertical angle is β_1, such that Eq.(3.18) is satisfied for various orders. If a parallel beam of monochromatic x-rays is incident on a perfect crystal, the diffraction can occur only when all the following three conditions are to be satisfied:

$$\cos \alpha_1 - \cos \beta_1 = n_1 (\lambda/a), \; n_1 = 1,2,3.....$$

$$\cos \alpha_2 - \cos \beta_2 = n_2 (\lambda/b), \; n_2 = 1,2,3.....$$

$$\cos \alpha_3 - \cos \beta_3 = n_3 (\lambda/c), \; n_3 = 1,2,3..... \qquad (3.19)$$

where a, b, c, are the lattice vectors. Eqs (3.19) are called the Laue equations.

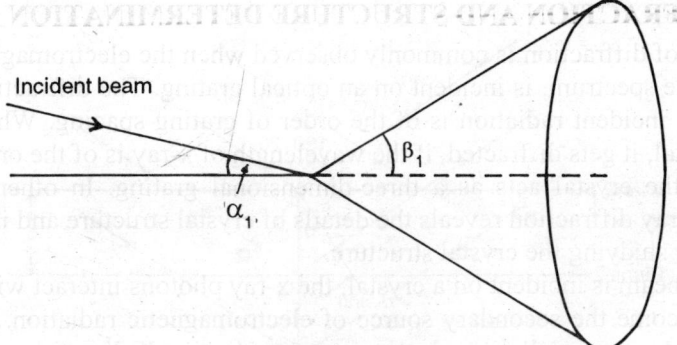

Fig. 3.28   Scattering from a one-dimensional lattice

Thus, the diffraction can only occur in three-dimensional lattice when the cones on a, b and c axes all intersect in one line, which is a possible direction of the diffracted beam. To obtain a diffracted beam, the wavelength of the incident x-ray or the position of the crystal is to be changed appropriately such that the three cones intersect in one line, Fig. 3.29.

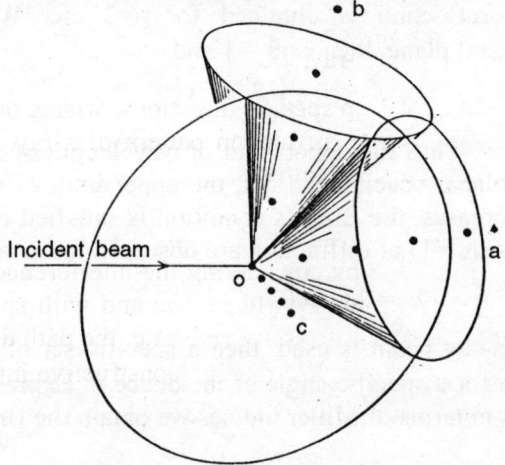

Fig. 3.29   Cones on three axes (a, b, c) defining possible diffraction directions (where three cones intersect in one line) from a three- dimensional lattice.

(ii) Bragg's Method

According to Bragg's method the diffraction process can be understood by considering reflection of x-rays by a set of parallel crystal planes, Fig. 3.30. Consider a beam of x-ray of wavelength λ is incident on a crystal plane making an angle θ with it. The crystal planes are semitransparent i.e. x-ray beam incident on a given plane gets partially reflected while rest of it passes through the plane and is incident on other crystal planes. The angle of incidence θ is equal to the angle of reflection.

The rays 1 and 2, reflected by crystal planes 1 and 2 will reinforce to give maximum intensity if the path difference between the rays is an integral multiple of the wavelength. According to Bragg's law,

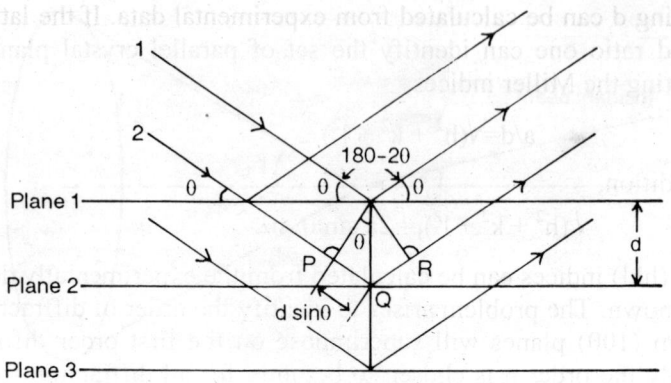

Fig. 3.30. Bragg diffraction.

$$PQ + QR = 2d\sin\theta = n\lambda \qquad (3.20)$$

where n is an integer and d is interplanar spacing. If n= 1, we obtain first order reflection. Similarly, second order reflection is obtained for n=2 etc. When the x-ray beam falls perpendicularly on the crystal plane, then $\sin\theta = 1$ and

$$\lambda = 2d \qquad (3.21)$$

for the first order reflection. Thus the upper limit of wavelength of x-ray, which can give rise to diffraction, is 2d. If interplanar spacing is 1.5Å, the upper limit of wavelength becomes 3Å. As the angle of incidence decreases, the Bragg's condition is satisfied for the x-rays having smaller wavelength. Higher orders (n > 1) of diffraction are observed for x-rays having wavelength

$$\lambda = 2d \sin\theta /n \qquad (3.22)$$

When monochromatic x-ray beam is used, then a specific set of parallel crystal planes (hkl) satisfies Bragg's condition for a specific angle of incidence θ. Expressing the interplanar spacing of the set of crystal planes in terms of Miller indices we obtain the Bragg's condition as

$$2\left[\frac{h^2}{a^2} + \frac{k^2}{b^2} + \frac{l^2}{c^2}\right]^{-1/2} \sin\theta = n\lambda \qquad (3.23)$$

for cubic, tetragonal and orthorhombic crystals. In case of cubic crystal the above expression reduces to

$$\frac{2a}{\sqrt{(h^2+k^2+l^2)}} \sin\theta = n\lambda \qquad (3.24)$$

In case the parallel set of crystal planes are represented by (111) and the order of diffraction is 1, the lattice parameter a can be expressed as

$$a = \sqrt{3}\,\lambda/\,2 \sin\theta \qquad (3.25)$$

The interplanar spacing d can be calculated from experimental data. If the lattice parameter a is known, then from a/d ratio one can identify the set of parallel crystal planes responsible for diffraction by calculating the Miller indices,

$$a/d = \sqrt{(h^2 + k^2 + l^2)} \qquad (3.26)$$

and from Bragg's condition,

$$\sqrt{(h^2 + k^2 + l^2)} = 2a \sin \theta / n\lambda \qquad (3.27)$$

Thus, the possible (hkl) indices can be calculated from the experimentally determined θ values provided a value is known. The problem arises to identify the order of diffraction e.g. the second order diffraction from (100) planes will superimpose on the first order diffraction from (200) planes. Thus, normally the order n is chosen to be unity for all diffraction from parallel sets of crystal planes such as (100), (200) etc., since higher order diffraction lines are usually of lower intensity. In a crystal it may turn out that there are no such crystal planes as (200) etc. It may imply that these imaginary planes do not pass through any lattice point. Then, in that case first order diffraction from (200) actually refers to as second order diffraction from (100) planes.

Structure determination

The crystal structure can be determined from the x-ray diffraction pattern using any of the following techniques:

(i)Spectrometric method

Crystal structure was first determined by Bragg using a spectrometer. This spectrometer is now called as Bragg's x-ray spectrometer. It consists of three parts (Fig. 3.31)

- Source of x-ray beam
- Spectrometer table holding the crystal, which is graduated (provided with vernier scale) to find the position of the crystal with respect to the x-ray beam.
- Device to measure the intensity of x-ray (ionization chamber and electrometer).

The collimated x-ray beam is incident on the crystal mounted on the spectrometer table (S), which can rotate about a vertical axis so that suitable position of the crystal can produce the

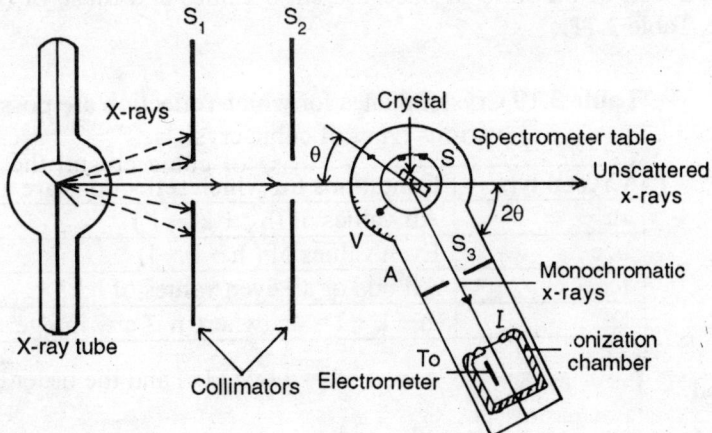

Fig. 3.31 Bragg's x-ray spectrometer showing X-ray absorption.

diffracted beam. The angle of diffraction can be measured by using the vernier provided with the graduated table. The reflected rays enter the ionization chamber I, which is mounted on the arm A, after suitable adjustment of the arm. The arm A can be rotated about same vertical axis as that of the table S.

The intensity of x-ray is maximum at certain glancing angles as indicated by sharp peaks. The corresponding 2θ values can be used in the Bragg's condition to find d, the interplanar spacing of the set of parallel planes responsible for maximum of x-ray diffraction. The structure determination of a crystal involves the following steps:

1) Determination of microscopic symmetry,
2) Finding out space lattice and its parameters,
3) Atomic arrangement within the unit cell.

Measurement of the density of the crystal and chemical composition can give useful information about the structure. The information about basis i.e. atomic arrangement within unit cell can be obtained only from the actual measurement of the intensity of the reflected x-ray beam. In case of monoatomic cubic crystal the procedure becomes simplified. Let us illustrate the method of structure determination in this case. We know that the Bragg's condition (3.24) for cubic crystal can be expressed as

$$\{\lambda^2/4a^2\} (h^2 + k^2 + l^2) = \sin^2\theta \tag{3.28}$$

where n is taken to be 1. In case of monochromatic x-radiation λ can be determined easily by using powder method for a known crystal structure. To determine the Miller indices of the reflecting planes of a given cubic lattice, we calculate $(h^2+k^2+l^2)$ for all possible combinations of h, k, l values for a measured value of θ. These h, k, l values are arranged in the increasing order. If the crystal structure of the given crystal is same as that of the assumed lattice, $\sin^2\theta$ is found to be proportional to $(h^2 + k^2 + l^2)$.

One can determine the type of cubic lattice by finding out the sets of parallel planes (hkl) corresponding to which certain reflection lines will be missing although Bragg conditions are satisfied. Table 3.10 gives crystal planes (hkl) for which reflection lines are obtained. Now we write down in Table 3.11 the ratio of $(h^2+k^2+l^2)$ values for allowed reflections for different cubic crystals. A comparison of the ratio of observed $\sin^2\theta$ values and those of $(h^2+k^2+l^2)$ reveals the crystal structure, Table 3.11.

Table 3.10 Crystal planes for which reflection are possible for various types of cubic crystals.

Crystal type	Conditions for which reflections are allowed
sc	All values of $(h^2 + k^2 + l^2)$
bcc	Even values of $(h + k + l)$
fcc	All odd or all even values of h,k,l
DC	$h + k + l = 4n$, where n is any integer

(ii)Laue method

In Laue method, a single crystal of the sample is kept in the path of white x-ray beam. A specific set of crystal planes will diffract x-radiation of discrete values of λ, if the incidence angle θ

Table 3.11 Ratio of $(h^2+k^2+l^2)$ for allowed reflections of cubic crystal

$h^2+k^2+l^2$	{h k l}	Ratio of allowed $(h^2+k^2+l^2)$ values sc	bcc	fcc	DC
1	1 0 0	1	-	-	-
2	1 1 0	2	1	-	-
3	1 1 1	3	-	3	-
4	2 0 0	4	2	4	-
5	2 1 0	5	-	-	-
6	2 1 1	6	3	-	3
8	2 2 0	8	4	8	4
9	2 2 1 3 0 0	9	-	-	-
10	3 1 0	10	5	-	5
11	3 1 1	11	-	11	-
12	2 2 2	12	6	12	-
13	3 2 0	13	-	-	-
14	3 2 1	14	7	-	-
16	4 0 0	16	8	16	8
17	3 2 2 4 1 0	17	-	-	-
18	3 3 0 4 1 1	18	9	-	-
19	3 3 1	19	-	19	-
20	4 2 0	20	10	20	-
21	4 2 1	21	-	-	-
22	3 3 2	22	11	-	11
24	4 2 2	24	12	24	12
25	4 3 0 5 0 0	25	-	-	-

satisfies the Bragg's Law. In this method a collimated beam of x-ray from a suitable source is incident on the single crystal. The diffracted beam from the crystal falls on a flat photographic plate. The diffraction pattern, consisting of a series of spots, is obtained on the photographic plate. The symmetry of the Laue spots (Fig 3.32) determines the symmetry of the crystal. No

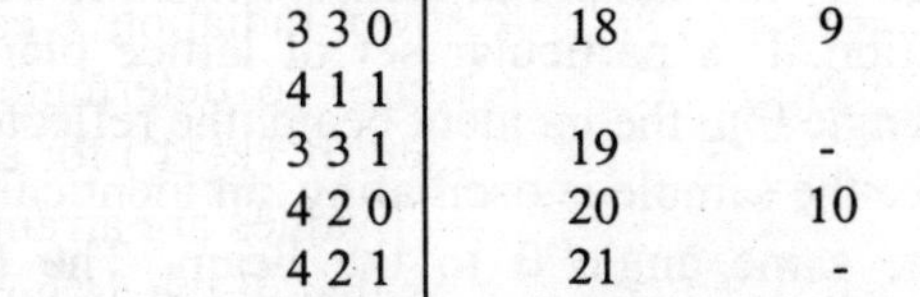

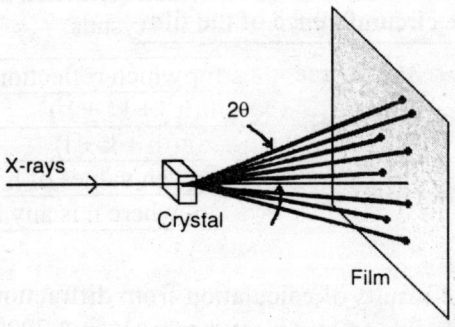

Fig.3.32. Laue Pattern.

other technique can give so much of information about the symmetry of the crystal as the Laue photography. This method is often used to orient the crystal appropriately for carrying out experiments involving condensed matter.

(iii) Powder method

In powder method a monochromatic x-ray of a specific wavelength is diffracted with a large number of possible θ values, so that the reflection occurs at the right combination that satisfies the Bragg's Law. This is achieved by using specimen in powder form consisting of an aggregate of a large number of tiny crystallites. This method is commonly used in laboratories to find a crystal structure since many a time the specimen is not available in the form of single crystal of appropriate size and also all possible orientations of lattice planes are present in the powder sample.

Debye-Scherrer camera is used in powder method to determine the structure of unknown compounds. The sample can be in the form of cylindrical thin wire, or powdered specimen can be placed inside a capillary tube, or a thin fibre can be coated with the powdered specimen. The sample is placed on the sample holder at the centre of the cylindrical Debye-Scherrer camera. The film with two punched holes diametrically opposite of each other is loaded onto the cylindrical surface of the camera. The sample is provided with oscillatory motion. The camera is attached to the x-ray tube mounting such that x-ray beam enters the camera through a pinhole and punched hole of the film and gets diffracted by the sample at centre. Principle of powder method can be illustrated by using Bragg reflection. If a particular set of lattice planes in one crystallite is oriented at the appropriate Bragg angle θ to the incident beam, the reflected beam makes an angle 2θ with the undeviated beam. Since the sample is oscillating, an identical set of planes in another crystallite can be oriented at the same angle θ to the beam. The reflected beams, which correspond to these orientations, outline a cone, which is coaxial with the incident beam and has a half apex angle 2θ. The oscillation of the sample gives rise to dense reflection cones. Each reflection cone intersects the film in a pair of arcs and all the cones generated by the sample are represented on the film. The front reflection arcs are curved about the hole of the film facing the pinhole through which x-ray beam enters the camera, the curvature decreasing to zero at $2\theta = 90°$. The back reflection arcs are curved in the opposite direction, the curvature increasing with increasing 2θ angles. The distance $2S$ between a pair of arcs on the film bears a simple relation to apex solid angle 4θ of the corresponding reflecting cone,

$$\frac{2S}{4\theta} = \frac{\text{The circumference of the film}}{360 \text{ degree}}$$

So,

$$\theta = \frac{2\pi S}{4\pi r} = \frac{S}{2r} \tag{3.29}$$

where r is the radius of the camera.

A typical experimental data and results of calculation from diffraction pattern of fcc iron are tabulated in Table 3.12 where monochromatic x-ray beam, $\lambda_v = 2.2909$ Å of chromium target using vanadium filter is used and the radius of the camera is 28.65 mm.

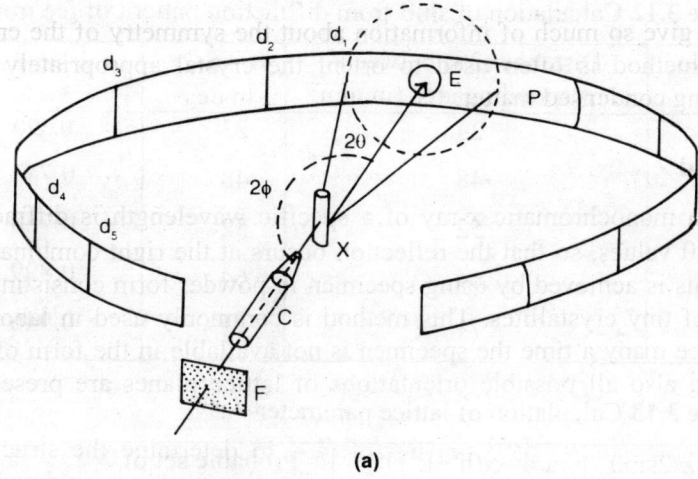

(a)

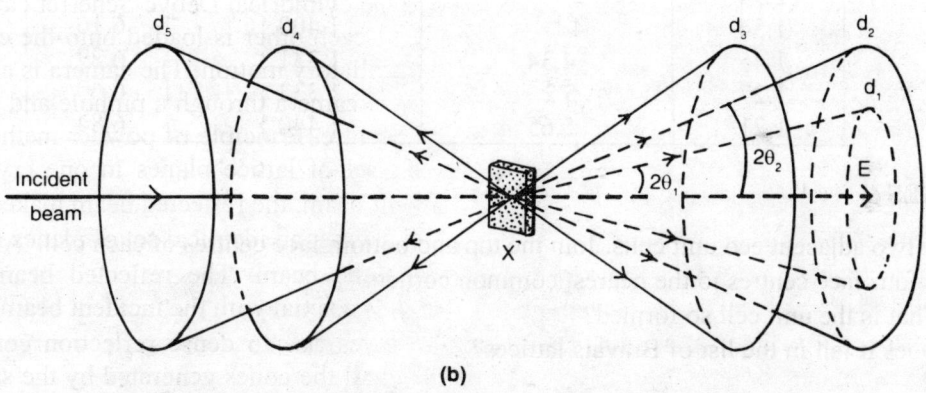

(b)

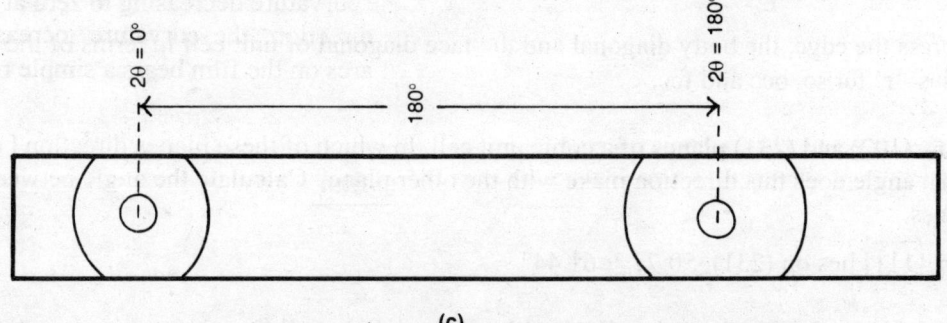

(c)

Fig. 3.33 Diffraction pattern using powder method.
(a) Arrangement in Debye-Scherrer Camera.
(b) Reflection cones.
(c) Film with punched holes.

Table 3.12 Calculation of sinθ from diffraction pattern of fcc iron.

S.No	Distance (2S) between corresponding arcs (in mm)	θ = S/2r in deg	sinθ
1.	23	23	0.390
2.	48	48	0.743
3.	52	52	0.788
4.	64	64	0.899
5.	71	71	0.946

Table 3.13 Calculation of lattice parameter 'a'

$d = \lambda/2\sin\theta$ (Å)	$a/d = \sqrt{(h^2+k^2+l^2)}$ (Approx.)	Probable set of diffraction planes	a (Å)
2.94	2	{ 200 }	5.88
1.54	4	{ 400 }	6.16
1.45	4.34	{ 331 }	6.29
1.27	5.2	{ 333 }	6.60
1.21	5.65	{ 440 }	6.84

EXERCISES

3.1 Draw two adjacent eco unit cells. Join the top and bottom face centres of each cell. Also join these four face centres to the nearest common corners
(i) What is the unit cell so formed?
(ii) Does it fall in the list of Bravais lattices?

3.2 Explain why there is no ecc lattice in the Bravais lattice but there is eco lattice.

3.3 Express the edge, the body diagonal and the face diagonal of unit cell in terms of the atomic radius 'r' for sc, bcc and fcc.

3.4 Show (102) and (231) planes of a cubic unit cell. In which of these planes direction $[\bar{1}1\bar{1}]$ lies? What angle does this direction make with the other plane? Calculate the angle between these planes.

Ans: $[\bar{1}1\bar{1}]$ lies on (231), 50.77 °; 61.44°

3.5 Sketch (112) plane in the unit cell of a cubic crystal. Show all the <110> directions that lie on this plane, giving the Miller indices of each one of them.

3.6 Sketch a $(\bar{1}1\bar{1})$ plane in the unit cell of a cubic crystal. Draw all the <110> directions that lie on this plane, giving the Miller indices of each one of them.

Ans: [110], [011], [101], $[\bar{1}0\bar{1}]$, $[\bar{1}\bar{1}0]$, $[0\bar{1}\bar{1}]$

3.7   Find the conditions for the crystal to be rhombohedral.

Ans: $a = b = c$; $\alpha = \beta = \gamma \neq 90°$

3.8   Given the plane (220) in a monoatomic fcc with lattice constant, $a = 2\text{Å}$. Find the planar density.

Ans: $3.536 \times 10^{19}/m^2$

3.9   Find the planar density of atoms and lattice points in $(0\overline{1}12)$ plane in a hexagonal unit cell where $c/a = 1$.

Ans: $0.432/a^2$

3.10 Draw $(10\overline{1}2)$ and $(11\overline{2}0)$ planes of a hexagonal unit cell. Draw the direction of the line of intersection of these two planes. Also write down MBI of other planes belonging to crystal forms $\{10\overline{1}2\}$ and $\{11\overline{2}0\}$.

3.11 Given in figure below, 'I' & 'J' are mid points of FD and EC respectively. Find the angle between AI and AJ.

Ans: 41.81°

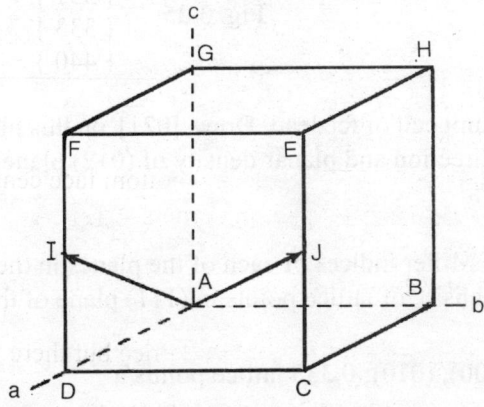

Fig. 3.34

3.12 The position of the origin in the unit cell of Fe (fcc) is chosen at one of the tetrahedral interstitial positions. Specify the coordinates of equivalent points in the unit cell.

Ans: $(\frac{1}{4},\frac{1}{4},\frac{1}{4})$; $(\frac{1}{4},\frac{3}{4},\frac{3}{4})$; $(\frac{3}{4},\frac{1}{4},\frac{3}{4})$; $(\frac{3}{4},\frac{3}{4},\frac{1}{4})$

3.13 Calculate the radius of an atom of a metal X (fcc) with density $= 22.4 \times 10^3$ kg/m³ and an atomic weight of 192.2 g/mole.

Ans: 1.36 Å

3.14 A hypothetical metal X has hcp unit cell with c/a ratio equal to 2, atomic radius 1.78 Å, and atomic weight $= 150.36$ g/mole, calculate (i) the volume of the hexagonal unit cell (ii) density of X.

Ans: (i) 2.34×10^{-28} m³; (ii) 6388 kg/m³

3.15 Find the Miller-Bravais indices of a plane that makes intercepts in a_1, a_2 and c axes equal to 2Å, 2 Å and 3.2 Å in a hexagonal crystal unit cell with c/a ratio 1.6. Draw the plane obtained in the hexagonal unit cell.

Ans: $(11\bar{2}1)$

3.16 Find the Miller - Bravais indices of the shaded plane ABCD.

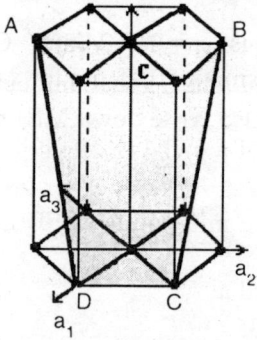

Fig. 3.35

Ans: $(10\bar{1}1)$

3.17 Draw (012) plane in unit cell of fcc lead. Draw $[02\bar{1}]$ on this plane. Find the linear density of lattice points in this direction and planar density of (012) plane.
Ans: 0.447/a; 0.8944/a²

3.18 In bct crystal find the Miller indices of each of the planes in the family {100} If c/a = 2, calculate the planar density of lattice points in (111) plane of the crystal.

Ans: (100), (010), $(\bar{1}00)$, $(0\bar{1}0)$; 0.333 lattice points/a²

3.19 (a) How many atoms per mm² are there on the (111) plane of Ir (fcc)? where a = 3.84 Å
(b) Calculate the linear density of atoms and lattice points in DC along [111] if a =3.57 Å.
(c) Calculate the interplanar spacing between (111) planes in Cr (bcc).
(d) Show [112] & [110] in simple cubic unit cell and calculate the angle between them.
Ans: (a) 1.566 x 10¹⁹ atoms/m²; (b) 3.234 x 10⁹ atoms/m, 1.617 x 10⁹ lattice points/m; (c) 8.314 x 10⁻¹¹ m; (d) 54.74°

3.20 The interplanar spacing d_{321} in a fcc metal is 0.085 nm. What is its lattice constant 'a'?
Ans: 3.18 Å

3.21 Calculate the planar density in atom/nm² for the (0001) plane in hcp zinc, which has a = 0.26649 nm and c/a = 1.856.
Ans: 1.626 x 10¹⁹ atoms/m²

3.22 The planar density of (111) in Ag (fcc) is 1.392 x 10¹⁹ atoms/m². Find
(i) Atomic diameter of Ag atom

(ii) Density of the material (At. wt. of Ag = 107.87 amu)

(iii) The angle between [001] and the normal of the above plane.

Ans: (i) 2.88 Å; (ii) 10600 kg/m³; (iii) 54.74°

3.23 Find the planar density of (100) and (110) planes in vanadium (bcc) if the lattice parameter is 'a'.

Ans: $1/a^2$; $\sqrt{2}/a^2$

3.24 The lattice parameter of Ni(fcc) is a = 0.3524 nm. Calculate the values of θ corresponding to the first 5 diffraction lines, using λ= 0.15 nm for x-ray wavelength.

3.25 Calculate the unit cell volume of a metal β which has an atomic radius of 0.0214 nm and a packing factor of 0.67.

Ans: 0.1225 Å³

3.26 Diffraction peaks are determined at the following values of 2θ for a sample of a cubic metal (either fcc or bcc): 25.062°, 35.698°; 44.116° and 51.406°. Calculate the wavelength of x-ray if the lattice constant a of the unit cell is 0.502 nm. Determine the crystal structure.

3.27 When a sample of a cubic metal (bcc) was placed in a diffractometer using x-ray of λ = 0.154 nm, diffraction from {310} planes was obtained at 2θ = 101.502°. Calculate the value of a, considering first order diffraction.

3.28 The first diffraction line of copper from a plane (111) is at 2θ = 43°. Calculate the lattice constant and atomic radius of copper.

3.29 What are the values of 2θ for first five diffraction lines of bcc iron when x-ray wavelength is 0.58Å?

3.30 Let the radii of ions in NaCl structure be r_1 and r_2 ($r_2 > r_1$). Calculate the length of the face diagonal in the unit cell and hence show that the structure is stable only when $r_1 > 0.414\, r_2$.

3.31 Prove Bragg's law in your own way without referring back to the text.

3.32 (a) Copper (fcc) crystal has lattice parameter a = 3.61Å at 0°K. Find the bond length in Å.

(b) Considering the potential energy expression for above material to be

$$W = - (A/r) + (B/r^9),$$

find the value of B/A ratio of the material at equilibrium.

(c) Find the bond energy of the above material, if the energy required atomizing it is : 338.4 kJ/mole.

4

Solids and Their Structure

4.1 NONMETALS

Nature is partial about metals and that is why metals are in abundance. Nonmetals are nearly 20% of the total elements so far discovered. Nearly half of the nonmetals are gaseous substances. In group zero of the periodic table, all the elements are inert gases. These elements have completed outermost shell i.e. they have eight electrons in these shells (octet structure). Atoms in these elements may have weak binding due to existence of van der Waals bonding at low temperature and high pressure. All the elements in group VIIA are nonmetals and are called halogens. Atoms of these elements have one electron short to fill their outermost p orbitals. Any two similar atoms of this group form diatomic molecules through covalent bonding e.g. F_2, Cl_2, Br_2, I_2 etc. At appropriate temperature and pressure these diatomic molecules form only weak secondary bonding. Fluorine and chlorine remain in the gaseous state whereas bromine remains in liquid form and iodine in crystalline form at room temperature.

Elements in the sixth group are also nonmetals. Atoms of these groups have two electrons short in the outermost p subshell to form an octet. Oxygen forms diatomic molecule in which two atoms are bonded by two covalent bonds by sharing of two partially filled p orbitals. Some of these elements like sulphur exist in several allotropic forms. The common form of sulphur is orthorhombic sulphur, which is composed of S_8 molecules. The eight atoms in the molecule are bonded through covalent bonding in a puckered ring, each sulphur atom forms two covalent bonds with two other sulphur atoms, Fig. 4.1(a). Plastic sulphur is formed when liquid sulphur is poured into water. This has molecule in the form of zigzag long chains of sulphur, Fig. 4.1 (b).

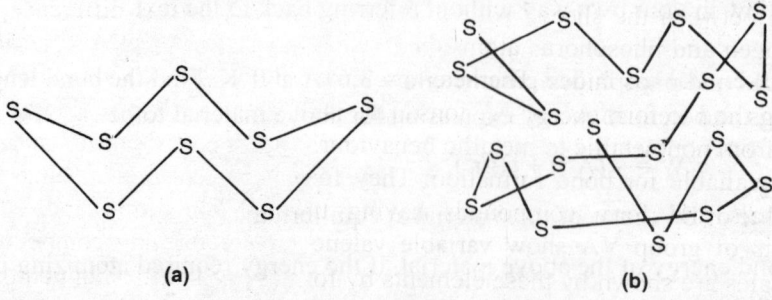

(a)   (b)

Fig. 4.1 (a) Puckered rings of sulphur molecule S_8, (b) Zigzag chains of sulphur

These chains or ring molecules of sulphur form three-dimensional structure with very weak bonding between them. The other form of sulphur is monoclinic. Selenium also has allotropes, namely rhombic, monoclinic and gray forms. The rhombic variety has Se_8 molecules with atoms in an open ring whereas in the monoclinic variety Se_8 molecules have the form of puckered ring as is found in case of S_8. The gray form has regularly arranged spirals of Se atoms. The allotropes of tellurium are also of the type of selenium.

The two bonding electrons of VIA group can also form small molecules with other elements, namely H_2O, H_2S etc. These molecules are also bonded together by weak secondary forces when they form crystalline structures. Water molecule forms hexagonal ice crystal with hydrogen bonds between the water molecules, Fig. 4.2. In VA group, the elements P, As, Sb and Bi have three half-filled p orbitals. Hence each atom forms three covalent bonds with other three atoms. Their structure consists of puckered sheets. In these sheets, the orientations of the p orbitals are nearly unaltered. The puckered sheets are held together by van der Waals forces such that in the solid form these nonmetals are mostly noncrystalline.

Fig. 4.2 (a) Hydrogen bonding between H_2O molecules in ice crystals; each oxygen atom is bonded to two hydrogen atoms through covalent bonding in a water molecule and is bonded to two hydrogen atoms in other H_2O molecules by hydrogen bonds.
(b) Hexagonal symmetry of ice is shown.

The covalent bond length between two atoms in the puckered plane is smaller than the secondary bond length between the sheets. As we go down group VA, this difference in bond length decreases. Nitrogen and phosphorus atoms, being small and electronegative, accept three electrons to form nitrides and phosphides. The heavier elements of this group have the tendency of giving away electrons and to form A^{5+} or A^{3+} ions. This is the reason why the elements of this group show transition from nonmetallic to metallic behaviour. The heavier elements in this group have empty d orbital available for bond formation. They form covalent bonds using sp^3d and sp^2d^2 hybridized orbitals and form compounds having trigonal bipyramid and octahedral structure. The elements of group VA show variable valency while forming compounds. The maximum oxidation states are shown by these elements by forming pentoxides. In pentoxides all the five electrons (two s electrons and three p electrons) take part in bond formation. At room temperature nitrogen is a gas, phosphorous is a lusterless soft waxy solid, arsenic and bismuth are semimetals having luster like metals. Antimony and arsenic are brittle whereas bismuth is less brittle than antimony and arsenic. The electrical conductivity of the elements increases as we go down VA group elements.

Molten antimony and bismuth form nearly close-packed structures characteristic of a typical molten metal. In this molten state the coordination number becomes as large as 10 or 11. When these elements crystallize, the atoms form three covalent bonds with its neighbour giving rise to much smaller coordination number. So these elements expand on cooling and solidification. This property of Sb, Bi, is utilized in type casting for reproducing the details of the mould accurately by filling up minute cavities. Nitrogen also produces molecules of NH_3 and N_2 ($N \equiv N$).

In group IVA carbon is an insulator, whereas silicon and germanium are semiconductors and all other elements of group IVA are metals. Carbon exists in nature mainly in two allotropic forms: (a) graphite and (b) diamond.

Graphite

In graphite, carbon atom forms three sp^2-hybridized bonds. Each carbon atom is bonded to three other carbon atoms in a plane, bond angle being $120°$. Thus, the carbon atoms form hexagonal network in a plane called sheets of graphite, Fig. 4.3.

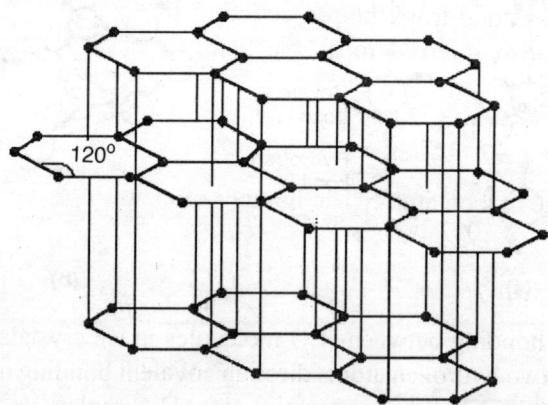

Fig. 4.3 Sheets of graphite held together by van der Waals bonding to form crystal.

The sheets are held together by van der Waals bonds to form crystal. Thus, only three of the four valence electrons of carbon form covalent bonding whereas the fourth bonding electron of carbon is nearly free and resonates between three sp^2 bonds. The weak bonding between the sheets gives softness to graphite. The delocalized electrons are mobile and contribute to the good electrical and thermal conductivity of graphite in the direction parallel to the sheets as compared to the conductivity perpendicular to the sheet.

Thus, the conductivity property of graphite is directional. The sheets are usually longer distance apart as compared to the bond length in the sheet and can slide over one another due to presence of weak van der Waals forces between the sheets. That is why one can observe lubricating property of graphite even at high temperatures. At high temperatures, the graphite does not evaporate due to presence of strong covalent bonding in the sheet. The graphite, mixed with clay in various proportions, are used in the lead of pencils. The graphite sheets, aligned parallel to the fibre axis give strength to carbon fibre. The graphite is also used as a resistance in heating applications.

Diamond

Diamond has either fcc or hcp structure. Structure of cubic diamond is also called diamond cubic (DC). Each carbon atom in diamond is bonded to four neighbouring carbon atoms, through (sp^3) covalent bonding. The basis of DC consists of two carbon atoms at (0,0,0) and (¼,¼,¼)) locations with respect to each lattice point of fcc. Another way of visualizing the DC structure is by considering the unit cell to be divided into eight cubes of edge a/2 and the carbon atoms are located at the body centre positions of the alternate cubes along all the axes, besides the carbon atoms at corner and face centre lattice points of the unit cell. Thus there are four carbon atoms within the DC unit cell, each bonded to the carbon atom at the nearest corner and those at three adjacent face centres forming tetrahedron, Fig. 4.4(b). The structure of a DC unit cell is open. DC unit cell can also be imagined as two interlocking fcc unit cells of carbon displaced by √3a/4 along the body diagonal. DC crystal has coordination number equal to four in contrast to the monoatomic fcc crystal which has coordination number 12. This is because the monoatomic fcc crystals have predominantly metallic bonding whereas DC has covalent bonding. The directional property of covalent bonding restricts the number of nearest neighbours. Low packing efficiency in DC arises due to smaller value of coordination number. Low packing efficiency and covalent bonding result in low density and hardness in diamond. The positions of carbon atoms in diamond unit cell may be well understood from the projection of the unit cell on the basal plane as shown in Fig. 4.4(a), which is also referred to as the plan view. The distance between the nearest neighbours in DC is given by

$$2\,r = \sqrt{3}a/4$$

or
$$r = \sqrt{3}a/8 \qquad\qquad (4.1)$$

where r is the radius of the carbon atom. The number of carbon atoms per unit cell of DC is 8.

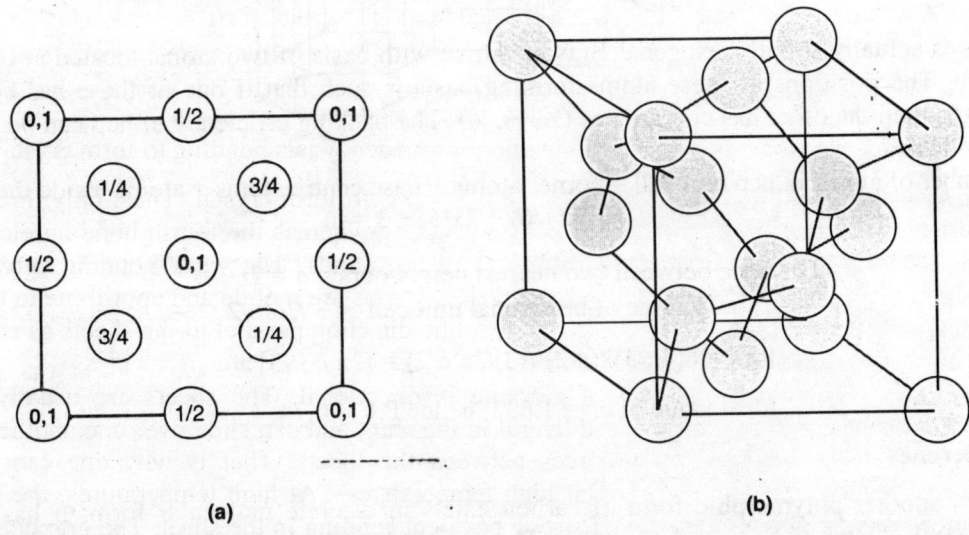

(a) (b)

Fig. 4.4 (a) Plan view of the atoms in DC unit cell, showing height from the base of the unit cell in the unit of lattice parameter a , (b) Unit cell of DC.

Packing efficiency

The packing efficiency of a crystal is defined as the ratio of the volume occupied by the atoms in the unit cell to that of the unit cell.

Example 4.1

Calculate the packing efficiency of diamond.

Solution:

In diamond unit cell, the volume occupied by 8 atoms is given by
$$V_{atoms} = 8 \times (4\pi/3) \, r^3 = 8 \times (4\pi/3) \, (\sqrt{3}a/8)^3 = 0.34 \, a^3 \text{ and P.E.} = V_{atoms}/V_{unit \, cell} = 0.34$$

There are a large number of compounds and elements, which have DC structure. Diamond can also exist in another form, which has hexagonal close-packed (hcp) structure. The hexagonal close-packed unit cell has carbon atoms located at the corners, at the centres of basal planes and also at the positions (⅓, ⅔, ½), (−⅔, −⅓, ½) and (⅓, −⅓, ½), Fig. 4.9(a). Projection of the hcp unit cell on the basal plane shows the position of the atoms in the unit cell, Fig. 4.5.

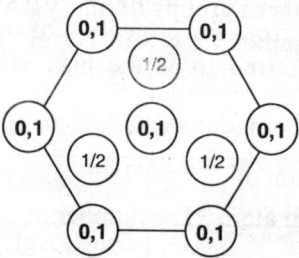

Fig. 4.5 Plan view of hcp unit cell.

This is actually simple hexagonal Bravais lattice with basis of two atoms located at each lattice point. The positions of these atoms forming basis is such that if one of these has coordinates (0,0,0) then the other has coordinates (⅓, ⅔, ½). The packing efficiency of hcp can be calculated as follows:

Number of atoms in hcp unit cell = corner atoms + base centre atoms + atoms inside the unit cell
$$= 12 \times (1/6) + 2 \times ½ + 3 = 6$$

Distance between two nearest neighbours $= a = 2r$
Volume of hexagonal unit cell $= 3\sqrt{3}a^2c/2$

$$P.E. = 6(4\pi/3) \times (a/2)^3/(3\sqrt{3}a^2c/2) = [2\pi / 3\sqrt{3}] \, a/c \qquad (4.2)$$

Ideal value of c/a= 1.633, hence P.E.= 0.74

Fullerenes

C_{60} – another polymorphic form of carbon exists in discrete molecular form. It has a hollow spherical cluster of sixty carbon atoms. These carbon atoms, which are bonded to one another, form 20 hexagons and 12 pentagons. They are joined in such a way that no two pentagons share a common side. The molecular surface looks like a soccer ball. R. Buckminister Fuller invented this geodesic dome. The molecule C_{60} is a molecular replica of such a dome, which is also called buckyball. The class of materials, which are composed of this type of molecules, is called Fullerenes.

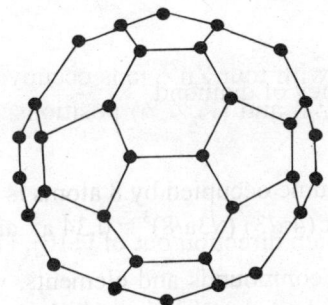

Fig. 4.6 The structure of a buckyball

Compounds with DC structure

The other elements of Group IV A like silicon, germanium and gray tin also have DC structure. Compounds formed of elements of groups IB and VIIA in equal atomic proportions also have DC or hexagonal structure like diamond e.g. CuCl, AgI. Table 4.1 gives more examples of similar compounds.

Table 4.1 Compounds having diamond structure

Combination of elements of groups	Examples of compounds
I B & VII A	CuCl, AgI
II B & VI A	ZnO, ZnS, CdS
III A & V A	AlP, AlAs, GaP, GaAs
IV A & IV A	SiC

The structure of ZnS is similar to that of diamond in which sulphur atoms occupy corners and face centre positions of cubic unit cell and zinc atoms occupy alternate tetrahedral voids and vice versa. The plan view of ZnS unit cell is shown in Fig. 4.7.

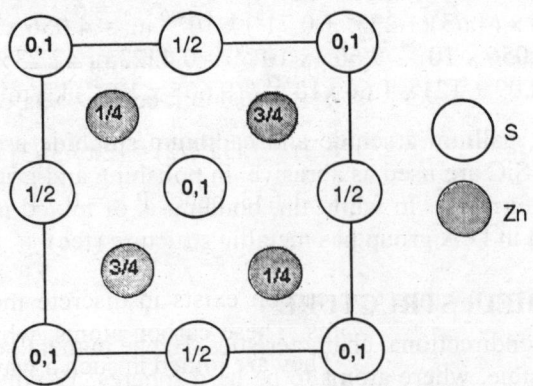

Fig. 4.7 Plan view of ZnS showing the positions of Zn and S atoms. The c coordinates are shown at each point

Example 4.2

Show analytically that in a unit cell of ZnS structure the four Zn atoms, which lie fully within the unit cell, do not lie on a plane.

Solution:

The S^{2-} ions form fcc structure with four Zn^{2+} ions occupying alternate tetrahedral voids, which are at $(\frac{1}{4},\frac{1}{4},\frac{1}{4})$, $(\frac{1}{4},\frac{3}{4},\frac{3}{4})$, $(\frac{3}{4},\frac{3}{4},\frac{1}{4})$ and $(\frac{3}{4},\frac{1}{4},\frac{3}{4})$ positions. Hence, they do not lie on the same plane.

Example 4.3

In Ge (DC) crystal, calculate which direction out of [110], [100] and [111] have maximum linear density?

Solution:

Given :   Directions, A = [110],   B=[100] and C =[111]
Linear density along A = Effective number of atoms / length

i.e. $L_d (A) = [2\times\frac{1}{2}$ (at corners) + 1 (at centre)] $/ [a \sqrt{(1^2 + 1^2 + 0^2)}] = \sqrt{2}/a$

Similarly $L_d (B) = [2 \times\frac{1}{2}$ (at corners)] $/[a\sqrt{(1^2+ 0^2 + 0^2)}] = 1/a$

and $L_d (C) = [2 \times \frac{1}{2}$ (at corners) + 1 (at tetrahedral void)] $/ a\sqrt{3} = 2/\sqrt{3}a$.

Hence $L_d (A): L_d (B): L_d (C):$: 1.414 : 1 : 1.155 i.e maximum linear density is along A.

Example 4. 4

Calculate the packing efficiency and density (in kg/m³) for SiC from the following data: covalent radii of Si and C are 1.273 Å and 0.71 Å, respectively. The centre-to-centre distance between Si and C is given by $r_{Si}+ r_C$.

Solution:

SiC has DC structure with Si atoms occupying corners and face centres and C atoms occupying alternate tetrahedral voids. Hence

$$a = 4(1.273 + 0.71)/\sqrt{3} \text{ Å } = 4.58 \text{ Å}$$
$$V_{unit\ cell} = a^3 = 9.605 \times 10^{-29} \text{ m}^3$$
$$V_{atoms} = 4 \times (4\pi/3)(1.273^3 + 0.71^3) \times 10^{-30} \text{ m}^3 = 4.056 \times 10^{-29} \text{ m}^3$$
$$P.E. = 4.056 \times 10^{-29} /9.605 \times 10^{-29} = 0.4223 \text{ or } 42.23\%$$
Hence $\rho = 4(28.09 + 12) \times 1.66 \times10^{-27} / (9.605 \times 10^{-29}) \text{ kg/m}^3 = 2771 \text{ kg/m}^3$

Silicon, germanium, gallium arsenide and cadmium sulphide are used in making solid-state devices. Diamond and SiC are used as abrasive, in polishing and grinding operations. SiC is used as heating element in furnaces. In white tin, bonding is of mixed nature, partially covalent and partially metallic. Lead in IVA group has metallic structure (fcc).

4.2 METALS AND THEIR STRUCTURE

Metallic bonding has nondirectional characteristics. Hence in metal, each atom surrounds itself by as many atoms as possible, where atoms to be hard spheres. 12 equal size atoms are required to surround an atom in order to have close-packing.

Packing in crystals

Close packing in crystal structure can be obtained by the following arrangements of atoms:

(i) Consider that the atoms are arranged in layer A where each atom is labeled A and is surrounded by six atoms in close packing. This layer has valleys, which are alternately named as B and C. The second layer of atoms can be arranged in close packing either over B valleys or over C valleys, Fig. 4.8(a). If atoms are arranged over B(C) valleys, the layer of atoms is to be called B(C) layer. In the third layer, atoms can be placed over C(B) valleys or the layer may be exactly parallel to A layer. If the third layer is parallel to A layer, then repetition of such a close-packing is called ABABAB.... stacking; this type of close-packing is characteristic of hcp structure, Fig. 4.8(b). (ii) If after A and B layers, atoms are arranged in close-packing over the C valleys, then repetition of such close packing is termed as ABCABCABC......stacking. Such close packing occurs in fcc structure, Fig. 4.8(c).

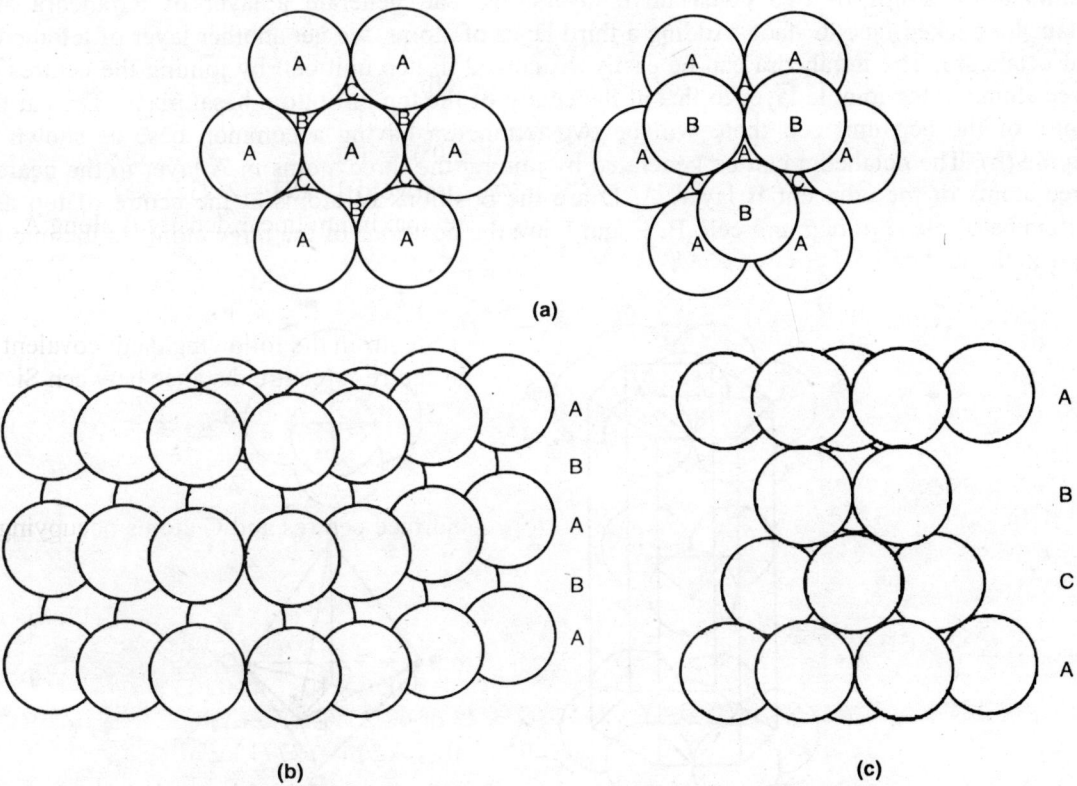

(a)

(b)   (c)

Fig. 4.8 (a) and (b) Showing ABAB.... stacking in hcp structure;
(c) Showing ABCABC.... stacking in fcc structure.

The effective number of lattice points in hcp unit cell is three and a basis of two atoms is associated with each lattice point. Hence the effective number of atoms is six in hcp unit cell. The coordination number in any close-packed structure is 12. There are other arrangements of layers, which can give rise to close packing. There are structures in which stackings are random such as ACBCABAC... It occurs in small crystallites of cobalt at room temperature. Cobalt also has two different crystal structures – hcp below 400°C and fcc above 400°C. Sometimes random stacking is termed as crystallinity in two dimensions since there is no regular pattern in third dimension. In a close-packing of layers of atoms one has to remember that no two consecutive layers should have parallel arrangement of atoms e.g. ABBA.... or ABCCBA... etc. since these arrangements cannot give rise to close-packed structure. Sometimes, crystal is found to have stacking sequences with a long repeat unit. This is called polyptism. SiC is an example of polyptism in which there are 45 known stacking sequences. The polyptype of SiG, known as 393R, has a= 3.079 Å and c=989.6 Å, the longest known repetitive distance. SiC also has two other possible crystal structures – cubic ZnS and hexagonal ZnS.

Polyhedra in close-packed structure

In an ideal hcp crystal the atoms in ABAB.... stacking touch each other. Joining the centres of the neighbouring atoms of two consecutive layers, we can generate a layer of tetrahedra and octahedra packed face-to -face. Adding a third layer of atoms, we get another layer of tetrahedra and octahedra. The tetrahedra can be easily visualized in hcp unit cell by joining the centres of three atoms in the middle layer to that at the centre of the top or bottom basal plane. Thus at the centre of the hcp unit cell there will be two tetrahedra having a common base as shown in Fig. 4.9(b). The octahedra can be generated by joining the three atoms in A layer to the nearest three atoms of the adjacent B layer. A, D are the positions of atoms at the centre of top and bottom basal plane of hcp unit cell. B, C and E are the positions of the three atoms in the middle layer in the unit cell.

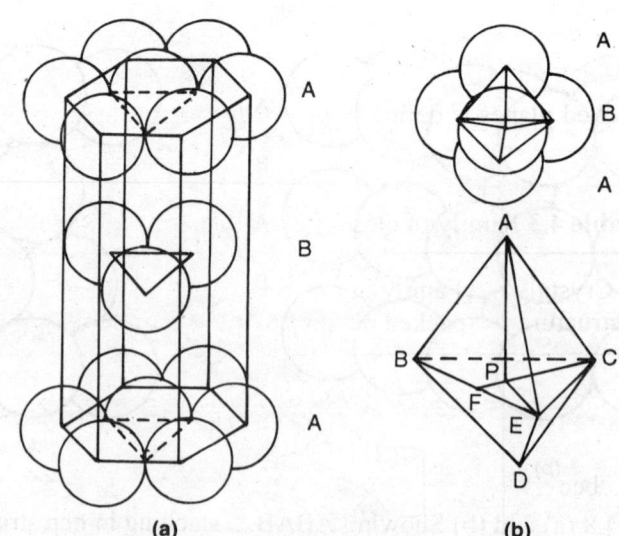

(a) (b)

Fig. 4.9 Three-dimensional view of (a) hcp structure; (b) Tetrahedra in hcp

One can calculate c/a ratio, which is given by AD/BE using simple geometrical consideration. CF is perpendicular to BE and AP is perpendicular to BCE plane and it cuts CF at P, Fig. 4.9b.

Now,
$$CP : PF = 2 : 1, \quad CP = \tfrac{2}{3} CF = \tfrac{2}{3}\sqrt{[\ CE^2 - FE^2\ ]} = a/\sqrt{3}$$
$$AD = 2\ AP = 2\sqrt{(AC^2 - CP^2)} = 2\sqrt{[a^2 - (a/\sqrt{3})^2]} = 2\sqrt{\tfrac{2}{3}}\ a$$

We can write
$$BE = AC = a$$

Hence one can show that $\quad c/a = (2\sqrt{⅔}) = 1.633$ $\qquad\qquad$ (4.3)

This is ideal c/a ratio since in this packing atoms in A and B layers are touching each other i.e. the packing is closest possible.

Most metals crystallize as monoatomic fcc or hcp structures. Since these structures have efficient packing and hence low energy. Nearly half of the elements in solid state have fcc structure. All these close-packed crystals expand on heating since heating increases energy as well as disorder and thus decreases packing efficiency. Calculation of packing efficiency shows all close-packed structures have 74% of unit cell filled up with the atoms and 26% of space remains unoccupied. The packing efficiency of other structures can also be calculated similarly. Table 4.2 shows the packing efficiency of various crystal structures.

Table 4.2 Packing efficiency and other related parameters.

Crystal structure	Atomic radius in terms of a	Coordination number	P.E. %
sc	a/2	6	52
bcc	$\sqrt{3}a/4$	8	68
fcc	$a/2\sqrt{2}$	12	74
hcp	a/2	12	74
DC	$\sqrt{3}a/8$	4	34

Close-packed plane

A family of close-packed planes is defined as that which has maximum planar density in the unit cell, Table 4.3.

Table 4.3 Family of close-packed planes and directions.

Crystal structure	Family of close-packed planes	Family of close-packed directions
sc	{100}	<100>
bcc	{211}	<$\bar{1}11$>
	{101}	<$\bar{1}11$>
fcc	{111}	<$\bar{1}10$>
hcp	{0001}	<$11\bar{2}0$>
	{10$\bar{1}$0}	<$11\bar{2}0$>

Close-packed directions

Family of close-packed directions in a crystal contains the directions having maximum linear density. The close-packed directions in a crystal lie in close-packed planes. For example, say the close-packed direction [h k l] lies in close-packed plane (101) of bcc, hence h.1+k.0+ l.1 = 0 i.e. h = − l. So the close-packed direction that lies on (101) plane is

$$[\bar{1}11].$$

4.3 VOIDS IN CRYSTAL STRUCTURE

Three-dimensional packing of atoms in a crystal always leaves some void that is why packing efficiency can never be 100%. When the centers of neighbouring atoms are joined, we obtain cube or tetrahedron or octahedron. Centres of these polyhedra are interstitial voids. Thus in case of three-dimensional crystal structure, the interstitial voids can be mainly

> (i) tetrahedral (T-void)
> (ii) octahedral (O-void)
> (iii) cubic (C-void)

depending upon the arrangement of atoms surrounding the voids. In case of fcc crystal, the unit cell has eight tetrahedral voids. When the atom at one of the corners of a unit cell is joined to the face centre atoms of three adjacent faces, a tetrahedron is generated at the centre of which is the tetrahedral void. Similarly, octahedral voids are there at the edge centres and also at the body centre of the fcc unit cell. The edge centre voids are shared by four unit cells hence the effective contribution of each edge centre void is ¼ to each unit cell. One octahedral void is at body centre position, which is surrounded by six face centre atoms. Thus the number of octahedral voids, in fcc unit cell is ¼ x 12 +1 = 4. The effective number of lattices points per unit cell is four. In other words there are two tetrahedral and one octahedral voids per lattice point in a close-packed structure like fcc and hcp.

Example 4.5

When 20% of the octahedral voids of fcc iron is filled up with impurity atoms of radii 0.46 Å, find the change in the packing efficiency and density if the molecular weight of impurity is 10.81 amu.

Solution:

> Radius of Fe = 1.27 Å, Radius of impurity = 0.46Å
> P.E. = $4 \times (4\pi/3)[1.27^3+0.2 \times 0.46^3]/ (2\sqrt{2} \times 1.27)^3 = 0.747$, Change in P.E. = 0.007
> Density: $\rho_{Fe} = \{4 \times 55.85 \times 1.66 \times 10^{-27}\}/[(2\sqrt{2} \times 1.27)^3 \times 10^{-30}] = 8002$ kg/m^3
> $\rho_{\text{impure Fe}} = 4(55.85 + 0.2 \times 10.81) \times 1.66 \times 10^{-27}/[(2\sqrt{2} \times 1.27)^3 \times 10^{-30}] = 8312$ kg/m^3
> Change in density = (8312 − 8002) kg/m^3 = 310 kg/m^3

In simple cubic, there is a cubic void at the centre of the unit cell. In bcc, there are twelve tetrahedral voids, one at (½,¼,0) and others at equivalent locations. Each of these voids is surrounded by two adjacent corner atoms and body centre atoms of the two neighbouring unit cells, Fig. 4.10(a). Each face of the unit cell has four such voids. Since such void contributes ½ to each of the neighbouring unit cells, the effective number of tetrahedral voids to bcc unit cell becomes 6x4x½=12. In bcc, there are octahedral voids (irregular ones, since the distances between the centre of the void and its surrounding atoms are not equal) at the edge centres and face centres, Fig. 4.10(b). Total number of octahedral voids in bcc is 6 x ½ + 12 x ¼ = 6. The volume of the voids can be determined from simple geometrical considerations.

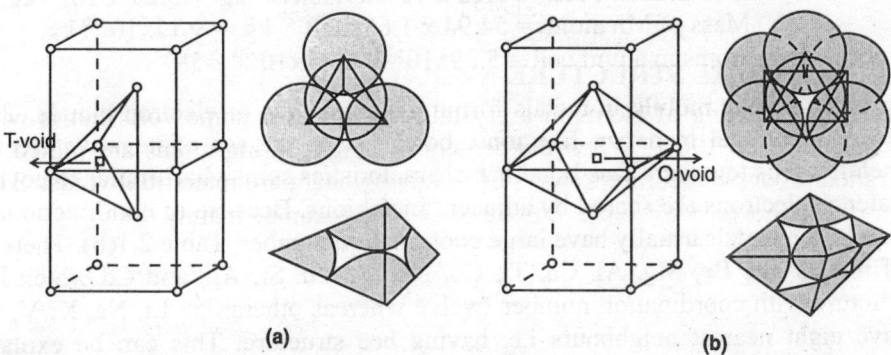

T-void

O-void

(a)   (b)

Fig. 4.10 Voids in bcc unit cell,(a) tetrahedral voids, (b)octahedral voids.

Polymorphic phase transformation

Certain metals undergo changes in crystal structure with temperature. For example, the crystal structure of iron is bcc at temperatures below 910°C and above 1490°C. It becomes fcc in between these two temperatures. Similarly, the crystal structure of titanium undergoes change from hcp (α phase) to bcc (β phase) on heating above 880°C, Fig. 4.11. Such phase transformation is called polymorphic phase transformation. The phases of a material with different crystal structures occurring at different temperatures are called polymorphs. Sometimes element exists in nature in different phases having different crystal structures. These phases are called allotropes e.g. graphite and diamond. The polymorphic phase transitions are associated with free energy changes. This transformation takes place at a fixed temperature, beyond which a specific crystal structure possesses lowest free energy, Fig. 4.11.

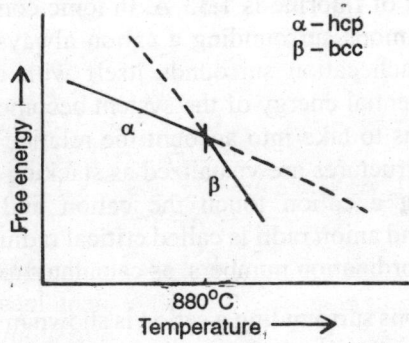

α – hcp
β – bcc

α

β

Free energy →

880°C
Temperature →

Fig. 4.11 Polymorphic phase transition of titanium at 880°C

Example 4.6

Manganese has a polymorph possessing complex cubic structure with lattice parameter = 8.93 Å and density = 7.43×10^3 kg/m^3. Find the number of atoms per unit cell.

Solution:

$$\text{Volume of unit cell} = 7.12 \times 10^{-28} \text{ m}^3$$
$$\text{Mass of atoms in a unit cell} = 7.12 \times 10^{-28} \times 7.43 \times 10^3 \text{ kg} = 5.29 \times 10^{-24} \text{ kg.}$$
$$\text{Mass of Mn atom} = 54.94 \times 1.66 \times 10^{-27} \text{ kg.} = 9.12 \times 10^{-26} \text{ kg.}$$
$$\text{Number of Mn atoms in a unit cell} = 5.29 \times 10^{-24} / 9.12 \times 10^{-26} = 58$$

Valence electrons are mobile in metals forming electron gas or electron clouds. The metallic bonds are nondirectional in nature like ionic bonds, since the sea of negative electrons holds positive metallic ions together. This bond has characteristics somewhat similar to covalent bonds because valence electrons are shared by adjacent metal ions. Because of nondirectional nature of metallic bonds, the metals usually have large coordination number, Table 2.1(b). There are a large number of metals like Be, Mg, Al, Ca, Ti, Co, Ni, Cu, Zn, Sr, Ag, and Cd which have close-packed structure with coordination number twelve whereas others like Li, Na, K, V, Cr, Mo, W and Fe have eight nearest neighbours i.e. having bcc structure. This can be explained using following arguments:

(i) The bcc structure is more open than fcc or hcp crystal structures. A number of alkali metals whose melting temperature is close to room temperature have bcc crystal structure at room temperature thereby allowing larger amplitudes of vibrations for atoms. The larger amplitudes of vibrations increase disorder i.e. entropy and hence lower the free energy of the system.

(ii) The transition metals possess partial covalent character hence many of them have bcc structure at low temperatures.

At low temperature silicon, germanium and tin have only four nearest neighbours since they have sp^3 hybridized covalent bonding.

4.4 IONIC SOLIDS AND PACKING

Nondirectional nature of ionic bonding results in close-packing in ionic solids. Normally radius of cation (r_c) is smaller than that of anion (r_a) in an ionic compound, since cation loses electrons whereas anion gains electrons, leaving apart some exceptional cases like RbF where ionic radius of rubidium is 1.49 Å and that of fluorine is 1.33 Å. In ionic compounds anions and cations are considered as hard spheres. Anions surrounding a cation always touch cation to maximize the attractive coulomb energy. Each cation surrounds itself with as many number of anions as possible such that the total potential energy of the system becomes minimum. While considering packing in ionic solids, one has to take into account the relative sizes of cations and anions. In ionic compounds, the crystal structures are visualized as stacking of close-packed planes of ions. When the anions surrounding a cation touch the cation and also touch one another, the corresponding ratio of cation and anion radii is called critical radius ratio. This radius ratio will be different for different anion coordination numbers, as calculated here for a few cases:

(i) Linear coordination of anions surrounding a cation is shown in Fig. 4.12, where $r_c/r_a < 0.155$. If $r_c/r_a \geq 0.155$, triangular coordination is favoured.

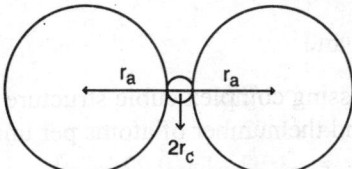

Fig. 4.12 Linear coordination of anions surrounding a cation

(ii) Triangular coordination of anions is shown in Fig. 4.13.

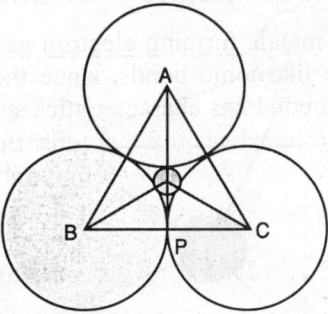

Fig. 4.13 Triangular coordination of anions surrounding a cation

The critical radius ratio r_c/r_a can be calculated as follows:

$$AB = 2r_a$$
$$AO = (2/3) AP = r_a + r_c$$

From Fig. 4.13, $r_a + r_c = (2/3) 2r_a \sin 60° = (2/\sqrt{3}) r_a$

or $- r_a + (2/\sqrt{3})r_a = r_c$

$$r_c/r_a = (2/\sqrt{3}) -1 = 0.155 \qquad (4.4)$$

If the radius ratio is less than 0.155, the triangular coordination of anions is not possible since the cation rattles inside the triangular packing of anions. This decreases the coulomb attractive energy and hence increases the energy of the system. If r_c/r_a is greater than the critical value triangular coordination is possible where cation touches all the three neighbouring anions but anions do not touch one another. If r_c/r_a becomes large enough such that the four anions can become nearest neighbours to cation, tetrahedral coordination is favoured.

(iii) Tetrahedral coordination of anions is shown in Fig. 4.14. To calculate the critical radius ratio r_c / r_a, tetrahedral coordination is visualized as follows: Let us consider two anions at the end positions of a face diagonal and the other two anions are at the ends of the other face diagonal in the parallel faces of a cube, such that the directions of the face diagonals are perpendicular to each other. If we join one of the anions with other three anions, which are also joined, a regular tetrahedron is obtained. The cation is at the body centre position of this tetrahedron. It touches all the four surrounding anions and anions also touch one another in the critical packing of anions. In this geometry,

$$2r_a = \sqrt{2}a, r_a = a /\sqrt{2}$$

where a is the side of the cube and

$$r_a + r_c = (\sqrt{3}/2)a$$
$$1 + r_c/r_a = \sqrt{3}/\sqrt{2}$$

i.e. $\qquad r_c/r_a = \sqrt{3}/\sqrt{2} - 1 = 0.225 \qquad\qquad\qquad (4.5)$

If $r_c/r_a > 0.225$, tetrahedral coordination is still possible until r_c/r_a is equal to 0.414. A ligancy of five anions is not possible in ionic compound since it does not result in a stable configuration in the long-range order and does not satisfy minimum energy condition.

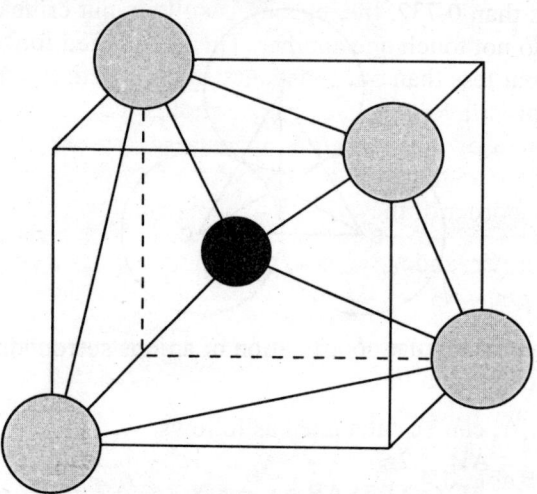

Fig. 4.14 Tetrahedral coordination of anions

(iv) If $0.732 > r_c/r_a \geq 0.414$, the ligancy becomes six. This is called octahedral coordination of anions. To calculate the critical radius ratio r_c/r_a for this case, consider four anions are located at the corners of a square such that the anions along the sides of the square are touching and two more anions one above and the other below the centre of the square touching the other four anions, thus forming an octahedron if we join the centres of anions. The centre of the octahedron is occupied by cation, which touches all the six anions and all these six anions touch one another, Fig. 4.15.

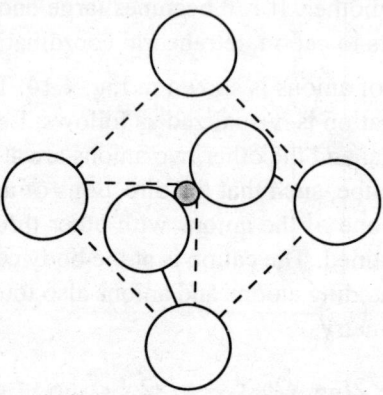

Fig. 4.15 Octahedral coordination of anions

Each side of octahedron is equal to $2r_a$ and the distance of the centre of the octahedron to its corner is $r_a + r_c$. Hence $\quad r_a + r_c = \sqrt{2}\, r_a$

$$r_c = 0.414\, r_a$$

So, critical radius ratio $r_c / r_a = 0.414$ \hfill (4.6)

Example of octahedral coordination of anions around a cation is found in NaCl, Fig. 4.16. In NaCl unit cell, considering Na^+ at the body centre position, nearest Cl^- ions are located at the six face centre positions; thus, Na^+ is at the centre of a regular octahedron formed by Cl^- ions. The radius ratio $r_{Na+} / r_{Cl^-} = 0.541$ which is greater than the critical value 0.414. If $r_c / r_a > 0.414$ but less than 0.732, the legacy is still six but critical packing condition is not satisfied i.e. anions do not touch one another. This is satisfied for NaCl.

(v) When $r_c / r_a \geq 0.732$ but less than one, a legacy of eight satisfies minimum energy condition i.e. cubic coordination is favoured. The critical radius ratio in case of cubic coordination of anions can be calculated by considering the anions at the corners of a cube of side 's' such that the anions along the edge touch each other and the cation at the body centre touches all the eight anions at the corner of the cube. From the geometry it follows that

$$\text{Side of cube } s = 2r_a \text{ and } r_c + r_a = \sqrt{3}s/2 = \sqrt{3}\, r_a$$
$$r_c/r_a = \sqrt{3} - 1 = 0.732 \hfill (4.7)$$

(vi) In case of fcc or hcp packing $r_c / r_a = 1$.

If the radius ratio is greater than the critical value corresponding to a specific ligancy, packing geometry remains the same till the radius ratio attains critical value for higher ligancy. The geometry of local packing can be determined from the radius ratios for a large number of ionic compounds as shown in Table 4.4. We observe quite a few ionic compounds, (refer Table 4.5) whose ligancies are determined from their radius ratio. The coordination numbers 5,7,9,11 do not obey all the rules of ionic compound formation. However, there are also ionic compounds for which ligancy rule is not followed because of the presence of mixed nature of bonding.

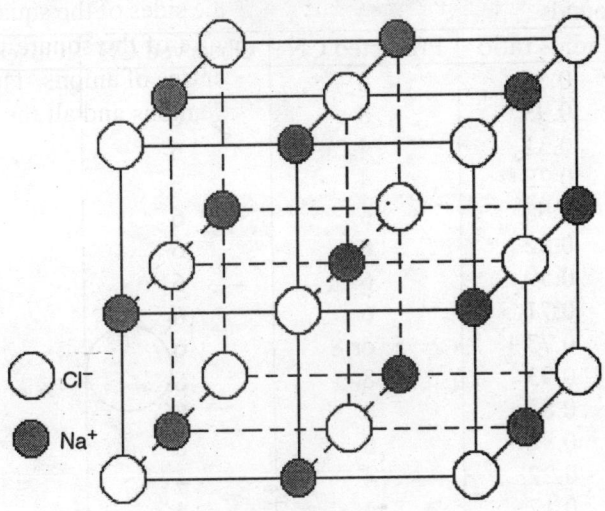

Fig. 4.16 Sodium chloride structure

Table 4.4 Geometry of ionic packing and legancy

CN or legancy	Radius ratio r_c/r_a	Packing geometry
2	$r_c/r_a < 0.155$	Linear
3	$0.155 \leq r_c/r_a < 0.225$	Triangular
4	$0.225 \leq r_c/r_a < 0.414$	Tetrahedral
6	$0.414 \leq r_c/r_a < 0.732$	Octahedral
8	$0.732 \leq r_c/r_a < 1.0$	Cubic
12	1.0	fcc or hcp

4.5 CERAMICS AND THEIR STRUCTURE

Ceramic materials are ionic compounds, abundant in earth crust as compared to pure metals since they are more stable than metals at higher temperatures and also less reactive with environment. Archaeologists have found ceramics (e.g. glass, bricks etc.) to be amongst the earliest man made materials. Ceramics have more compressional strength than tensile strength due to presence of ionic bonding. Ceramics are dielectrics and are mostly electrical and thermal insulators. They undergo polymerization and form linear and sheet molecules because they contain nonmetals or semimetals, which share electrons. The glasses are highly polymerized.

Ceramics are more rigid than metals or organic compounds but have less plasticity than organic compounds. In some of these materials, cations have two or more positive charges, hence the electrostatic repulsive energy becomes more and as a result the cation positions in the neighbouring planes are staggered such that the mutual repulsive energy of cations is reduced. The bond energy of ceramics with multivalent ions is quite high, hence the melting points of these compounds are also high.

Table 4.5 Predicted and observed coordination numbers (CN) of certain ionic compounds.

Compound	Radius ratio	Predicted CN	Observed CN	Observed structure
NaCl	0.54	6	6	NaCl
AgF	0.93	8	6	NaCl
LiBr	0.31	4	6	NaCl
LiI	0.28	4	6	NaCl
KBr	0.68	6	6	NaCl
RbCl	0.82	8	6	NaCl
MgO	0.59	6	6	NaCl
CaO	0.71	6	6	NaCl
KCl	0.73	5 or 8	6	NaCl
CsCl	0.93	8	8	CsCl
CsBr	0.87	8	8	CsCl
RbCl	0.82	8	8	CsCl*
BeO	0.22	4	4	ZnS
BeS	0.17	4	4	ZnS
CaF_2	0.73	6 or 8	8	CaF_2

.*At high pressure

These compounds have a wide range of applications because they produce hard crystals. They also have excellent insulating properties. Ceramics can have varieties of crystal structures:

a) Rock salt or NaCl type structure, Fig. 4.16
b) Cesium chloride structure, Fig. 3.8
c) Zinc blende structure, Fig. 4.17
d) Wurtzite structure, Fig. 4.18
e) Corundum structure
f) Rutile structure, Fig. 4.19
g) Fluorite structure, Fig. 3.10
h) Perovskite structure, Fig. 4.20
i) Spinel structure, Fig. 4.21

The coordination number, structure and packing in various ionic compounds are given in Tables 4.5 and 4.6. Their applications are given in Table 4.7.

Table 4.6(a) Anion packing in ionic compounds.

Names of the compounds	Anion packing	Structure	Fractional occupancy of cation sites
NaCl, KCl, MgO, CaO, SrO, BaO, LiF, KBr, CdO, VO, MnO, FeO, CoO, NiO	fcc	NaCl	All octahedral sites
$CdCl_2$, MgF_2, TiO_2	fcc	NaCl	Half of octahedral sites (alternate layers)
K_2O, Rb_2O, Li_2O, Na_2O	fcc	antifluorite	All tetrahedral sites
$Zn(CN)_2$	fcc	anticuprite	One fourth of upright tetrahedral and one fourth of inverted tetrahedral sites
ZnS, BeO, SiC	fcc	ZnS	All upright tetrahedral sites
ZnS, ZnO, SiC	hcp	ZnS	All upright tetrahedral sites
NiAs, FeS, FeSe	hcp	NiAs	All octahedral sites
Al_2O_3, Cr_2O_3	hcp	Corundum	Two thirds of the octahedral sites
CdI_2	hcp	---	One half of the octahedral sites

Table 4.6(b) Cation packing in ionic compounds.

Names of the Compounds	Cation packing	Fractional occupancy of anions
W_2N, MO_2N Mn$_4$N	fcc fcc	Half of the octahedral sites One fourth of the octahedral sites

(a) Rock salt or NaCl type structure

In NaCl type structure, Fig. 4.16, Cl^- ions occupy corners and face centre positions and Na^+ ions occupy edge centres and body centre position of the cubic unit cell or vice versa. Thus, there are 4 Na^+ and 4 Cl^- ions in each unit cell. An ionic compound has NaCl type structure if it has

   i) equal number of cations and anions,
   ii) the radius ratio r_c/r_a between 0.414 and 0.732, and
   iii) coordination number as six both for cations and anions.

These compounds have fcc Bravais lattice with basis of one cation and one anion separated by a distance of a/2 at each lattice point. Thus, NaCl type crystal structure can be visualized as two interpenetrating fcc lattices one of cations and the other of anions such that one of them is displaced along the edge by a/2 with respect to the other. Some of the ionic compounds of this type are MgO, FeO, MnS etc.

Table 4.7 Applications of ionic compounds.

Compound	Applications
MgO	Refractory
Al_2O_3	Substrate for building integrated circuits, spark plugs in automobiles, abrasives.
Fired clay Alumina mixture	Electrical insulator
Substitutional compound made by replacing a fraction of Al^{3+} ions by other trivalent ions like Fe^{3+}, Cr^{3+}	Gemstones, e.g. solid solution, Ruby (Cr^{3+} added to Al_2O_3) used in making devices like LASER, sapphire (Fe^{3+} added to Al_2O_3) and ruby are used in jewelry and cutting tools.
$BaTiO_3$	Ferroelectrics
SiO_2(quartz)	Piezoelectric transducers
Spinels	Magnets

Example 4.7

Find the volume of unit cell and density of NaCl crystal if the radii of Na^+ and Cl^- ions are 0.98Å and 1.81Å, respectively. Find the distance between the surfaces of the nearest chloride ions.

Solution:

a= 2(0.98+ 1.81) = 5.58 Å;   Volume of unit cell = a^3 = 173.74 x 10^{-30} m^3
Density = 4 x (23+ 35.5) x 1.66 x 10^{-27} kg /173.74 x 10^{-30} m^3 = 2235.75 kg/ m^3
Assuming chlorine ions are located at corners and face centres, distance between corners and nearest face centre = $a/\sqrt{2}$ = 3.946 Å.
So the distance between the surfaces of chlorine ions is =(3.946 – 3.62) Å = 0.326 Å

Example 4.8

FeO is an ionic solid and has a cubic unit cell.
        (a) Determine its density in kg/m^3.
        (b) Calculate its packing efficiency.
Atomic weight of iron and oxygen are 55.85 amu and 16 amu, respectively. Ionic radii of Fe^{2+}
and O^{2-} are 0.83 Å and 1.32 Å, respectively.

Solution:

The radius ratio of Fe^{2+} and O^{2-} is 0.83:1.32 =0.629 > 0.414. Hence anions form octahedral configuration around cations. Oxygen ions form fcc packing, with Fe^{2+} ions occupying all the octahedral
voids. Thus, FeO has NaCl type structure having unit cell with lattice parameter

$$a = 2 \times (r_{Fe2+} + r_{O2-}) = 4.30 \text{ Å}$$

and has four Fe^{2+} and four O^{2-} ions having masses as:
      Mass of Fe ion = 55.85 x 1.66x 10^{-27} kg
      Mass of O ion = 16.0 x 1.66x10^{-27} kg
      Vol. of unit cell = a^3 = 7.95 x 10^{-29} m^3
      Density ρ = 4x(55.85+ 16.0) x 1.66x10^{-27} /a^3 = 6001 kg/m^3
    Packing efficiency P.E. = Vol. of ions/Vol. of unit cell
        Vol. of ions = 4x (4π/3) x (0.83^3 + 1.32^3) Å^3
                 = 16.76x(0.57 + 2.30) x 10^{-30} m^3 = 48.12 x 10^{-30} m^3
Therefore   P.E. = 60.52 %

b) Cesium chloride structure

In cesium chloride crystal anions and cations both have coordination number 8, Fig. 3.8. The
radius ratio r_c/r_a lies between 0.732 and 1. Anions are located at the corners and a cation at the
body centre position of the cubic unit cell. Interchange of cation and anion positions produces the
same structure. It has simple cubic Bravais lattice with a basis of a cation and an anion separated
by a distance of √3a/2, attached to each lattice point. This structure is often mistaken as a bcc
lattice, which is not true since the body centre position of the unit cell is occupied by a cation,
which is different from the anions, which are at the corners. Thus either corners or body centre
position should be treated as lattice points but not both.

Example 4.9

What is the name of the Bravais lattice to which CsCl belongs? Write down the basis for this crystal
structure. What is the coordinate of Cl$^-$ considering nearest Cs$^+$ ion at origin? Calculate the packing
efficiency of ions in CsCl.

Solution:

Bravais lattice of CsCl: simple cubic.
The basis for this structure is Cl$^-$ at (0,0,0) and Cs$^+$ at (½,½,½); r_{Cs}^+ =1.65 Å and r_{Cl}^- =1.81 Å.
There is one Cs$^+$ ion and one Cl$^-$ ion per unit cell.
      Lattice parameter a = 2(r_{Cs+} + r_{Cl-})/√3 = 3.995 Å, volume of unit cell = 6.376 x 10^{-29} m^3

Volume of ions = $(4\pi/3)(1.65^3 + 1.81^3) \, 10^{-30} = 4.366 \times 10^{-29} \, m^3$
P.E.= $4.366 \times 10^{-29}/6.376 \times 10^{-29} = 0.6847 = 68.47\%$

c) Zinc blende structure

This structure is also called sphalerite or diamond cubic structure, Fig. 4.17. The coordination number is 4 for ZnS type structure. This is a fcc structure where the corners and face centre positions are occupied by anions and each cation is tetrahedrally bonded to four anions through sp^3 hybridized bonds. The four cations occupy alternate tetrahedral voids in the fcc unit cell. An equivalent structure results if cation and anion positions are interchanged. The radius ratio r_{Zn2+}/r_{S2-} is 0.48, which predicts octahedral coordination as per ligancy rules. Yet we observe four-fold coordination in this type of compounds due to presence of covalent character of bonding. The lattice points of this type of fcc lattice are attached with a basis consisting of one anion S^{2-} and one cation Zn^{2+} displaced by $\sqrt{3}a/4$ along the body diagonal. This structure can be visualized as two interpenetrating fcc lattices, one of anions and other of cations displaced by one fourth of the body diagonal. Examples of these type ionic compounds are ZnS, ZnTe, BeO, SiC etc.

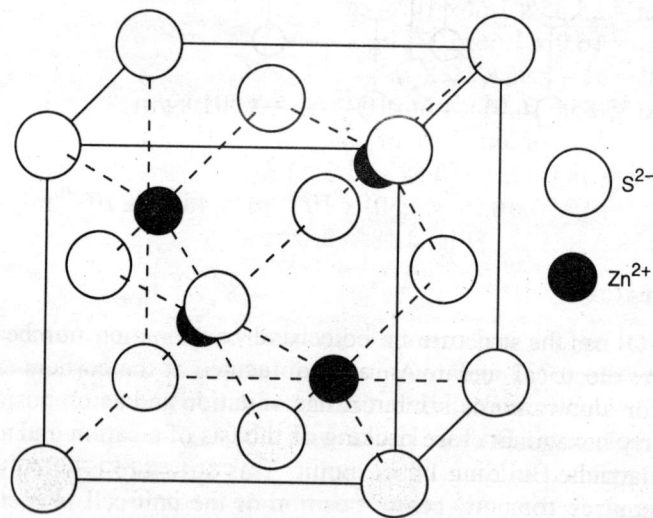

S^{2-}

Zn^{2+}

Fig. 4.17 Zinc blende structure

Example 4.10

Calculate the packing efficiency for ZnS crystal where r_{Zn2+} = 0.83 Å, r_{S2-} = 1.74 Å.

Solution :

Given that r_{Zn2+} = 0.83 Å and r_{S2-} = 1.74 Å ; $\sqrt{3}a/4 = r_{S2-} + r_{Zn2+} = 2.57$ Å
a = 5.935 Å and volume = $2.091 \times 10^{-28} \, m^3$
Number of Zn^{2+} and S^{2-} ions per unit cell = 4 each
Vol. of Zn^{2+} ions = 4 x (4/3)π x $(r_{Zn2+})^3$ = $9.58 \times 10^{-30} m^3$
Vol. of S^{2-} ions = 4 x (4/3)π x $(r_{S2-})^3$ = $8.83 \times 10^{-29} \, m^3$
P.E. = vol. of ions / a^3 = $9.788 \times 10^{-29} / 2.091 \times 10^{-28}$ = 46.81%

d) Wurtzite structure

This structure has hexagonal packing of anion with half the tetrahedral interstitial voids filled with cations, such that cation-cation separation maximizes, Fig. 4.18. Its Bravais lattice is simple hexagonal with basis of one anion and one cation. The radius ratio is between 0.225 and 0.414. In case of beryllium oxide, this ratio is 0.25. ZnS also possesses this structure.

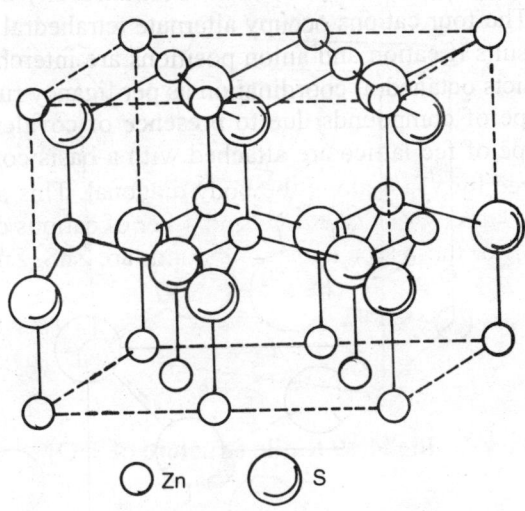

Fig. 4.18 Wurtzite structure

e) Corundum structure

Naturally available Al_2O_3 has the structure called corundum. Al_2O_3 is a prime constituent of spark plug insulators, other electrical ceramics and abrasion-resistant materials. The preferred coordination number for aluminium is six but it has valency three, so the bond strength is only half. This leads to nearly hexagonal close packing of the oxygen ions and aluminium ions filling ⅔ rd of the octahedral voids. Building up of similar layers leads to maximum spacing of Al^{3+} ions. Such packing requires four Al^{3+} ions adjacent to each O^{2-}. Cr_2O_3 also has got Al_2O_3 structure.

f) Rutile structure

TiO_2 has rutile structure, Fig. 4.19. The coordination number for Ti^{4+} is six and it has valency 4, which leads to a bond strength of two-thirds. It requires three-fold coordination of Ti^{4+} adjacent to O^{2-}. The fcc close-packing of O^{2-} ions in rutile structure is distorted due to filling up of cations in half of the octahedral voids. This can be explained as follows. The radius of the octahedral void is

$$0.414 r_{O2-} = 0.5465$$

and radius of cation Ti^{4+} is 0.64. So filling up of octahedral voids with Ti^{4+} causes distortions. GeO_2, PbO_2, SnO_2, MnO_2 and many other oxides possess rutile structure.

g) Fluorite structure

In this type of ionic compounds, the number of anions is twice as many as that of cations since the charge on cation is +2 and that on anion is −1; an example of this type crystal is CaF_2 where $r_{Ca2+}/r_{F-} = 0.73$. In this compound, Fig. 3.11, the coordination number for cation is 8 and that of anion is 4. Ca^{2+} ions occupy corners and face centre positions. F^- ions occupy all the eight tetrahedral voids of the fcc unit cell. Ca^{2+} ions occupy half of the body centre positions of the cubes formed by joining the eight F^- ions. Examples of such compounds are UO_2, ThO_2, TeO_2 etc. This structure of UO_2, a nuclear fuel, helps to accumulate fission fragments from nuclear reactions in its vacant sites without introducing much strain in the solid.

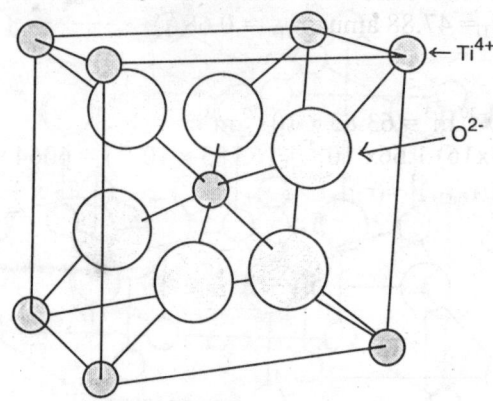

Fig. 4.19 Rutile structure of TiO_2

Example 4.11

Describe the Bravais lattice and basis of CaF_2 crystal. Find the coordination number of cations as well as of anions. Calculate the density of CaF_2.

Solution:

This crystal has fcc structure with Ca^{2+} ions located at corners and face centres, F^- ions are at all the tetrahedral voids. The coordination number of cation is eight since each Ca^{2+} ion is coordinated to eight F^- ions whereas that of anion is four since each F^- ion is coordinated to four Ca^{2+} ions. There are four cations and eight anions per unit cell. Since $r_{Ca+} = 1.06$ Å and $r_{F-} = 1.33$ Å we get $a = 4(1.06 + 1.33) / \sqrt{3} = 5.52$ Å, Volume $= 1.68 \times 10^{-28}$ m^3, Density $= 4(40.08 + 2 \times 19) \times 1.66 \times 10^{-27}$ kg $/1.68 \times 10^{-28}$ m$^3 = 3086$ kg/ m^3

h) Perovskite structure

Perovskite contains more than one type of cations. The coordination number of cations depends upon their charges. For example, consider the cubic $CaTiO_3$ structure, where cations are Ca^{2+} and Ti^{4+} and O^{2-} is anion. This structure does not have a centre of symmetry. Ca^{2+} ions occupy corner positions of the unit cell, O^{2-} ions are located at face centre positions and Ti^{4+} ions occupy the octahedral sites at the centre of the cube, Fig. 4.20. The structure of $CaTiO_3$ changes to tetragonal below 120°C. Other ionic compounds like $BaTiO_3$, $SrTiO_3$, $LaAlO_3$ and many other oxides also possess this structure. These crystals show piezoelectric effect and are used for phonograph cartridges and pressure transducers. Other applications of perovskites are the following: $BaTiO_3$ in multilayer capacitor, $Pb(Zr, Yi) O_3$ in piezoelectric transducer, $BaTiO_3$ in thermistor, (Pb, La) (Zr,

Ti) O_3 in electrooptical modulator, $LiNbO_3$ in switch, $BaZrO_3$ in dielectric resonator, $BaRuO_3$ in resistor, Ba (Pb, Bi) O_3 layered cuprates in superconductor, $GdFeO_3$ in magnetic bubble memory, (Ca, La) MnO_3 in ferromagnet and $LaCoO_3$ in refractory electrode.

Example 4.12

Calculate the density and packing efficiency of $BaTiO_3$, where the unit cell dimensions are 3.98 Å, 3.98 Å, and 4.03 Å. The following data is given: M_{Ba}= 137.33 amu, r_{Ba2+}=1.36 Å, M_O= 16 amu, r_{O2-}= 1.40 Å, M_{Ti}= 47.88 amu, r_{Ti4+} = 0.68Å.

Solution:

$V_{unit\ cell}$= 3.98 x 3.98 x 4.03x 10^{-30} m^3 = 63.83 x 10^{-30} m^3

Density = [(137.33 + 47.88 + 3x16) 1.66x 10^{-27}]/ (63.83 x 10^{-30}) = 6064.99 kg/m^3

Packing Efficiency = (4/3)π x [$(r_{Ba2+})^3$ +$(r_{Ti4+})^3$+3 x $(r_{O2-})^3$] m^3/ 63.83 x 10^{-30} m^3 =0.726

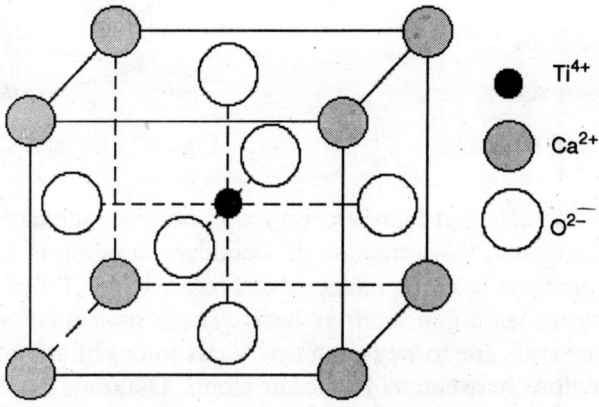

Fig. 4.20 Perovskite structure

i) Spinel structure

Spinel structure, Fig. 4.21, is found in nature in certain compounds of the type $A_mB_nC_p$, e.g. magnesium aluminate, $MgAl_2O_4$. In this structure O^{2-} ions form fcc lattice, Mg^{2+} ions fill up tetrahedral voids and Al^{3+} ions occupy octahedral void positions. Magnetic ceramics or ferrites have a crystal structure, which is a slight variation of the spinel structure. The occupancy of the tetrahedral void positions decides the magnetic characteristics of the material.

4.6 MOLECULAR CRYSTALS

Molecular crystals are obtained from molecular compounds. When these compounds in the liquid form are cooled slowly to a temperature below the freezing point, the molecules arrange in regular three-dimensional array to form a crystal. The molecular crystal has a specific space lattice with a molecule as basis at each lattice point. Normally the atoms are bonded by primary bonding to form molecules, and the molecules are bonded to each other through secondary bonding. The secondary bonding is also referred to as molecular bonding. When the configuration

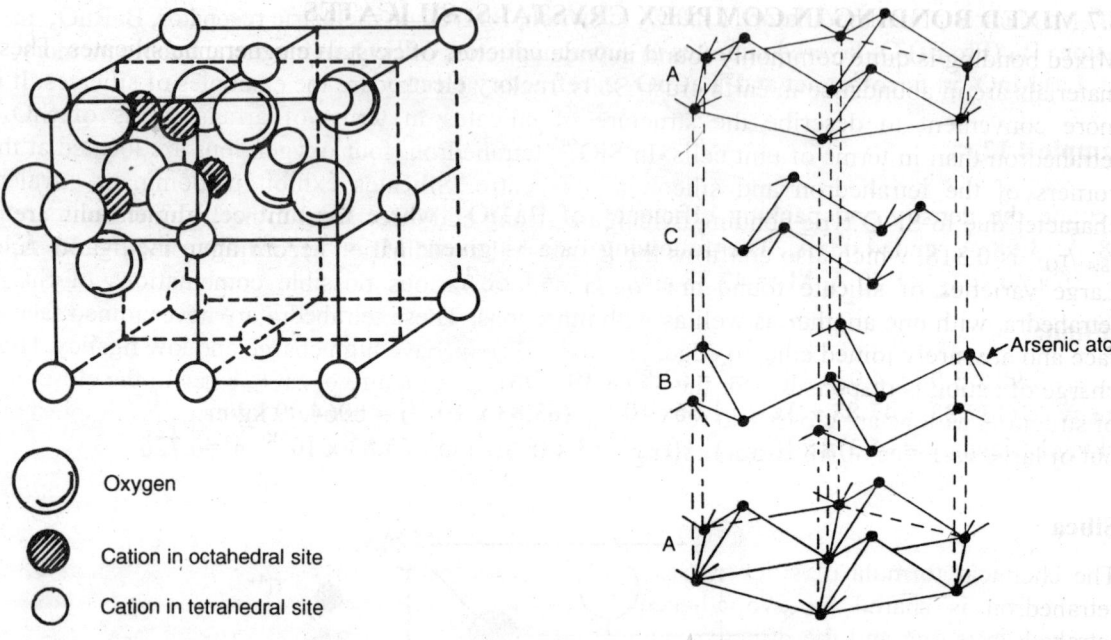

Oxygen

Cation in octahedral site

Cation in tetrahedral site

Fig. 4.21 Spinel structure

Arsenic atom

Fig. 4.22 Crystal structure of Arsenic

of atoms or molecules is such that there are very less number of electron transfers taking place at the time of bond formation, the formation of secondary bonding is favoured. For example, the molecular crystal of methane has CH_4 molecules as basis, which forms an fcc crystal at –183°C; the fcc crystals of argon has argon atom as basis. These molecular bonds are usually van der Waals bonds since they arise due to weak van der Waals forces of attraction arising due to dipolar or higher polar interactions between molecules or atoms. Diatomic molecules like halides, N_2, O_2 do not form cubic structure on crystallization at very low temperature due to the specific shape of molecules. Linear molecules like tellurium, polyethylene etc. do not crystallize completely due to the problem of aligning the linear molecules in same direction so as to generate a regular three-dimensional structure. Thus, molecular crystals having basis of linear molecules vary markedly in their structure. The linear molecules are bonded by weak van der Waals attractions.

There are molecular crystals formed of sheet molecules e.g. the puckered sheets of arsenic stacked one above another with repeating structure after every third layer, Fig. 4.22. Graphite is also a molecular crystal with sheet molecules stacked one above another with the sequence repeating after every second layer, Fig. 4.3. The sheets are bonded to one another due to the presence of weak van der Waals forces. Ice is another example of a molecular crystal, which has hexagonal structure, Fig. 4.2. In an ice crystal hydrogen bonds are formed between the unshielded proton of hydrogen of one H_2O molecule and the unshared electrons of oxygen in the neighbouring water molecules. This type of bonding is often referred to as hydrogen bridge and is responsible for low packing factor in ice. Hence the water expands on freezing to ice state. As ice melts the hydrogen bonds break down and contraction occurs. The hydrogen bridges remain even after melting of ice till the temperature reaches 4°C, at which the density of water is a maximum. Molecular crystals are generally good insulators and transparent to light. These compounds have low melting point due to weak bonding between molecules.

4.7 MIXED BONDING IN COMPLEX CRYSTALS: SILICATES

Mixed bonding is quite commonly found in wide varieties of crystals e.g. ceramic silicates. These materials are in abundance in earth crust. Soils, rocks, and clay are the examples of silicates. It is more convenient to describe the structure of silicates in terms of arrangements of SiO_4^{4-} tetrahedron than in terms of unit cells. In SiO_4^{4-} tetrahedron, four oxygen ions are located at the corners of the tetrahedron and silicon at its centre. Silicates exhibit predominant covalent character due to Si–O type bonding, which are strong and directional in nature. The radius ratio r_{Si4+}/r_{O2-} is 0.318, which also confirms the tetrahedral coordination according to the ligancy rule. Large varieties of silicate found are found due to various possible combinations of SiO_4^{4-} tetrahedra, with one another as well as with other ions. These tetrahedra are never joined face to face and are rarely joined edge to edge, because Si^{4+} ions have high charge and low ligancy. High charge of cation is responsible for the formation of open structure of SiO_2, since in the close type of structure, cations are closer and result in increase of potential energy. Silica is the simplest one out of large varieties of silicate compounds.

Silica

The chemical formula of silica is SiO_2. In this compound oxygen ion at the corner of every tetrahedron is shared by two adjacent tetrahedra. The effective number of Si^{4+} ions per tetrahedron is one and the effective number of O^{2-} ions are 4 x ½ = 2, because each corner is shared by two tetrahedra. So the ratio of Si^{4+} to O^{2-} is 1: 2 as given by formula SiO_2. When these tetrahedra are arranged in regular manner, three primary polymorphic crystalline forms of silica are obtained:

a) Quartz
b) Cristobalite
c) Tridymite

Quartz has stable forms at low temperatures as well as in the temperature range from 573°C to 867°C. The crystal structure of β-quartz is hexagonal having two spiral chains wound around a hexagonal prism. Quartz is composed of connected networks of silica tetrahedra and has quite symmetrical structure at high temperature but gets distorted at low temperatures. The most stable forms of silica are low quartz below 573°C , high quartz (573°C to 867°C), high tridymite (867°C to 1470°C) and high cristobalite (1470°C to 1710°C) and liquid above 1710°C at atmospheric pressure. High quartz has closer packing and greater density (2.65 Mg/m³) than high tridymite (density= 2.26 Mg/m³) and high cristobalite (density = 2.32 Mg/m³). Structures of high cristobalite and high tridymite are shown in Figs. 4.23 and 4.24, respectively. The structure of cristobalite is same as that of sphalerite in which silicon ions occupy both the positions of Zn^{2+} as well as S^{2-} and oxygen ions are located between the silicon ions. In other words the cristobalite has diamond cubic structure. Tridymite may similarly be compared with wurtzite structure.

Fused silica or vitreous silica

Fused silica is silica glass having high degree of atomic randomness. Silica can exist in both crystalline and noncrystalline form as shown in Fig. 3.2 schematically. The crystallization occurs in silica or in other oxides like B_2O_3 and GeO_2 only with difficulty upon very slow cooling from liquid state. When basic oxides CaO, Na_2O are added to silica, the cations from these oxides fit into the network in such a way that the formation of glassy substance becomes easier.

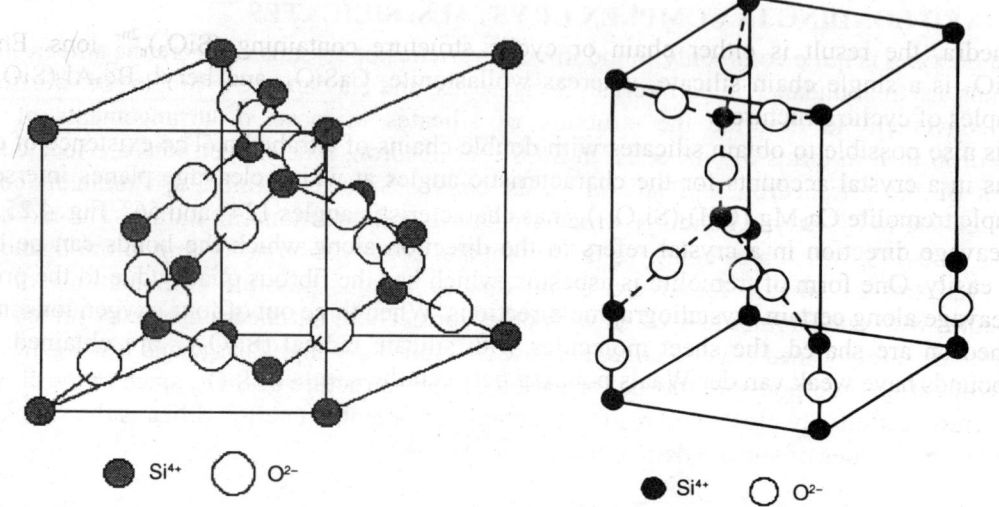

Fig. 4.23 Structure of cristobalite.   Fig. 4.24 Structure of tridymite.

Silicates

As we have discussed earlier the silicates can be formed by various possible arrangements of SiO_4^{4-} tetrahedron. Sometimes in certain silicates one, two or three of the corner oxygen atoms are shared by other tetrahedra giving rise to complex structures having ions of various types like,

   a) SiO_4^{4-}

   b) $Si_2O_7^{6-}$ in which two tetrahedra share one corner

   c) $Si_3O_9^{6-}$ in which two corners of tetrahedron are shared to form varieties of ring and chain structures, Fig. 3.2.

Orthosilicates

Orthosilicates groups of compounds contain SiO_4^{4-} tetrahedra. In these compounds SiO_4^{4-} tetrahedra are bonded to positive ions and are not directly joined to one another. One of the positive ions like Mg^{2+}, Fe^{2+} etc. is in between two tetrahedra. There are effectively two Mg^{2+} ions for each SiO_4^{4-} ion resulting in the formation of compounds like olivine silicate e.g. Mg_2SiO_4, Fe_2SiO_4, are formed. Every Mg^{2+} has six O^{2-} ions as nearest neighbours. Garnets are formed with two positive ions and SiO_4^{4-} tetrahedra e.g. $Mg_3Al_2(SiO_4)^3$, $Fe_3Cr_2(SiO_4)^3$. These compounds are also called island silicates. In olivine, oxygen atoms form distorted hcp structure with only one eighth of the tetrahedral interstitial sites filled up by silicon atoms and half of the octahedral sites filled up by Mg or Fe atoms. Mullet $(3Al_2O_3 - 2SiO_2)$ is also a compound of this type and is a common constituent of fire clay products.

Pyrosilicates

Pyrosilicates contain $Si_2O_7^{6-}$ ions and are less stable. Akermanite $Ca_2MgSi_2O_7$ is an example of pyrosilicate.

Metasilicates

These silicates are formed when the two corners of each tetrahedron are shared with other tetrahedra, the result is either chain or cyclic structure containing $(SiO_3)_n^{2n-}$ ions. Enstatite $MgSiO_3$ is a single chain silicate, whereas wollastonite, $CaSiO_3$, and beryl, $Be_3Al_2(SiO_3)_6$ are examples of cyclic structure.

It is also possible to obtain silicates with double chains of tetrahedra. The existence of double chains in a crystal accounts for the characteristic angles at which cleavage planes intersect for example tremolite $Ca_2Mg_5(OH)_2(Si_4O_{11})_{12}$ has characteristic angles 124° and 56°, Fig. 4.25.

Cleavage direction in a crystal refers to the direction along which the bonds can be broken most easily. One form of tremolite is asbestos, which has the fibrous quality due to the presence of cleavage along certain crystallographic directions. When three out of four oxygen ions in every tetrahedron are shared, the sheet molecules with silicate radical $(Si_2O_5)^{2-}$ are obtained. These compounds have weak van der Waals bonding between the sheets.

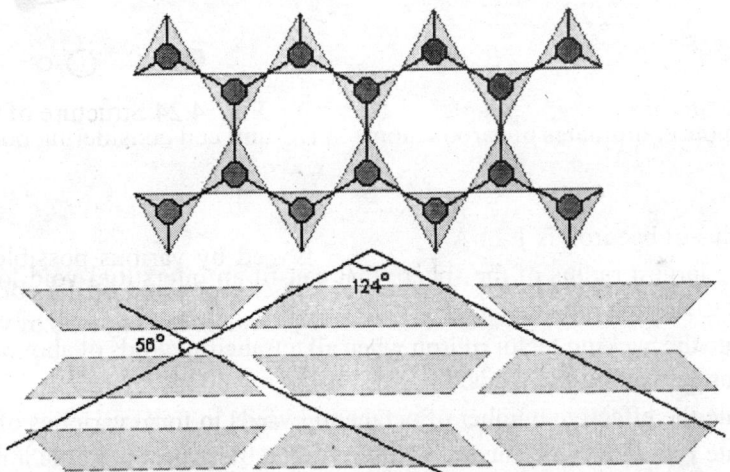

Fig. 4.25 Arrangement of double chains in tremolite

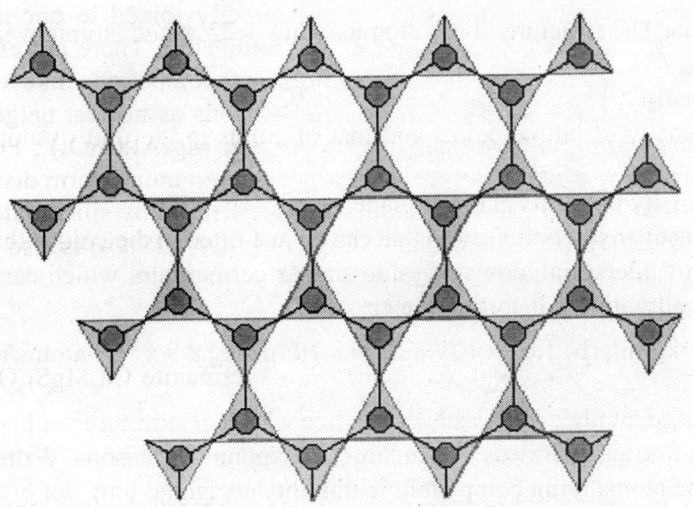

Fig. 4.26 Arrangement of silicate tetrahedra in a sheet silicate.

Examples of this type of silicates are:

Clay : $Al_2(OH)_4(Si_2O_5)$ (Kaolinite)
Talc : $Mg_3(OH)_2(Si_2O_5)_2$
Mica : $KAl_2(OH)_2(AlSi_3O_{10})$

The clay (Kaolinite) sheet molecules are polar in nature and hence attract water molecules through dipolar interactions, Fig. 4.26. That is why clay absorbs water. Talc is soft like clay but does not absorb water. When all the four oxygen ions of SiO_4^{4-} are shared, the three-dimensional network silicate compounds are obtained, Fig. 4.26. SiO_2 and fused silica are examples of these network silicates. When Al atoms replace some of the Si atoms in silica, the feldspar is produced. The silica and feldspar constitute around 50% of the earth crust.

EXERCISES

4.1   Write down the coordinates of carbon atoms in DC unit cell considering body centre position as origin.

4.2   Atomic radius of bcc iron is 1.24 Å.
   i)   Find the largest radius of the sphere that can fit an interstitial void located at $(0, ½, ¼)$ without distorting the crystal.
   ii) Calculate the packing factor of iron when all tetrahedral voids of above type are occupied by atoms just fitting the voids.
   iii) Calculate the effective number of octahedral voids in unit cell.
   iv) Calculate P.E. if 50% octahedral voids are filled by largest possible impurity atoms such that there are no distortions in shape or size of the unit cell.

   Ans: (i) 0.36 Å; (ii) 0.78; (iii) 6; (iv) 0.686

4.3   Germanium has DC structure. It has atomic radius 1.22 Å and atomic weight 72.6 amu. Find the following:
   (a) Mass density.
   (b) Linear density of lattice points and that of atoms in the unit of atoms per meter along [111] direction.
   (c) Planar density of atoms in (110) plane.
   (d) The radius of an interstitial atom that can be just fitted in the void at the centre of unit cell.
   (e) Number of interstitial atoms of same size as germanium, which can be accommodated per unit cell without distorting the crystal.

   Ans: (a) 5411 kg/m³; (b) 1.02 x 10⁹/m, 2.04 x 10⁹/m; (c) 8.9 x 10¹⁸ atoms/m²; (d) 1.22 Å; (e) 8

4.4   Although a large number of crystals crystallize with hcp structure yet hcp structure does not figure among fourteen Bravais space lattices. Explain the reasons. Write down the coordinates of all the atoms lying completely within the hexagonal unit cell of Zn (hcp).

Ans: hcp is simple hexagonal having a basis of two atoms – one at (0,0,0) and the other at (⅓,⅔,½)--attached to each lattice point (⅓,⅔,½), (⅓,–⅓,½), (–⅔,–⅓,½).

4.5 An alloy of iron (bcc structure) and carbon contains 0.05 atomic percent of carbon. Find the number of carbon atoms per m^3 of the alloy. Given that the atomic radius of iron is 1.24 Å.

Ans: $4.26 \times 10^{25}/m^3$

4.6 Calculate the number density of atoms and packing efficiency of fcc iron if 0.05 atomic percent of its octahedral voids are filled by impurity atoms each of radius $0.414r_{Fe}$, where r_{Fe} = 1.27 Å.

Ans: 8.63×10^{28} atoms/m^3; 0.7405

4.7 Find the planar density of ions (considering both types of ions) on the plane (110) of NaCl. Given: Ionic radius of Na^+ = 0.98 Å and that of Cl^- = 1.81 Å.

Ans: 9.08×10^{18} ions /m^2

4.8 Quartz (SiO_2) has a density of 2.65 Mg/m^3. How many silicon and oxygen atoms are there per unit volume? Calculate the value of P.E., if r_{Si} = 0.38Å and r_o= 1.17 Å.

Ans: $2.66 \times 10^{28}/m^3$, $5.32 \times 10^{28}/m^3$; 0.36

4.9 Consider an ionic compound which on heating changes its cation to anion radius ratio from 0.4 to 0.6. Consequently, its structure is changed from diamond cubic to NaCl type. Find the percentage change in its packing efficiency and in density if the lattice parameter of fcc is 0.5% larger as compared to that of DC structure

Ans: 18%; –1.49%

4.10 On the basis of ionic charge and ionic radii, predict the crystal structures for the following ionic compounds: (a) CsI, (b) NiO,(c) KI, (d) MgO and (e) NiS with proper explanation.

4.11 The parameter 'a' of CdS unit cell is 0.58 nm. If the density of CdS having certain number of vacancies is 4.82 Mg/m^3, calculate P.E. of the crystal structure.

Ans: 0.501

4.12 Titanium has an hcp unit cell and its density is 4.51 Mg/m^3. What is the volume of the unit cell in cubic meters?

Ans: 105.78×10^{-30} m^3

4.13 Calculate the percentage change in density, as iron transforms from fcc to bcc. The atomic radii are 1.270 Å and 1.241Å, respectively. Given that mol. wt. of Fe = 56 amu.

Ans: –2.14%

4.14 Draw the plan view of Cd^{2+} for $CdCl_2$ unit cell, clearly indicating positions and height from base. Take a = 1 unit.

4.15 Consider an ordered alloy of 25% Cu-75% Ni in which the Cu atoms are occupying the corners and Ni atoms the face-centres of the unit cell. Find the volume of the largest atom that would fit in the interstitial void. Given that $r_{Cu} = 1.28$ Å and $r_{Ni} = 1.25$ Å.

Ans: 0.656 Å^3

4.16 What is the percentage change in void space of sc unit cell if the cubic void is filled up by largest possible impurity atom such that there is no distortion in size or shape of the unit cell.

Ans: –43.1%

4.17 Calculate the planar density of equivalent points per Å^2 for Sn (bct) crystal (a = 5.82 Å and c = 3.175 Å) on (021) plane. This unit cell has four Sn atoms at locations (0, 0, 0), (½, ½, ½), (½, 0, ¼) and (0, ½, ¾).

Ans: 1.995×10^{18}/ Å^2

5

Natural and Synthetic Polymers

5.1 INTRODUCTION

In the last century, new materials have contributed immensely to industrial and technological development. The list of materials includes a large number of metals, alloys, composites, ceramics and polymers. Out of these polymers have made an important contribution to this developmental process. What is a polymer? 'Poly' is a Greek word meaning many and 'mer' means unit. Polymer is a macromolecule consisting of a repetitive unit, which may be a chemical group or a small molecule. Small molecules, which interact to form polymer, are called monomers whereas repetitive units in polymer is called mer. These macromolecules may contain hundreds or even tens of thousands of atoms. One of the most common examples of polymer is polyethylene having monomer as ethylene molecule, C_2H_4. Mer in polyethylene is represented as

$$\begin{array}{c} \text{H} \quad \text{H} \\ | \quad\;\; | \\ -\text{C}-\text{C}- \\ | \quad\;\; | \\ \text{H} \quad \text{H} \end{array}.$$

The number of repeating units in the macromolecule is called the degree of polymerization (D_p). The product of degree of polymerization and the molecular weight of monomer (M_m) is the molecular weight of the polymer (M_w) i.e.

$$M_w = M_m D_p \qquad (5.1)$$

Polymers with high degree of polymerization are called high polymers and those with low degree of polymerization (e.g. 500 – 600 amu) are called oligomers. The name of the polymer is derived from the name of monomer with a prefix of "poly" attached to it. Polymer may contain identical monomers or monomers having different formula or structure. In the former case the polymer is termed as homopolymer and in the later case, the polymer is heteropolymer or mixed polymer e.g. cellulose is a homopolymer of glucose whereas peptidoglycan is a heteropolymer with mers of N-acetyl glucosamine and muramic acid.

Natural polymers are also synthesized in laboratoies for industrial applications e.g. cellulose, starches, proteins, rubber, enzymes etc. Some of the polymeric substances are used directly as are available in nature e.g. wood, wool, silk, cotton etc., whereas in some cases materials are reconstituted of naturally occurring polymeric substances e.g. rayon and cellophane are reconstituted from cellulose. There are other synthetic polymers like dacron, polyethene, nylon, plastic fibers etc. which are used in everyday life.

5.2 CLASSIFICATION OF POLYMERS

Polymers can be classified based on their behaviour with respect to the change of temperature, chemical composition, and structural formula.

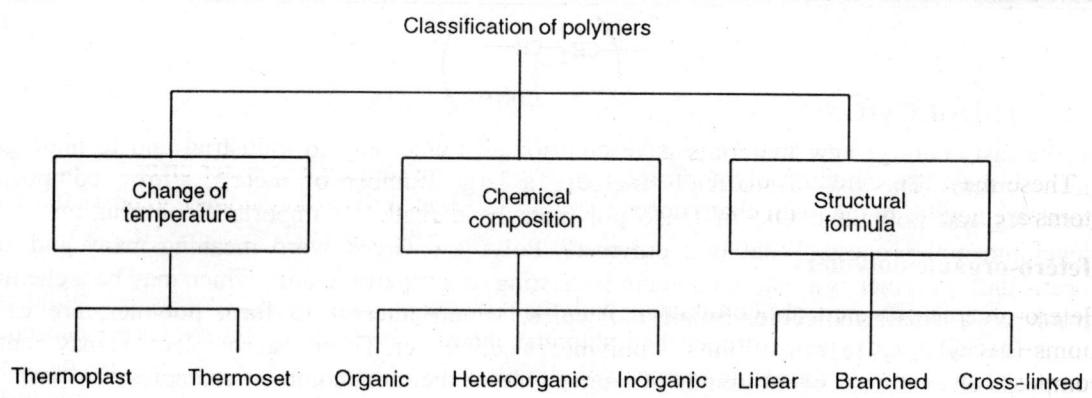

Fig. 5.1. Classification of polymers.

a) Classification based on behaviour with change of temperature

Polymers can be broadly grouped into two types as per their behaviour with change of temperature: thermoplasts and thermosets.

Thermoplasts

Thermoplasts are usually linear chain polymers held together by secondary forces. On application of heat, the secondary bonding between the polymeric chains breaks down and as a result, plasticity of the material increases making the polymers mouldable. The temperature at which a polymer becomes soft is known as softening temperature. As the temperature is increased beyond the softening temperature the plastic melts and hence these materials are not suitable for high temperature applications. These materials usually melt at a few hundred degree Celsius.

Thermosets

Thermoset polymers have three-dimensional network type structure where bondings in three dimensions are primary. Hence they are hard and rigid at room temperature. As the thermal energy increases, these materials become harder due to formation of more primary bonding between the molecules. Ultimately it decomposes at a specific temperature instead of melting in contrast to what happens in case of thermoplasts. Of course the thermoplasts can be treated such that these become hard and suitable for high temperature applications. Scrap plastics can be recycled but thermoset polymers cannot be reused since thermoset polymers are degraded when heated to an elevated temperature e.g. bakelite.

b) **Classification based on chemical composition**

According to their chemical composition, polymers can be grouped into organic, hetero-organic and inorganic polymers.

Organic polymers

Organic polymers fall into the classes of aliphatic and aromatic compounds. These can further be classified as hydrocarbons, halogen derivatives, alcohols, acids, ethers, nitriles, aldehydes, ketones etc. These are homochain polymers. Organic polymers contain compounds of carbon, hydrogen, nitrogen, oxygen, sulphur and halogens e.g. polyacrylamide

$$\left(\!-CH_2\!-\!\underset{\underset{CONH_2}{|}}{CH}\!-\!\right)_n$$

These may also contain polymeric molecules consisting of atoms of other elements but these atoms are neither in the main chain nor are directly connected to carbon atoms e.g. polyacid salts.

Hetero-organic polymers

Hetero-organic polymers are compounds with main chain composed of carbon atoms and hetero-atoms (except for oxygen, nitrogen and sulphur). These compounds can also have main chain composed of atoms of inorganic elements with side groups containing carbon atoms connected directly to the chain or they can also have main chain composed of carbon atoms and side group containing hetero-atoms (except for nitrogen, sulphur, oxygen and halogen atoms) connected directly to carbon atoms of the chain e.g. polysiloxanes where R is vinyl group.

$$-\underset{\underset{R}{|}}{\overset{\overset{R}{|}}{Si}}-O-\underset{\underset{R}{|}}{\overset{\overset{R}{|}}{Si}}-$$

Inorganic polymers

Inorganic polymers do not contain carbon atoms e.g. polysilanes

$$-\underset{\underset{H}{|}}{\overset{\overset{H}{|}}{Si}}-\underset{\underset{H}{|}}{\overset{\overset{H}{|}}{Si}}-\underset{\underset{H}{|}}{\overset{\overset{H}{|}}{Si}}-$$

(c) **Classification based on structural formula**

Polymers can be classified as linear chain, branched, and cross-linked or three-dimensional polymers based on structural formula.

Linear chain polymers

Linear chain polymers are formed from micromolecules (monomers), which contain bifunctional groups or double bonds like vinyl compounds. These groups can react with at least two neighbouring monomers so that the long chain molecule can be built up in both the directions along the chain. Examples of vinyl compounds are given in Table 5.1.

Table 5.1 Organic molecules

Vinyl compounds

	R	Compound
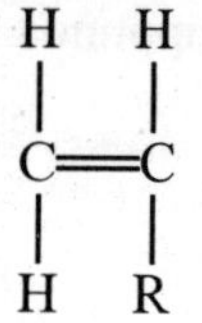 $\begin{array}{c} H \quad H \\ \vert \quad\; \vert \\ C{=}C \\ \vert \quad\; \vert \\ H \quad R \end{array}$	H	Ethene
	Cl	Vinyl chloride
	OH	Vinyl alcohol
	$OCOCH_3$	Vinyl acetate
	C_6H_5	Vinyl benzene or Styrene

Vinylidene compounds

	R'	R"	Compound
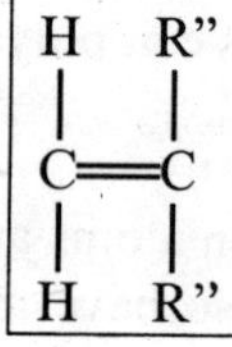 $\begin{array}{c} H \quad R" \\ \vert \quad\; \vert \\ C{=}C \\ \vert \quad\; \vert \\ H \quad R" \end{array}$	CH_3	CH_3	Isobutylene
	Cl	Cl	Vinylidene chloride
	$OCOCH_3$	CH_3	Methyl methacrylate

Tetrafluoroethylene :
$$\begin{array}{c} F \quad\; F \\ \vert \quad\;\; \vert \\ C{=}C \\ \vert \quad\;\; \vert \\ F \quad\; F \end{array}$$

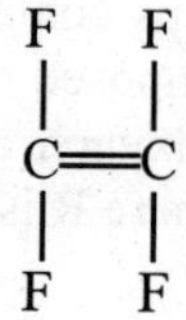

Some common functional groups, which enter into polymerization reaction, are given below:

	Group		Group
HO–	Hydroxyl	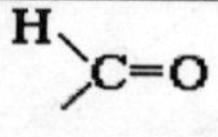 $\begin{array}{c} H \\ \diagdown \\ \quad C{=}O \\ \diagup \end{array}$	Aldehyde
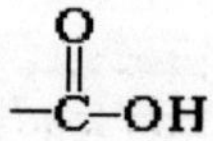 $\begin{array}{c} O \\ \parallel \\ {-}C{-}OH \end{array}$	Carboxyl	–O–	Ether
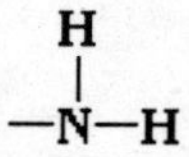 $\begin{array}{c} H \\ \vert \\ {-}N{-}H \end{array}$	Amine	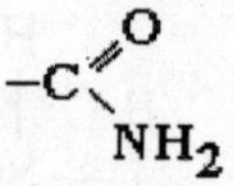 $-C\diagup^{\displaystyle O}_{\displaystyle NH_2}$	Amide
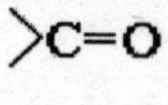 $>C{=}O$	Ketone	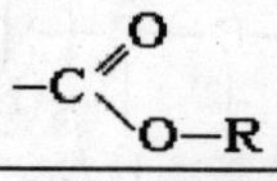 $-C\diagup^{\displaystyle O}_{\displaystyle O{-}R}$	Ester

Branched polymers

During addition polarization process, propagation may cause sometimes irregularities due to which branching of polymer may occur. Termination by disproportion can also cause branching. For example,

(i)

$$X-(mer). + C=C \rightarrow X-(mer)-C. \qquad (5.2)$$

where side radicals (H—C—H) is formed instead H adding to the linear chain.

(ii)

$$X-\left(\begin{matrix} H & H \\ | & | \\ -C-C- \\ | & | \\ H & H \end{matrix}\right)_m \bullet + \begin{matrix} H & H \\ | & | \\ C=C\bullet \\ | & | \\ H & H-C-H \\ | \\ H-C-H \\ | \\ X \end{matrix}_n \longrightarrow X-\left(\begin{matrix} H & H \\ | & | \\ -C-C- \\ | & | \\ H & H \end{matrix}\right)_m \begin{matrix} H & H \\ | & | \\ -C-C\bullet \\ | & | \\ H & H-C-H \\ | \\ H-C-H \\ | \\ X \end{matrix}_n \qquad (5.3)$$

where the combination of propagating polymeric chain with a second product having a double bond, has given rise to branching instead of extension of linear chain in a specific direction. Thus polymer chains can be branched and normally such branching may occur only at one out of a few thousand mers. A general type of branching is shown below:

The propagation may start at a number of places in polymerization process, where a free radical or an initiator is available. Sometimes termination by disproportion can lead to branching:

$$X-\left(\begin{matrix} H & H \\ | & | \\ C-C \\ | & | \\ H & H \end{matrix}\right)_m + \begin{matrix} H & H \\ | & | \\ C=C \\ | & | \\ H & H-C-H \\ | \\ H-C-H \\ | \\ X \end{matrix}_n + \left(\begin{matrix} H & H \\ | & | \\ C-C \\ | & | \\ H & H \end{matrix}\right)_p-X \longrightarrow X-\left(\begin{matrix} H & H \\ | & | \\ C-C \\ | & | \\ H & H \end{matrix}\right)_m \begin{matrix} H & H \\ | & | \\ C-C \\ | & | \\ H & H-C-H \\ | \\ H-C-H \\ | \\ X \end{matrix}_n \left(\begin{matrix} H & H \\ | & | \\ C-C \\ | & | \\ H & H \end{matrix}\right)_p-X \qquad (5.4)$$

Cross-linked or three-dimensional polymers

Cross-linked or three-dimensional types of polymers form primary bonding in three dimensions. Sometimes linear chain molecules may be bonded to each other through cross-links (via covalent bonds). This has been discussed in detail in Sec. 5.11.

5.3 SYNTHESIS OF POLYMERS

Two most common types of polymerization reactions produce linear polymers synthetically:

i) Addition polymerization

Monomers with multiple bonds polymerize through addition polymerization process. This consists of three main steps -- activation or initiation, propagation and termination.

Activation or initiation

During activation process, the initiator monomer forms free radicals. This process requires application of energy, which may be in the form of thermal energy, ionizing irradiations (α, β, γ rays), light energy or through chemical reactions. For example, the initiators hydrogen peroxide, benzoyl peroxide readily form free radicals when heated,

$$C_6H_5COO\text{-}O\text{-}COC_6H_5 \rightarrow C_6H_5\cdot + C_6H_5COO\cdot + CO_2 \tag{5.5}$$

$$HO\text{-}OH \rightarrow 2HO\cdot \tag{5.6}$$

Sometimes ions also act as initiators e.g. HCl is used as an ionic initiator for isobutene polymerization. Both H^+ and Cl^- serve as initiators and terminators. The initiation reaction can be given by

$$H^+ + \begin{array}{c} H \quad CH_3 \\ | \quad\; | \\ C=C \\ | \quad\; | \\ H \quad CH_3 \end{array} \longrightarrow \begin{array}{c} H \quad CH_3 \\ | \quad\; | \\ H-C-C^+ \\ | \quad\; | \\ H \quad CH_3 \end{array} \tag{5.7}$$

Initiation can be activated by decomposition of an organic peroxide e.g.$(CH_3)_3CO\text{-}OC(CH_3)_3$. Initiation involves formation of free radicals. Interaction of free radicals of $\cdot(CH_3)_3CO\text{-}OC(CH_3)_3$ with ethylene molecules can be expressed as

$$(CH_3)_3CO\text{-}OC(CH_3)_3 \rightarrow (CH_3)_3CO\cdot + \cdot OC(CH_3)_3 \tag{5.8}$$

$$(CH_3)_3CO\cdot + \begin{array}{c} H \quad H \\ | \quad | \\ C=C \\ | \quad | \\ H \quad H \end{array} \rightarrow (CH_3)_3CO- \begin{array}{c} H \quad H \\ | \quad | \\ C-C\cdot \\ | \quad | \\ H \quad H \end{array} \tag{5.9}$$

Propagation

Activated monomers combine with other monomers and the propagation process gives rise to the formation of chain from each activated monomer as follows:

$$
\begin{array}{c}
\text{H}\quad\text{CH}_3 \quad \text{H}\quad\text{CH}_3 \qquad\qquad \text{H CH}_3\text{H CH}_3 \\
\text{H}-\text{C}-\overset{+}{\text{C}} \;+\; \text{C}=\text{C} \;\longrightarrow\; \text{H}-\text{C}-\text{C}-\text{C}-\overset{+}{\text{C}} \\
\text{H}\quad\text{CH}_3 \quad \text{H}\quad\text{CH}_3 \qquad\qquad \text{H CH}_3\text{H CH}_3
\end{array}
\tag{5.10}
$$

Similarly,
$$
\begin{array}{c}
\text{H}\quad\text{CH}_3 \qquad\qquad \text{H}\quad\text{CH}_3 \\
\text{Cl}^- + \text{C}=\text{C} \;\longrightarrow\; \text{Cl}-\text{C}-\text{C}^- \\
\text{H}\quad\text{CH}_3 \qquad\qquad \text{H}\quad\text{CH}_3
\end{array}
\tag{5.11}
$$

$$
\begin{array}{c}
\text{H}\quad\text{CH}_3 \qquad \text{H}\quad\text{CH}_3 \qquad\quad \text{H CH}_3\text{H CH}_3 \\
\text{Cl}-\text{C}-\text{C}^- \;+\; \text{C}=\text{C} \;\longrightarrow\; \text{Cl}-\text{C}-\text{C}-\text{C}-\text{C}^- \\
\text{H}\quad\text{CH}_3 \qquad \text{H}\quad\text{CH}_3 \qquad\quad \text{H CH}_3\text{H CH}_3
\end{array}
\tag{5.12}
$$

More generally, we obtain two types of polymer chains as,

$$\text{Cl} - (\text{mer})_m^-$$

and
$$\text{H} - (\text{mer})_n^+$$

This is also called ionic polymerization, where active centres are ions that initiate chains. This may proceed in the presence of catalysts and is then called catalytic polymerization. In case of free radical initiators, the chains can be represented as $X-(\text{mer})_m$, where '.' indicates the reaction center i.e. unpaired electron which can easily form bond with another electron.

Termination

When the supply of monomers decreases drastically, then the termination process starts. Termination mechanism can be either due to coupling of two polymer chains or due to disproportionation. Examples are shown below:

Termination by coupling

$$
\begin{array}{c}
\left(\begin{array}{cc}\text{H} & \text{H} \\ \text{C} & \text{C} \\ \text{H} & \text{H}\end{array}\right)_m \quad
\left(\begin{array}{cc}\text{H} & \text{H} \\ \text{C} & \text{C} \\ \text{H} & \text{H}\end{array}\right)_n \quad
\left(\begin{array}{cc}\text{H} & \text{H} \\ \text{C} & \text{C} \\ \text{H} & \text{H}\end{array}\right)_{m+n}
\end{array}
$$

$$X - \text{.} + \text{.} - X \longrightarrow X - X \tag{5.13}$$

In case ionic initiators are used, oppositely charged polymeric chains are generated which combine due to coulomb attraction and thus, termination occurs.

$$H-\left(\begin{matrix} H & H \\ | & | \\ C-C- \\ | & | \\ H & H \end{matrix}\right)_m^+ + \ ^-\left(\begin{matrix} H & H \\ | & | \\ C-C- \\ | & | \\ H & H \end{matrix}\right)_n Cl \longrightarrow H-\left(\begin{matrix} H & H \\ | & | \\ C-C- \\ | & | \\ H & H \end{matrix}\right)_{m+n} Cl \qquad (5.14)$$

Termination by disproportionation

$$X-\left(\begin{matrix} H & H \\ | & | \\ C-C- \\ | & | \\ H & H \end{matrix}\right)_m \cdot + \cdot \left(\begin{matrix} H & H \\ | & | \\ C-C \\ | & | \\ H & H \end{matrix}\right)_n X \longrightarrow X-\left(\begin{matrix} H & H \\ | & | \\ C-C \\ | & | \\ H & H \end{matrix}\right)_m H + C=C-\left(\begin{matrix} H \\ | \\ C \\ | \\ H \end{matrix}\right)\left(\begin{matrix} H & H \\ | & | \\ C-C \\ | & | \\ H & H \end{matrix}\right)_{n-1} X \qquad (5.15)$$

To summarize, we can say that the molecules with multiple bonds take part in chain polymerization process in which the micromolecules form at least two new bonds but no specific group for polymer linkage. Polymer molecule grows rapidly till termination occurs. Sometimes in chain polymerization more than one type of molecule may be involved which is known as copolymerization. This is dealt in detail in Sec. 5.9.

Example 5.1

0.2 w/o of H_2O_2 was added to ethylene prior to polymerization. What would be the average number of mers in the chain if all the H_2O_2 molecules were used as terminals for the molecules?

Solution:

Molecular weight of H_2O_2 is 34 and that of single mer of ethylene is 28. Let n be the average number of mers present in polyethylene molecule, then considering one mole of H_2O_2 is added before polymerization,

$$\frac{34}{n \times 28} = \frac{0.2}{100}$$

Hence the number of moles of mers n = 607. The number of polymeric chains in the sample equals Avogadro number since each polyethyene chain has two −OH radicals, one at each end i.e. one H_2O_2 molecule is associated with each polymeric chain. Hence the degree of polymerization is 607.

ii) Condensation (stepwise) polymerization

Different micromolecules may interact and form special group for linkage during polymerization process. This type of polymerization is called condensation or stepwise polymerization. In stepwise polymerization the following points are to be remembered:

- Usually two different molecules are involved.
- Reactions advance in steps and no termination occurs till the reacting molecules are exhausted.
- Small molecules are normally produced as byproducts.
- The reactant molecules are mostly bifunctional so that each one reacts to extend the polymeric chain in two directions.

Consider the reaction of ethylene glycol $HO-CH_2-CH_2-OH$ reacting with adipic acid to form polyester

$$HO-(CH_2)_2-OH + HO-\underset{O}{\overset{O}{C}}-(CH_2)_4-\underset{O}{\overset{O}{C}}-OH \longrightarrow$$

$$HO-(CH_2)_2-O-\underset{O}{\overset{O}{C}}-(CH_2)_4-\underset{O}{\overset{O}{C}}-OH + H_2O \qquad (5.16)$$

and

$$HO-(CH_2)_2-O-\underset{O}{\overset{O}{C}}-(CH_2)_4-\underset{O}{\overset{O}{C}}-OH + H_2O + HO-(CH_2)_2-OH \longrightarrow$$

$$HO-(CH_2)_2-O-\underset{O}{\overset{O}{C}}-(CH_2)_4-\underset{O}{\overset{O}{C}}-O-(CH_2)_2-OH + 2H_2O \qquad (5.17)$$

$$HO-(CH_2)_2-O-\underset{O}{\overset{O}{C}}-(CH_2)_4-\underset{O}{\overset{O}{C}}-O-(CH_2)_2-OH + 2H_2O + HO-\underset{O}{\overset{O}{C}}-(CH_2)_4-\underset{O}{\overset{O}{C}}-OH$$

$$\longrightarrow HO-(CH_2)_2-O-\underset{O}{\overset{O}{C}}-(CH_2)_4-\underset{O}{\overset{O}{C}}-O-(CH_2)_2-O-\underset{O}{\overset{O}{C}}-(CH_2)_4-\underset{O}{\overset{O}{C}}-OH + 3H_2O$$

$$(5.18)$$

Thus, if m molecules of ethylene glycol interact with equal number of adipic acid molecules, we obtain polyester such that all the molecules of reactants are fully exhausted then the number of water molecules obtained as byproduct is 2m − 1,

$$m(HO-(CH_2)_2-OH) + m(HO-\underset{O}{\overset{O}{C}}-(CH_2)_4-\underset{O}{\overset{O}{C}}-OH) \longrightarrow$$

$$H\left[O-(CH_2)_2-O-\underset{O}{\overset{O}{C}}-(CH_2)_4-\underset{O}{\overset{O}{C}}\right]_m OH + (2m-1)H_2O \qquad (5.19)$$

If we write this reaction in terms m moles of adipic acid interacting with m moles of ethylene glycol, a single chain of polyester is formed, which consists of 'm' moles of mers shown within bracket. The byproduct in this reaction is [(2m) moles–1 molecule] $\approx$ 2m moles of water. This reaction proceeds in both the directions to form linear molecule of polyester. In these reactions the –OH group of acid reacts with H^+ ion of glycol to form the byproduct H_2O and simultaneously forming polymer linkage, which is a polar group to extend the polyester chain as given by

$$-O-\underset{\underset{O}{\|}}{C}-$$

This type of reaction is also called polycondensation. The polar group linkages are responsible for secondary bonding through hydrogen bridges from adjacent polyester molecules. Polycondensation or stepwise reaction can also occur when molecules of similar type interact e.g. polycondensation of dihydric alcohols results in linear polyethers as given by

$$m(HO-R-OH)+n(HO-R'-OH)\rightarrow ..-R-O-R'-O .. + [(m+n)\text{moles} -1 \text{ molecule}]\, H_2O$$
$$(5.20)$$

Polycondensation reactions occur in any number of pairs, which may react again to form polymeric chain with only restriction of interaction between specific pairs. Thus step reactions can proceed without termination. Examples of other polycondensation are given below:

Dacron, terylene or mylar film are produced during polycondensation reaction of ethylene glycol and dimethyl terephthalate. As the step reaction proceeds, longer chain of macromolecule and more number of molecules of methyl alcohol are produced. If m moles of dimethyl terephthalate interact with m moles of ethylene glycol, the byproducts are [(2m) moles–1 molecule] i.e. approximately 2m moles of methyl alcohol.

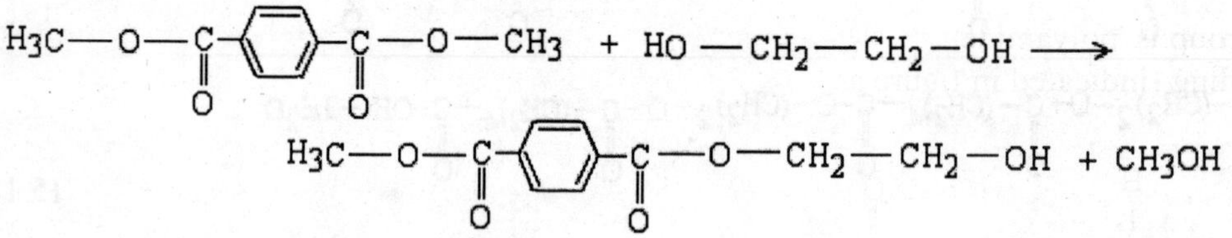

$$H_3C-O-\underset{\underset{O}{\|}}{C}-\bigcirc-\underset{\underset{O}{\|}}{C}-O-CH_3 \; + \; HO-CH_2-CH_2-OH \longrightarrow$$

$$H_3C-O-\underset{\underset{O}{\|}}{C}-\bigcirc-\underset{\underset{O}{\|}}{C}-O-CH_2-CH_2-OH \; + \; CH_3OH \qquad (5.21)$$

$$m\left(H_3C-O-\underset{\underset{O}{\|}}{C}-\bigcirc-\underset{\underset{O}{\|}}{C}-O-CH_3\right) + m(HO-CH_2-CH_2-OH) \rightarrow$$

$$H_3C-O-\left(-\underset{\underset{O}{\|}}{C}-\bigcirc-\underset{\underset{O}{\|}}{C}-O-CH_2-CH_2-O-\right)_m H \; + (2m-1)CH_3OH \quad (5.22)$$

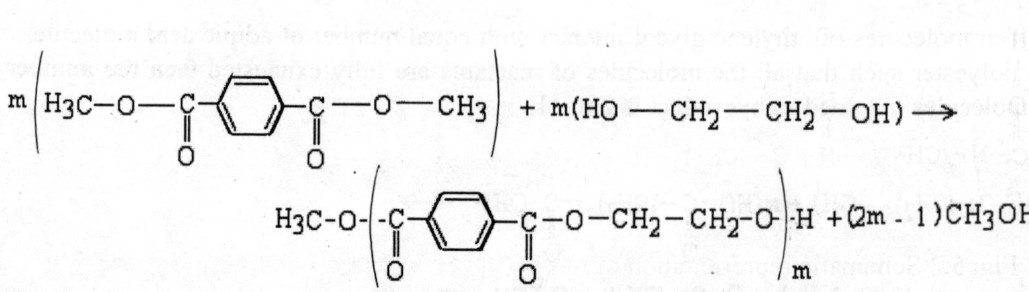

The dacron macromolecule thus produced has degree of polymerization equal to m times Avogadro number if all the reactant molecules produce only one polymeric chain of dacron. The condensation reaction of dicarboxylic aromatic acids with aromatic diamines results in heat resistant polymers.

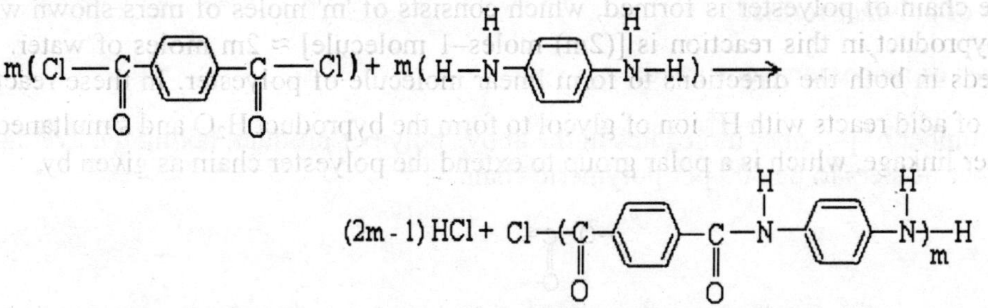

$$m(Cl-\underset{O}{\underset{\|}{C}}-\text{⟨benzene⟩}-\underset{O}{\underset{\|}{C}}-Cl)+m(H-\underset{H}{\overset{H}{N}}-\text{⟨benzene⟩}-\underset{H}{\overset{H}{N}}-H)\longrightarrow$$

$$(2m-1)\,HCl+Cl-\left(\underset{O}{\underset{\|}{C}}-\text{⟨benzene⟩}-\underset{O}{\underset{\|}{C}}-\underset{H}{\overset{H}{N}}-\text{⟨benzene⟩}-\underset{H}{\overset{H}{N}}\right)_{m}-H \qquad (5.23)$$

These polymers are used to form fibre e.g. isophthalic acid chloride is polycondensed with m-phenylediamine to produce polyphenylone. The polar group in this macromolecule is polyamide and the byproduct is HCl. The condensation reaction of adipic acid $HOOC-(CH_2)_4-COOH$ and hexamethylne diamine, $H_2N-(CH_2)_6-NH_2$ produces nylon-66 and is given by

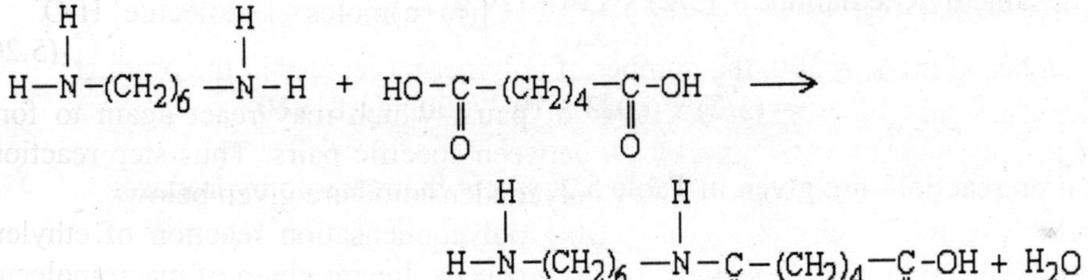

$$H-\underset{H}{\overset{H}{N}}-(CH_2)_6-\underset{H}{\overset{H}{N}}-H \; + \; HO-\underset{O}{\underset{\|}{C}}-(CH_2)_4-\underset{O}{\underset{\|}{C}}-OH \longrightarrow$$

$$\qquad (5.24)$$

$$H-\underset{H}{\overset{H}{N}}-(CH_2)_6-\underset{H}{\overset{H}{N}}-\underset{O}{\underset{\|}{C}}-(CH_2)_4-\underset{O}{\underset{\|}{C}}-OH \; + \; H_2O$$

where polar group is 'polyamide' $-\underset{O}{\overset{H}{\underset{\|}{\overset{|}{N}}}}-C-$. In nylon, the molecular chains are held together by hydrogen bonding (indicated in figure as '|') to form crystal as shown in Fig. 5.2 below.

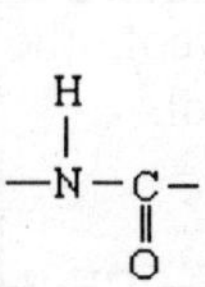

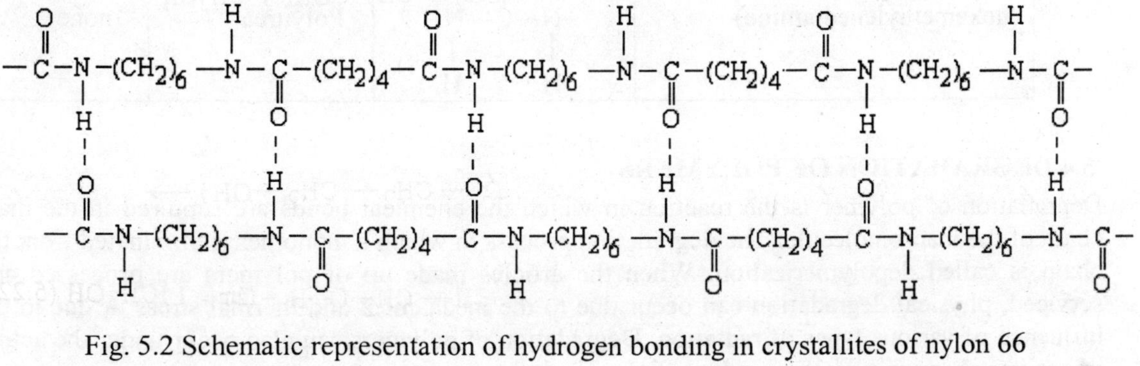

Fig. 5.2 Schematic representation of hydrogen bonding in crystallites of nylon 66

Example 5.2

54g of water is formed during the preparation of nylon-66. Assuming all the functional groups are consumed, determine the following:

(i) Mass of polymer formed.

(ii) Masses of monomers consumed.

(iii) The number of polymer molecules in the above polymer product, taking the average number of mers to be 500 per polymeric chain.

Solution:

The condensation reaction for the formation of nylon-66 is given in Eq. (5.24).
The amount of water formed = 54 g. Number of moles of water = (54/18) = 3
Number of moles of mers present in the polymeric product = 3/2 moles

(i) Amount of polymer = (3/2) x 226 = 339 g , where gmole of mer = 226 g

(ii) Amount of adipic acid = (3/2) x 146 = 219 g
Amount of hexamethylenediamine = (3/2) x 116 = 174 g

(iii) If average number of mers = 500, the number of polymer molecules in the product
$$= (3/2) \times 6.023 \times 10^{23} / 500 = 1.81 \times 10^{21}$$

Other condensation reactions are given in Table 5.2, where there are no byproducts.

Table 5.2

Reactions	Polar groups	Final product	Byproduct
O=C=N-(CH$_2$)$_4$-N=C=O (di-isocyanate) + HO-(CH$_2$)$_2$-OH	$\begin{matrix} O \\ \| \\ -O-C-N- \\ \| \\ H \end{matrix}$	Polyurethane	none
O=C=N-(CH$_2$)$_4$-N=C=O +H$_2$N-(CH$_2$)$_6$-NH$_2$ (hexamethylenediamine)	$\begin{matrix} O \\ \| \\ -N-C-N- \\ \| \quad\quad \| \\ H \quad\quad H \end{matrix}$	Polyurea	none

5.4 DEGRADATION OF POLYMERS

Degradation of polymer is the reaction in which the chemical bonds are ruptured in the main chain of the macromolecule. The degradation process in which a monomer is eliminated from the chain is called depolymerization. When the articles made up of polymers are processed and serviced, physical degradation can occur due to the mechanical and thermal stress or due to the influence of various types of radiation. Degradation of polymers can also occur under the action of various chemicals such as acids, amines, alcohols, oxygen, water etc.

5.5 MOLECULAR WEIGHT

Molecular weight of a polymer is normally quite high due to presence of thousands of monomers in a single macromolecule. In a sample there are macromolecules of different sizes i.e. the number of monomers in different molecules are different. Thus to estimate the molecular weight of a sample we have to consider only average molecular weight M. The average molecular weight can be defined as "number average" molecular weight, M_n, which is given by

$$\overline{M_n} = \frac{\Sigma_i N_i Mw_i}{\Sigma_i N_i} \qquad (5.25)$$

where N_i is the number of molecules in the ith range having mean molecular weight Mw_i. Similarly we can also define "weight average" molecular weight, which is given by

$$\overline{M_w} = \frac{\Sigma_i W_i Mw_i}{\Sigma_i W_i} \qquad (5.26)$$

where W_i is the weight fraction of the molecules having mean size Mw_i.

Example 5.3

Consider a sample of polyvinyl chloride which has the following molecular size distribution.

Size range (kg/mole)	5-10	10-15	15-20	20-25	25-30	30-35	35-40
Number N_i (10^{21} molecules/kg)	9.0	8.6	7.5	6.2	3.5	2.4	1.5

Calculate number average molecular weight.

Solution:

Size range (kg/mole)	Mean Size (kg/mole)	Number N_i (10^{21} molecules/kg)	Product $(N_i)(Mw_i)$ (10^{22} molecules/mole)
5 – 10	7.5	9.0	6.75
10 – 15	12.5	8.6	10.75
15 – 20	17.5	7.5	13.12
20 – 25	22.5	6.2	13.95
25 – 30	27.5	3.5	9.62
30 – 35	32.5	2.4	7.80
35 – 40	37.5	1.5	5.62
Total		38.7	67.61

$\overline{M_n} = 67.61 \times 10^{22} / 38.7 \times 10^{21} = 17.47$ kg/mole

Example 5.4

Find the "weight average molecular weight" and degree of polymerization of a polyethylene $(C_2H_4)_n$ using the following data table.

Molecular Weight (kg/mole)	7	12	17	22	27	32	Total
Weight fraction	0.22	0.15	0.33	0.18	0.10	0.02	1.00

Solution:

$$\Sigma_i \, W_i \, Mw_i = 1.54 + 1.8 + 5.61 + 3.96 + 2.7 + 0.64 = 16.25$$
$$\Sigma_i \, W_i = 1$$

So,　　　$M_w = 16.25$ kg/mole

Average weight of a molecule= $16.25/6.023 \times 10^{23}$ kg
The degree of polymerization D_p is given by

$$D_p = \frac{\overline{M_w}}{M_w \text{ of monomer}} = \frac{16.25/6.023 \times 10^{23}}{28 \times 1.66 \times 10^{-27}} = 580.46$$

This means average size macromolecule in the given sample has 580 monomers.

5.6 CONFIGURATIONS AND LENGTH OF POLYMERS

A long chain polymer is like an earthworm, which can remain in straight, coiled or kinked configurations. The ratio of true length and diameter of a polymer chain is quite high. A polymer chain assumes various configurations due to thermal motion such that the end-to-end distance of the chain is usually much less than the true length. If polyvinyl chloride chain is straightened, except for 109.5° bond angle, then a molecule having D_p equal to 2000 will have true length, Fig. 5.3,

$$L = 2000 \times 2 \times 1.5 \text{ Å} \times \sin(109.5°/2) = 4900 \text{ Å}$$

since each carbon-carbon bond is 1.5Å.

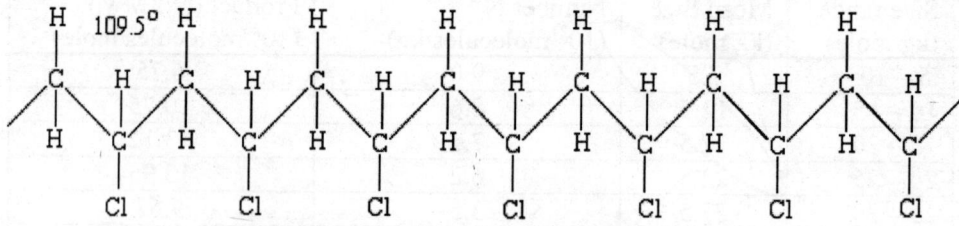

Fig. 5.3 Linear chain of polyvinyl chloride

Using the above method one can easily calculate the length of a molecule containing one or more monomers. The length of butane mer is 3.7 Å for trans-isomers whereas it is 2.5 Å for cis-isomers. The 109.5° bond angle across each carbon atom is free to rotate unless restrained, because all molecules are moving due to thermal energy.

A polymer chain can be considered consisting of m independent segments, Fig. 5.4. If there are S mer units per segment, then

$$m = D_p / S \qquad (5.27)$$

The vector length of the polymer can be expressed as

$$\mathbf{L} = \mathbf{l}_1 + \mathbf{l}_2 + \mathbf{l}_3 + \dots + \mathbf{l}_m \qquad (5.28)$$

where l_1, l_2, l_3, etc. are the lengths of various segments. The mean square of end-to-end length, L, is given by

$$\overline{L}^2 = ml^2 + 2l^2 \Sigma \cos \theta \qquad (5.29)$$

assuming each segment length to be l and is equal to C-C bond length. We have also assumed that the segments of an unrestrained molecule have equal probabilities for all orientations θ in space and hence the last term of the above equation becomes zero when average value is considered. Hence we get

$$\overline{L}^2 = ml^2 \qquad (5.30)$$

Hence the root mean square length is

$$\overline{L} = l \sqrt{m} \qquad (5.31)$$

Since there are two C-C bonds per mer of polyvinyl chain, hence $m = 2D_p$, and

$$\overline{L} = l \sqrt{(2D_p)} \qquad (5.32)$$

Hence the root mean square length of a polyvinyl chloride chain of D_p equal to 2000 is given as 94.87 Å. The noncrystalline polymeric molecules usually remain coiled and kinked; hence the mean length of such a molecule is always much less than the extended length since such a molecule can be extended considerably when tensile stress is applied. When the stress is removed they kink and coil again.

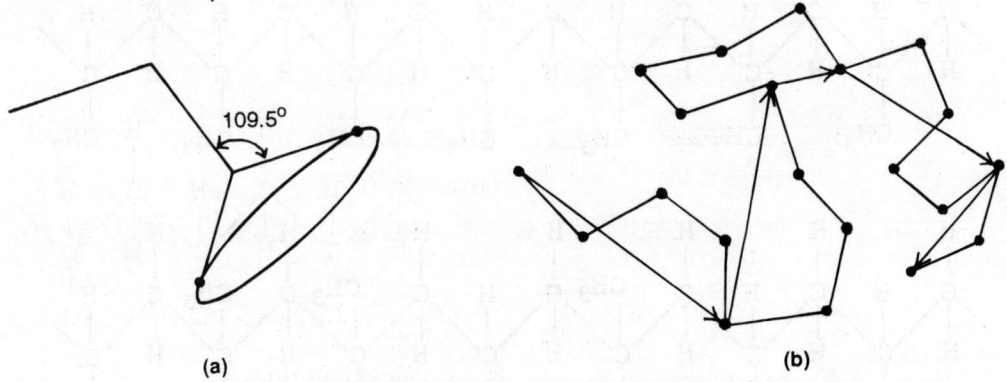

(a) (b)

Fig. 5.4 Segments of polymeric chain

5.7 STEREOSPECIFIC POLYMERIZATION

Stereospecific polymerization is a form of addition polymerization in which spatial orientation of a monomer can be decided in growing polymeric chains with the help of proper choice of catalysts. This type of polymerization becomes important whenever there is a side group or a large atom present e.g.

The various types of arrangements of mers in the polymeric chain can be classified as follows:

- The adjacent mers may have same orientation in polymeric chain. This type of configuration is referred to as head to head (i.e. HHHH...) or tail to tail (i.e. TTTT...) configurations or any combination of them (e.g. THTHHTT...). As an example, such chain irregularities are there in polyvinyl chloride,

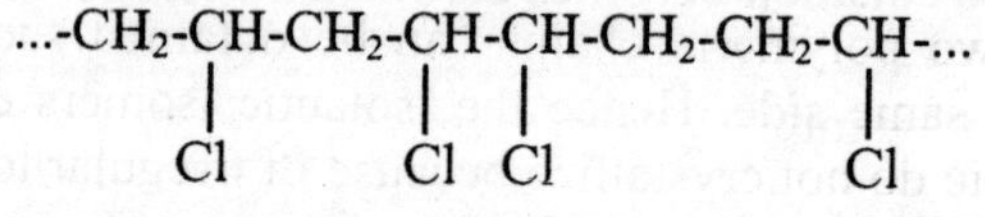

$$...\text{-CH}_2\text{-CH-CH}_2\text{-CH-CH-CH}_2\text{-CH}_2\text{-CH-}...$$
$$\qquad\quad |\qquad\quad |\ \ |\qquad\qquad\quad |$$
$$\qquad\quad \text{Cl}\qquad\ \text{Cl Cl}\qquad\qquad\ \text{Cl}$$

Fig. 5.5 Chain irregularities in polyvinyl chloride

- Each successive pair of mers may have same pattern in the polymeric chain. This is designated as head to tail addition polymerization or HTHT... In many polymerization processes monomers combine in this way.

The possibility of crystallization of these polymers becomes feeble since van der Waals forces will not come into play due to lack of close-packing. The above arrangements can exist in various configurations. To illustrate these configurations, let us consider the example of polyisoprene, which has three stereoisomerism, Fig. 5.6, whose configurations are given below:

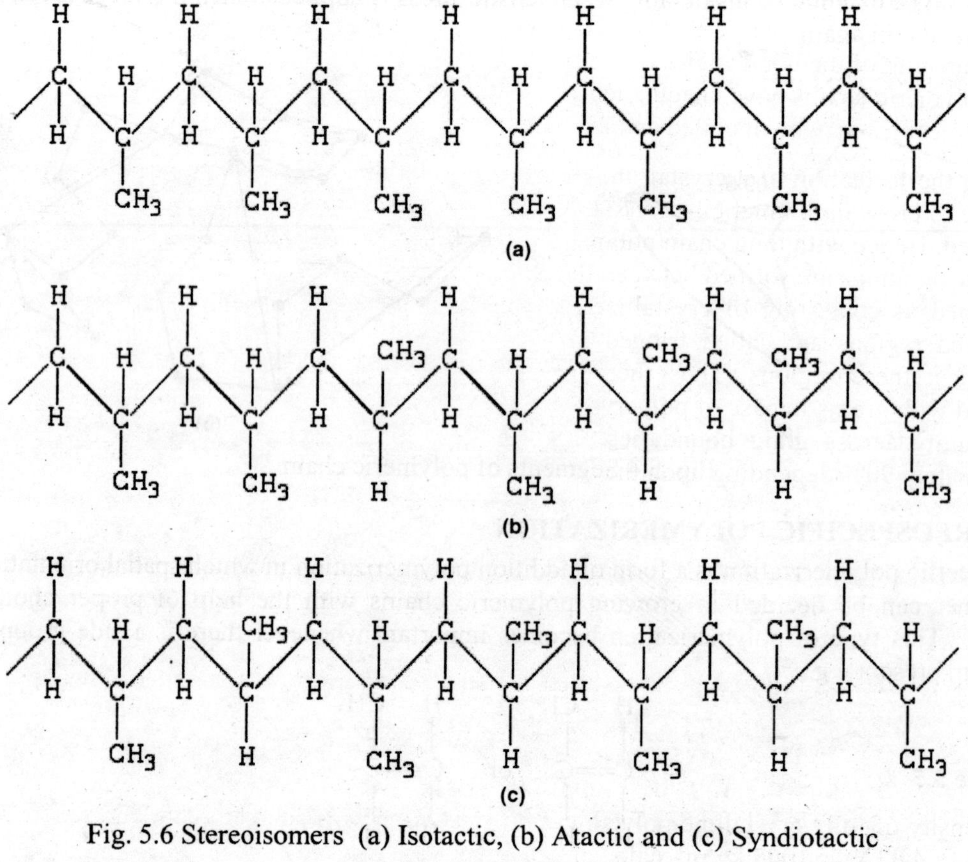

Fig. 5.6 Stereoisomers (a) Isotactic, (b) Atactic and (c) Syndiotactic

a. Isotactic isomer has side groups CH_3 always located on one side of the zigzag polyprene chain i.e. HHH... or TTT... arrangement.

b. Atactic isomer has CH_3 groups randomly located on the two sides of the chain e.g. HTHHT..

c. Syndiotactic isomer has CH_3 groups on alternate sides of the chain i.e. HTHT...

Since the isotactic polymeric chain has side groups on one side of the zigzag chain, the matching of bond angle of two polymeric chains can be obtained such that their side groups should not be located on the same side. Hence the isotactic isomers crystallize comparatively easily. Polymers that are atactic do not crystallize because of irregular locations of $.CH_3$ group in the chain of polyprene. Atactic polymers form a waxy material at room temperature while isotactic polymers form hard plastic. Syndiotactic polymers have more chance of forming crystal as compared to atactic. The two-$.CH_3$ groups tend to overlap in the planar configuration. However, the repulsion of $.CH_3$ groups prevents them to occupy the same space. Thus the chain gets twisted out of the planar zigzag form after getting rotated about C-C bonds. This is called hindered rotation. Mechanical properties are different for different forms of stereoisomers. Plastics become more rigid if hindered rotation is present in the sample. The helical coils pack well and readily crystallize, but steric hindrance prevents free rotation of the chain.

5.8 CRYSTALLIZATION OF POLYMERS

Complete crystallization of long chain polymers cannot be achieved easily due to the following reasons:

    i) Matching of the bond angles of the adjacent molecules may not be complete.

    ii) Polymeric chains have unequal lengths.

    iii) Molecules may be in coiled and kinked conditions.

During the formation of a crystal, the molecules move to suitable lattice positions, for which they have to cross high kinetic barriers. Owing to its large length, movement of a polymer chain is hindered. Hence with long chain polymers (e.g.$(C_2H_4)_n$) crystallization can only be local where secondary bondings are formed between the molecules. Thus, a partially crystallized polymer can be obtained as consisting of crystallized regions interconnected by amorphous region. These crystallized regions are called fringed micelles, Fig. 5.7, since threads of molecules mostly interconnect these regions. Thus there is some continuity between micelles, which can be compared with grains in case of polycrystalline substances, the only difference being the absence of continuity across grain boundaries. The crystallization in polymers may vary from 0 to approximately 90% depending upon the factors described above.

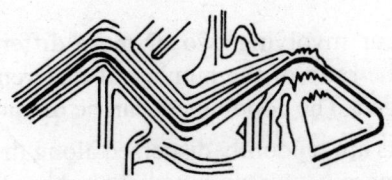

Fig. 5.7 Fringed micelles

Example 5.5

If the density of fully crystalline polyethylene (unit cell is orthorhombic with dimensions 2.53Å x 7.40 Å x 4.93Å) is 1005 kg/m³, calculate the number of mers per unit cell.

Solution:

$$\text{Density} = \frac{\text{Mass/ unit cell}}{\text{Vol./ unit cell}}$$

i.e. $\quad 1005 = \dfrac{n \times \text{mass of mer}}{92.30 \times 10^{-30}} = \dfrac{n \times 28 \times 1.66 \times 10^{-27}}{92.30 \times 10^{-30}}$

$$n = 2 \text{ mers per unit cell}$$

Steric hindrance

The presence of large side radicals or atoms may put spatial interference to the matching of bond angles of the adjacent polymeric chains since these adjacent molecules will not be able to come nearer to form bonding. As the radicals provide spatial interference it is called steric hindrance. In the absence of crystallization the mechanical and thermal properties are changed to a large extent. Such polymeric substances melt at lower temperature as compared to crystalline polymers. They are usually soft materials. The irregularities arising due to side groups can be understood from the reaction given by Eq.(5.33) which causes branching of polymers and hence cause steric hindrance but such reactions are not always energetically favoured.

The polymer like polytetrafluoroethylene (PTFE) has large fluorine atoms in place of the four hydrogen atoms of polyethylene. Large fluorine atoms cause steric hindrance and prevent zigzag arrangements as we observe in case of polyethylene. PTFE chain takes a helical configuration in which fluorine atoms cannot come one above another; these atoms are arranged along helical curve. As a result a PTFE chain is much stiffer than polyethylene. It is crystalline below 30°C. It is soft and has a melting point 330°C.

$$
\begin{array}{c}
\quad\ \text{H}\ \ \text{R}\quad\ \ \text{H}\ \ \text{R}\qquad\qquad \text{H}\ \ \text{R}\ \ \text{R}\\
\quad\ |\ \ \ |\quad\ \ \ |\ \ \ |\qquad\qquad |\ \ \ |\ \ \ |\\
\text{H}^{+}\!-\!\text{C}\!-\!\text{C.} + \text{C}\!=\!\text{C} \rightarrow \text{H}\!-\!\text{C}\!-\!\text{C}\!-\!\text{C.}\\
\quad\ |\ \ \ |\quad\ \ \ |\ \ \ |\qquad\qquad |\ \ \ |\ \ |\\
\quad\ \text{H}\ \ \text{H}\quad\ \ \text{H}\ \ \text{H}\qquad\qquad \text{H}\ \ \text{H}\ \ |\\
\qquad\qquad\qquad\qquad\qquad\qquad\qquad\ \ \text{H}\!-\!\text{C}\!-\!\text{H}\\
\qquad\qquad\qquad\qquad\qquad\qquad\qquad\qquad |\\
\qquad\qquad\qquad\qquad\qquad\qquad\qquad\qquad \text{H}
\end{array}
\qquad (5.33)
$$

5.9 COPOLYMERIZATION

The polymerization can also occur involving two or more different monomers. This is called copolymerization. By adjusting the ratio of the number of different monomers, the properties of synthesized polymer can be changed. The copolymers can be of various types:

(a) When two different mer units are randomly dispersed along the polymeric chain, the polymer is called a random copolymer, e.g. styrene-butadiene rubber that is used in automobile tyre, vinyle acetate-vinyl chloride copolymer, and nitrile-butadiene. The gasoline hoses are made of copolymer of nitrile and butadiene. A triple polymer of acrylonitrile, butadiene and styrene is also an example of random copolymer. The copolymerization of polyvinyl acetate and polystyrene produces textile fibre, Fig. 5.9.

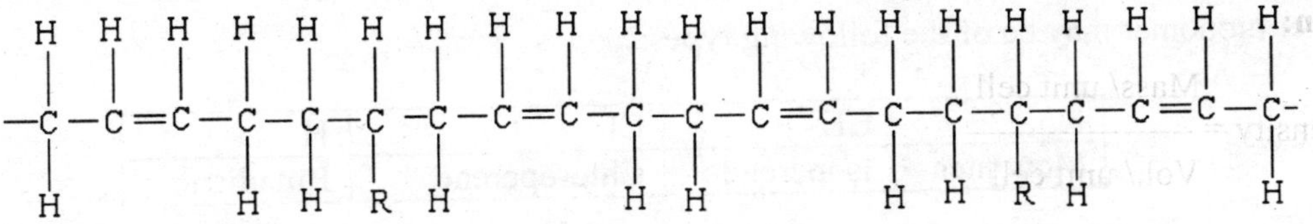

Fig. 5.8 Copolymer of styrene-butadiene where 'R' is benzene ring (C_6H_5)

(b)　In an alternating copolymer, the two mer units occupy alternate chain positions.

(c)　When a block of identical mers occur in polymeric chain, the copolymer called block copolymer.

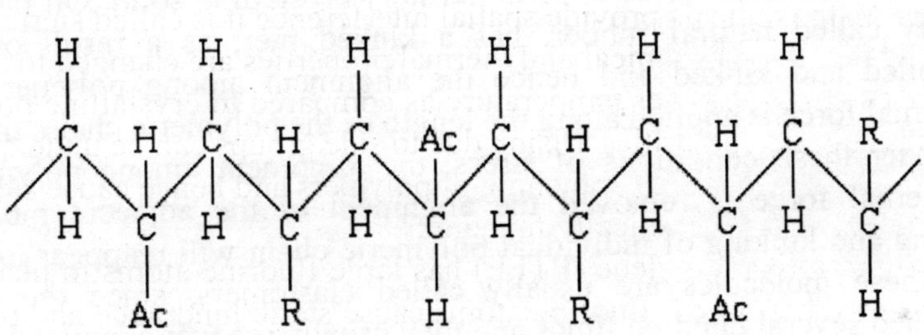

Fig. 5.9 Copolymer of polyvinyl acetate and polystyrene

where Ac is acetate radical i.e.

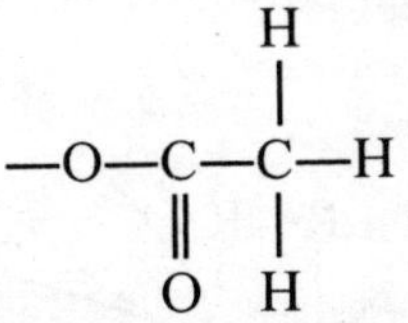

and R is benzene ring (C_6H_5)

The copolymers have lower melting points and lower glass points as compared to its polymer components.

5.10 STEREOISOMERISM IN RUBBER: ELASTOMERS

The unsaturated polymers have double bonds and show stereoisomerism. If tetrafunctional monomer (containing two double bonds) like isoperene, is polymerized, an unsaturated polymer is generated which shows cis- or trans-isomerism. In this reaction, out of the two double bonds one is consumed for chain polymerization and the other double bond is shifted at the center of mer.

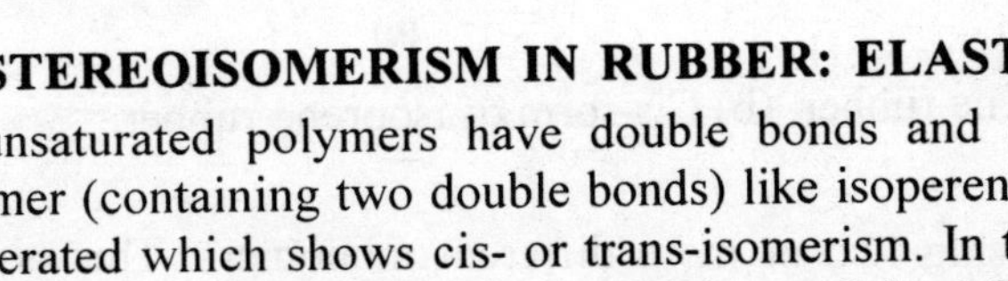

(5.34)

The rubber monomer may be of the following types:

R	CH_3	Cl	H
Monomer	Isoperene	Chloroperene	Butadiene

The bond angle C==C—C is 124.7° and C==C bond cannot be rotated like a single bond. Cis- and trans-isomers are two distinct isomers of rubber. The cis- or trans-configurations, Fig. 5.10 depend on whether the vacant sites are on the same side or on the opposite sides of the chain. In other words in the trans-form of isomers, CH_3 and H attached to double bonded carbon atoms, reside on the opposite sides of chain. Trans-isomers cannot be converted to cis-isomers by a double bond rotation since the double bond is extremely rigid. The trans-isoprene, commonly called gutta purcha has a mer, which has zigzag structure. Hence the polymeric chain of gutta purcha can align with adjacent chain so as to form partially crystalline solid. On the other hand cis-isoprene, commonly called natural rubber, has a kinked mer, as a result of which the polymeric chain is coiled and kinked and hence the alignment among polymeric chains is difficult. When an external force is applied along the length of the polymeric chain, the chains are partially unkinked. Under these conditions of stress, the alignment among polymeric chains results. When the external force is removed the alignment of the adjacent molecules gets disturbed and the coiling and kinking of individual polymeric chain will reappear just as if they are free molecules. These molecules are usually called elastomers, since the cis-isoprene molecules can be stretched several hundred times and they usually recover completely, soon after the removal of stress.

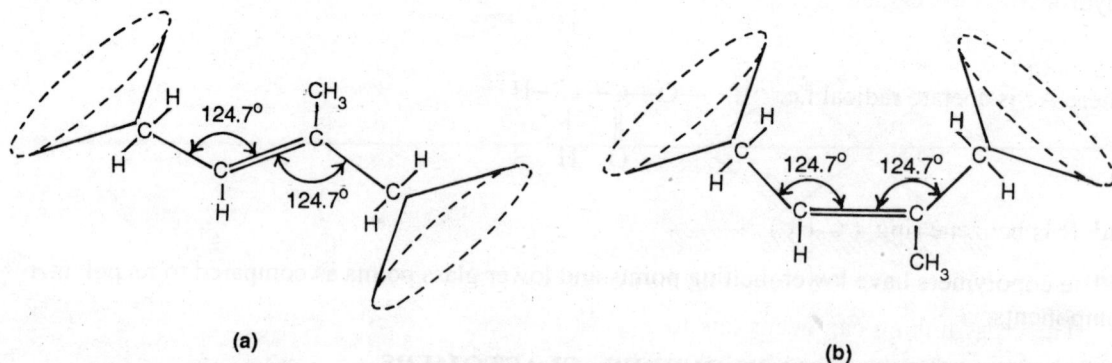

(a)   (b)

Fig. 5.10   (a) Trans-form of isoprene rubber. (b) Cis-form of isoprene rubber.

5.11 CROSS-LINKING

Cross-linking between linear chain polymers take place during the synthesis or processing of polymers. Cross-linking during synthesis is undesirable, as insoluble or non-melting products are produced, which are difficult to separate. The cross-linking reactions are desirable, while producing goods from polymers. These reactions are commonly called vulcanization in the rubber industry. Cross-linking of polymers using ionizing radiations are also desirable since the materials thus produced have increased thermal stability and insolubility in organic solvents. This process is known as radiation cross-linking. In radiation cross-linking, a polymer molecule generates free radicals under the action of ionizing radiation. These free radicals in turn react with molecules or with a macroradical, to form a branch or a cross-link e.g. silicone rubbers are vulcanized by radiation cross-linking, Eq (5.35). Silicone plastics are cross-linked to form hard silicone rubber.

$$
\begin{array}{ccc}
\begin{array}{cc}
CH_3 & CH_3 \\
| & | \\
\cdots Si-O-Si-O\cdots \\
| & | \\
CH_3 & CH_3 \\
\\
CH_3 & CH_3 \\
| & | \\
\cdots Si-O-Si-O\cdots \\
| & | \\
CH_3 & CH_3
\end{array}
&
\xrightarrow{h\nu}
\begin{array}{cc}
CH_3 & CH_3 \\
| & | \\
\cdots Si-O-Si-O\cdots \\
| & | \\
\cdot CH_2 & CH_3 \\
\\
\cdot CH_2 & CH_3 \\
| & | \\
\cdots Si-O-Si-O\cdots \\
| & | \\
CH_3 & CH_3
\end{array}
&
\longrightarrow
\begin{array}{cc}
CH_3 & CH_3 \\
| & | \\
\cdots Si-O-Si-O\cdots \\
| & | \\
CH_2 & CH_3 \\
| & \\
CH_2 & CH_3 \\
| & | \\
\cdots Si-O-Si-O\cdots \\
| & | \\
CH_3 & CH_3
\end{array}
\end{array}
\qquad (5.35)
$$

The silicones are heat resistant and good electrical insulators. When silicone rubber caps are stretched over steel containers and the temperature is reduced to $-30°C$, the caps no longer act as seal. This can be explained because when the rubber is stretched the molecules tend to crystallize. Lowering the temperature reduces the thermal motion of the chains and hence the resistance to the stretching force decreases. The silicone treatment for waterproofing is done on polar substrates like glass, paper etc. to increase the surface electrical resistivity. Ordinarily the polar substrates adsorb water, which decreases the surface resistivity. By treating the surface with silicone, Fig. 5.11, the surface electrical resistivity can be improved. The treated surface is now hydrocarbon in character.

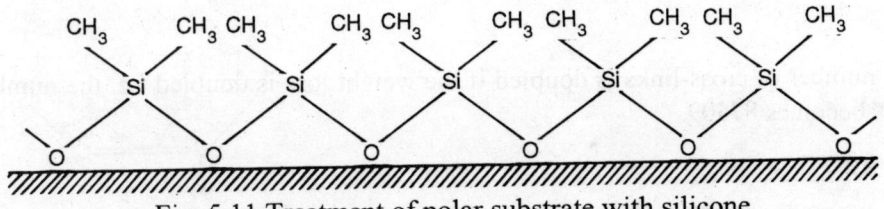

Fig. 5.11 Treatment of polar substrate with silicone

The cross-linking can occur due to the presence of some elements called vulcanizing agents, e.g. S, Se, Te and O_2. In case of polyisoprene (natural rubber), the sulphur bridges are formed between two macromolecules during vulcanization.

$$
\begin{array}{ccc}
\begin{array}{c}
H \quad CH_3 \; H \; H \\
| \quad\; | \quad\, | \; | \\
-C-C=C-C- \\
| \qquad\qquad | \\
H \qquad\qquad H \\
\\
H \quad CH_3 \; H \; H \\
| \quad\; | \quad\, | \; | \\
-C-C=C-C- \\
| \qquad\qquad | \\
H \qquad\qquad H
\end{array}
&
+\ 2S\ \rightarrow
&
\begin{array}{c}
H \quad CH_3 \; H \; H \\
| \quad\; | \quad\, | \; | \\
-C-C=C-C- \\
| \qquad\qquad | \\
H \qquad\qquad H \\
| \qquad\qquad | \\
H \;\; S \qquad S \; H \\
| \qquad\qquad | \\
-C-C=C-C- \\
| \quad\; | \quad\, | \; | \\
H \;\; CH_3 \; H \; H
\end{array}
\end{array}
\qquad (5.36)
$$

In this vulcanization process sulphur bridges are formed at the point of opening of double bonds. If the number of cross-links is small, the final product is soft and flexible. The stiffness of the polymeric material increases with the density of cross-links. When the sulphur content in rubber is as high as 32 weight percent, the hard product is called ebonite.

Example 5.6

During radiation cross-linking the weight of silicone rubber is reduced by 0.7%. If degree of polymerization is 17×10^4 then compute number of (i) cross links, (ii) cross-links if weight loss is doubled.

Solution:

i) The weight of the polymeric chain (before cross-linking) is given by

$$17 \times 10^4 \times 74.08 \text{ amu} = 12.6 \times 10^6 \text{ amu}$$

where weight of a mer is 74.08 amu [CH_3-Si-CH_3]
$$\overset{\|}{O}$$

During radiation cross-linking, silicone rubber loses two hydrogen atoms per cross-linking, Eq(5.35). When the weight of the polymer is reduced by 0.7%, the weight of the hydrogen atoms left the polymer is $12.6 \times 10^6 \times 0.007 = 88200$ amu i.e. $88200/1.0079 \sim 87509$ number of hydrogen atoms have left during radiation cross-linking. So the number of cross-links formed is $87509/2 \sim 43754$.

ii) The number of cross-links is doubled if the weight loss is doubled i.e. the number of cross-links becomes 87509.

Example 5.7

The vulcanization of polybutadiene is different from that of natural rubber where sulphur interacts at the point of opening of the double bonds of the vinyl side groups, Eq (5.37). In the process of vulcanization of polybutadiene, sulphur bridges are formed at the vinyl side groups.

i) If the weight percent of sulphur added is 16, find the fraction of mers cross-linked.

ii) Calculate the exact amount of sulphur required for 100% cross linking of a polybutadiene sample weighing 324 g.

Solution:

When four mers of polybutadiene are cross-linked fully, Eq (5.37), two sulphur atoms are used i.e. one sulphur atom per two mers is required for full cross-linking. Now weight of a mer of butadiene is 54 amu. Thus, sulphur to be added for cross-linking is $(32/108) \times 100 = 29.63\%$ of the polybutadiene sample

$$
\begin{array}{c}
-CH_2-CH-CH_2-CH- \\
\quad\quad | \quad\quad\quad | \\
\quad\quad CH \quad\quad CH \\
\quad\quad \| \quad\quad\quad \| \\
\quad\quad CH_2 \quad\quad CH_2 \\
\\
+2S \longrightarrow \\
\\
-CH_2-CH-CH_2-CH- \\
\quad\quad | \quad\quad\quad | \\
\quad\quad CH \quad\quad CH \\
\quad\quad \| \quad\quad\quad \| \\
\quad\quad CH_2 \quad\quad CH_2
\end{array}
\qquad
\begin{array}{c}
-CH_2-CH-CH_2-CH- \\
\quad\quad | \quad\quad\quad | \\
\quad\quad CH \quad\quad CH \\
\quad\quad | \quad\quad\quad | \\
\quad\quad CH_2 \quad\quad CH_2 \\
\quad\quad | \quad\quad\quad | \\
\quad\quad S \quad\quad\quad S \\
\quad\quad | \quad\quad\quad | \\
\quad\quad CH_2 \quad\quad CH_2 \\
\quad\quad | \quad\quad\quad | \\
\quad\quad CH \quad\quad CH \\
\quad\quad | \quad\quad\quad | \\
-CH_2-CH-CH_2-CH-
\end{array}
\qquad (5.37)
$$

(i) If only 16% of sulphur is added the cross-linking is (16/29.63) x100%= 54%

ii) The weight of polybutadiene sample is 324 g = 324/54 moles = 6 moles of mers. Each mer requires one sulphur atom for cross-linking, so 3 moles of sulphur are required for full cross-linking i.e. 96 g of sulphur is to be added.

Example 5.8

Consider a natural isoprene rubber, which is 50% cross-linked with sulphur. How much more sulphur has to be added to one mole of above rubber such that ebonite is produced. Also calculate the increase in the number of cross- links.

Solution:

When one mole of isoprene rubber is 50% cross-linked, one has to add ½ mole of sulphur i.e. 16 g of sulphur more so that it is fully cross-linked which is called ebonite. Number of cross-link increased = $0.5 \times 6.023 \times 10^{23} = 3.0115 \times 10^{23}$

Example 5.9

A rubber contains 54.0 w% butadiene (Mol. wt. 54), 34.0 w/o isoprene (Mol.wt. 68) and 12.0 w% sulphur (Mol.wt. 32). What fraction of the possible cross-links is used in cross-linking?

Solution:

The number of possible cross-links in the polymer= Number of moles of mers of butadiene and isoprene since each butadiene mer can link with one 'S' atom and each isoprene mer with one 'S' atom. Consider 100g of polymer. Then 54g of butadiene and, 34g of isoprene are present. Amount of sulphur present = 12g.
Number of moles of butadiene = 54/54 = 1
Number of moles of isoprene = (34/68) = 0.5
Number of moles of sulphur = (12/32) = 0.375
Number of moles of sulphur required = 1 + 0.5 = 1.5
Fraction of cross linking = (0.375/1.5) = 0.25

Example 5.10

48 g of sulphur is added to 1.36 kg of polyisoprene. If the degree of polymerization is 1000, find how many chains can be fully cross-linked.

Solution:

48 g of sulphur = 3/2 N_A number of sulphur atoms. For full cross-linking of a pair of polymeric chains ($D_p = 1000$), one requires to add $1000 \times 2 = 2000$ sulphur atoms, Eq (5.37). So the number of fully cross-linked polymeric chains can be

$$2 \times (3/2) \times 6.023 \times 10^{23} / 2000 \approx 9.0345 \times 10^{20}$$

Cross-linking reactions can occur through substitution reaction without opening of double bond e.g. vulcanization of polychloroprene using metallic oxide.

$$(5.38)$$

Each cross-link produces a $ZnCl_2$ molecule.

Example 5.11

As a result of vulcanization of polychloroprene, through ZnO substitution reaction, the weight of zinc chloride formed is 681.9 g.

  i) Compute the number of cross-links if all the zinc chloride generated is due to cross-linking reaction.

  ii) Compute the amount of oxygen in moles that is used up in cross-linking a sample of 531g.

  iii) Also find the weight of the sample after cross-linking.

Solution:

  i) Mol.wt.of $ZnCl_2$ = 136.38

    Amount of $ZnCl_2$ formed = 681.9 g = 681.9/136.38 = 5 moles. From Eq. (5.38), it is evident that as many $ZnCl_2$ molecules are formed, that many cross-links are also formed. Thus, the number of cross-links formed are $5 \times 6.023 \times 10^{23} = 3.0115 \times 10^{24}$

  ii) A polychloroprene sample of 531 g (= 531/88.5 moles =6 moles of mers) requires 3 moles of oxygen which is equal to 48g for full cross-linking

  iii) Weight of the sample after cross-linking

    = 531g – weight of chlorine formed $ZnCl_2$ + weight of the oxygen formed cross-linking

    = $(531 - 6 \times 35.5 + 48)$ g = 366 g.

Cross-linking may result from the recombination of two previously formed polymeric chains e.g. heating of polyacrylonitrile at elevated temperature results in the intermolecular reactions and formation of cycles or network structure. Cycles of polyacrylonitrile are shown in Eq. (5.39).

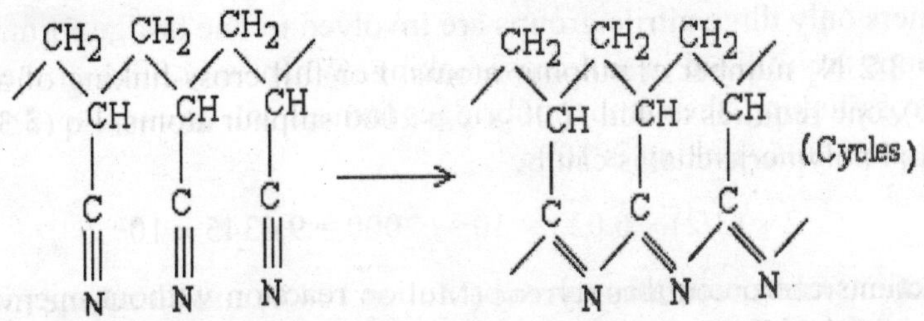

$$(Cycles) \qquad (5.39)$$

Although intramolecular ring formation does not decrease the solubility in polymers, but in general polyacrylonitrile becomes less soluble due to formation of three-dimensional network when the temperature is increased, Fig. 5.12.

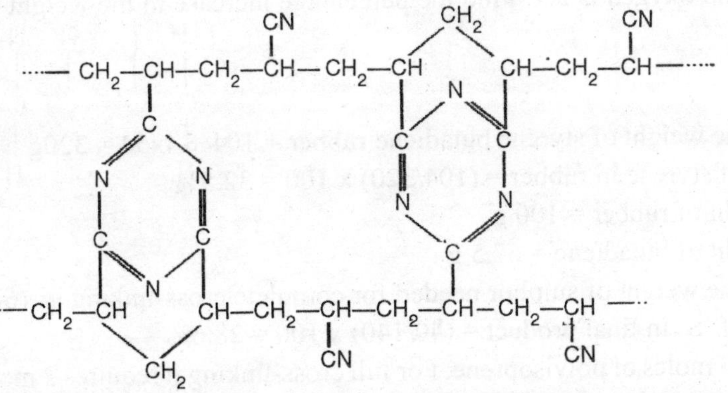

Fig. 5.12 Three-dimensional network of polyacrylonitrile.

Example 5.12

At an elevated temperature the nitrile groups of polyacrylonitrile (with average degree of polymerization 5×10^6) forms a network structure with nitrile bridges.

   i) Compute the number of cross-links.

   ii) Find the number of nitrile groups involved in bridge formation.

   iii) If 30% of nitrile groups are only involved in cross-linking then compute the decrease in the number of cross-bridges.

Solution:

   i) Corresponding to every five mers of each of the polymeric chain there are two nitrile bridges i.e. corresponding to total five mers of two linking polymeric chains of polyacrylonitrile there is one bridge, Eq (5.39). When two polymeric chains, with average degree of polymerization 5×10^6 are fully cross-linked, there will be 2×10^6 nitrile bridges and hence 2×10^6 cross-links.

   ii) The number of nitrile groups involved in one nitrile bridge is 3, hence there are 6×10^6 nitrile groups involved for complete cross-linking.

iii) Out of five mers only three nitrile groups are involved in one bridge formation i.e. only 60% of nitrile groups are utilized in full cross-linking. When only 30% of the nitrile groups are involved in cross-linking, the number of bridges formed will be equal to $2 \times 10^6 \times 0.3/0.60 = 10^6$ So the decrease in cross-linking is 10^6.

Example 5.13

(a) Styrene-butadiene rubber contains styrene (Mol wt.104.0) and butadiene (Mol. wt 54.0) in 1:4 mole ratio.
  i)   Find w/o of styrene in rubber
  ii) If 100 g of rubber is completely cross-linked by sulphur, calculate the weight of sulphur in cross-linked product.
  iii) Find w/o of sulphur in final product of (ii).
(b) A cis-polyisoprene rubber is fully cross-linked with sulphur and oxygen. The atomic ratio of sulphur and oxygen is 2:1. Find the percentage increase in the weight of isoprene (Mol. wt. 68.0).

Solution:

(a) (i) Let the weight of styrene butadiene rubber $= 104 + 4 \times 54 = 320g$
       w/o of styrene in rubber $= (104/320) \times 100 = 32.5\%$
    (ii) Weight of rubber $= 100$ g.
       Weight of butadiene $= 67.5$ g
       and the weight of sulphur needed for complete cross-linking is $(67.5/54) \times 32 = 40g$
    (iii) w/o of 'S' in final product $= (40/140) \times 100 = 28.6\%$
(b) Consider 3 moles of polyisoprene. For full cross-linking it requires 2 moles of sulphur and one mole of oxygen.
    After cross-linking the weight of the polymer will be
    $3 \times 68 + 2 \times 32 + 16 = 284$ g.
    So the percentage increase in weight $= (80/204) \times 100\% = 39.2\%$

Example 5.14

1% by weight of divinyl benzene (Mol.Wt.= 130) is added to 10 g of styrene (Mol. Wt.= 104). Calculate the maximum possible number of cross-links in the polymer formed.

Solution:

The polymer is 10 g. Then, the weight of divinyl benzene is 0.1gm. Then, the number of cross-links is $(0.1/130) \times N_A$, where N_A is Avogadro number. So, the maximum possible number of cross-links is $(N_A/130)/10 = 4.63 \times 10^{20}/g$

During the synthesis of polystyrene if divinyl benzene is added to styrene, divinyl benzene molecule can polymerize as a bridge or a cross-link between two polystyrene chains Divinyl benzene is a tetrafunctional monomer. Each of these molecules joins four molecules of styrene. Thus the number of cross-links will be equal to the number of bridges formed by divinyl benzene molecules.

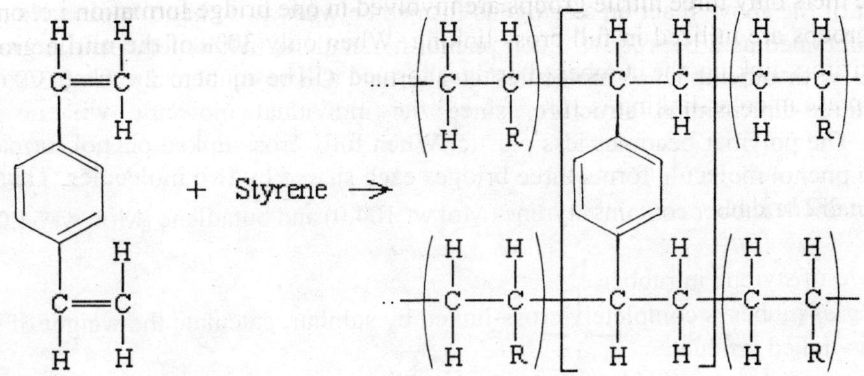

$$+ \text{Styrene} \rightarrow \qquad (5.40)$$

where R is benzene ring C_6H_5.

Example 5.15

Consider a cross-linked polymer having cross-links equal to 5×10^{20}/kg of product containing 50 kg of polystyrene. Find the amount of divinyl benzene mixed up with the polystyrene so that the above product is obtained.

Solution:

The number of cross-links $= 5\times10^{20}$ / kg
The number of cross-links in 50 kg product $= 50 \times 5\times10^{20} = 25\times10^{21}$
The amount of divinyl benzene mixed with the polystyrene $= 25\times10^{21}\times130$ amu $= 5.418\times10^{-3}$kg

The vulcanization of rubber results in hard polymeric substances. With the rise of temperature cross-linking reactions in polymers increase and ultimately give rise to thermo-hardening or thermosets polymers. The process of converting thermoplastics to thermoset polymers by mixing suitable vulcanizing agents at an elevated temperature is called curing. There are certain polymer-vulcanizing agent systems which cure polymeric substances rapidly even in cold. This process is called cold curing or cold vulcanization. Hard rubber combs are examples of vulcanized rubber. The cross-links produced during the vulcanization of rubber are also called anchor points. When cycle tubes are punctured, vulcanization is done to cure the tubes. In plastic industries, hard articles are produced by adding curing agent to polycondensed linear polymers at high temperature.

5.12 FRAMEWORK STRUCTURES

Polycondensation of organic molecules with three or more functional groups can occur if special reagents (hardeners) are added. These polymers assume three-dimensional structure. The three-dimensional network of phenol formaldehyde is an example of three-dimensional cross-linking where the following reaction takes place:

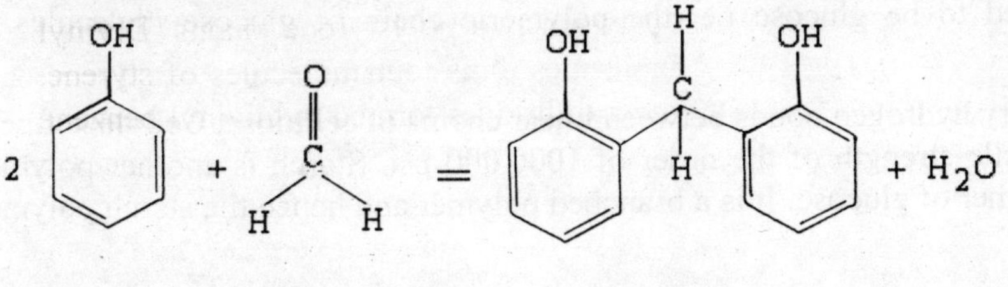

$$+ H_2O \qquad (5.41)$$

As a result of the above reaction a bridge is formed between two phenol molecules and a water molecule is obtained as a byproduct. Each phenol molecule is trifunctional and hence can form three bridges with other phenol molecules at alternate C-site of benzene ring. Cross-linkings produce a three-dimensional structure, since the individual molecule will no longer be independent. The polymer becomes less plastic. When fully cross-linked phenol formaldehyde is formed, each phenol molecule forms three bridges each shared by two molecules. Thus, these are on an average 3/2 bridges

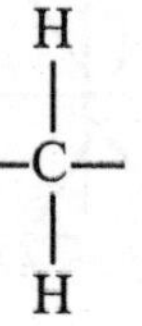

$$
\begin{array}{c}
H \\
| \\
-C- \\
| \\
H
\end{array}
$$

corresponding to each phenol molecule in fully cross-linked phenol formaldehyde plastic. When one mole of phenol is fully cross-linked by formaldehyde, 3/2 moles of water molecules are obtained as a byproduct. The resultant three-dimensional framework structure is a noncrystalline giant single molecule. It is also called phenolic or bakelite. The dark brown light switch is a single molecule of phenolic! The long-range order in a phenol formaldehyde resin is not possible due to the framework structure. When phenol formaldehyde is heated, more and more cross-links are formed. As a result of this it becomes harder and decomposes at higher temperature. So phenol formaldehyde is a thermoset polymer. The other important network polymers are melamine and epoxies. The possible structure of one type of epoxy polymers is

$$
\overset{OH}{\underset{|}{\cdots -O-R-O-CH_2-CH-CH_2-\cdots}} \quad (\text{R= aromatic or aliphatic radical})
$$

R can be

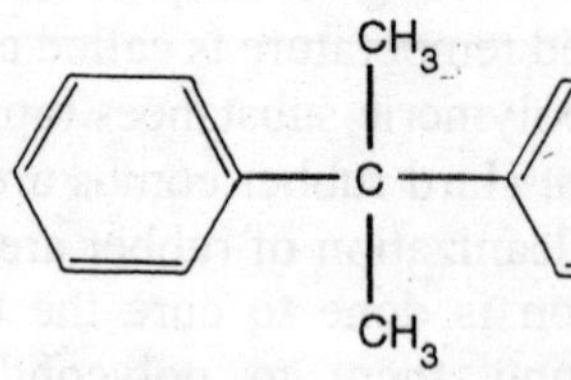

Fig. 5.13 Epoxy polymers

The epoxies are useful adhesives having high strengths at relatively high temperatures.

5.13 POLYMERS IN LIVING MATTER

Cellulose is an ideal example of polymer in the living systems. Cellulose is present in plants. skin, hair, toenails, muscle and tendons of animals are also made of polymers. The monomer of cellulose can be considered to be glucose i.e. the polymeric chain of glucose molecules is cellulose, Fig. 5.14.

There are strong secondary hydrogen bonds between linear chains of cellulose. Cellulose fibers are very strong having tensile strength of the order of 1000,000 psi. Starch is another polymer formed from a different isomer of glucose. It is a branched polymer and hence the starch polymer cannot be crystallized easily. When water is added to the starch, it swells since the packing is poor in it. One of the first commercial plastics used for toy making was celluloid which is nitrocellulose plasticized camphor.

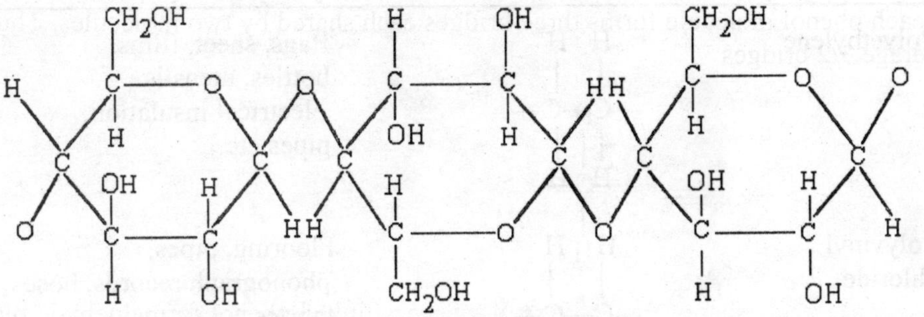

Fig. 5.14 Cellulose as a polymeric chain of glucose

The polypeptide chain in living system is formed due to the condensation polymerization of α-amino acids having the general structure,

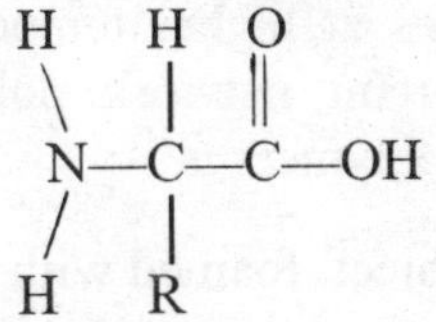

The polymerization reaction is as follows:

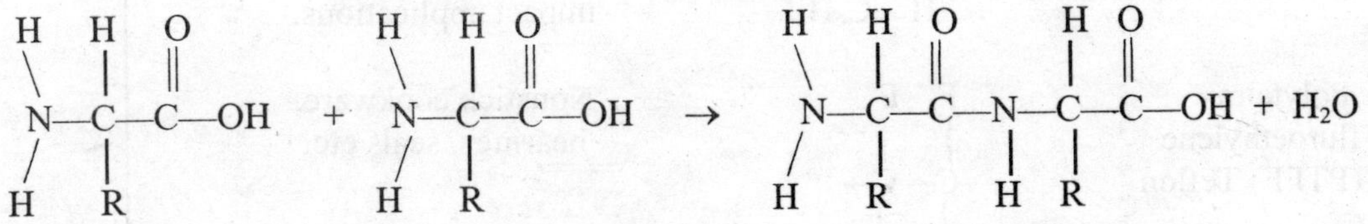

$$\text{(5.42)}$$

A protein is a polypeptide chain where α-amino acids are arranged in specific order. Excellent hydrogen bondings are there between the polypeptide chains if R is hydrogen i.e. if the chains are polyglycines. If R is an alkyl group the hydrogen bond formation is hindered.

5.14 GLASSY NATURE OF POLYMERS

Polymers usually exist in amorphous form. The interference in stereoregularity arises due to atacticity, cross-linking, unequal lengths of polymeric chains and branching. This hinders crystallization and the melting point of polymer is therefore blurred. The liquid polymers solidify over a range like a glass. The slower the rate of cooling, the molecules have more time to pack efficiently which results in higher polymer density. The volume of polymer melts decreases linearly until the glass transition temperature is reached. At higher temperature the melt has free volume available for molecules to execute their motion. Below glass temperature the free volume reduces to zero, there by leaving no scope for molecular motion.

5.15 USES OF POLYMERS

The uses of various types of polymers are given below:

Name of polymer	Mer	Uses
Polyethylene	$\begin{array}{c} \text{H} \quad \text{H} \\ | \quad | \\ -\text{C}-\text{C}- \\ | \quad | \\ \text{H} \quad \text{H} \end{array}$	Bags, sheet, films, bottles, utensils, electrical insulation, pipes etc.
Polyvinyl chloride	$\begin{array}{c} \text{H} \quad \text{H} \\ | \quad | \\ -\text{C}-\text{C}- \\ | \quad | \\ \text{H} \quad \text{Cl} \end{array}$	Flooring, pipes, phonograph records, hoses, fabrics
Polypropylene	$\begin{array}{c} \text{H} \quad \text{H} \\ | \quad | \\ -\text{C}-\text{C}- \\ | \quad | \\ \text{H} \quad \text{CH}_3 \end{array}$	Stiffer tubing, bottles, sheets etc.
Polystyrene	$\begin{array}{c} \text{H} \quad \text{H} \\ | \quad | \\ -\text{C}-\text{C}- \\ | \quad | \\ \text{H} \quad \text{C}_6\text{H}_5 \end{array}$	Moulded object, foamed with CO_2 to form insulating packaging, toughened with butadiene used in high impact applications.
Polytetra-fluroethylene (PTFE) Teflon	$\begin{array}{c} \text{F} \quad \text{F} \\ | \quad | \\ -\text{C}-\text{C}- \\ | \quad | \\ \text{F} \quad \text{F} \end{array}$	Nonstick cookware, bearings, seals etc.
Polymethylmetha-crylate	$\begin{array}{c} \text{H} \quad \text{CH}_3 \\ | \quad | \\ -\text{C}-\text{C}- \\ | \quad | \\ \text{H} \quad \text{COOCH}_3 \end{array}$	Aeroplane windows, sheets, windscreen etc.
Polyester	$-\text{O}-(\text{CH}_2)_2-\text{O}-\overset{\displaystyle \text{O}}{\overset{\displaystyle \|}{\text{C}}}-(\text{CH}_2)_4-\overset{\displaystyle \text{O}}{\overset{\displaystyle \|}{\text{C}}}-$	Fibre

(Contd.)

Name of polymer	Mer	Uses
Polyethylene terephthalate	$-O-(CH_2)_2-O-\overset{\displaystyle O}{\overset{\displaystyle \|}{C}}-(C_6H_5)-\overset{\displaystyle O}{\overset{\displaystyle \|}{C}}-$	Dacron fibre, mylar insulation film
Polyamide Nylon	$-\overset{\displaystyle O}{\overset{\displaystyle \|}{C}}-(CH_2)_4-\overset{\displaystyle O}{\overset{\displaystyle \|}{C}}-\underset{\displaystyle H}{\overset{\displaystyle \|}{N}}-(CH_2)_6-\underset{\displaystyle H}{\overset{\displaystyle \|}{N}}-$	Fibre ropes, machine parts
Phenol formaldehyde (bakelite amorphous)	Eq.(5.40)	Electrical insulation, switches
Epoxy(amorphous)	Fig. 5.13	Adhesives, fibreglass
Polybutadiene	$\begin{array}{cccc} H & H & H & H \\ \| & \| & \| & \| \\ -C & -C & =C & -C- \\ \| & & & \| \\ H & & & H \end{array}$	Tyres, moulds
Polyisoprene	$\begin{array}{cccc} H & CH_3 & H & H \\ \| & \| & \| & \| \\ -C & -C & =C & -C- \\ \| & & & \| \\ H & & & H \end{array}$	Gaskets, tyres
Polychloroprene	$\begin{array}{cccc} H & Cl & H & H \\ \| & \| & \| & \| \\ -C & -C & =C & -C- \\ \| & & & \| \\ H & & & H \end{array}$	Oil resistant rubber for seal
Silicone rubber	$\begin{array}{cc} CH_3 & CH_3 \\ \| & \| \\ -O-Si & -O-Si-O- \\ \| & \| \\ CH_3 & CH_3 \end{array}$	Thermal and electrical insulation components, foam rubber, coatings

5.16 POLYMER PROCESSES

Polymer processes are important technologies, which produce varieties of objects that rival even metals in a number of applications. Thermoplastic polymer forming processes can be described in terms of the following operations:

(i) Production of the polymer in a powder, granular or sheet form
(ii) Basic pattern of heating to soften
(iii) Mechanical deformation to obtain desired form
(iv) Cooling to harden.

During the production process the polymers are mixed with suitable additives in the form of solid or liquid in order to have the finished material with the required properties. In case of thermoset polymers, solid additives like chalk, carbon black, cork dust, paper pulp etc. are added to reduce the brittleness of the material. The flow characteristics are improved by adding liquid additive during processing. Gas additives are used to produce foam plastic components. In thermoset polymer, the curing is done in the mould to form three-dimensional network structure and then cooling is done. The various machine operations that produce the ultimate product are broadly classified as:

a) Mixing
b) Extrusion
c) Moulding
d) Sheet formation
e) Casting
f) Calendering
g) Machining
h) Grinding
j) Heat Welding
k) Slitting
l) Adhesive bonding.

Thermoplastic materials can be softened by heating and reused indefinitely provided the temperature is not that high which causes decomposition of the material. Extrusion and injection moulding can readily process thermoplast. Thermosetting materials cannot be softened by application of heat. These materials undergo chemical changes when heated and become more rigid. So reusing of these materials is not possible. Moulding and casting are the processes used for such material.

Mixing

Mixing is the first step in shaping the plastics. The polymer and various additives like fillers, plasticizers, dyes etc. are mixed intimately in an open-roll mill. The two rolls in open-roll mill are kept at different temperatures and they rotate at slightly different speed. The components of the plastic are blended due to the shearing force acting in the nip region between the rolls. The other method of blending uses drums with internal rotors and blades. The inert atmosphere is needed during these processes since polymers may oxidize during the shearing action.

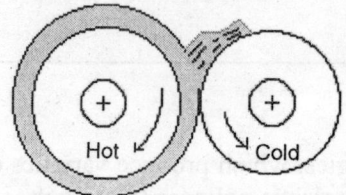

Fig. 5.15 Schematic representation of Mixing process

Extrusion

In this process the molten polymer, mixed with additives, is forced through a die. This is usually a continuous process. Granules of polymer mixture are hopper fed into the rear of the cylinder where extrusion process is carried out. The polymer mixture is passed through a heated zone by means of screw mechanism. As the screw rotates the polymer granules are compacted, mixed, heated, forwarded and eventually forced through an open-ended die. This process produces plastic pipes, plastic sheets or any other product, which has constant cross-sectional profile. There are external heaters surrounding the compression cylinder, which create heated zone in the cylinder.

When thin film or sheet is to be produced by this process, an extruded cylinder is produced first using a suitable die. This hot cylinder is inflated by compressed air to give a sleeve of thin film. Fiber, curtain rails, household guttering, polybag etc. are examples of the products obtained from this process. Hollow containers like plastic bottles etc. can be produced by extrusion flow moulding.

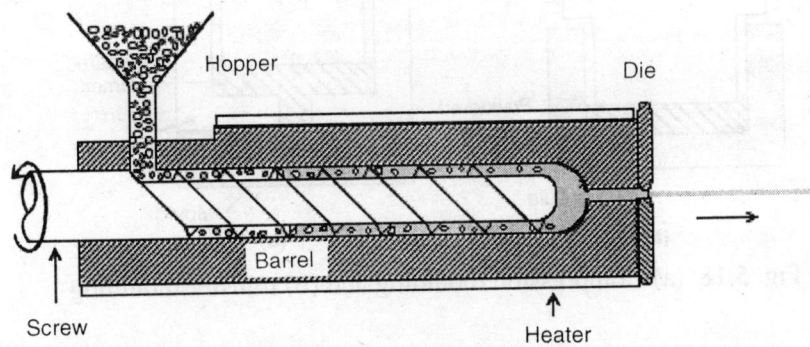

Fig. 5.16 Extrusion

Injection moulding

In this process, the polymer is melted and then forced into a mould. Products having complex shapes with inserts, thread etc. can be produced at a fast rate using this method. Some typical products are parts for electronic equipments, tool handles, toys, crates etc. Foam plastic components can also be produced by this method. Inert gases are dissolved in the molten polymer. The gases while coming out of the melt expand to form a cellular structure during cooling. When the hot polymer comes in contact with the cold surface of the mould, a solid surface appears. In case of thermoplast moulding, the mould is water-cooled whereas for thermosets, the material must be put into a heated mould to complete polymerization reaction.

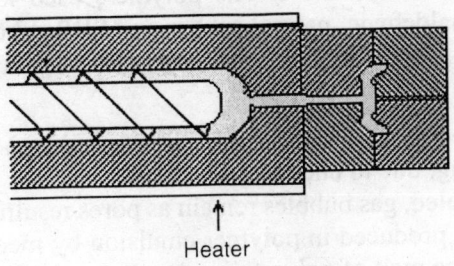

Fig. 5.17 Injection moulding

Compression moulding and transfer moulding

Compression moulding and transfer moulding are the methods usually adopted for shaping thermoset polymers, Fig. 5.18. The powdered polymer is compressed between the two parts of the mould and also heated during the compression process. In case of transfer moulding, the powdered polymer is heated in a chamber before transferring it into a mould by using plunger.

Cold setting

Cold setting is a casting process, generally used for encapsulating small electrical components. In this method short molecular chains mixed with required additives are polymerized in a mould where solidification takes place.

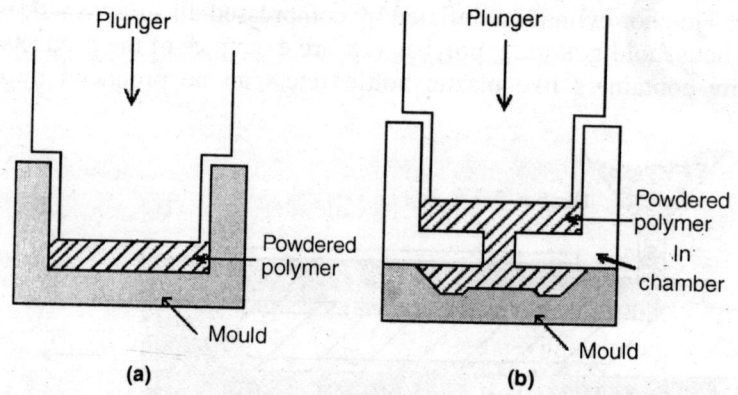

Fig. 5.18 (a) Compression moulding and (b) transfer moulding

Calendering

Calendering consists of three or more heated rollers. The heated polymer input is added in the gap between the first pair of rollers, and the output is a plastic sheet, Fig. 5.19. During forming process, the heated sheet is pressed into or around a mould. Such process is often called thermoforming. Pressing may also be done by the application of air under pressure on one side of the sheet, which is called pressure forming. Vacuum formation between the sheet and the mould, also results in forming, which is called vacuum forming.

Foamed polymers

This form of plastics and elastomers has wide applications in cushions with foamed fillings, packaging, thermal insulation materials etc. The polymers used in this way are polyurethane, polyvinyl chloride, urea formaldehyde, natural rubber and SBR rubber. Thermosetting materials, elastomers and thermoplastics can be foamed. The processes involved in the formation of foamed polymers are the following:

i) Gas bubbles are produced throughout the polymeric liquid by adding a blowing agent which liberates gas upon heating, due to chemical reactions.

ii) When this material is cooled, gas bubbles remain as pores resulting in a sponge like structure. Gas bubbles can also be produced in polymer emulsion by mechanical agitation or by bubbling inert gas through the melt of polymeric substance or by heating polymeric liquid or by reducing pressure in it.

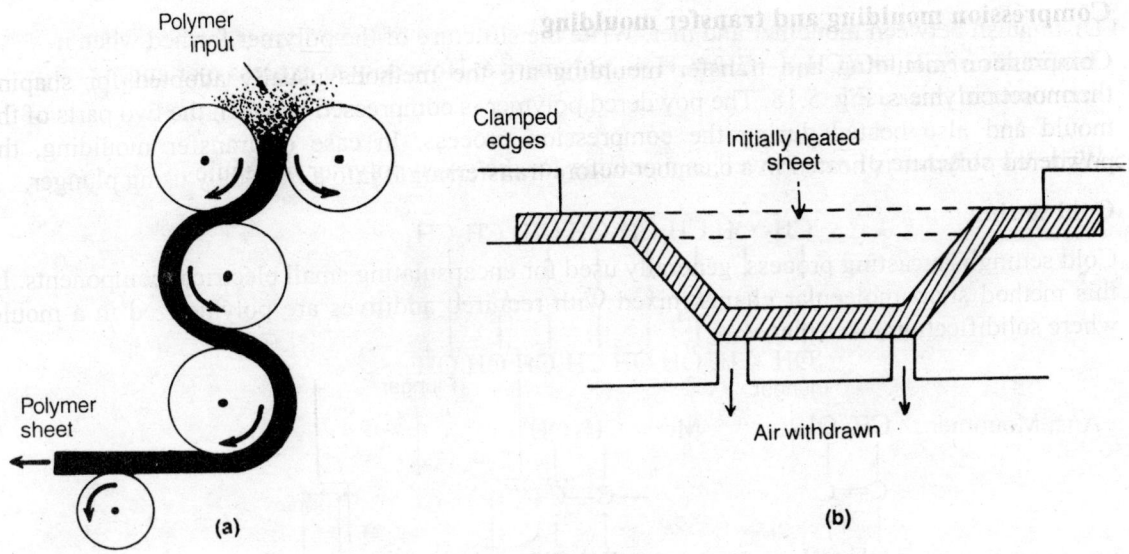

Fig. 5.19 Calendering

EXERCISES

5.1 What are the weight-average molecular weight and degree of polymerization of a polybutadiene sample as per data given below:

$$—(CH—CH_2)_n—$$
$$|$$
$$CH$$
$$||$$
$$CH_2$$

Molecular wt. (kg/mole)	5.4	16.2	21.6	32.4	37.8	48.6
Weight fraction	0.18	0.18	0.35	0.15	0.08	0.06

Ans: 22248 g/mole, 412

5.2 Distinguish between number-average and weight-average molecular weight and find the ratio of the two, if the following data is given for polyacrylonitrile:

Range kg/mole	0–5.3	5.3–10.6	10.6-15.9	15.9-21.2
Weight percent	15	35	40	10

Also find out the degree of polymerization.
Ans: 7342 g/mole, 10335 g/mole, 0.71, 195

5.3 Five hundred polyvinyl chloride molecules are selected, out of which 100 have 20 mers, 200 have 40 mers, 100 have 70 mers and 100 of them have 100 mers. What are the values of $\overline{Mw}$ and $\overline{Mn}$?

Ans: 4282 g/mole, 3375 g/mole

5.4 Distinguish between monomer and mer. Write the structure of the polymer formed when n-molecules of $CH_2=C(CH_3)—CH=CH_2—$ polymerize. How do you express mer and monomer in this case?

5.5 Write the structure of mer as well as monomer for the given polymer molecule

$$\begin{array}{ccccccccc} CH_3 & OH & CH_3 & CH_3 & OH & CH_3 & CH_3CH_3 \\ | & | & | & | & | & | & | & | \\ —C— & C— & C— & C— & C— & C— & C— & C— \\ | & | & | & | & | & | & | & | \\ OH & CH_3 & OH & OH & CH_3OH & OH & OH \end{array}$$

Ans: Monomer : $\begin{array}{cc} CH_3 & OH \\ | & | \\ C{=}C \\ | & | \\ OH & CH_3 \end{array}$ Mer: $\begin{array}{cc} CH_3 & OH \\ | & | \\ —C—C— \\ | & | \\ OH & CH_3 \end{array}$

5.6   0.1 mole of HCl is added to 1.02 kg of isoprene $CH_2=C(CH_3)-CH=CH_2$,. The efficiency of HCl is 40%. If whole of HCl and isoprene are used in the polymerization reaction, find the degree of polymerization and the molecular weight of polymer.
Ans: 375, 25500 g/mole

5.7   If HBr is to be used as an ionic initiator for isoprene polymerization, how much HBr has to be added to 500 kg of isoprene to produce an average molecular weight of 9000 g/mole. Assume 20% efficiency of HBr. Both H^+ and Br^- ions serve as initiators as well as terminators.
Ans: 22.5 kg

5.8   (a) Draw the spread out structure of the cross-linked mer in the styrene-divinylbenzene polymer.
       (b) Draw the spread out structure of polyvinyl chloride, which crystallizes easily.

5.9   Write down the polymerization reaction for the following three cases, in which one of the reactants is,
       i)   bifunctional
       ii)   trifunctional
       iii) tetrafunctional

5.10 a) Cross-linked cis-isoprene rubber contains 5 w/o oxygen and 10 w/o sulphur. What fraction of this polymer is cross-linked ?
       b) 5g of divinyl benzene is added to 500 g of styrene. What are the maximum possible cross-links per gram of product?
       Ans: (a) 0.5; (b) 4.633×10^{19}/g

5.11 a) Write the polymerization reaction and show the mer by a bracket when m-molecules of ethylene glycol react with m-molecules of dimethyl terephthalate.

b) What type of bonding is there for the crystallization of above polymer?

c) If the degree of polymerization for the above condensation reaction in part a)is 1000, calculate how many molecules of polymer are there in 10 g of polymer.

$$-[\overset{\text{O}}{\overset{\|}{C}}-C_6H_4-\overset{\text{O}}{\overset{\|}{C}}-O-CH_2-CH_2-O]_n-$$

d) Find the number of byproduct molecules formed in part c) for 10g polymer.

Ans: (c) 3.14×10^{19} ; (d) 6.249×10^{22}

5.12 Write down the formula for condensation reaction between urea and formaldehyde. What is the byproduct in this case ?

5.13 If rubber is formed from copolymerization of equal number of chloroprene and butadiene molecules.

a) What is the weight fraction of each of the component?

b) How many grams of oxygen must be added to 1 kg of this rubber to cross-link 50% of all the mers?

Ans: (a) chloroprene 0.621, butadiene 0.379; (b) 112 g of O_2

5.14 If the degree of polymerization of polypropylene is 735, find the molecular weight of polypropylene.

Ans: 30870 g/mole

5.15 Plastic foams are used for insulation, floating devices and packing. They contain large volumes of entrapped gas that can reside in either isolated closed cells or in interconnected open cells. If the specific gravity of polymer is 1.21 and its density becomes 0.05 g/cm³ after foaming, by what factor the volume is increased during foaming?

Ans: 23.2 times

5.16 A rubber contains 27 w/o butadiene,68 w/o isoprene and 5 w/o sulphur. What fraction of cross-links will be formed during vulcanization?

Ans: 0.104

5.17 The specific volume v(ml/g) of a polymer is given as a function of temperature (T in °C)

$$v(L) = 1.135 + 1.2 \times 10^{-4} T$$

for amorphous solid polymer at low temperature and

$$v(H) = 1.150 + 8.0 \times 10^{-4} T$$

for the polymer melt at higher temperature

a) What is the glass transition temperature?

b) Find the density of the polymer at 30°C.

Ans: (a) –22.06 °C; (b) 0.85 g/cc

5.18 What fraction of chloroprene is cross-linked if the product contains 3.0 w/o S?
Ans. 0.086

5.19 A product of polystyrene contains 26 w/o DVB calculate the number of cross-links per kg of the sample
Ans. 1.2046×10^{24} cross-links per kg.

5.20 Write the condensation reaction for the polymerization of hexamethylenediamine and adipic acid. Encircle and name the polar group. 54g of byproduct are formed assuming all the functional groups to be consumed, determine
(i) Mass of polymer formed
(ii) Average number of mers, if the molecular wt. of polymer is 798006.

6
Crystal Imperfections

6.1 INTRODUCTION

Most of the crystals, which are found in nature or are grown in laboratories, are not perfect. The imperfections may arise due to the presence of impurities or vacancies in the crystals. Crystals grow from the melt of materials during cooling under specific conditions. During the process of formation of crystal, impurity atoms, dust particles, vacancies, atoms and molecules at the surface of the container provide the centers around which crystals start nucleating. Thus, the crystallites are formed around these centers in the melt. These crystallites are also called grains, which grow and ultimately touch along the boundary surfaces giving rise to grain boundaries. This is called polycrystalline solid. Crystals are mostly found in polycrystalline form.

Crystal planes and directions are discontinuous along grain boundaries. So, the anisotropic properties also change abruptly at the grain boundaries. The crystal surfaces and grain boundaries are surface imperfections. In general, one can define the imperfections or defects in a crystal as the discontinuity or absence of periodicity in certain region of the space lattice.

6.2 TYPES OF IMPERFECTIONS

Imperfections in materials change the properties. The presence of impurities or defects may change the properties of metals e.g. strength of steel increases with the presence of alloying elements like carbon, manganese, chromium etc. The doped or extrinsic semiconductors show improved conductivity and are used in semiconductor devices. In vulcanization process sulphur or oxygen atoms are introduced in natural rubber to make it harder. Thus, imperfections are sometimes introduced to improve upon the properties of material and make it suitable for specific applications. Sometimes the defects are accidental and undesirable e.g. undesirable defects in glass increase its brittleness; large additions of micromolecules in rubber hosepipe make it harder and brittle. Imperfections can be classified as follows, Table 6.1:
   (i) Zero dimensional or point imperfections
   (ii) One dimensional or line imperfections
   (iii) Two dimensional or surface imperfections
   (iv) Three dimensional or volume imperfections.

6.3 POINT IMPERFECTIONS

Imperfections of zero dimension or point imperfections are present in almost all the crystals. At the time of formation of crystals some vacancies are created accidentally i.e. some of the regular

Table 6.1 Classification of crystal imperfections

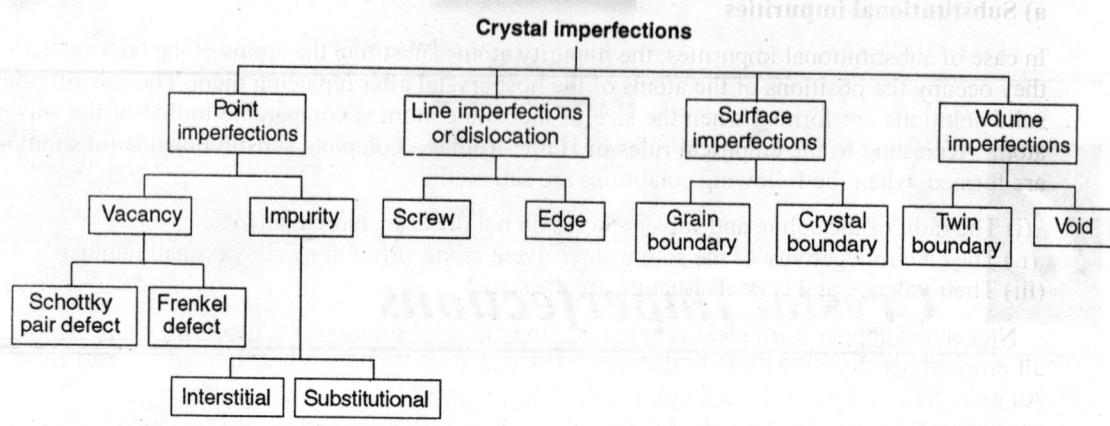

atomic sites in the crystal remain vacant. The vacancies may also be created during heat treatment of the crystal or as a result of thermal disorder, when some of the atoms at the regular lattice sites gain enough kinetic energy such that they can cross the kinetic barrier and diffuse to the surface, or grain boundaries of the crystal where vacancies are present.

Fig. 6.1 shows vacancy due to a missing atom. Sometimes, atoms or molecules of other materials are present or are introduced in a crystal as impurities. These are called solid solutions. The solid solutions are very much similar to the liquid solutions. In solid solutions, mixing of solute and solvent takes place on the atomic scale. The presence of impurity atoms as solute in solid solution can occur in two ways:

    a) Substitutional impurities
    b) Interstitial impurities

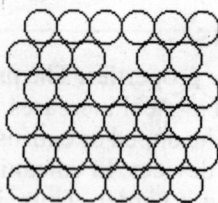

Fig. 6.1 Point imperfections as vacancies

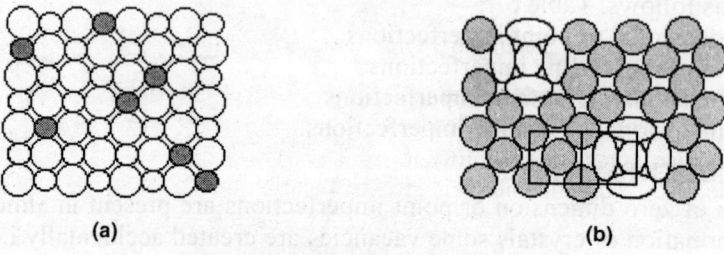

(a)   (b)

Fig. 6.2 (a) Substitutional impurities, (b) Interstitial impurities.

a) Substitutional impurities

In case of substitutional impurities, the impurity atoms substitute the atoms of the host crystal i.e. they occupy the positions of the atoms of the host crystal after replacing them. The substitutional solid solutions are formed when the size of the solute atom is comparable to that of the solvent atom. According to the empirical rules of Hume-Rothary, complete substitutional solid solutions are formed, when the following conditions are satisfied:

 (i) The radii of the solute and solvent atoms do not differ by more than15%.
(ii) The electronegativity of the solute and solvent atoms differ at most by a small number.
(iii) Their valency and crystal structure are the same.

   Nickel and copper form ideal couple for complete solid solutions i.e. these two elements mix in all proportions since they satisfy the conditions of Hume-Rothary rules. Similarly, Ge-Si and Ag-Au systems also form solid solutions in all proportions. Mostly we find different types of atoms are randomly distributed on the lattice sites in solid solutions e.g. brass containing 30 atomic percent (a/o) Zn and 70 a/o Cu. This type of solution is called a random substitutional solid solution. Extrinsic or impurity semiconductors like phosphorous doped silicon (n-type) or aluminium doped silicon (p-type) are also examples of solid solutions. At room temperature about 35% of Zn can dissolve in fcc Cu whereas the solubility of copper in zinc is only 1%. This is because Cu-Zn couple does not satisfy Hume-Rothary conditions fully as the valency of zinc is two where as that of copper is one. The crystal structure of zinc is hcp whereas that of copper is fcc. The extra bonding electrons of zinc can easily be accommodated in the host matrix of fcc copper but lack of bonding electrons of copper cannot easily be matched in the solvent of zinc The electrical properties of brass are similar to that of copper but it is cheaper than copper since it contains cheaper element zinc. Brass alloy has many compositions. Substitutional solid solutions can also be ordered, as happens in case of β-brass, which has bcc type crystal structure, Fig 6.3, containing equal number of copper and zinc atoms. The Bravais lattice of β-brass is sc.

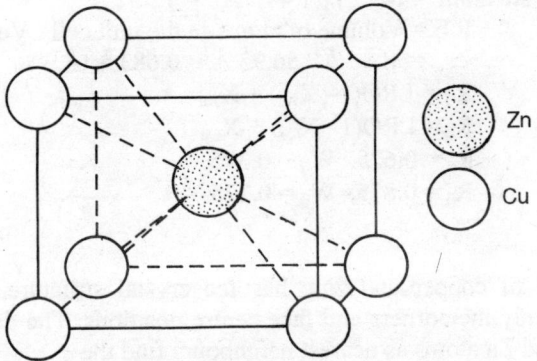

Fig. 6.3 Ordered solid solution of β-brass

The arrangement of Cu and Zn atoms in the unit cell of β-brass is such that zinc atom is at the centre of cubic unit cell surrounded by eight copper atoms as nearest neighbours positioned at the corners or vice versa. The stability in the structure of β-brass can be explained as follows: the magnitude of Cu-Zn bond energy is greater than that of Cu-Cu as well as Zn-Zn bond i.e.

$$E_{Cu-Zn} > \frac{1}{2} (E_{Cu-Cu} + E_{Zn-Zn}) \tag{6.1}$$

Hence, copper-zinc bonding is more favoured as compared to the copper-copper and zinc-zinc bonding and hence copper and zinc atoms become nearest neighbours. β-brass forms an intermediate structure, a system of limited solid solutions, which is also called intermetallic compound. Iron carbide (Fe_3C) is also an intermetallic compound having complex crystal structure with orthorhombic lattice. It is hard and brittle due to presence of 6.67 w/o carbon. If the free energy of the solution is less than the sum of the free energy of the pure components, the components readily form the solid solution.

Example 6.1

Consider an fcc gold-copper solid solution of having 25 a/o gold and 75 a/o Cu i.e. $AuCu_3$ where the corner sites are designated as α and face centre sites as β. In a perfect long-range ordered alloy all α-sites are occupied by Au and all β-sites by Cu atoms. Find the packing efficiency and density of the alloy. This alloy becomes disordered at higher temperatures. Indicate the rightly and wrongly occupied sites as a function of long-range order parameter (LRO). Calculate these values, if LRO =0.5. Given that r_{Cu} = 1.28 Å, r_{Au} = 1.44 Å.

[Hint: LRO is expressed as $(R_\alpha - X_{Au})/(1 - X_{Au})$ where R_α indicates the fraction of α-sites rightly occupied, W_α (= 1− R_α) indicates the fraction of α-sites wrongly occupied. and X_{Au} is atomic fraction of Au atoms.]

Solution:

In the unit cell of the alloy there are three atoms of Cu and one atom of Au. Assuming Cu and Au atoms touch along the face diagonal

$$\sqrt{2}a = 2 \times (r_{Cu} + r_{Au}) = 5.44 \text{ Å}$$
$$a = 3.847 \text{ Å, Volume of unit cell} = 56.93 \text{ Å}^3,$$

Mass of atoms in unit cell = $3 \times 63.54 + 197 = 387.62$ amu $= 6.47 \times 10^{-25}$ kg

$$\text{Density} = 6.47 \times 10^{-25} /56.93 \times 10^{-30} = 11365 \text{ kg/m}^3$$

Volume of three Cu atoms = $3 \times (4\pi/3) \times (1.28)^3 \text{ Å}^3 = 26.35 \text{Å}^3$

Volume of one Au atom = $(4\pi/3) \times (1.44)^3 \text{ Å}^3 = 12.51 \text{ Å}^3$

$$\text{P.E.= Volume of atoms in the unit cell / Volume of unit cell}$$
$$=38.86 \text{ Å}^3/ 56.93 \text{ Å}^3 = 0.683 = 68.3 \%$$
$$R_\alpha = LRO(1- X_{Au}) + X_{Au}$$
$$R_\beta = LRO(1- X_{Cu}) + X_{Cu}$$

For LRO = 0.5
$$R_\alpha = 0.625, \ W_\alpha = 0.375$$
$$R_\beta = 0.875, \ W_\beta = 0.125$$

Example 6.2

A solid solution of copper and zinc has fcc crystal structure, with copper and zinc atoms occupying randomly the corners and face centre positions. The density of the alloy is 8665 kg / m^3. Taking Cu and Zn atoms as nearest neighbours find the
(a) atomic percent of copper present in the solid solution,
(b) fraction of the unit cell volume of the solid solution that is not occupied by any of the atoms?

Solution:

(a) Calculation of 'a'
$$a = \sqrt{2} (R_{Cu} + R_{Zn}) = \sqrt{2} (1.28 \text{ Å} +1.31 \text{ Å}) = 3.66 \text{ Å}$$
Calculation of atomic fraction f of Cu

$$\rho = 4 [M_{Cu} (f) + M_{Zn} (1-f)]/ a^3 \times 1.66 \times 10^{-27}$$

$$8665 \text{ kg/m}^3 = \frac{4\,[(f\,)63.54 + (1-f\,)65.37\,]}{(3.66)^3 \times 10^{-30}} \times 1.66 \times 10^{-27} \text{ kg/m}^3$$

$$f = 0.76$$

Atomic percent of copper $= 76\,\%$

(b) P.E. $= (4\pi/3)\,[4(1.28^3 \times 0.76 + 1.31^3 \times 0.24)] \times 10^{-30} \text{ m}^3/\,(3.66 \times 10^{-10} \text{ m})^3 = 0.73$

Fraction of the volume not occupied by any atom $=0.27$

b) Interstitial impurities

Interstitial impurities are the impurity atoms, which occupy interstices in the host crystal. These atoms are usually smaller in size as compared to the host atoms. If the size of the impurity atom is less than or equal to the size of the interstices, then a large number of impurity atoms can be doped in the host crystal under proper experimental conditions. The solubility of interstitial impurity is reduced in case the size of the impurity atom is larger than the void size. The large size impurity atom causes distortion in the region around it. So the energy of the system increases. Therefore such distortions are less probable and are unstable. This explains why the solubility of carbon in bcc iron is poorer than that in fcc iron. The interstitial solid solution of carbon-steel contains carbon atoms, which occupy the octahedral voids in the fcc iron. The maximum solubility of carbon in fcc iron is 2 w/o. At room temperature solubility of carbon in bcc iron is only 0.008 w/o. Poor solubility of carbon in iron is due to the fact that in fcc iron, the octahedral void has radius $0.414\ r_{Fe}$ i.e. 0.53Å whereas the radius of carbon atom is 0.71 Å. Therefore, compressive stress field is present around the carbon atoms in the octahedral voids. As a result, the energy of the crystal increases and the introduction of more carbon impurities is not possible near the compressive stress field until and unless large amount of work is done. Thus, other carbon atoms can be introduced only at locations away from this compressive stress field, which is responsible for poor solubility of carbon atoms in fcc iron. Similarly, one can easily explain the poor solubility of carbon in bcc iron. The largest interstitial void of the bcc unit cell is tetrahedral void, located at $(\frac{1}{2},\frac{1}{4},0)$ and other equivalent sites. There are twelve such sites per unit cell, Table 6.2. Carbon atoms occupy tetrahedral voids that have radii $0.29\ r_{Fe}$ i.e. 0.365 Å. This void size is even smaller than 0.53 Å. Thus, the solubility of carbon in bcc iron is poorer than that in fcc iron. Table 6.2 gives the details of the voids present in various crystal structures where radius ratio $\gamma = r_v/r_a$, r_v being the radius of the largest void and r_a, the radius of the atom in the host crystal.

The atoms from host crystal can also be trapped in voids or interstices, which are called self-interstitials. The self-interstitials also cause large distortions in the region around the interstitial sites and increase the energy of the system, so the solubility of self-interstitials is also low.

Interstitial solid solutions are obtained from metals when small atoms such as hydrogen, boron, carbon, nitrogen and some times oxygen dissolve in metal. In case of random solid solution, random distribution of solute (impurity) atoms in the host crystal gives rise to an increase in configurational entropy S. The increase in entropy can be expressed as

$$dS = S_{final} - S_{initial} = k \ln P \tag{6.2}$$

where k is the Boltzmann constant and P is the total number of distinct configurations arising due to the presence of impurity and is given by

$$P = \frac{(N_A + N_B)!}{N_A!\ N_B!} \tag{6.3}$$

where N_A, N_B are the numbers of solute and solvent atoms respectively. As a result of change in entropy, there is change in the free energy dG:

$$dG = dH - TdS - SdT \tag{6.4}$$

Table 6.2 Voids in crystal structure

Crystal Structure	Cubic			Octahedral			Tetrahedral		
	Equivalent Position	No. of voids	γ	Equivalent Position	No. of voids	γ	Equivalent Position	No. of voids	γ
sc	(½,½,½)	1	0.732	—			—		
bcc	—			(½,½,0) (½,0,½) (0,½,½)	3	0.154	(½,¼,0) (¼,½,0) (¾,½,0) (½,¾,0) (0,½,¼) (0,¾,½) (0,½,¾) (0,¼,½) (½,0,¼) (¾,0,½) (½,0,¾) (¼,0,½)	12	0.29
fcc	—			(½, 0, 0) (0, ½, 0) (0, 0, ½) (½,½,½)	4	0.414	(¼,¼,¼) (¼,¼,¾) (¼,¾,¼) (¾,¼,¼) (¾,¼,¾) (¾,¾,¼) (¼,¾,¾) (¾,¾,¾)	8	0.225

When the free energy of the solid solution decreases and attains the minimum value, the solid solution attains stable configuration. The change in free energy depends on the changes in the enthalpy dH and that of the entropy of the solution. The entropy change in the solution can be expressed in a simpler form as

$$dS = k [\ln (N_A + N_B)! - \ln N_A! - \ln N_B!] \tag{6.5}$$

Using Stirling's formula $\ln N! \approx N \ln N - N$ for large N, $\tag{6.6}$

we get

$$dS = k [(N_A + N_B) \ln (N_A + N_B) - (N_A + N_B) - N_A \ln N_A + N_A - N_B \ln N_B + N_B]$$
$$= - k (N_A + N_B) [f_A \ln f_A + f_B \ln f_B] \tag{6.7}$$

where f_A, f_B are fractions of solute and solvent atoms, respectively, present in solution. If we take one mole of solution, the change in entropy is

$$dS = - R [f_A \ln f_A + f_B \ln f_B] \tag{6.8}$$

where $R = Nk$ and $N = N_A + N_B$ is Avogadro number. In case of random solid solutions increase in entropy may lead to decrease in free energy. The enthalpy of the solid solution may or may not change as compared to the sum of enthalpies of solute and solvent. According to this enthalpy change, the solution can be classified into the following categories:

(i) Ideal solution

In case of an ideal solution of components A and B, the enthalpy value of a specific composition is linearly interpolated value between H_A and H_B, Fig. 6.4. Thus, there is no change in enthalpy as compared to the sum of enthalpies of A and B. Increase in entropy due to formation of solid solution of various composition is also shown in Fig. 6.4(a). The change in free energy of this solution can easily be represented by

$$dG = -TdS \qquad (6.9)$$

i.e G decreases as S increases.

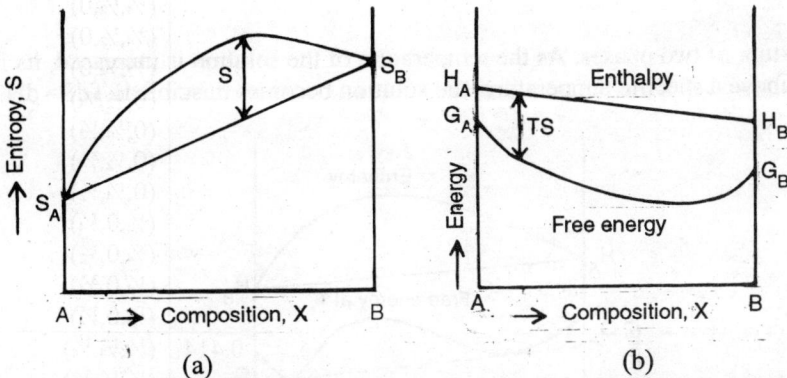

(a)   (b)

Fig. 6.4   (a) Entropy, (b) Enthalpy and free energy versus composition for ideal solution.

(ii) Negative deviation from ideality

The enthalpy of the solution may have negative deviation from ideality i.e. dH is negative. In such cases, the solid solution may become ordered one, as happens in case of β-brass below 470°C. If the solid solution is ordered one, the entropy decreases. Hence the term $(-TdS)$ becomes positive. The free energy of the solid solution decreases if the magnitude of dH is less than that of TdS, Fig. 6.5.

The solid solution can also be random one and in that case the entropy of the solid solution increases. The free energy of such solution decreases for all compositions and thus the solution becomes stable. Such reactions are always spontaneous. The reactions, in which the solutions have negative deviation from ideality, are exothermic reactions. Free energy considerations favour solubility for all such solution compositions.

(iii) Positive deviation from ideality

The enthalpy of the solid solution may also have positive deviation from the ideal value of enthalpy. In that case dH is positive, Fig. 6.6. The entropy of such solution cannot decrease as compared to the initial entropy of components i.e. the solid solution cannot be ordered one. Free energy of such solid solution decreases, if $TdS > dH$ and it increases when $TdS < dH$, gives rise to endothermic reactions. The free energy of the intermediate solution may be more than that of mixtures for certain compositions of solutions, at a specific temperature, Fig. 6.7. At such compositions, minimum free energy favours phase separation. So the solution becomes

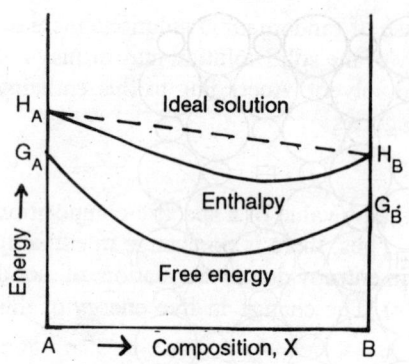

Fig. 6.5 Negative deviation from ideality:
Exothermic solution

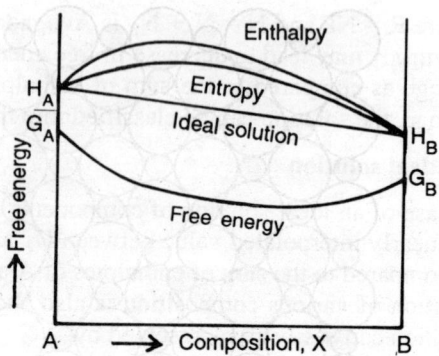

Fig. 6.6. Positive deviation of enthalpy from
ideality: Endothermic solution

immiscible mixture of two phases. As the temperature of the solution is increased, its free energy decreases and above a specific temperature, the solution becomes miscible if $TdS > dH$.

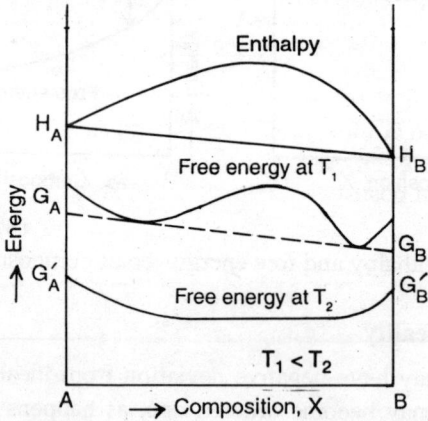

Fig. 6.7 Positive deviation from ideality, giving immisible mixture.

(c) Vacancies in ionic crystals

The vacancies in ionic crystals occur in such a way that the overall electrical neutrality is maintained. The imperfection due to vacancy is called Frenkel defect, if an ion is displaced from a regular site and moves to an interstitial site thereby creating a vacancy in the regular site. Since the cations are small so they have more tendency to get displaced and to create Frenkel defects. Such defects are produced by heating effect or by bombardment with neutrons, electrons and γ-rays. The vacancies and interstitials can be observed using field ion microscopy or scanning tunneling microscopy. The Frenkel defects are commonly observed in silver halides and CaF_2.

Another type of defect that is observed in NaCl type ionic crystal at high temperature is Schottky defect. When cation and anion vacancies occur in pairs, it is called Schottky defect. These vacancies in pairs maintain the charge neutrality. A vacant cation site is effectively having negative charge and that of anion site is positively charged.

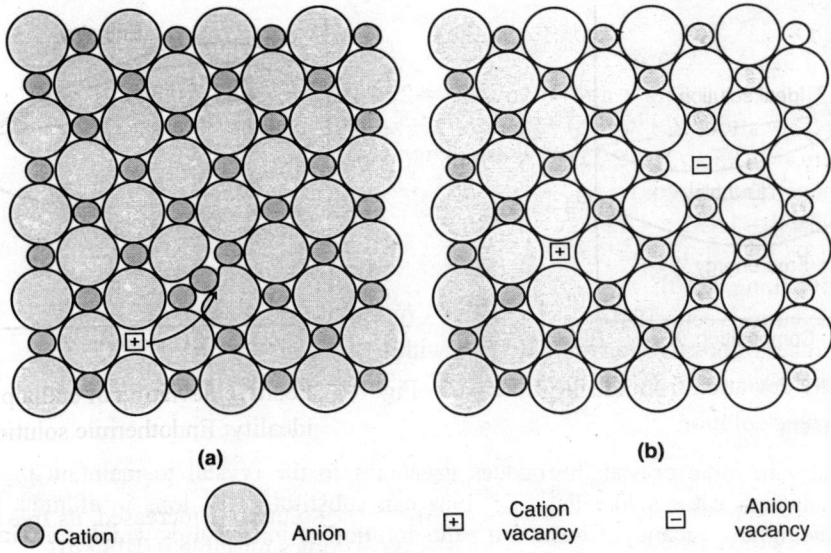

(a)

(b)

○ Cation   ⬭ Anion   ⊞ Cation vacancy   ⊟ Anion vacancy

Fig. 6.8 (a) Frenkel defect and (b) Schottky defect in ionic crystal

Example 6.3

CaO has NaCl type structure, Fig. 6.9. Find the area of (102) plane in the unit cell and hence the planar density of ions in (102) plane. Find the radius of the tetrahedral void in the above crystal structure. Find the linear density of positive ions along [210] direction.

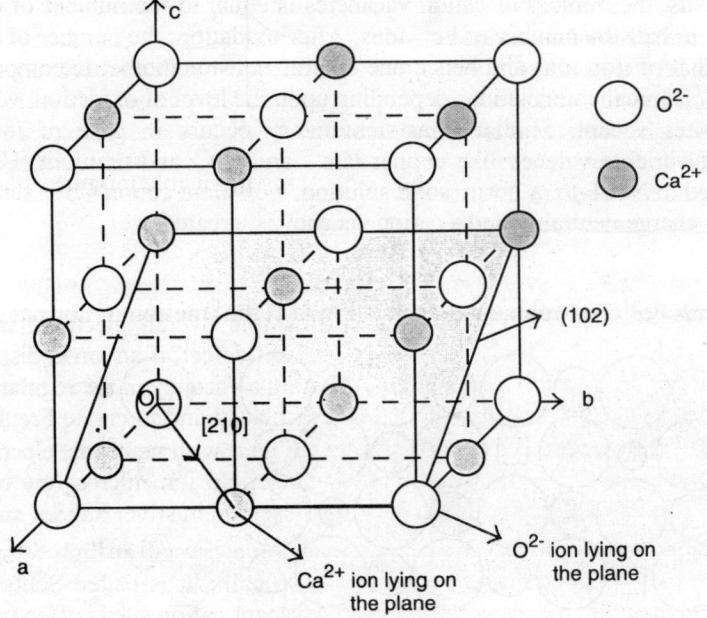

○ O^{2-}

⬤ Ca^{2+}

(102)

[210]

b

a

Ca^{2+} ion lying on the plane

O^{2-} ion lying on the plane

Fig. 6.9 CaO crystal

Solution:

a = 2(0.94 + 1.32) Å = 4.52 Å

Area of (102) plane = a $\sqrt{(a^2 + a^2/4)}$ = $\sqrt{5}$ a^2/2 = 22.84 Å^2

No. of ions = 1 cation(¼ + ¼ + ½) + 1 anion(¼ + ¼ + ½) = 2 ions

Planar density = 2/($\sqrt{5}$ a^2/2) = 2/22.84 Å^2 = 0.08756 ions/ Å^2

Radius of the tetrahedral void = r$_v$

r$_v$ + r$_{O2-}$ = $\sqrt{3}$a/ 4 = 1.732 x 4.52/4 Å = 1.957 Å

r$_v$ = 1.957 Å – 1.32 Å = 0.637 Å

Linear density along [210]:

Length of the lattice vector [210]= a$\sqrt{(1^2 + (½)^2 + 0)}$ = a$\sqrt{(5/4)}$ = a$\sqrt{5}$/ 2

Effective number of positive ion on [210] = ½ within unit cell

Linear density = ½ / (a$\sqrt{5}$/ 2) = 1/ (a$\sqrt{5}$) = 1/(4.52 $\sqrt{5}$) Å = 0.09894 positive ions Å^{-1}

Nonstoichiometry

Nonstoichiometry in ionic crystal, introduces vacancies in the crystal to maintain the charge balance. The trivalent cations like Fe^{3+}, Cr^{3+}ions can substitute Al^{3+} ions in alumina crystals without introducing any vacancy. Consider a solid solution of ionic solids which contain anions of same valency, but the solute has cations with valency greater than that of cations in solvent, then the cation vacancies are created to maintain charge balance, Fig. 6.10. Similar effects are there, if ferrous oxide is oxidized to ferric i.e.

$$3Fe^{2+} + ½ O_2 \rightarrow 2Fe^{3+} + 1 \text{ cation vacancy} + FeO \qquad (6.10)$$

As Fe^{2+} ions are oxidized to Fe^{3+}, the charge balance is disturbed. One can also explain the above process as follows. FeO has NaCl structure i.e. each lattice point is associated with a basis consisting of one cation Fe^{2+} and one anion O^{2-}. During oxidation, each oxygen ion entering the host lattice adds a cation vacancy since no Fe^{2+} ion joins the lattice. In this process two negative charges are added to the host matrix; hence two Fe^{2+}ions convert into two Fe^{3+} ions to maintain charge neutrality. Thus, the number of cation vacancies is equal to the number of oxygen ions added and also equal to half the number of Fe^{3+} ions. After oxidation, the number of oxygen ions becomes more than that of iron ions and hence one obtains nonstoichiometric compound $Fe_{1-x}O$ which has a range of chemical compositions depending upon the level of oxidation, where x is the fraction of cation sites vacant. Similar nonstoichiometry occurs in case of ionic crystals containing cations of variable valency like copper (Cu^{2+} and Cu^+) and uranium (U^{3+} and U^{4+}). When $CdCl_2$ is added to NaCl to a form solid solution, a divalent cation Cd^{2+} substitutes two Na^+ ions to maintain charge neutrality and a cation vacancy is created.

Example 6.4

A sample of $Fe_{1-x}O$ has Fe^{3+}/Fe^{2+} ratio equal to 1/3. Find (a) the fraction of normal cation sites

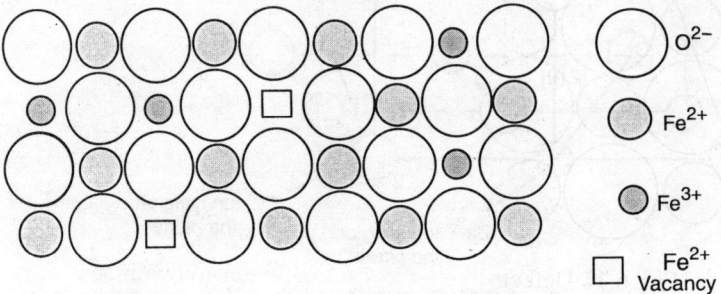

Fig. 6.10 Defects in FeO due to nonstoichiometry

vacant, (b) weight percent (w/o) oxygen in the sample, (c) Fe/O ion ratio, (d) formula of the compound.

Solution:

(a) Assume there are 300 Fe^{2+} ions, so there are 100 Fe^{3+} ions and 50 vacant cation sites.
   Total number of cation sites $= 300 + 100 + 50 = 450$
   Fraction of cation vacancy $= 50/450 = 0.1111$

(b) To maintain charge neutrality there are 300 O^{2-} ions corresponding to 300 Fe^{2+} ions and there are 150 O^{2-} ions corresponding to 100 Fe^{3+} ions. Hence

   Number of Fe ions $= 400$
   Weight of iron $= 400 \times 55.8 /N_A = 22320/N_A$
   Number of O^{2-} ions $= 450$
   Weight of oxygen $= 450 \times 16.0/N_A = 7200/N_A$
   Weight fraction of oxygen $= 7200 / (22320 + 7200) = 0.2439$

(c) Fe/O ratio $= 400/450 = 0.8889$
   So, $x = 1 - 0.8889 = 0.1111$

(d) Formula of the compound: $Fe_{0.8888}O$

Example 6.5

If in a NaCl crystal 25 atomic percent of the NaCl molecules are substituted by $CdCl_2$, what should be the chemical formula of the final product?

Solution:

Consider initially there are 1000 NaCl molecules. So after substitution, there are 750 NaCl molecules and 125 $CdCl_2$ molecules. Each Na+ ion, replaced by Cd^{2+} creates a cation vacancy. Now number of Cd^{2+} ions is equal to 125 and number of cation vacancies is equal to 125. The chemical formula for the solid solution can be written as $Na_{0.75} Cd_{0.125}Cl$.

Example 6.6

When ZnO is heated in zinc vapour, the excess cations occupy interstitial voids forming nonstoichiometric compound $Zn_{1.11}O$, Fig. 6.11. Find the w/o Zn occupying interstitial positions in the final product.

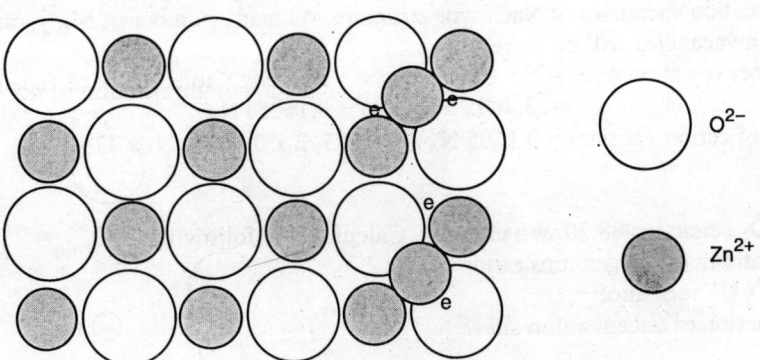

Fig. 6.11 Defects in zinc oxide due to nonstoichiometry

Solution:

According to the given formula of nonstoichiometric compound 1.11 mole of Zn combines with 1 mole of oxygen such that 0.11mole of Zn occupies interstitial positions.

Weight of the compound = 65.37 x 1.11 + 16 = 88.56 g

w/o Zn in interstitial positions = [(65.37x0.11) / 88.56] x 100 = 8.12 %

When an ionic compound is reduced it may become nonstoichiometric. As a result, anion vacancies are created to maintain the charge neutrality of the compound. For example, a compound like UO_2 becomes nonstoichiometric UO_{2-x}, when some of U^{4+} ions become U^{3+} ions due to reduction. When a pair of U^{4+} ions are replaced by a pair of U^{3+} ions, there is deficiency of two positive charges and this reduction in positive charges is neutralized by creation of an anion (O^{2-}) vacancy. The anion vacancies contribute to the increased mobility of anions.

Solid solution of two or more compounds has either cation or anion vacancies depending upon the formula of the compound and the valencies of cations and anions. If $Fe_{1-x}O$ is dissolved in MgO under oxidizing conditions, one cation vacancy is created for the substitution of every three Mg^{2+} ions by two Fe^{3+} ions. Such substitutions maintain the charge balance. No vacancies are produced if Fe^{2+} ions replace Mg^{2+} ions since they have the same ionic charges. Similarly when MgF_2 is dissolved in LiF, each Mg^{2+} ion replaces two Li^+ ions to maintain charge neutrality resulting in a cation vacancy.

Example 6.7

Consider a ceramic compound containing 10 w/o MgF_2 and 90 w/o LiF forms a solid solution having NaCl type structure. F^- ions occupy all anion sites. Determine

(i) the ratio of Mg^{2+} to Li^+

(ii) fraction of cation sites vacant.

Solution:

Let us consider 100 gms of sample which contains 10 gms of MgF_2 and 90 gms of LiF.

One mole of MgF_2 = 62.31 g

10 g of MgF_2 contains (10/62.31) N_A molecules

So number of Mg^{2+} ions = 0.1605 N_A,

where N_A is Avogadro number. Similarly, 90g of LiF contains (90/25.94) N_A number of Li^+ ions, which is equal to 3.4695 N_A numbers of Li^+ ions.

Ratio of Mg^{2+} and Li^+ ions = 0.04626

Since all anion sites are occupied, and each Mg^{2+} is forming compound with two anions (F^- ions), there are cation vacancies in NaCl type structure. As many number of Mg^{2+} ions are present that many cation vacancies will be there.

Here total number of cation sites = No. of Li^+ ions + No. of Mg^{2+} ions + No. of vacancies
= (3.4695 + 0.1605 + 0.1605) N_A

So fraction of cation vacancy = 0.1605 N_A / (3.4695+2 x 0.1605) $N_A \approx$ 423/1000

Example 6.8

A sample of UO_{2-x} contains 88.20 w/o uranium. Calculate the following:

(i)   Uranium to oxygen ions ratio;

(ii) U^{3+}/ U^{4+} ion ratio;

(iii)   Fraction of vacant anion sites.

Solution:

Consider 100 g of sample, out of which 88.20 g is uranium.

(i) Number of U ions present in the sample = $(88.20/238)\,N_A = 0.3706\,N_A$
Number of oxygen ions = $(11.80/16)\,N_A = 0.7375\,N_A$
Ratio of Uranium ions to O^{2-} = $0.3706/0.7375 = 0.5025$

(ii) In a stoichiometric UO_2 compound ratio of U^{4+} ions to O^{2-} ion is 0.5
Number of cation sites is $0.3706\,N_A$ and the number of anion sites is $0.7412\,N_A$
Out of these anion sites, only $0.7375\,N_A$ sites are occupied by O^{2-} ions, so the number of
vacant sites = $(0.7412 - 0.7375)\,N_A = 0.0037\,N_A$

If two U^{4+} ions are replaced by two U^{3+} ions, one anion vacancy is created so that charge
neutrality is maintained. Hence in a sample UO_{2-x}, the number of U^{3+} ions is twice the number
of anion vacancy i.e. equal to $0.0074\,N_A$. So, the number of U^{4+} ions is equal to
$(0.3706 - 0.0074)\,N_A = 0.3632\,N_A$
Therefore the ratio U^{3+}/U^{4+} ions = $0.0074\,N_A/0.3632\,N_A \approx 0.02$

(iii) Fraction of vacant anion sites = $0.0037\,N_A/0.7412\,N_A \approx 0.005$

Example 6.9

Consider a nonstoichiometric cubic crystal having chemical formula $(Zr_{0.85}Ca_{0.15})O_{1.85}$ where 4
cation sites per unit cell are all filled up but out of 8 anion sites some are vacant. Calculate
    i) number of anion vacancies per 1000 unit cells,
    ii) ion ratio (Zr^{4+}/Ca^{2+}),
    iii) w/o of O^{2-} ions in the sample.

Solution:

(i) Consider that the above sample contains 1000 unit cells. So there are 4000 cation and
8000 anion sites. From the formula it follows that out of 2 anion sites 0.15 anion sites are
vacant i.e. number of anion vacancies per 1000 unit cells is equal to $0.15 \times 8000/2 = 600$.

ii) Ion ratio $(Zr^{4+}/Ca^{2+}) = 0.85/0.15 = 5.67$

iii) w/o of oxygen = $\dfrac{1.85 \times 16}{0.85 \times 91.22 + 0.15 \times 40.08 + 1.85 \times 16} \times 100\% = 26.16\%$

Example 6.10

Analysis of a sample of otherwise pure iron (bcc) showed that 0.008 w/o carbon atoms was
present in the interstices among iron atoms. How many carbon atoms are there per 10,000
unit cells?

Solution:

Consider the sample of 100 g. It contains 99.992 g of iron and 0.008 g of carbon.
Number of iron atoms = $(99.992/55.8)\,N_A = 1.792\,N_A$
Number of carbon atoms = $(0.008/12)\,N_A = 6.667 \times 10^{-4}\,N_A$
In 10,000 unit cells there are 20,000 iron atoms.
Hence in 10,000 unit cells there are $20000 \times 6.667 \times 10^{-4}/1.792$ carbon atom = $7.44 \approx 7$
carbon atoms.

6.4 FORMATION OF POINT IMPERFECTIONS

During the formation of point imperfections, the elastic stress-strain field is created around the
defect site irrespective of the type and size of the imperfection. Thus, the formation of defects
results in distortion in the crystal. When vacancies are formed, certain bonds are broken. When
solute atoms or molecules are introduced in the host material substitutionally or interstitially,
some bonds are broken and some are formed. The stress fields are created due to mismatch in the

sizes of the host and the solute atoms. A larger impurity atom introduces compressive stress-strain field whereas smaller size impurity introduces tensile stress-strain field. When the solute atoms are introduced in the host matrix, these atoms diffuse and create distortions in the crystal. Similarly, interstitial atoms produce stress around the voids they are occupying, if there is size mismatch. All these processes require energy.

Even when there is no mismatch in size, the energy is required for the impurity atoms to diffuse and to occupy appropriate sites, after crossing kinetic barriers. Similarly, the energy is required for the formation of a vacancy. The energy required for the formation of a point imperfection is called the enthalpy of formation of point imperfection (ΔH_f). This is expressed in eV/point imperfection or kJ mole^{-1}. The enthalpy of formation of vacancies is given in Table 6.3 for a few elements. The presence of solutes increases configurational entropy of a solid solution. In general, formation of any type of point imperfections in crystal increases configurational entropy, which is given by

$$\Delta S = k [N_A \ln(N_A) - (N_A-n) \ln(N_A-n) - n \ln(n)] \qquad (6.11)$$

where n is the number of point imperfections present in one mole of a crystal.

Table 6.3 Enthalpy of formation of vacancy

Crystal	Ag	Al	Au	Cd	Cu	Kr	Mg	Ni	Pb	Zn
ΔH_f in kJ/mole	106	68	95	38	120	7.7	56	168	48	49

The change in free energy is given by

$$\Delta G = n \Delta H_f - T \Delta S = n \Delta H_f - kT [N_A \ln(N_A) - (N_A-n)\ln(N_A-n) - n \ln(n)]$$

ΔG will be minimum, if

$$\frac{d(\Delta G)}{dn} = 0 = \Delta H_f - kT [\ln(N_A-n) + (N_A-n)/(N_A-n) - \ln(n) - 1] = \Delta H_f - kT [\ln((N_A-n)/n)]$$

i.e. $\ln((N_A-n)/n) = \Delta H_f/kT$

Since $N_A/n \gg 1$ hence $n/N_A \approx \exp{-(\Delta H_f/kT)}$

i.e. $\qquad n = N_A \exp{-(\Delta H_f/kT)}$

If we express ΔH_f in kJ/mol, the above expression gets modified as

$$n = N_A \exp{-(\Delta H_f/RT)} \qquad (6.12)$$

Example 6.11

Find the ratio of equilibrium concentration of vacancies in copper at 500°K and at 1000°K, where enthalpy of formation of point imperfection is 120 kJ mole^{-1}.

Solution:

$$\frac{n_{500}}{N} = \exp\left[-\frac{120 \times 1000}{8.314 \times 500}\right] = \exp(-28.86)$$

$$\frac{n_{1000}}{N} = \exp\left[-\frac{120 \times 1000}{8.314 \times 1000}\right] = \exp(-14.43)$$

$$\frac{n_{500}}{n_{1000}} = \exp(-14.43) = 5.4 \times 10^{-7}$$

In case of Schottky defects, ΔH_{fSch} is the enthalpy of formation of one mole of pair of vacancies (both anion and cation vacancies). Hence, the enthalpy of formation for one mole of each type of vacancies is $\Delta H_{fSch}/2$. Hence the concentration of Schottky imperfections in thermal equilibrium at temperature $T°K$ is given by

$$n_{Sch} = N_A \exp-(\Delta H_{fSch}/2RT) \qquad (6.13)$$

Example 6.12

When the temperature of sodium chloride crystal is increased from room temperature 300°K to 800°C, Schottky defects increase by a factor of 3.98×10^{12}. Find the enthalpy of formation for the crystal.

Solution:

The ratio of defects at room temperature and at 800°C is given by

$$\frac{n_{300°K}}{n_{1073°K}} = \exp -\left[\frac{\Delta H_{fSch} \times 1000}{2 \times 8.314} \times (1/300 - 1/1073) \right]$$

$$\ln\frac{1}{3.98 \times 10^{12}} = -\frac{\Delta H_{fSch} \times 1000}{2 \times 8.314} \times 0.0024014 = -\Delta H_{fSch} \times 0.1444$$

$$\Delta H_{fSch} = 201 \text{ kJ/mole}$$

Density of defect structure

The vacancies as well as impurities change the density of the crystal. The presence of vacancies decreases the density of the crystal.

Example 6.13

Find the expression for percentage change in length due to thermal expansion and due to formation of vacancies in the crystal as the temperature is increased from 0°K to $T°K$.

Solution:

$$\text{Thermal expansion} = \Delta l_{ther} = l_o \, \alpha \, T$$

where α is the coefficient of linear expansion. Now let us consider expansion due to vacancy, formation.

$$\frac{\Delta V_{vac}}{V_o} = \frac{n}{N} = \exp(-\Delta H_f/RT),$$

where ΔV_{vac} = increase in volume due to vacancies

$$\Delta l_{vac} = l_o \, \tfrac{1}{3}\exp(-\Delta H_f/RT)$$

where Δl_{vac} = change in linear dimension due to vacancy formation.

$$\Delta l_{total} = \Delta l_{ther} + \Delta l_{vac}$$

Percentage change in length due to thermal expansion as compared to the total change in length,

$$\frac{\Delta l_{ther}}{\Delta l_{total}} \times 100\%$$

and percentage change in length due to vacancies as compared to the total change in length,

$$\frac{\Delta l_{vac}}{\Delta l_{total}} \times 100\%$$

Example 6.14

In a sample of copper crystal, one site is vacant for every 250 unit cells at a temperature 727°C. If these vacancies remain in the copper when it is cooled to 27°C, what will be the density of copper?

Solution:

The number of vacancies at 727°C is one out of 250 x 4 lattice sites i.e.

$$\frac{n}{N} = \frac{1}{1000}$$

The density of copper before introduction of vacancies

$$= \frac{M_{Cu} \times 4}{a^3} = \frac{63.54 \times 1.66 \times 10^{-27} \times 4}{(1.278 \times 2\sqrt{2})^3 \times 10^{-30}} = 8.94 \times 10^3 \text{ kg/m}^3$$

where M_{Cu} is the mass of copper atom and a^3 is the volume of unit cell. After introduction of vacancies,

$$\text{Density of copper} = \frac{M_{Cu} \times 999}{250a^3} = 8.94 \times 10^3 \times \frac{999}{1000} = 8.93 \times 10^3 \text{ kg/m}^3$$

Example 6.15

A material having fcc structure has 4 vacancies/1000 unit cells at room temperature. When it is heated, it is found that the lattice parameter has increased by 0.6% due to thermal expansion, while the density has decreased by 2.5%. Find the number of vacancies per 1000 unit cells at the higher temperature.

Solution:

Total fractional change in density = – 0.025
Fractional change in length due to thermal expansion = 0.006
Hence fractional change in volume = 0.018
Fractional change in density due to thermal expansion = – 0.018
Fractional change in density due to vacancies formed = – 0.025 – (– 0.018) = – 0.007
Number of vacancies created during heating per 1000 unit cells = 0.007 x 4 x1000 =28
Total number of vacancies at higher temperature = 28+4= 32 per 1000 unit cells

6.5 DISLOCATIONS

The line imperfection, which is commonly called dislocation, is geometrically one-dimensional defect. These defects may arise due to formation of incomplete crystal plane or due to plastic deformation. To visualize this let us consider a crystal is undergoing shear. Due to shear stress some part of the crystal is displaced in the direction of shear by one or fractional lattice translation. The boundary between slipped and unslipped region is the line of imperfection or dislocation. The dislocation can be of two types:

      a) Edge dislocation

      b) Screw dislocation

Edge dislocation

To visualize the edge dislocation, let us compare a perfect crystal with a crystal having edge dislocation. We can represent the crystal consisting of vertical planes parallel to one another and also to side faces. Each such crystal plane composed of parallel horizontal lines of atoms matching with each other the plan view of which is shown in Fig. 6.12.

If one of these vertical planes is incomplete, the boundary of incomplete plane within the crystal is called a dislocation. The dislocation line is the line of mismatched atoms within the crystal. The dislocation line is perpendicular to the plain of the figure and passes through the atom at point A, Fig. 6.12b. Dislocation can arise at the time of crystal formation or due to shear

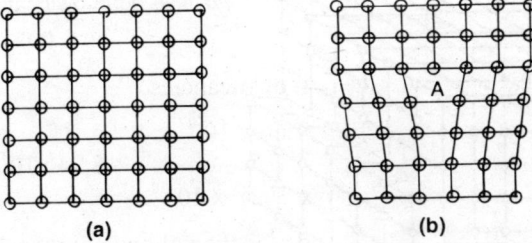

      (a)   (b)

Fig. 6.12 Schematic two-dimensional representation of a crystal (a) without dislocation;(b) with dislocation

stress. Consider the upper part of the crystal above the hatched area PQRS, is displaced to the left by one smallest lattice vector as compared to the lower part of the crystal due to shear, Fig. 6.13. As a result of this displacement, the first vertical crystal plane from the right side of the crystal face over hatched area is going to match with the second vertical plane below the hatched area. The second plane above the hatched area matches with the third plane below the area and so on till we reach the fifth half plane TRSV. This plane does not match with any vertical plane below the hatched area. This plane is termed as extra half-plane or incomplete plane. The line of mismatched atoms along the boundary RS of this extra half-plane is the edge dislocation line. The horizontal plane, on which the upper part of the crystal is slipped, is called a slip plane. The edge dislocation line is the boundary between slipped and unslipped parts of the crystal. It lies on the slip plane. If the shear is increased, the upper part of the crystal displaces further to the left and hence the extra half-plane moves to match with the next vertical plane i.e. sixth plane below the slip plane. This results in the motion of the dislocation line to the left along the slip plane. When the extra half-plane lies in the upper part of the crystal above the slip plane, the edge dislocation is positive and symbolically represented as ($\perp$). The edge dislocation can also occur when the lower part of the crystal below the hatched area PQRS, is displaced to the left by one smallest lattice vector as compared to the upper part of the crystal. In this case, the incomplete plane appears in the lower part of the crystal and the edge dislocation is called negative edge dislocation. It is represented as (⊤).

Thus, the edge dislocation line (⊥ or ⊤) moves along slip plane in the direction of shear, which is perpendicular to the dislocation line. This type of motion is called glide or slip of dislocation line. Gradual increase of shear causes the dislocation line to glide along slip plane till it reaches the crystal boundary to the left, as a result the upper part of the crystal slips. In this situation there is no dislocation line in the crystal.

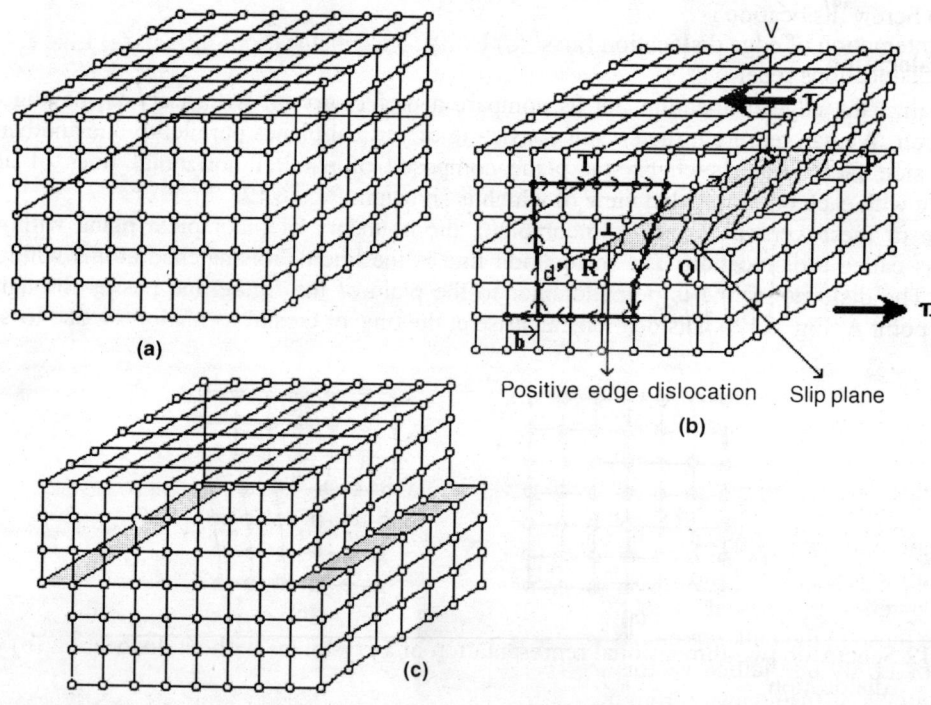

Fig. 6.13(a)Perfect crystal (b) Dislocation created by shear (c) Slip of the upper part of the crystal

The atoms just above the positive edge dislocation line are in a state of compression and those below the dislocation line are in a state of tension since the atoms are pulled apart in this region. In case of negative edge dislocation, atoms just below the dislocation line are in a state of compression and those above the slip plane are in a state of tension. Thus, stress-strain fields are of opposite nature in case of negative and positive edge dislocations. When these dislocations lie in the same slip plane, they attract and annihilate each other after application of suitable shear stress, Fig. 6.14 (a). The annihilation of dislocations causes removal of dislocations. When they are not in the same plane, then if these dislocations approach each other due to applied force, regions of compressive and tensile strain may partially overlap but annihilation does not take place. The dislocations of similar type lying in the same slip plane repel each other so they cannot easily approach nearer even when external force is applied. This results in extra stress- strain field.

The edge dislocations are usually good sources or sinks for vacancies. Vacancies can be created or destroyed at jogs on dislocation lines. Jogs are positions where dislocation moves from one slip plane to another, Fig. 6.15.

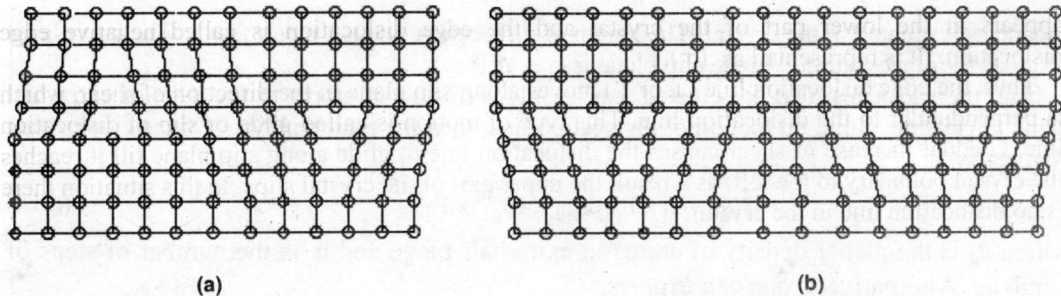

(a)   **(b)**

Fig. 6.14 Interaction of edge dislocation lines: (a) Unlike edge dislocations attract. (a) Like edge dislocations repel

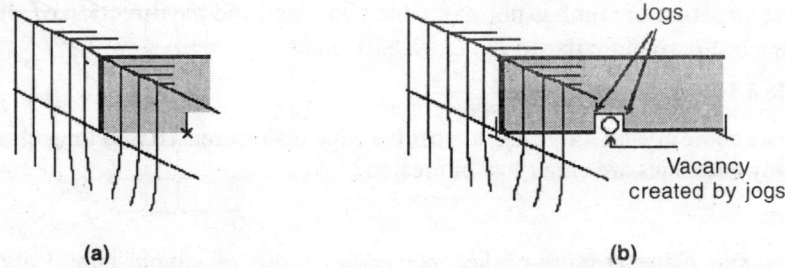

(a)   **(b)**

Fig. 6.15 (a) At jogs, dislocation line moves from one slip plane to another.
(b) Vacancy created by jogs

Dislocation lines can climb on a plane perpendicular to the glide plane. If the crystal is heated, the atoms at the boundary of the extra half-plane can easily gain enough energy to move to some vacant sites within the crystal or to the surface of the crystal, since these atoms have higher energy than those in the bulk region. As a result, the positive edge dislocation climbs up by a step i.e. it climbs up by one lattice vector in a direction normal to the slip plane. If there is larger number of atoms diffusing away from the edge of the incomplete plane, the plane shrinks further i.e. the dislocation climbs up by more than one step.

Positive edge dislocation climbs down, when atoms diffuse to the edge of the extra half-plane, there by adding rows of atoms from other parts of the crystal. Similarly, the negative edge dislocation climbs up or climbs down depending upon the addition or subtraction of rows of atoms at the edge of the extra half-plane. The negative edge dislocation climbs up, if the atoms diffuse on to the edge of extra half-plane, creating equal number of vacancies in the crystal or at the surface. Atoms may diffuse from the negative edge dislocation to the crystal where they annihilate some of the existing vacancies. As a result the negative edge dislocation climbs down. Heating facilitates these processes. Climb motion is said to be nonconservative since it involves addition or subtraction of rows of atoms at the edge of incomplete plane, whereas glide motion is conservative, since it does not involve addition or subtraction of atoms at the boundary of the incomplete plane.

Calculation of number of steps of climb up or climb down of dislocation line

Consider climbing up motion of a negative edge dislocation line lying on a slip plane whose Miller indices are (hkl). The climb up motion adds rows of atoms to the negative edge dislocation, there by creating vacancies elsewhere in the crystal. Consider that N_v number of vacancies is created in the crystal. The length of dislocation line is 1 meter (say). The interplanar distance for planes parallel to the slip plane is $d_{[hkl]}$ Å (say).

If all the vacancies, that are created, cause only diffusion of atoms to the edge of extra half plane to generate climb up motion of dislocation line, then the total number of vacancies created is given by

$$N_v = n_s \times 1 \times d_{[hkl]} \times 10^{-10} \times P_d \qquad (6.14)$$

where P_d is the planar density of atoms in extra half-plane and n_s is the number of steps of climb up. Alternatively, one can express,

$$N_v = L \times 1 \times P_d,$$

where L is the distance through which climb up has taken place. Similarly, one can calculate the number of vacancies annihilated during climb down motion of negative edge dislocation. The direction of climb is normal to the slip plane and the direction of slip is parallel to the slip plane but perpendicular to the extra half-plane.

Example 6.16

In a simple cubic crystal (a= 2Å), a positive edge dislocation 10 mm long climbs down by 1μm. How many vacancies are either lost or created?

Solution:

Since the slip plane is close-packed plane (say (100) of simple cubic) and slip direction is perpendicular to climb down direction, the planar density of extra half plane is equal to that of one of the close-packed planes in this case. Thus,

$$P_d = 1/a^2 = 0.25 \times 10^{20} \text{ atoms /m}^2$$

and

$$N_v = 10^{-2} \times 10^{-6} \times 0.25 \times 10^{20} = 2.5 \times 10^{11}$$

Burgers vector and burgers circuits

The magnitude and direction of displacement due to shear can be defined in terms of Burgers vector **b**. The Burgers vector of a dislocation line can be determined by drawing Burgers circuit around the dislocation line. The Burgers circuit is, in general, rectangular and is drawn on the crystal plane such that starting from a lattice point, one should go n_1 steps up, n_2 steps to the right, n_1 steps down and then n_2 steps to the left. If the end lattice point coincides with starting lattice point, then the region surrounded by the Burgers circuit is perfect without any dislocation or imperfection crossing the crystal plane, Fig. 6.16(a). If the Burgers circuit drawn on a crystal plane enclosed a region with line imperfection intersecting the plane, then the end point of the Burgers circuit does not coincide with the starting point, Fig 6.16 (c) (where $n_1=1$ and $n_2=3$). Thus, incomplete Burgers circuit indicates the existence of line imperfection. A vector is drawn to join the end point T of the Burgers circuit to the starting point P. **TP** is called Burgers vector **b**.

The Burgers vector is perpendicular to the edge dislocation line and is parallel to slip plane. Thus the cross product of direction indices of Burgers vector and that of dislocation line **d** gives the direction indices of normal **n** to the slip plane, which is also the Miller Indices of the slip plane. The Miller indices of slip plane (or glide plane) can be calculated by taking the cross product of **b** and **d** vectors. Let us say MI of b is [$h_1k_1l_1$] and that of **d** is [$h_2k_2l_2$], then the Miller indices (hkl) of the glide plane for pure edge or mixed dislocations are given by

$$\begin{bmatrix} h \\ k \\ l \end{bmatrix} = \begin{matrix} k_1l_2 - k_2l_1 \\ l_1h_2 - l_2h_1 \\ h_1k_2 - h_2k_1 \end{matrix} \qquad (6.15)$$

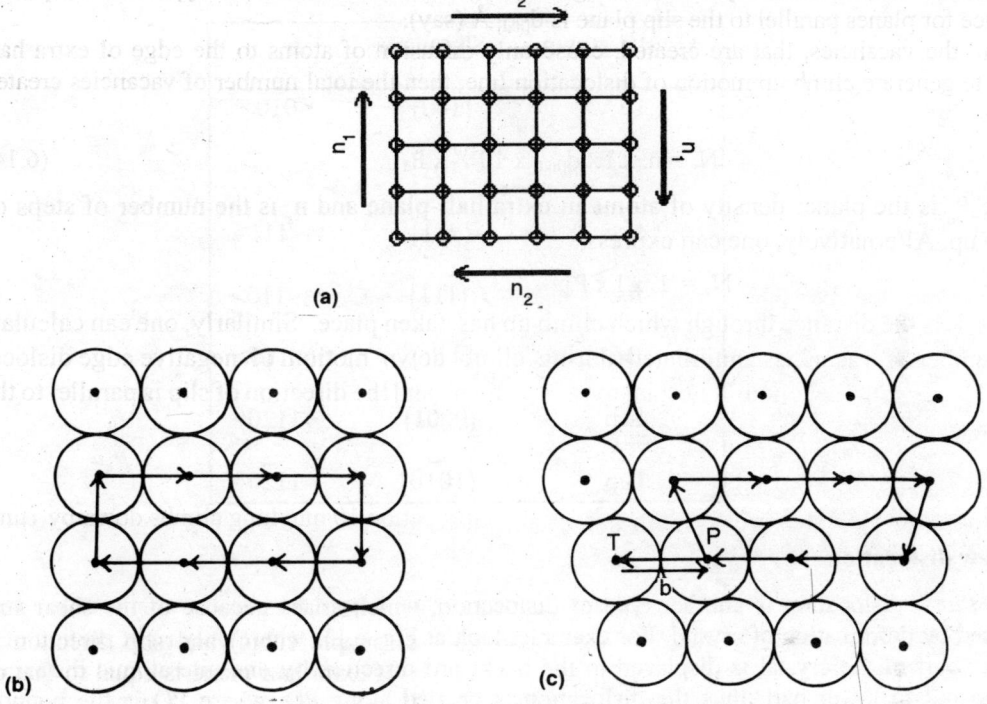

Fig. 6.16 Burgers circuit and Burgers vector

The Miller indices of extra half-plane is $(h_1k_1l_1)$ in case of pure edge dislocation since **b** is perpendicular to the extra half-plane. The direction of climb is [hkl] which is normal to slip plane.

Example 6.17

Consider an fcc crystal having Burgers vector $\frac{1}{2}[10\bar{1}]$ and the edge dislocation is [121]. Find the slip plane.

Solution:

$$\text{Normal to slip plane} = \frac{1}{2}[10\bar{1}] \times [121] = [1\bar{1}1]$$

$$\text{Hence MI of slip plane are } (1\bar{1}1) \text{ or } (\bar{1}1\bar{1})$$

The direction of the Burgers vector depends on whether the Burgers circuit is clockwise or anticlockwise. The direction of the Burgers circuit is determined in the following way. The direction of edge dislocation line is denoted by unit vector **d**. The direction vector **d** is tangential to the dislocation line at the point around which the Burgers circuit is drawn. Then the right hand screw (RHS) convention has to be applied to draw the Burgers circuit such that the direction of the movement of the end of the screw should coincide with the direction **d** and the direction of Burgers circuit should be same as the direction of rotation of the screw. If the direction **d** is pointing into the crystal plane, the Burgers circuit is clockwise. The slip plane is usually the close-packed plane and the direction of slip is parallel or antiparallel to **b**. Table 6.4 gives the possible slip planes and Burgers vectors for various crystallographic groups.

Table 6.4 Slip plane and **b** for various crystal systems

Crystal system	Slip plane	b
sc	{100}	<010>
bcc	{101}	½<11$\bar{1}$>
bcc	{211}	½<$\bar{1}$11>
fcc	{111}	½<$\bar{1}$10>
DC	{111}	½<$\bar{1}$10>
hcp	{0001}	<11$\bar{2}$0>
hcp	{10$\bar{1}$0}	<11$\bar{2}$0>

Screw dislocation

The screw dislocation is another type of dislocation, which arises because of the shear stress caused by deformation of crystal. For example, look at Fig. 6.17, where only right portion of the upper part of the crystal is displaced to the backward direction by one interatomic distance as compared to lower part, thus the dislocation is created along PQ, where PQ is the boundary between displaced and undisplaced part of the crystal. The name screw dislocation derives from the fact that the lattice points next to the dislocation lie on a helical or spiral path, which is traced around the dislocation line, Fig 6.18.

The plane on which slipping of the crystal occurs, is called the slip plane. Let **d** be the screw dislocation vector such that it is normal to the front face and points out of the crystal. Using RHS convention, the Burgers circuit is drawn in a anticlockwise sense. As defined earlier, incomplete Burgers circuit is completed by joining the end point to the starting point of circuit with the help of a vector called Burgers vector. In case of screw dislocation, the Burgers vector as well as screw dislocation vector lies on the slip plane but these two vectors are parallel or antiparallel as is quite clear from Fig 6.17.

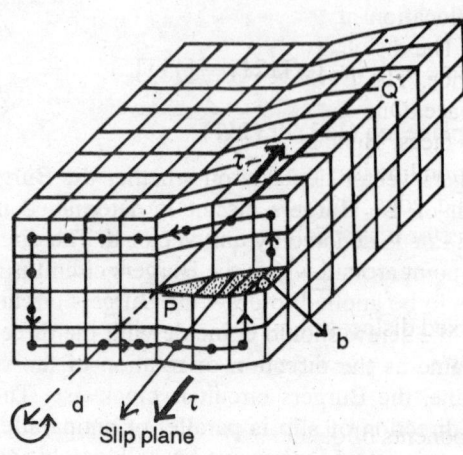

Fig. 6.17 Screw Dislocation

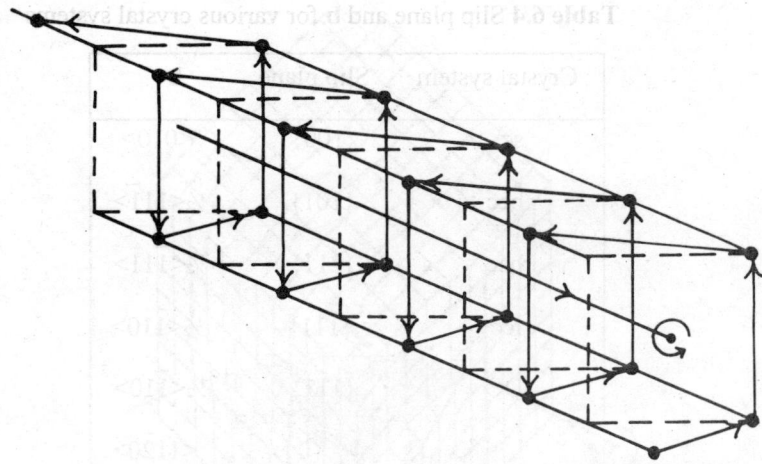

Fig 6.18 Helical path of atoms around screw dislocation

The screw dislocations are symbolically represented by ↺ or ↻ depending on whether **b** and **d** vector are either parallel or antiparallel. These two types of screw dislocations are also called positive and negative screw dislocations, respectively. Screw dislocations can also be observed if the lower part of the crystal is displaced similarly, with respect to the upper part due to shear stress. The glide or slip motion of a screw dislocation is perpendicular to the stress direction, in contrast to edge dislocation whose glide motion is parallel to the direction of shear stress. That is why the screw dislocation slips or glides in the direction perpendicular to **b**. There is no extra half plane in case of screw dislocation in contrast to the case of edge dislocation. Hence there is no climbing up or climbing down motion in case of screw dislocation.

Most of the dislocations that are observed, are mixed type i.e. they are combinations of screw and edge dislocations. Consider, for example, the dislocation line in Fig 6.19, which is mixed type in nature. This dislocation is shown as curved line, which forms boundary of the hatched area of the slip plane. The Burgers circuit, drawn using RHS convention on the right face, clearly shows that the Burgers vector is perpendicular to the dislocation at this face, hence the dislocation is pure edge, whereas on the front face the Burgers vector is antiparallel to the dislocation line, hence it is pure screw dislocation at this point. The dislocation line has mixed character in between these two points. The dislocation line points inside the crystal as we see from front and it points out of the right face. The Burgers circuit is clockwise on the front face, whereas it is anticlockwise in the right face. One can observe that the Burgers vector is having same direction and magnitude in both the faces. Hence one can conclude that the Burgers vector **b** is invariant along a continuous dislocation line irrespective of the changing character of dislocation. Thus one can find out the screw and edge components of dislocation as projection of **d** parallel and perpendicular to Burgers vector **b**, respectively.

Example 6.18

The Burgers vector of a mixed dislocation line is ½[110]. Consider the dislocation line lies along the [112] direction. Find the

(a) slip plane.
(b) edge and screw components of dislocation.

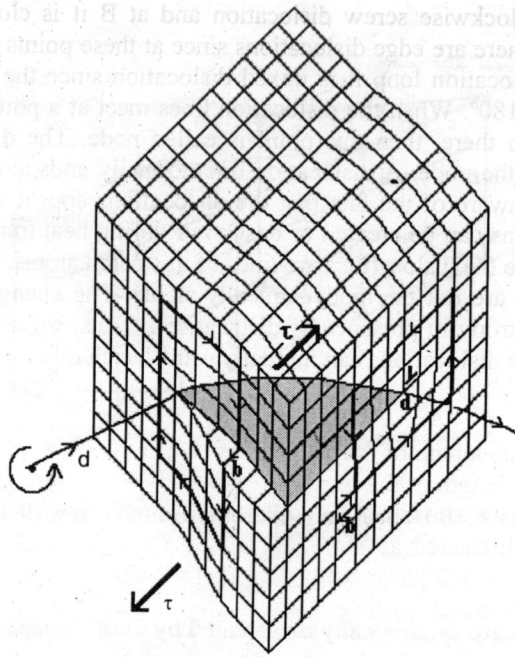

Fig. 6.19 Burgers vector for mixed dislocation

Solution:

Slip plane : ½[110] x [112] = (1$\bar{1}$0)
  d = [112], |d| = $\sqrt{6}$ i.e.   unit vector along **d** = [112] / $\sqrt{6}$
  b = ½ [110], |b| = 1/$\sqrt{2}$ i.e.   unit vector along **b** = [110] / $\sqrt{2}$

Unit vector perpendicular to **b** but parallel to slip plane = ½[110] x [1$\bar{1}$0] = [00$\bar{1}$] or [001]
Component of **d** parallel to **b** i.e. magnitude of screw component of dislocation = |d| cosθ

$$= \sqrt{6} \ (1 \times 1 + 1 \times 1 + 0) \ /(2\sqrt{3}) = \sqrt{2}$$

where θ is the angle between **d** and **b**. So the screw component of dislocation vector

$$= \sqrt{2} \ [110] \ x \ 1/\sqrt{2} \ = [110]$$

The edge component of **d** is perpendicular to **b** and is given by

$$|d| \ \sin\theta = |d| \sqrt{(1 - \cos^2\theta)} = \sqrt{6}\sqrt{(\frac{2}{3})} = 2$$

Hence the edge component of dislocation vector = 2[001]

The direction of the dislocation line at any given point is denoted by unit vector along d, which is tangential to the dislocation line, and it is continuous around a dislocation loop. Thus the dislocation vector **d** has opposite direction on the opposite sides of a loop. When the displacement of one part of the crystal has occurred on a slip plane over the hatched area bounded by the elliptical dislocation loop, the character of the dislocation changes along the loop. The plan view of the dislocation loop is shown in Fig 6.20.

The Burgers vector **b** is invariant along the dislocation loop but vector **d** is changing continuously along the loop. At point A, **d** and **b** are antiparallel whereas they are parallel at B.

So at A there is anticlockwise screw dislocation and at B it is clockwise screw dislocation. Similarly at C, and D there are edge dislocations since at these points **b** is perpendicular to **d**. At other points of the dislocation loop it is mixed dislocation since the angle between **b** and **d** is other than 0°, 90° and 180°. When the dislocation lines meet at a point such that the sum of the Burgers vectors is zero there, then this point is called node. The **d** vectors must either point towards or away from the node. A dislocation line normally ends at a node or at the surface or forms a loop. We are aware of the fact that the dislocations appear at the time of formation of crystal. Also dislocations can be created or destroyed during heat treatment or by application of mechanical stress. Since the dislocations are lines of mismatch atoms, which are very much prone to react, so these lines are not thermodynamically stable. The change in enthalpy occurs more rapidly than that of entropy in presence of dislocations. Thus, by simple application of heat or mechanical stress, these dislocations can mutually cancel each other or move out of the crystal.

Dislocation density

Dislocation density is the length of the dislocation line per unit volume of the crystal, but it is not possible to measure the length of the dislocation line physically. So how to calculate the length of the dislocation line? Dislocations can be observed with the help of an electron microscope, since the electron beam is diffracted as it passes through the lattice distortions created due to the dislocations. Chemical technique is also used to observe dislocations. In chemical technique, etchant is used on the polished surface of the crystal. The etchant forms a pit where a dislocation line emerges through the surface. The etch pits are formed at the dislocation sites because atoms surrounding a dislocation possess strain energy and react more readily with the enchant. The density of dislocations in a crystal can be estimated by finding out the average number of etch pits per unit area on random cross-sections of the crystal. The dislocation density varies from crystal to crystal. It is of the order of 10^8–10^{10}/m^2 in an annealed crystal. Consider a cube of volume 1 m^3 and assume that the dislocation lines pass the unit cube perpendicular to one of its face starting from or ending at the etch pits on the face. In that case, the length of each dislocation line is 1 meter and the number of dislocation lines is equal to the number of etch pits on one of the faces of the crystal on which the dislocation lines are either starting or terminating. So we can as well say dislocation density is 10^8 – 10^{10} m/m^3 in the above case.

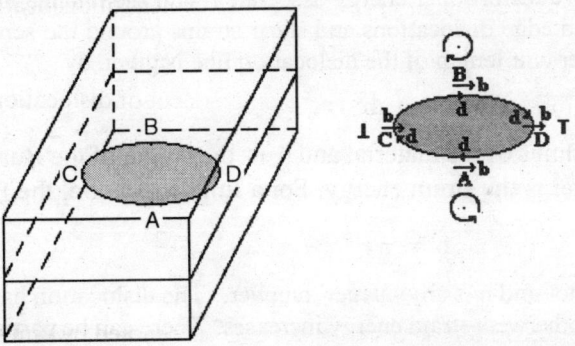

Fig 6.20 Dislocation loop

Slip in compounds

In case of compounds (specifically ionic compounds), usually a specific type of atom or ion has unlike nearest neighbours. If during glide motion of the dislocation, like ions become nearest neighbours, the energy of the system increases, since the binding energy of the unlike ions is higher than like ones in such compounds. Let us consider the example of CsCl crystal. If the Burgers vector is ½<111> for the edge dislocation in a CsCl crystal, which is also the slip vector, then the extra half plane will have excess electrical charges since this results in bringing the like charged ions nearer. There will be excess energy in the crystal since the magnitude of binding energy

$$E_{Cs^+ - Cl^-} > E_{Cs^+ - Cs^+} \text{ and } \qquad E_{Cs^+ - Cl^-} > E_{Cl^- - Cl^-}$$

Hence such a shear is resisted. The configurations with like charges, as nearest neighbour is not a stable one. The Burgers vectors are <100> directions in this case. The shear stresses on some crystal planes are sufficiently high so that slip is nearly impossible. This property decreases the ductility of the material and that is why quite a few ceramic materials like MgO, $Ni_8Fe_{16}O_{24}$ etc. are brittle at room temperature. Similarly the Burgers vector for NaCl crystal cannot be the vector joining nearest Na+ and Cl⁻ ions, rather it is ½<110>. Hence these restrictions in ionic compounds have resulted in a larger Burgers vector. In case of fcc iron the Burgers vector of a full dislocation is 2.54 Å, whereas the Burgers vector is 3.95 Å for NaCl. The related data for other ionic compounds are given in Table 6.5.

Table 6.5 Burgers vectors for compounds

Structure	Example	Burgers Vector
NaCl	LiF,MgO,MnS,TiC	½<110>
CsCl	CsCl	<100>
Al₂O₃	Al₂O₃	<1120>
DC	ZnS	½<110>

Distortional energy of dislocations

The dislocation lines have distortional energy associated with them since there are compressive and tensile strains around edge dislocations and shear strains around the screw dislocations. The elastic strain energy E per unit length of the dislocation line is given by

$$E = (½) \mu b^2 \qquad (6.16)$$

where ì is the shear modulus of the material and b is the length of the Burgers vector. Larger the Burgers vector, larger is the strain energy. For a full dislocation, the Burgers vector is

$$b = n r \qquad (6.17)$$

where r is the lattice vector and n is any integer number. The dislocation has a tendency to have smallest Burgers vector otherwise strain energy increases. There can be partial dislocation also, if the Burgers vector is a fraction of a lattice translation.

Interactions and dissociations of dislocations

As we have discussed earlier a dislocation line can end at the free surface of the crystal or form a loop. Another exceptional situation in which the dislocation line may end inside a crystal i.e. at a node, where more than one dislocation lines meet. A node is a point inside a crystal at which say two dislocation lines with Burgers vectors b_2 and b_3 combine to produce a resultant dislocation line with Burgers vector b_1 such that

$$\mathbf{b}_2 + \mathbf{b}_3 = \mathbf{b}_1 \tag{6.18}$$

Normally we observe the full dislocation lines. The dislocations with Burgers vector greater than one lattice spacing are unstable due to energy considerations and hence dissociate into two or more dislocations with Burgers vector of lower magnitude. The criterion for dissociation of a dislocation or combination of two or more dislocations to occur is that the strain energy of the system should decrease. In case of combinations of dislocation lines with Burgers vectors $\mathbf{b}_2$ and $\mathbf{b}_3$ to form a dislocation line with Burgers vector $\mathbf{b}_1$, the criterion for occurrence is given by

$$(b_2^2 + b_3^2) > b_1^2$$

Similarly, a dislocation reaction $\mathbf{b}_1 \rightarrow \mathbf{b}_2 + \mathbf{b}_3$ is possible if

$$b_1^2 > (b_2^2 + b_3^2)$$

A unit dislocation is a dislocation with Burgers vector of magnitude, one lattice vector length. It is also called dislocation of unit strength. If the Burgers vector is parallel to the direction of closest packing in the lattice, the energy associated with unit dislocation is minimum. Mostly the crystal slips in the close-packed directions. Dislocations with Burgers vector equal to fraction of lattice translations are also possible, especially in case of close-packed lattice.

Example 6.19

Consider the family of slip plane {111} of fcc lattice. The vector $\mathbf{b}_1 = \frac{1}{2}[10\bar{1}]$ defines the Burgers vector. Prove that an atom in a B plane moves easily by a zigzag path $\mathbf{b}_2 + \mathbf{b}_3$ rather than along $\mathbf{b}_1$ direction.

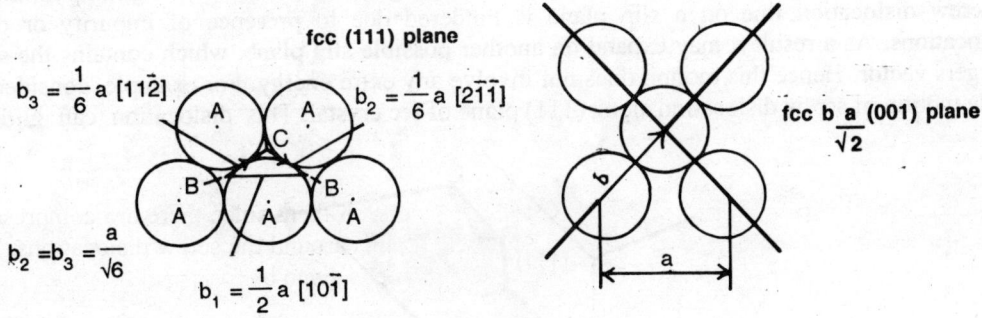

Fig. 6.21 Easy path of slip

Solution:

The dislocation reaction is given by $\mathbf{b}_1 \rightarrow \mathbf{b}_2 + \mathbf{b}_3$
As an example consider the reaction,

$$\tfrac{1}{2}[10\bar{1}] = 1/6[2\bar{1}\bar{1}] + 1/6[11\bar{2}]$$

The vector equation can be checked by comparing separately the equality of the sum of x,y and z components of Burgers vectors on both the sides of equality sign. The energetically favourable reaction should move from a state of higher strain energy to lower strain energy

$$b_1^2 = 1/2$$
$$b_2^2 + b_3^2 = 1/3$$

Hence the strain energy decreases in the above reaction, Eq. (6.16) i.e. zigzag path has lower energy than the path with Burgers vector b_1.

Example 6.20

Prove that an energetically possible dislocation reaction is given by

$$\tfrac{1}{2}[10\bar{1}] + \tfrac{1}{2}[011] = \tfrac{1}{2}[110]$$

Solution:

First let us check the vector equality:

	L.H.S.	R.H.S.
Sum of x components	$\tfrac{1}{2}+0 = \tfrac{1}{2}$	$\tfrac{1}{2}$
Sum of y components	$0 + \tfrac{1}{2} = \tfrac{1}{2}$	$\tfrac{1}{2}$
Sum of z components	$-\tfrac{1}{2}+\tfrac{1}{2} = 0$	0

Hence L.H.S. = R.H.S.

Strain energy before reaction $=\tfrac{1}{2}\mu[(\tfrac{1}{2})^2+(\tfrac{1}{2})^2+(\tfrac{1}{2})^2+(\tfrac{1}{2})^2] = \tfrac{1}{2}\mu$

Strain energy after reaction $=\tfrac{1}{2}\mu[(\tfrac{1}{2})^2+(\tfrac{1}{2})^2] = \tfrac{1}{4}\mu$

Hence the reaction is favourable since the strain energy decreases.

Cross slip

In case of screw dislocation **b** vector is either parallel or antiparallel to **d**. Therefore there is no unique glide plane for screw dislocations. The screw dislocation can therefore glide through any crystal plane parallel to **b** and **d**. This property of screw dislocation gives rise to another kind of motion for dislocation line, which is called cross slip, Fig. 6.21. Sometimes the gliding motion of a screw dislocation line on a slip plane is hindered due to presence of impurity or other dislocations. As a result it may expand on another possible slip plane, which contains the same Burgers vector. Hence this motion does not involve any extra energy. For example, consider the glide motion of screw dislocation along (111) plane of fcc crystal. This dislocation can glide to

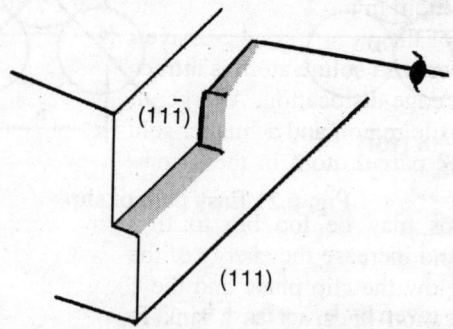

Fig. 6.22 Schematic representation of cross slip.

another plane $(11\bar{1})$ without any loss of energy since both have the same Burgers vector $[1\bar{1}0]$ Thus, screw dislocation can cross slip from one glide plane to another provided the Burgers vector of dislocation lies in both the planes i.e. Burgers vector is the line of intersection of these slip planes. One should remember that the slip planes are usually close-packed planes and slip directions are close-packed directions. The screw dislocation may cross slip more than once during glide motion. The cross slip does not involve any diffusion so it can occur more easily than climb at low temperature.

Example 6.21

If the slip plane of a bcc lattice is (110), find the possible slip directions for screw dislocation on this plane. Find MI of all the planes on which screw dislocation can cross slip.

Solution:

The slip directions are close-packed directions. The slip directions that lie on (110) plane are

$$[\bar{1}11], [1\bar{1}1], [1\bar{1}\bar{1}] \text{ and } [\bar{1}1\bar{1}]$$

Slip direction	Screw dislocation cross slips on planes from (110) plane
$[1\bar{1}1]$	$(011), (0\bar{1}1), (\bar{1}01), (10\bar{1})$
$[\bar{1}11]$	$(01\bar{1}), (0\bar{1}1), (101), (\bar{1}0\bar{1})$
$[1\bar{1}\bar{1}]$	$(01\bar{1}), (0\bar{1}1), (101), (\bar{1}0\bar{1})$
$[\bar{1}1\bar{1}]$	$(011), (0\bar{1}1), (\bar{1}01), (10\bar{1})$

Interactions of dislocations with point imperfections

Sometimes dislocations interact with point imperfections to decrease the strain energy of the system. The experimental evidence shows that vacancies are created during plastic deformation, which increases strain energy. The climb up motion of positive edge dislocation involves diffusion of atoms from the edge dislocation. These atoms annihilate the vacancies and hence strain energy decreases.

A substitutional solute atom in a solid solution is normally of different size as compared to the parent atom. As a result of this, a local compressive or tensile strain field is created around the solute atom. Hence an oversized solute atom is attracted to the tensile region and repelled by the compressive region of an edge dislocation. A larger atom can reduce strain energy by displacing the parent atom in the tensile region and a smaller solute atom can reduce the strain energy of the crystal by substituting the parent atom in the compressive strain field above slip plane of the positive edge dislocation.

Interstitial solute atoms may be too big to fit in the interstitial void and hence create compressive stress field and increase the energy of the crystal. The positive edge dislocation line has tensile stress field below the slip plane and the compressive stress field lies above the slip plane. Thus, these dislocation lines act as a sink for solute atoms. Dislocation lines capture diffusing atoms.

Dislocation lines can act as sink for a long period of time till they are saturated with the solute atoms. Such interaction is particularly strong in mild steel, where interstitial carbon atoms concentrate around the edge dislocation in bcc iron. The atmosphere around a dislocation condenses into a single line of carbon atoms parallel to the dislocation line below the core of positive edge dislocation.

Example 6.22

Consider a sample of mild steel (bcc) containing edge dislocation of density 10^{10} m^{-2}. Calculate w/o carbon to be added in one m^3 of volume to fill all the core sites of edge dislocations.

Solution:

Iron has bcc structure. So slip plane can be $(1\bar{0}1)$. Burgers vector (close-packed direction) can be chosen as $[111]$ and dislocation line d is cross product of normal to slip plane and Burgers vector i.e.

$$d = [111] \times [\bar{1}01] = [1\bar{2}1].$$

The length of the dislocation line is 10^{10} m per m^3.The number of mismatched atoms (N_m) along dislocation line per m^3

$$= \frac{\text{length of the dislocation line}}{\text{length of the lattice vector along dislocation line}} = \frac{10^{10}}{\sqrt{6}a} = \frac{10^{10} \times \sqrt{3}}{\sqrt{6 \times 4 \times 1.24 \times 10^{-10}}} = 1.425 \times 10^{19}$$

where 1.24×10^{-10} m is the atomic radius of Fe. The number N_m is also equal to number of core sites along edge dislocation. If all these sites are filled up by carbon atoms, w/o of carbon atoms are to be added

$$= (1.425 \times 10^{19}) \times 12 \times (1.66 \times 10^{-27}) \times 100 / 7870 = 3.607 \times 10^{-9} \%$$

where density of iron is 7870 kg/m^3

Example 6.23

Consider a simple cubic crystal with edge dislocation density 10^{10}m per m^3. The crystal is heated from 0° to 1000°K and as a result vacancies are created. 50% of these displaced atoms join the core sites of dislocation lines. How much would these edge dislocations climb down, if the enthalpy of formation of vacancies is 100 kJ mole^{-1}. The lattice parameter is 2.5 Å and molar volume of the crystal is 5.5×10^{-6}m^3.

Solution:

At equilibrium, there are no vacancies in the crystal at 0°K. The equilibrium concentration of vacancies at T°K is given by

$$n = N \exp(-\Delta H_f/RT)$$
$$= 6.023 \times 10^{23} \exp\{-(100 \times 1000)/(8.314 \times 1000)$$
$$= 3.6 \times 10^{18} \text{ mole}^{-1}$$
$$= (3.6 / 5.5) \times 10^{24} \text{ m}^{-3} = 6.55 \times 10^{23} \text{ m}^{-3}$$

One step of climb down = 2.5 Å

No. of atoms along 10^{10} m of dislocation line = $10^{20}/2.5 = 4 \times 10^{19}$
This dislocation line is located in a volume of 1 m^3 and we have already observed 6.55 x10^{23} vacancies or displaced atoms are there in 1 m^3 volume of crystal. Thus average number of steps of climb down = 6.55 x10^{23} x 0.5 /(4 x 10^{19}) = 8188, where we have considered only 50% of displaced atoms take part in climbing down process. The distance through which the climb down takes place is equal to

$$2.5 \times 8188 \times 10^{-10} \text{ m} = 2.047 \times 10^{-6} \text{ m}$$

6.6 SURFACE IMPERFECTIONS

The imperfections in two dimensions are called surface imperfections e.g. free surfaces of crystal, grain boundaries. The surface imperfections are regions of distortions that lie on the surfaces or grain boundaries having a layer of thickness of a few angstroms. Surface atoms are bonded to less number of nearest neighbours as compared to coordination number and hence are in a higher energy state as compared to the atoms in the interior. The bonds of these surface atoms, that are broken, give rise to surface energy. Hence the surface energy is dependent on (i) the number of bonds that are broken or absent at the surface and (ii) the planar density of atoms at the surface both of which depend on the type of surface.

To calculate the surface energy one should estimate the number of bonds broken per atom at the surface. To do this, one has to find out the number of nearest neighbours (n_p) of an atom lying in the lattice plane in question. Hence the number of broken bonds (n_b) for a surface atom can be approximately written as

$$n_b = (CN - n_p)/2$$

where CN is the coordination number for the crystal. As we have discussed earlier two atoms share each bond. Hence surface energy per atom

$$\varepsilon_a = \frac{n_b}{2} \times \text{bond energy per bond}$$

and

$$\varepsilon_s = \varepsilon_a \times P_d. \tag{6.19}$$

Table 6.6 gives the surface energy for different lattice planes of various crystal structures. Consider the close-packed plane (111) of fcc lattice to be the surface of the crystal. Any atom in fcc has CN = 12 out of which six are in the (111) plane, three below the plane and three above the plane. However, an atom on the surface plane (111) has nine nearest neighbours, three above the plane being absent. Thus, the surface energy per atom is

$$\varepsilon_a = (3/2) \times \text{bond energy per bond}$$

The surface energy per unit area corresponding to a particular crystal plane is given by

$$\varepsilon_{(111)} = \varepsilon_a \times P_d = \varepsilon_a \times 4/(\sqrt{3}a^2)$$

6.7 SPECIAL BOUNDARIES

Another type of surface imperfection is the grain boundary. During the process of crystallization of material from the melt, the nucleation takes place at various points of the melt. As the crystallites grow around various nuclei in the melt, their boundaries impinge on each other. Thus engineering materials may contain many small crystals with the boundaries of mismatch where the crystallites come into contact. Solid materials are usually found in polycrystalline form in nature. The crystallites are also called grains. The boundaries of mismatch where grains come into contact are called grain boundaries.

The grains are randomly oriented with respect to each other. The interfacial region between the two grains has thickness of only a few atomic diameters. The atoms in the grain boundary of a grain experience opposing forces from neighbouring grains. The orientations of crystal planes change sharply at the grain boundary. When the change in orientation is greater than $10° - 15°$, the grain boundaries are called high angle boundaries. In this region the atomic packing is imperfect. The average number of nearest neighbours of an atom in the grain boundaries is always less than that of an atom in the interior region of the crystal. This imperfect nature of structure of grain boundaries makes them visible through microscopes because they may scatter light in a transparent crystal. Since the atoms in grain boundaries have less number of nearest neighbours, hence less binding and so such boundaries are more prone to chemical etching. Sometimes thermal grooves as shown in Fig. 6.23 are formed under special conditions of vacuum at high temperature. The grooves are also formed to study the grain boundaries by using a solvent, which removes atoms from the grain boundaries. When the metal surface is observed through the microscope, the light reflects from the polished surface of the metal but grooves on the surface absorb light hence appear as black lines on the otherwise white background.

Table 6.6 Surface energy for different crystal planes of various crystal structures:

Crystal structure	CN	BE (kJ/mol)	Crystal surface	P_d $\times 10^{19} m^{-2}$	n_b	ε_a $\times 10^{-19} J$	ε_s Jm^{-2}
fcc Cu	12	56.4	(111)	1.77	3	1.405	2.49
			(100)	1.53	4	1.873	2.87
			(101)	1.08	5	2.342	2.53
			(012)	0.69	6	2.809	1.94
bcc Fe	8	104.6	(111)	0.037	4	3.473	0.128
			(100)	0.064	4	3.473	0.221
			(101)	0.089	2	1.737	0.154
			(012)	0.028	4	3.473	0.096
Diamond	4	347	(111)	1.78	2	5.761	10.255
			(100)	1.54	2	5.761	8.872
			(101)	1.63	1	2.881	4.696
hcp Mg	12	24.6	(0001)	2.41	3	0.613	1.477

There is no specific method to measure the surface energy of a grain boundary. The energy of a grain boundary can be measured experimentally. This energy depends on the following factors:
(a) Composition
(b) Orientation of the adjacent grains
(c) Area of the grain boundary

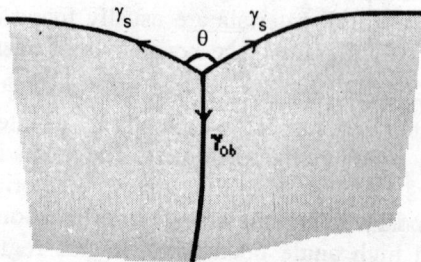

Fig. 6.23 Thermal grooves at the grain boundary.

Since the grain boundary has surface energy, it tries to minimize the surface area of the grain in order to minimize surface energy. The force which plays a role in this process of reduction of interfacial energy is called the surface tension, which is expressed either in units of N/m or as energy per unit area i.e. J/m².

As we know that the grain boundaries are chemically reactive, so certain impurity atoms have a tendency to align along these boundaries. This decreases the energy of the system. The total interfacial energy is lower for coarse-grained polycrystalline substance than for fine-grained one, since the total surface area of grains is more in the later case. The grains grow at elevated temperature to reduce the area of grain boundary and hence the boundary energy.

Twin boundary

A twin boundary is a surface imperfection, which separates atomic arrangements that are mirror images of each other, Fig. 6.24. Twin boundaries occur in pairs, such that the orientation change occurring at one boundary is restored at another. These boundaries are also called twinning planes. The volume of the material whose atomic arrangement is a mirror image of that of the matrix is called a twin. Twins may form during the growth of the crystal.

Shear parallel to twin boundary can produce a twin if slip is difficult to initiate. The region between the pair of twin boundaries is also referred to as twinned region. Twin boundaries are sometimes formed during annealing which are called annealing twins whereas those, which are formed during the process of deformation, are called deformation twins. Twin boundaries can be observed through optical microscopes. In case of materials with basis of two or more atoms like hcp, atomic readjustments occur after the twinning shear. Twinning affects a small fraction of total volume of the crystal. Twinning changes lattice orientations, as a result of which new slip systems may be placed in favourable orientation with respect to shear stress so that additional slip occurs. Thus, twinning increases the ductility of material. The metals with hcp structure are less ductile as compared to fcc or bcc metals since they have less number of slip systems. With the assistance of twinning mechanism, hcp metals like Zn and Mg can become more ductile. Twinning is observed in bcc metals like Fe, Mo, Cr and W when they are deformed at low temperatures. The fcc metal e.g. Cu shows twinning only when the stress level is high and the temperature is sufficiently low.

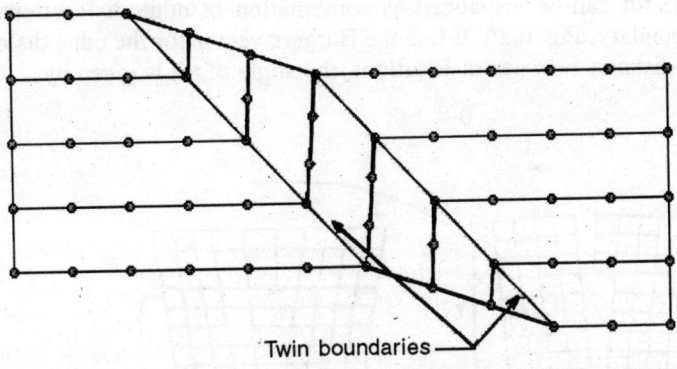

Twin boundaries

Fig. 6.24 Twinning process in crystal

Stacking fault

It is a surface imperfection that occurs due to the stacking of one layer of atoms out of sequence on another. The lattice on either side of the fault is perfect but is translated with respect to one another by a fraction of a lattice vector. We know that the stacking sequence in fcc is described as ABCABCABC..., Fig. 6.25(a). Suppose in this crystal, a stacking fault is present where the stacking sequence is ABCABABC.... This stacking fault is due to the A layer of atoms after the B layer instead of C layer, Fig. 6.25(b). Thus the stacking fault is a thin region of hcp. The stacking sequence ABCACBCA is referred as a twin-stacking fault. The defective layer ACB constitutes the twin, Fig 6.25(c). Stacking faults occur easily in fcc metals. The stacking fault in an fcc metal consists of a thin hexagonal region bounded by partial dislocations. The parallel dislocation lines tend to repel each other, which is balanced by the surface tension of the stacking fault, since the later tends to pull dislocations together.

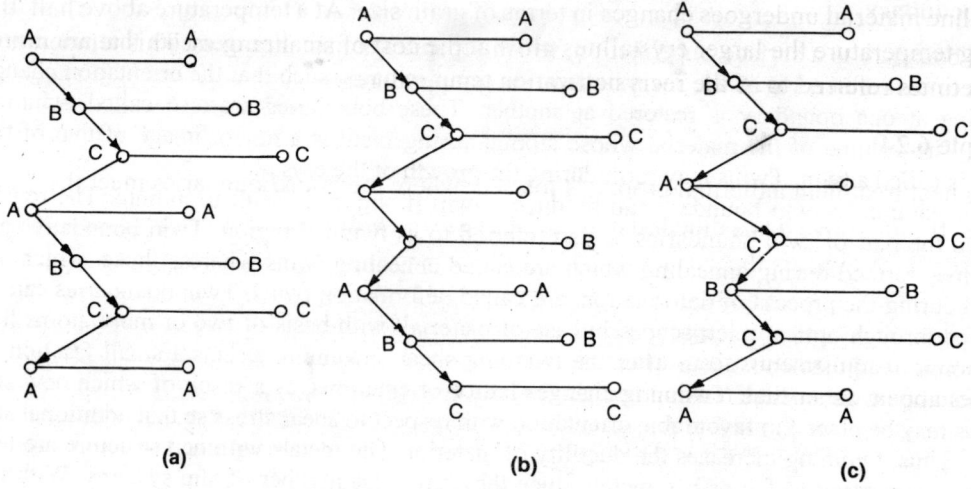

(a)   (b)   (c)

Fig. 6.25 Stacking faults: Stacking in a fcc crystal

Low-angle boundary

When the angular mismatch of the arrangement of lines of atoms in the grains of two sides of grain boundary is of the order of a few degrees, the grain boundary is called a low angle boundary. The low angle tilt can be visualized as combination of edge dislocations lying one above the other in the boundary, Fig. 6.26. If b is the Burgers vector for the edge dislocation and h is the average vertical distance between dislocations, the angle of tilt is given by

$$\theta = b/h \qquad\qquad (6.20)$$

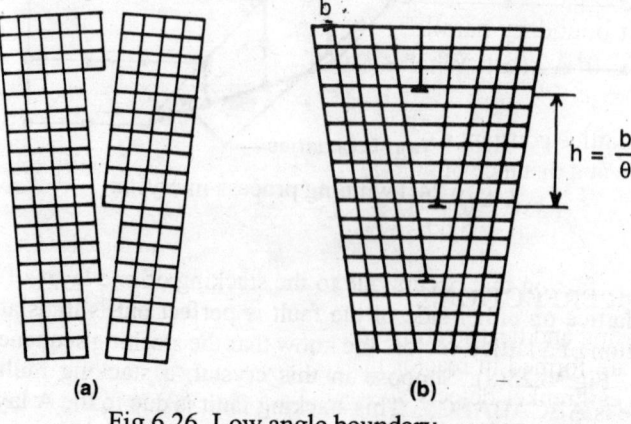

(a)   (b)

Fig 6.26 Low angle boundary

In case of low angle twist boundary, at least two sets of parallel screw dislocations lie in the boundary. In this case the angle of twist α is given by

$$\alpha = b/h \qquad (6.21)$$

where **b** and h refer to the screw dislocation.

In general, surface imperfections are not stable. As the thermal energy increases the poly-crystalline material undergoes changes in terms of grain size. At a temperature above half the melting temperature the larger crystallites grow at the cost of smaller ones. This temperature is sometimes referred to as the recrystallization temperature.

Example 6.24

During heat treatment a thermal groove is formed, where the grain boundaries meet the outer surface. If the groove has a dihedral angle ϕ of 138°, Fig 6.27 and the grain boundary energy γ_{ob} is 0.85 J/m^2, what is the surface energy γ_s of the material?

Solution:

At an elevated temperature, the surface tension forces come to equilibrium. Since the boundary energies appear as surface tension forces, at equilibrium we have

$$\gamma_s = 2\gamma_{ob} \cos(\theta/2) = 2(0.85) \cos 69° = 0.609 \text{ J/m}^2$$

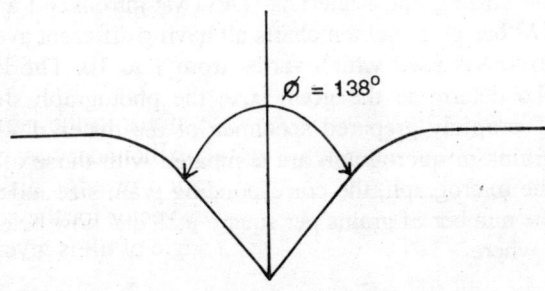

Fig. 6.27 Thermal groove

Example 6.25

The low-angle tilt boundary parallels the plane (110) in LiF and is generated by a series of etch, which are 5×10^4 Å apart. What is the tilt angle θ ?

Solution:

Lattice constant a of LiF = 4.2 Å
The etch pits are at a distance of 50,000 Å apart
$$b_{110} = (4.2 \text{ Å})/\sqrt{2} = 2.97 \text{ Å}$$
$$\theta = 2.97/50,000 = 5.94 \times 10^{-5} \text{ radian}$$

6.8 VOLUME IMPERFECTIONS

During the processing and heat treatment of materials the vacancies aggregate to form voids. The pores or bubbles are formed in reactor materials due to intense neutron irradiation. This type of irradiation produces Frenkel defects and Schottky pairs which aggregate to form pores and the fission gases occupy these pores, which are also called bubbles.

6.9 GRAIN-SIZE DETERMINATION

When we consider a single-phase polycrystalline material, each grain is a single crystal. The orientations and sizes are different for different grains. This structure of polycrystalline substances consisting of grains is referred to as microstructure since these structures can be revealed with the help of electron or optical microscopes. A single-phase microstructure has three basic features– grain shapes, grain size and grain orientation, which influences the properties of solid. The size of grains can vary very widely. The fine-grained materials are stronger than coarse-grained materials. Grain size can be estimated in terms of average grain volume, diameter or area. The grain size is often estimated from the two-dimensional section of the solid material. The photographic equipment along with microscope is commonly employed to take photograph of microscopic image of two-dimensional section of material. The photomicrographs can be used to estimate the grain size with the help of intercept method. In this method, straight lines of same length are drawn on several photomicrograph showing microstructures. Let us say N_l number of lines is drawn. Now the number of grains intersected by each line has to be counted. The number of grains intersection by the ith line is say n_i. Hence the average number of grains intersected by a line is

$$\frac{1}{N_l} \sum_{i=1}^{N_l} n_i$$

Now the length of the straight line is divided by the average number of grains which in turn is divided by the linear magnification to get the average grain diameter.

The American Society for Testing and Materials (ASTM) introduced a standard method to estimate the grain size. ASTM has prepared ten charts all having different average grain sizes. In this method grain size index (n) is used which varies from 1 to 10. The larger the index, the smaller is the grain size. To determine the grain size, the photograph should be taken at a magnification of 100x of a properly prepared specimen of the material which clearly reveals grain structure. Now the grains in micrograph are compared with those of ASTM charts. The chart which matches with the micrograph, the corresponding grain size index is assigned to the grains of the micrograph. The number of grains per square inch can now be estimated, at a linear magnification 100x to be N where

$$N = 2^{n-1}/in^2, \tag{6.22}$$

n being the grain size index. Thus if one knows the ASTM grain size, one can calculate the average area per grain.

Example 6.26

A steel sample has an ASTM grain size index to be 8. Find the average area observed for each grain in a surface.

Solution:

$$N = 2^{8-1} = 128 \text{ grains/ in}^2$$

at magnification x100. Hence at x1, the grain density is x 100 x 100/in^2
Average area for a grain = 7.9 x 10^{-7} in^2

EXERCISES

6.1 The fraction of sites vacant in Al at 900°K is 1.12×10^{-4} at equilibrium. Find the value of ΔH_f and also determine the ratio of the number of vacancies in equilibrium at 300°K to that produced by rapid quenching from 900°K to 300°K assuming all the vacancies generated at 900°K are frozen in the process of quenching.

Ans: 68.1 kJ, 1.24×10^{-8}

6.2 In an ionic compound the fraction of cation sites vacant is 10^{-12} at 0°C. Calculate the enthalpy of formation of vacancies for cation sites.

Ans: 62.68 kJ/mole

6.3 If there are 8 carbon atoms per 10000 unit cells of bcc iron, what w/o carbon occupies interstitial voids in iron? Find also the density of the sample.

Ans: 0.0086 w/o, 7889 kg/m³

6.4 In 1 cm³ of Cu crystal (fcc) find the following:
   (i) Calculate the number of vacancies at room temperature (27°C).
   (ii) How many nickel atoms would be required to fill 50% of the vacancies created at 1000°C? Given that $\Delta H_f = 120$ kJ/mole.

Ans: (i)106; (ii) 5×10^{17}

6.5 A solid solution contains three oxides, $Zn_{1+y}O$, $Fe_{0.8}O$ and $Ni_{1-z}O$ having NaCl type structure. The moles of iron (Fe^{2+}, Fe^{3+}) and nickel (Ni^{2+}, Ni^{3+}) ions in the sample are 0.8 and 0.9 respectively. $Zn_{1+y}O$ contains 1.2 moles of oxygen and 0.2 moles of interstitial Zn^{2+} ions. If the vacancy formation due to Fe^{3+} and Ni^{3+} is the same. Calculate the ion ratio of
   (i) Fe^{2+} and Fe^{3+}
   (ii) Ni^{2+} and Ni^{3+}
   (iii) interstitial Zn^{2+} and Zn^{2+} at regular sites.

Ans: (i) 1; (ii) 5/4; (iii) 1/5

6.6 Pure FeO is heated in oxygen vapor such that the oxygen content in the final sample is 25 w/o more than that in the original sample. Assuming that all the anion sites are occupied, calculate the fraction of cation sites vacant.

Ans: 0.2

6.7 A ceramic material containing 7.6 w/o Cr_2O_3 and 92.4 w/o MgO forms a solid solution with NaCl structure. All anion sites are occupied by O^{2-} ions. Calculate the
   (i) w/o Cr, Mg and O.
   (ii) fraction of the cation sites are vacant.

Ans: (i) 5.21 w/o, 55.71 w/o, 39.08 w/o; (ii) 0.02

6.8 In bcc crystal of molybdenum the slip plane is (211) and slip direction is along $[\bar{1}11]$.
   (i) Find the direction of glide motion of edge dislocation,
   (ii) Find the direction of climb motion of edge dislocation,
   (iii) Draw the slip plane in unit cell and show the direction of edge dislocation.

Ans: (i) $[\bar{1}11]$ or$[1\bar{1}1]$; (ii) [211]; (iii) $[0\bar{1}1]$

6.9 Compute the line energy of dislocations in fcc, diamond. The shear modulus of carbon is 462 GNm⁻². The diameter of carbon atom in diamond is 1.54 Å.

Ans: 1.46×10^{-8} J/m

6.10 (a) Calculate the ratio of surface enthalpies of bcc iron to that of fcc copper along (110)

   (b) As a result of shear stress, slip has occurred within a circular region on a slip plane as shown in Fig. 6.20. Indicate which part of the loop is pure edge dislocation land which part is pure screw dislocation.

   Ans: (a) 0.0609

6.11 (a) The lattice constant for lead (fcc) is 4.95 Å. If the density of a single crystal of lead is 11g/cc, what fraction of the atomic sites are vacant?

   (b) Compare the relative energies of edge dislocations with b_{111}, b_{100} and b_{110} as Burgers vectors in tungsten (bcc).

   Ans: (a) 0.0302; (b) 3:4:8

6.12 MgO, $Fe_{1-x}O$ and $Ni_{1-y}O$, have NaCl structure. Consider a solid solution having moles of MgO, iron (Fe^{2+} and Fe^{3+}) and nickel(Ni^{2+} and Ni^{3+}) ions as 1.2, 0.6 and 0.4, respectively. The total moles of O^{2-} ions in this sample are 2.29. Calculate the

   (a) total vacancy.

   (b) ratio Ni^{3+}/Ni^{2+} if vacancy formation due to Ni^{3+} is one-third to the total vacancy.

   (c) ratio Fe^{2+}/Ni^{2+}.

   Ans: (a) 5.42 x10^{22}; (b) 0.176; (c) 1.412

6.13 An ionic solid solution (having same structure as NaCl) of NaCl and $MgCl_2$ has 1 mole of NaCl and 0.125 moles of $MgCl_2$. Find the chemical formula of nonstoichiometric compound and density of material, if the lattice constant of solution is 10% more than the lattice constant of pure NaCl.

   Ans: 1688 kg/m³

6.14 In copper (fcc) at 1000 °C one out of every 500 lattice sites is vacant. If these vacancies remain in copper when it is quenched to 20°C, what will be the density of copper at room temperature? Lattice parameter of copper at room temperature = 3.6 Å.

   Ans: 8950 kg/m³

6.15 Two edge dislocations are shown in Fig. 6.14(b). Draw separate Burgers circuits for each of these dislocations and obtain their Burgers vectors. Draw a single Burgers circuit to include both the dislocations and show the corresponding Burgers vector.

6.16 There are 10^{11} m^{-2} of edge dislocations in a simple cubic crystal. When the crystal is heated from 0 to 700°C the dislocation climbs. Calculate the direction of climb and the average distance through, which climb has occurred. The enthalpy of formation of vacancies is 100 kJ/mole. The lattice parameter is 2.5 Å. The volume of 1 mole of the crystal is 5.55 x 10^{-6} m³.

   Ans: [100], 2.89 x 10^{-7} m

6.17 (a) Calculate the fractional change in energy during the following dislocation reaction:

$$\frac{a}{2}[110] \rightarrow \frac{a}{6}[110] + \frac{a}{6}[112] + \frac{a}{6}[1\bar{1}2]$$

   (b) The slip plane of a screw dislocation in a bcc crystal is (112). What are Miller indices of the direction of screw dislocation line?

(c) Positive edge dislocations in a simple cubic crystal (lattice parameter a = 2.5 Å) climb down by 2.5 Å along [001] direction when 5×10^{-9} atomic percent of vacancies are created. Calculate the density of edge dislocations.

Ans: (a) –2/9; (b) [11$\bar{1}$] or [$\bar{1}$11]; (c) 8×10^8 m/m^3 or 8×10^8 m^{-2}

6.18 Lattice constant of fcc aluminium is 4.05 Å. The shear modulus of elasticity of aluminium is 25 GN/m^2 and enthalpy of formation of vacancies is 68 kJ/mole. An edge dislocation is caused by slip in [$\bar{1}$10] direction on slip plane (111).

(a) Find the following,
  (i) direction indices of edge dislocation
  (ii) direction of glide motion of edge dislocation
  (iii) direction indices of the direction of climb up of edge dislocation
  (iv) Miller indices of the incomplete plane of edge dislocation

(b) Positive edge dislocation of length 1 cm climbs down by a distance of 1µm. Find the number of vacancies generated or annihilated, in this climbing down process. Calculate the change in energy of the system in this climbing down motion.

Ans: (a) (i) [11$\bar{2}$] or [$\bar{1}\bar{1}$2], (ii) [1$\bar{1}$0] or [$\bar{1}$10], (iii) [111], (iv) ($\bar{1}$10) or (1$\bar{1}$0);
     (b) 8.62×10^{10}, 9.734×10^{-9} J

6.19 The ratio of surface enthalpies of fcc iron to that of bcc iron along (110) is 3:2. The atomic radius of iron in fcc form is 1.27 Å. Assuming the bond energies to be same in bcc and fcc structures, Calculate the
  (i) radius of bcc iron, and
  (ii) ratio of density of fcc to that of bcc iron.

Ans: (i) 1.205 Å (result is approximate since bond energies are assumed to be same);
     (ii) 0.92

6.20 Compute the line energy of dislocation in fcc silicon. The shear modules of Si is 462GN/m^2. The diameter of atom is 2.36 Å. (b = ½ <110>)
Ans: 5.146×10^{-8} J/m

6.21 In a monoatomic sc crystal, the density of edge dislocation of smallest lattice translation is 10^{10}/m^2. If the crystal is heated to a particular temperature, each dislocation climbs down by 2.5×10^{-6} m. If this crystal is fcc and has the same density of edge dislocation, along [11$\bar{2}$] of smallest lattice translation, how much would each of them have climbed down, with the same number of vacancies created per unit volume? Let the atomic diameter be 2Å in both the systems.

Ans: 3.5×10^{-6}m

6.22 The copper and zinc mixed in equal atomic proportions form an ordered solid solution below 470°C, with all the copper atoms occupying the cube corners and the zinc atoms occupying body centers (or vice versa).

(i) Answer the following:
    (a) The shortest distance between two Zn atoms is ......
    (b) The shortest distance between two Cu atoms is ....
    (c) The shortest distance between Cu and Zn atoms is ...
    (d) The distance between two (111) planes is.......
(ii) Tick the correct answer:
    Plane (111) will contain only
        (a) Cu atoms   (b) Zn atoms   (c) Cu or Zn atoms
        (d) no atom   (e) Cu and Zn atoms
(iii) Coordinates for Zn atoms are
        (a) (0,0,0)   (b) (½,½,½)   (c) (0,0,0) or (½,½,½),
        (d) (0,0,0) or (¼,¼,¼),   (e) (0,0,0) or (¾,¾,¾)

Ans: (i) (a) $a = 2(r_{Cu}+r_{Zn})/\sqrt{3}$, (b) a, (c) $\sqrt{3}a/2$, (d) $a/(2\sqrt{3})$; (ii) Cu or Zn atoms;
    (iii) (½,½,½)

6.23 Calculate the surface energy of a bcc crystal when the external surface is (111) type. The bond energy is 50 kJ/mole and the planar density of atoms on (111) is $2 \times 10^{19}/m^2$.

Ans: 0.33 J/m^2

6.24 (a) What is the length of the smallest Burgers vector that a full dislocation may have in CsCl crystal. The radii of Cs^+ and Cl^- ions are 1.65 Å and 1.81 Å, respectively.

(b) An edge dislocation line lies along [121] direction on a slip plane in a cubic crystal. The Burgers vector for this dislocation $\mathbf{b} = \frac{1}{2}[10\bar{1}]$. Find all possible screw dislocation lines of the family <101> that can lie on the same slip plane.

(c) Sodium is a bcc metal having slip plane (211). Find the Miller indices of the directions of glide motion of edge dislocation.

Ans: (a) 3.995 Å; (b) [101] or $[\bar{1}0\bar{1}]$; (c) $[\bar{1}\bar{1}1]$ or $[1\bar{1}\bar{1}]$

6.25 In fcc copper crystal, the Burgers vector of a mixed dislocation line is given by $\mathbf{b} = a/2$ [110]. The mixed dislocation line $\mathbf{d}$ is along a [121]. Find the following:
(i) Miller indices of slip plane
(ii) Edge and screw dislocation components lying on the slip plane.
(iii) The direction indices of glide motion of screw dislocation line lying on the slip plane of part (i) above.
(iv) Miller indices of slip planes belonging to the family whose member is found in (i) above on which screw dislocation can cross-slip.

Ans: (i) $(1\bar{1}1)$ or $(\bar{1}1\bar{1})$, (ii) $a/2[\bar{1}\bar{1}2]$, $3a/2[110]$; (iii) $[\bar{1}12]$ or $[1\bar{1}\bar{2}]$; (iv) $(\bar{1}11)$ or $(1\bar{1}\bar{1})$

6.26 Calculate the indices of positive edge dislocation line, which has $\mathbf{b} = \frac{1}{2}[01\bar{1}]$ and lies in the (111) plane. What is the direction of climb up of this dislocation?

Ans: $[2\bar{1}\bar{1}]$, [111]

6.27 The Burgers vector of a mixed dislocation line is ½[111]. The dislocation line lies along [112] direction. Find the slip plane on which this dislocation lies.

Ans: $(\bar{1}10)$ or $(1\bar{1}0)$

6.28 What is the smallest Burgers vector that a dislocation may have in NaCl crystal with lattice parameter 5.6 Å.

Ans: 3.96 Å

6.29 Is the following dislocation reaction energetically favourable? Justify your answer.

$$a/2 \ [110] + a/2 \ [\bar{1}01] \rightarrow a/2 \ [011]$$

Ans: Reaction is energetically favourable since the final energy $\mu a^2/4 <$ initial energy $\mu a^2/2$.

6.30 The grains in a polycrystalline material can be approximated as cubic in shape with equal volume 10^{-15} m^3. Find out the grain boundary area per m^3 of the material.

Ans: 3×10^5 m^2/m^3

6.31 During annealing of iron, a thermal groove is formed where a grain boundary intercepts the surface. The angle at the bottom of groove is 141°. The grain boundary is symmetrical with the external surfaces at the groove. Calculate the ratio of grain boundary energy to surface energy.

Ans: 0.668

7 *Diffusion in Solids*

7.1 INTRODUCTION

Interior of a solid is dynamic. There is always movement of atoms or molecules inside the solid due to transportation of energy. They move in a direction opposite to the concentration gradient. If we place a drop of ink on a blotting paper, the ink drop spreads. Why does this happen? The molecules of ink drop move radially from the spot to other parts of the blotting paper. There is an equal probability for an ink molecule at the periphery to move in all possible directions outside the ink drop. The movement takes place from higher to lower concentration. In diffusion process, solute molecules or atoms move from higher to lower concentration.

If a jar of argon gas is connected to a jar of helium gas at the same temperature and pressure, after some time these gases diffuse into each other in such a way that a uniform composition is formed. Similar uniform composition can be obtained if water and alcohol are kept in contact since they diffuse into each other very easily. The solid solution of Cu and Ni is formed when the bars of Cu and Ni with intimate contact between surfaces are heated for a long period at elevated temperature. This phenomenon is due to the diffusion of Cu atoms in Ni and vice versa. However the rate of diffusion of nickel atoms down the concentration gradient in copper is faster than that of copper atoms in nickel. The above phenomenon can be explained as follows: During diffusion process, the bonds in the solvent are broken along the path of diffusion of solute atoms and hence this requires activation energy. We know that the melting point of copper is lower than that of nickel i.e. Cu-Cu bonds are weaker than Ni-Ni bonds, hence the activation energy for diffusion of nickel atoms in copper is less than that of copper atoms in nickel.

7.2 DIFFUSION MECHANISMS

Diffusion can take place in the presence of vacancies and interstitial voids of suitable size. During the diffusion process, the atom or vacancy moves from one lattice site or interstitial position to another. Diffusion of an atom from one lattice site to another vacant location can be interpreted as the diffusion of vacancy in the opposite direction. The atoms in the solid are vibrating about their equilibrium positions. A small fraction of these atoms diffuse from their equilibrium positions to other locations at a specific temperature, because they possess vibrational energy. Diffusion can take place provided,

- the diffusing atoms have enough energy to break the bonds with neighbouring atoms and create lattice distortions while moving through the lattice matrix and this energy is of vibrational type;
- there are vacant sites where the diffusing atoms can move.

In general the diffusion rate increases at elevated temperature. Various mechanisms of diffusion are described below:

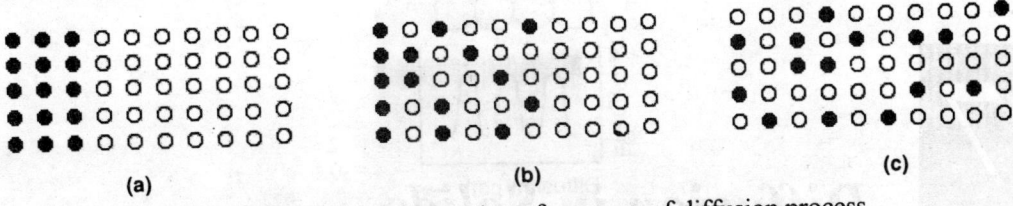

(a)   (b)   (c)

Fig.7.1 Schematic representation of progress of diffusion process.

Self diffusion

Self diffusion occurs in pure metals when the atoms exchanging positions are of the same type. The self diffusion process propagates through vacancy mechanism e.g. copper atom in copper matrix. The activation energy for self diffusion is the sum of activation energy for vacancy formation and the activation energy for movement of vacancy. Higher the melting temperature of a metal, higher is the activation energy.

Vacancy diffusion

An atom near a vacancy has a normal tendency to move into the vacancy, such that the vacancy is annihilated but a new vacancy is created at the position of the atom. After this, the new vacancy can receive another atom and thus the process continues. As a result, vacancies diffuse through metal and homogenization takes place, Fig. 7.2.

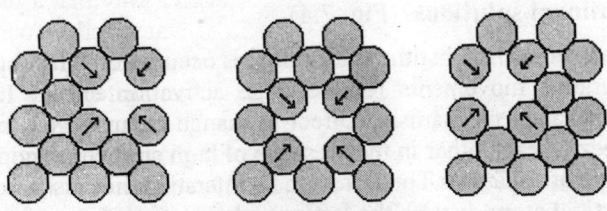

Fig. 7.2 Vacancy diffusion.

Interstitial diffusion

In this type of diffusion process, the atoms diffuse through solid via interstitial sites. A small solute atom in the interstitial site moves to a vacant neighbouring interstitial site without permanently displacing any of the atoms in its way through the host matrix and its movement from one interstitial site to a neighbouring site is called the unit step in interstitial diffusion.

During this process, interstitial atoms have to move through kinetic barrier (Fig.7.3) and hence require activation energy. Examples of such diffusions are found in case of inter-diffusion of hydrogen, carbon, oxygen etc, in metals. Consider carbon atoms as solute in iron. When the distribution of carbon atoms in the host matrix of iron is more or less uniform what we get is solid solution. Solid solution is formed due to interstitial diffusion of carbon atoms in α-iron (bcc) or γ-iron (fcc), during which carbon atoms squeeze between the iron atoms. This is so because the size of the largest interstitial void (octahedral void in fcc) in γ-iron is 75% of that of the carbon atom. So there will be distortion around the interstial site occupied by carbon atom. Similarly, the largest size void (tetrahedral void in bcc) in α-iron is 40% of that of the carbon atom. Hence the distortion in case of α-iron is even larger than that in case of γ-iron.

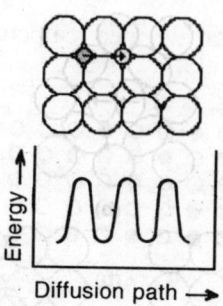

Fig. 7.3 Interstitial Diffusion

Hence the concentration of carbon in the steel is quite low. Consider that the concentration of carbon is 1.06 w/o in fcc iron. Hence in 100 g of the sample, iron is 98.94 g and carbon is 1.06 g. Now 98.94 g of iron = 1.77 moles of iron = $1.77N_A$ iron atoms = $1.77N_A$ interstitial voids. Since, we know that the number of iron atoms is equal to the number of interstitial voids (octahedral voids) in fcc unit cell.

$$1.06 \text{ g of carbon} = 0.088N_A \text{ carbon atoms.}$$

So, the fraction of interstitial sites occupied by carbon atoms $= 0.088N_A /1.77N_A \approx 0.05$. Thus, the probability that an interstitial site to be empty and be available for a diffusing atom to move into it from a neighbouring site is $1 - 0.05 = 0.95$. Thus, the probability of diffusion is higher in interstitial diffusion than in substitutional diffusion where the number of vacant sites is much less.

Diffusion in substitutional solutions (Fig. 7.4)

Diffusion of large solute atoms in substitutional solutions occurs through cooperative movements. Some of these cooperative movements require large activation energy like in case of ring diffusion and direct exchange mechanism. Direct exchange occurs when two atoms exchange their positions by squeezing each other in the presence of high strain energy. The ring diffusion is also possible in rare type of situations. The diffusion by interstitialcies is a mechanism of diffusion in which a large interstitial atom distorts the lattice and pushes to the nearest lattice site and the atom in this lattice site moves to the next interstitial site. This process of cooperative movement occurs in presence of compressive stresses.

7.3 FACTORS INFLUENCING DIFFUSION

The following factors influence diffusion:
  i) Diffusion rate is dependent on the temperature. In most of the cases diffusion becomes faster at higher temperatures.
 ii) Diffusion is faster in case of small interstitial solute atoms as compared to large substitutional solute atoms.
iii) Diffusion rate is faster in a solvent with a low melting point as compared to that in a solvent with a higher melting point. This can be explained as follows. During the diffusion of solute atoms, some of the atomic bondings in the solvent are broken. If the melting point is low, it is easier to break bonds between the atoms.
 iv) Diffusion of solute atoms takes place faster in solvent with lower packing efficiency than in a solvent with higher packing efficiency. In the later case more energy will be needed for the movement of the solute atoms through the solvent.
  v) The more the vacancies, the faster is the diffusion.

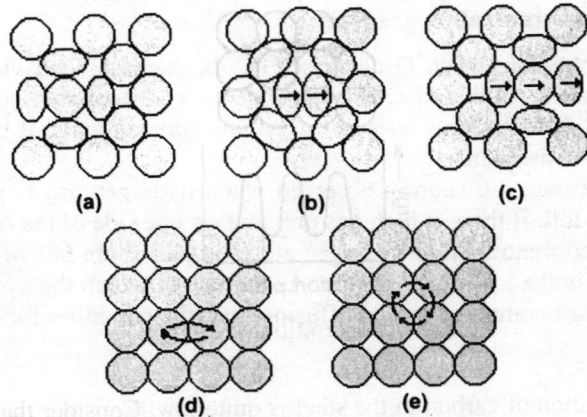

Fig.7.4 (a)-(c) Diffusion by interstitialcies, (d) Direct exchange mechanism, (e) Ring diffusion

7.4 LAWS OF DIFFUSION

Fick's first law

Fick's first law states that the net rate of flow, dN/dt, of solute atoms from the region of higher concentration to that of lower concentration is proportional to the concentration gradient of diffusing solute atoms i.e.

$$(dN/dt) \propto - (dC/dx) \tag{7.1}$$

where C is the concentration of diffusing solute atoms, x being the direction of diffusion. Also,

$$dN/dt \propto A, \tag{7.2}$$

where A is the area of cross-section perpendicular to the direction of diffusion. The negative sign appears in Eq. (7.1), since the solute atoms diffuse along negative concentration gradient. Thus,

$$dN/dt = - DA (dC/dx) \tag{7.3}$$

where D is a constant of proportionality. It is called the diffusion coefficient for the specific solvent-solute pair. The flux J is defined as the quantity of solute atoms crossing unit area in unit time. J is in the x-direction as diffusion is taking place in the x-direction. Thus, J can be expressed as

$$J = (1/A)(dN/dt) = - D(dC/dx) \tag{7.4}$$

This equation is known as Fick's first law. If J is expressed in moles per square meter per second, concentration C in moles per cubic meter, then D is expressed in meter square per second. Fick's first law becomes a linear relation in where D does not vary with C under steady state flow, and flux is independent of time and remains the same at any cross-sectional area perpendicular to the direction of diffusion. Hence,

$$J = constant \tag{7.5}$$

To keep J constant, D(dC/dx) should also remain constant. This leads to two different situations:

i) D is independent of concentration gradient

In this case, both D as well as dC/dx has to remain constant for steady diffusion flux.

ii) D is dependent on concentration gradient

Under steady state condition, when D decreases dC/dx increases and vice versa, to keep the product D(dC/dx) constant. In this case, D is dependent on concentration gradient. In such cases, the concentration of diffusing atoms does not vary with time although it has different values at different positions of the medium.

This type of steady state diffusion is observed when hydrogen gas is purified by passing it through the palladium foil. If there is high gas pressure on one side of the foil and a low pressure on the other side, the concentration of hydrogen atoms in palladium foil will be high in the high-pressure side and low in the low-pressure side. As the rate at which the hydrogen atoms enter or leave the foil is very fast compared to the diffusion rate, the condition for steady state diffusion prevails.

Example 7.1

Consider a steel tank of 7 mm thick sheet, containing hydrogen gas at a constant pressure of 20 atm, and with vacuum outside. If the diffusion coefficient of hydrogen in steel at room temperature is 10^{-9} m^{2}s^{-1} and diffusion flux is equal to 4 x10^{-6} kg m^{-2}s^{-1}, find the concentration gradient of hydrogen in the tank wall.

Solution:

The concentration gradient

$$dC/dx = J/D = 4x10^{-6} x10^9 \text{ kg m}^{-4} = 4000 \text{ kg m}^{-4}$$

Example 7.2

The diffusion coefficient for carbon in fcc iron are 4.4 x 10^{-5} m^2/s at 1000° K and 2.8 x 10^{-13} m^2/s at 500°K. At the surface of Fe crystals, there is one carbon atom per 10 unit cells and one millimeter below the surface there are two carbon atoms per 50 unit cells. Lattice parameter of iron is 3.65 Å.
i) Find the coefficient of diffusion of carbon in iron at 700°K.
ii) At 700°K, how many carbon atoms diffuse down the concentration gradient per second through each square meter area?

Solution:

(i)
$$\frac{D_{1000°K}}{D_{500°K}} = \frac{4.4 \times 10^{-5}}{2.8 \times 10^{-13}} = \exp[-Q(1/1000) - 1/500)/8.314]$$

$$Q = 1.569x10^5 \text{ J/mol}$$

$$D_o = 4.4 \times 10^{-5} \exp(1.569x10^5/8.314x10^3) = 6908 \text{m}^2/\text{s}.$$

$$D_{700°K} = 6908 \exp[-1.569x10^5/(8.314 \times 700)] = 1.352 \times 10^{-8} \text{m}^2/\text{s}$$

$$\frac{dC}{dx} = \frac{[-(1/10a^3) + (2/50a^3)]}{10^{-3}} = \frac{3}{50(3.65 \times 10^{-10})^3 x 10^{-3}} = -1.234 \times 10^{30} \text{ carbon atoms/m}^4,$$

(ii)
$$J = -D_{700°K}\frac{dC}{dx} = 1.668 \times 10^{22} \text{ atoms / m}^2 \text{ s.}$$

Example 7.3

A steel container has vacuum inside, has walls of thickness 0.5 mm and is placed in a room. Assuming that the air concentration on the outer surface is 15 kg m^{-3}, calculate the rate of air diffusion through the walls of the container, the diffusion coefficient of air in steel at room temperature being 10^{-8} m^2 s^{-1}.

Solution:

$$J = 10^{-8} [15/(5 \times 10^{-4})] = 3 \times 10^{-4} \text{ atoms/m}^4.$$

Example 7.4

Consider an fcc iron crystal having 25 carbon atoms per 1000 iron atoms present at its surface. One millimeter below the surface, there is 2 atomic percent of carbon present. Find the diffusion flux of carbon atoms per square meter at 927°C.
[Radius of γ-Fe atom is 1.27 Å. D for carbon atoms in γ-Fe at 927°C is 10^{-11} m^2/s.]

Solution:

The lattice constant of γ-iron = $2\sqrt{2}$ r = 3.59 Å. 25 carbon atoms per 1000 γ-Fe atoms means 25 carbon atoms per 250 unit cells (since each unit cell contains 4 atoms) and 20 carbon atoms per 1000 γ-Fe atoms means 20 carbon atoms per 250 unit cells.
Concentration of carbon atoms/ m^3 at the surface

$$= 25/[250 \times (3.59 \times 10^{-10})^3] = 2.16 \times 10^{27} \text{ atoms/m}^3$$

Concentration of carbon atoms/ m^3 at 1mm below the surface

$$= 20/[250 \times (3.59 \times 10^{-10})^3] = 1.73 \times 10^{27} \text{ atoms/m}^3$$
$$J = -D (C_2 - C_1)/(x_2 - x_1) = 4.3 \times 10^{18} \text{ carbon atoms/ m}^2.\text{s}$$

Fick's second law and nonsteady state flow

Fick's second law is an extension of Fick's first law to nonsteady state flow or time dependent diffusion process. In case of nonsteady state flow, the following considerations are taken into account:

- The flux is not same at different cross-sections perpendicular to the direction of diffusion;
- The flux through a given cross-section varies with time.

There is a net accumulation or depletion of diffusing species. This is shown in Fig.7.5, which illustrates the change in concentration profile with time in nonsteady state diffusion.

Consider a slab of area unity and thickness Δx at distance x from origin in the diffusing medium. The cross-sectional area of the slab is perpendicular to the direction of diffusion x. Under nonsteady state conditions, the flux of diffusing species J_x entering the slab is not equal to the flux $J_{x+\Delta x}$ coming out of the opposite face of the slab. This results in the depletion or accumulation of solute in the slab (whose volume is Δx). Hence one can express

$$J_x - J_{x+\Delta x} = (\partial C/\partial t)\Delta x$$
$$J_x - [J_x + (\partial J_x/\partial x) \Delta x] = (\partial C/\partial t)\Delta x \tag{7.6}$$

After simplification $\qquad \partial C/\partial t = -\partial J_x/\partial x \tag{7.7}$

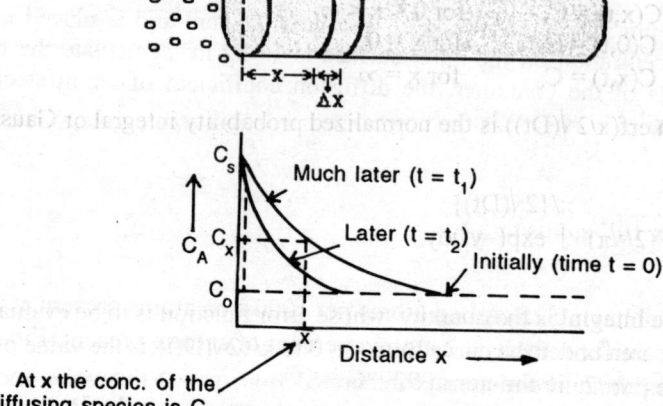

Fig.7.5 Variation of concentration profile with time for diffusing atoms

Using the expression of J from Fick's first law, one can get,

$$\partial C/\partial t = -\partial/\partial x \, (-D(\partial C/\partial x))$$ (7.8)

If D does not depend on concentration, Eq.(7.8) simplifies to

$$\partial C/\partial t = D(\partial^2 C/\partial x^2)$$ (7.9)

This is known as Fick's second law or parabolic diffusion equation. The Fick's second law in three dimensions can be expressed as

$$\partial C/\partial t = (\partial/\partial x)\,(D_x \partial C/\partial x) + (\partial/\partial y)\,(D_y \partial C/\partial y) + \partial/\partial z\,(D_z\,\partial C/\partial z)$$ (7.10)

where D_x, D_y and D_z are the coefficients of diffusion along x, y and z axes, respectively. These diffusion coefficients are equal for cubic crystals having higher symmetry. The crystals with lower symmetry may not have same D_x, D_y and D_z values.

The solution to Fick's second law for unidirectional diffusion from one medium to another across a common interface is given by

$$\frac{C_x - C_o}{C_s - C_o} = 1 - \text{erf}\,(x/\{2\sqrt{(Dt)}\})$$ (7.11)

where C_o is the base concentration, C_x is the concentration at distance x into the slab and C_s is the concentration at the surface. These are to be determined from the boundary conditions. The boundary conditions assumed are the following:

- Two media are considered to be semi-infinite having the common interface and other two ends of the media are at infinite distances.
- The initial uniform concentration of solute atoms in the two media are different having abrupt change in concentrations at the common interface.
- The origin of x-axis is at the common interface.
- The time is taken to be zero just before the diffusion process begins:

At t = 0 $\qquad$ $C(x,0) = C_o$ $\qquad$ for $0 < x < \infty$

At t ≥ 0 $\qquad$ $C(x,t) = C_x$ $\qquad$ for $0 < x < \infty$

$\qquad\qquad\qquad$ $C(0,t) = C_s$ $\qquad$ for $x = 0$

$\qquad\qquad\qquad$ $C(x,t) = C_o$ $\qquad$ for $x = \infty$ $\qquad\qquad$ (7.12)

In Eq.(7.11), the function $erf(x/2\sqrt{(Dt)})$ is the normalized probability integral or Gaussian error function which is given by

$$erf\frac{x}{2\sqrt{(Dt)}} = (2/\sqrt{\pi}) \int_0^{x/\{2\sqrt{(Dt)}\}} \exp(-y^2)dy \qquad (7.13)$$

where the upper limit of the integral is the quantity whose error function is to be evaluated. If we plot $\exp(-y^2)$ versus y, the area under the curve from y = 0 to $x/\{2\sqrt{(Dt)}\}$ is the value of the error function. The area becomes $\sqrt{\pi}/2$, if the area is integrated from y = 0 to y = ∞ and the area becomes $-\sqrt{\pi}/2$ if we integrate from y = 0 to y = – ∞. Hence,

$$erf(\infty) = 1 \qquad\qquad erf(0) = 0$$
$$erf(-\infty) = -1 \qquad\qquad erf(-z) = -erf(z). \qquad (7.14)$$

If the values of C_o, C_x and C_s are known then error function can be calculated using Eq.(7.11). The argument of error function $x/(2\sqrt{(Dt)})$ can be found out from the value of error function using the Table 7.1. Diffusion coefficient D can be calculated at a known value of x, and t from the value of the argument. If C_o, C_s and D are known for a specific material then C_x can be expressed as a function of x and t.

Table 7.1 Error function

z	erf(z)	z	erf(z)	z	erf(z)	z	erf(z)
0	0	0.40	0.4284	0.85	0.7707	1.6	0.9763
0.025	0.0282	0.45	0.4755	0.90	0.7970	1.7	0.9838
0.05	0.0564	0.50	0.5205	0.95	0.8209	1.8	0.9891
0.10	0.1125	0.55	0.5633	1.0	0.8427	1.9	0.9928
0.15	0.1680	0.60	0.6039	1.1	0.8802	2.0	0.9953
0.20	0.2227	0.65	0.6420	1.2	0.9103	2.2	0.9981
0.25	0.2763	0.70	0.6778	1.3	0.9340	2.4	0.9993
0.30	0.3286	0.75	0.7112	1.4	0.9523	2.6	0.9998
0.35	0.3794	0.80	0.7421	1.5	0.9661	2.8	0.9999

Example 7.5

The diffusion coefficients for metal A in metal B at 600°C and 800°C are $5 \times 10^{-14} m^2/s$ and $6 \times 10^{-13} m^2/s$, respectively. How much time will be taken to produce the same concentration of A at some specific point in B at 800°C, as produced by an 18 hour heat treatment at 600°C.

Solution:

It is desired that the composition in both the diffusion situations becomes same at the same position. Hence

$$Dt = constant$$

at both the temperatures where t is the time taken at a specific temperature. Thus,

$$(Dt)_{800} = (Dt)_{600}$$

t_{800} is the time taken to achieve the same concentration when the heat treatment was done at 800°C.

$$18 \text{ hr} \times 5 \times 10^{-14} = 6 \times 10^{-13} \times t_{800}$$

$$t_{800} = \frac{18 \times 5 \times 10^{-14}}{6 \times 10^{-13}} = 1.5 \text{ hr}$$

Example 7.6

An iron block is plated with nickel and heated to 900°C so that there is an atomic flux of 2.0×10^{15} Ni atoms/m²s. Find the concentration gradient of Ni-atoms. If concentration of Ni atoms is 40 a/o at the surface, what is the concentration at 0.2 mm below the surface? Given that D for nickel in γ-Fe is $2 \times 10^{-17} m^2/s$.

Solution:

Lattice parameter a of γ-Fe = 3.59 Å

$$(dC/dx) = -J/D = -10^{32} \text{ Ni atoms/m}^4$$

C_1 at surface = 40 Ni atoms per total of 100 atoms i.e. 25 unit cells

$$= 40/\{25 \times (3.59 \times 10^{-10})^3\} \text{atoms /m}^3$$
$$= 3.458 \times 10^{28} \text{ atoms /m}^3$$

$$C_2 \text{ at 0.2 mm} = (dC/dx) \times 0.2 \text{ mm} + C_1 \text{ below surface} = 1.458 \times 10^{28} \text{ atoms / m}^3$$

Let the concentration of diffusing atoms be C' at a distance x at time t (considering the concentration to be same C_o at any value of x at time t = 0). The distance x is called penetration depth. Now if the penetration depth is doubled such that C' is the concentration at 2x, the diffusion time will be four times, since $x/\{2\sqrt{(Dt)}\}$ is same for the two cases.

Carburization of steel

When steel bar is heated at an elevated temperature in the atmosphere rich in hydrocarbon gas, the concentration of carbon increases at the surface. This is called carburization of steel, Fig.7.6 (a). The carburization of steel is done to harden the surface of the steel compared to the interior of steel. The surface hardened steel is used for making certain machine parts like gear.

Example 7.7

The surface of a steel bar (0.1 w/o C - 99.9 w/o Fe) is exposed to a hydrocarbon gas at 950°C that maintains the surface composition to be 0.93 w/o of carbon. If the diffusion coefficient is $0.31 \times 10^{-6} cm^2/s$, find the time taken for the concentration to attain the value 0.5 w/o at depth of 0.137 mm.

Solution:

$$C_o = 0.1 \text{ w/o C}$$
$$C_s = 0.93 \text{ w/o C}$$
$$C_x = 0.5 \text{ w/o C}$$
$$x = 1.37 \times 10^{-4} m$$
$$D = 0.31 \times 10^{-10} m^2 / s.$$

Thus,

$$\frac{C_x - C_o}{C_s - C_o} = \frac{0.50 - 0.10}{0.93 - 0.10} = 1 - \text{erf} (1.37 \times 10^{-4} m)/ (2\sqrt{(3.1 \times 10^{-11} t)})$$

So, $\quad\quad\quad\quad$ erf$(12.3/\sqrt{t}) = 0.52$

Using Table 7.2, $\quad\quad$ $12.3/\sqrt{t} = 0.50$

$\quad\quad\quad\quad\quad\quad\quad\quad$ t = 605 s

Decarburization of steel

During decarburization process, carbon is lost from the surface layer of steel, due to oxidation of carbon, which reacts with atmosphere to form CO, CO_2 gases. If the heat treatment of steel is performed in a nonprotective atmosphere, the decarburization occurs. As a result of decarburization the concentration of carbon at the surface decreases below the initial carbon concentration of steel, Fig. 7.6(b), so the diffusion of carbon atoms takes place from inside to surface. When decarburization takes place during heat treatment, the machining of the surface is done such that certain level of carbon concentration can be maintained at the surface.

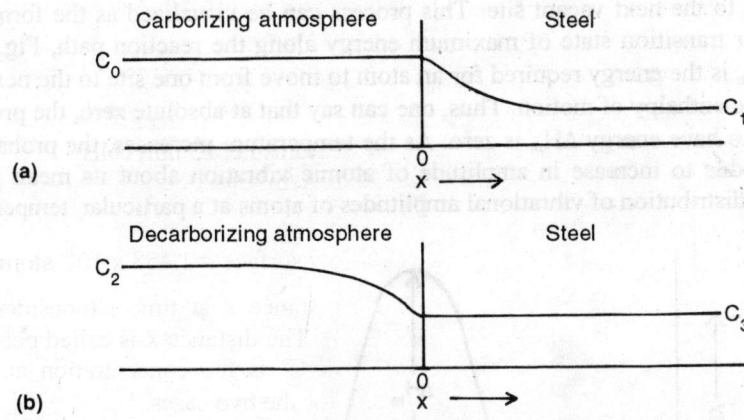

(a)

(b)

Fig.7.6 Concentration – distance profile for (a) carburization (b) decarburization

Example 7.8

A 0.6% carbon steel undergoes heat treatment at 800°C for duration of 6 hours in air. As a result of this, the surface concentration of carbon has become 0%. Find up to what depth post-machining should be done such that the carbon concentration at the surface becomes 0.4%.

Given that D_o for C in Fe (γ)= 0.7×10^{-4} m²/s, activation energy = 157 kJ/mole.

Solution:

We have $\quad$ D = 0.7×10^{-4} exp $(-157 \times 10^3/8.314 \times 1073) = 1.592 \times 10^{-12}$ m²/s.

$\quad\quad\quad\quad$ $C_o = C_{initial} = 0.6\%$

$\quad\quad\quad\quad$ $C_s = 0.0\%$

$\quad\quad\quad\quad$ $C_x = 0.4\%$

$\quad\quad\quad\quad$ t = 6hr = 21600s

The depth x is found out from

$$\frac{0.4 - 0.6}{0.0 - 0.6} = 1 - \text{erf}\,[x / \{2\sqrt{(Dt)}\}]$$

Since diffusion takes place from inside to surface, x will be negative, hence

$$- \text{erf} \, [x/2\sqrt{(Dt)}] = -0.667$$

Using Table 7.2, we get $x/\{2\sqrt{(Dt)}\} = -0.684$, $x = -2.54 \times 10^{-4}$ m

7.5 DIFFUSION COEFFICIENT AND ACTIVATION ENERGY

Diffusion processes are complex and require a series of unit steps to complete the process. These steps are simple. One such unit step is jumping of an atom to its neighbouring vacant site. As the diffusing atom moves from one site to another, it has to have enough energy to pass through the host matrix. While interstitial atom diffuses through host matrix, the outer electronic orbitals of diffusing atom may overlap with those of host atoms and overlap is maximum when it is midway between the site 1 and site 2. Other steps involve breaking up of bonds with neighbouring atoms at the initial site and formation of new chemical bonds at the final site, when the individual atom jumps from one site to the next vacant site. This process can be visualized as the formation of activated complex or transition state of maximum energy along the reaction path, Fig.7.7. The potential energy ΔH_m is the energy required for an atom to move from one site to the next vacant site. ΔH_m is called the enthalpy of motion. Thus, one can say that at absolute zero, the probability of a diffusing atom to have energy ΔH_m is zero. As the temperature increases, the probability of diffusion increases, due to increase in amplitude of atomic vibration about its mean position. There is a statistical distribution of vibrational amplitudes of atoms at a particular temperature T

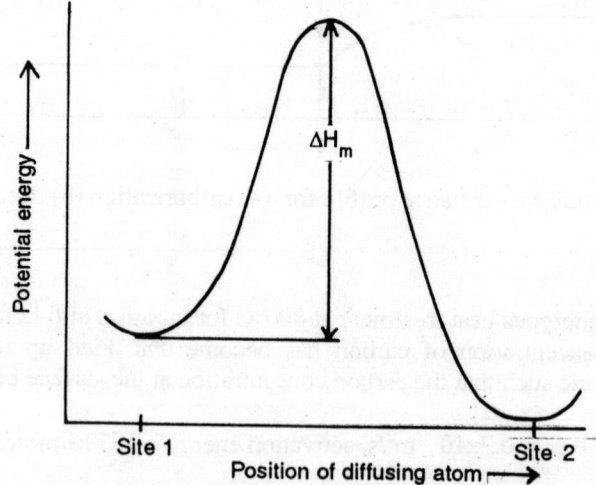

Fig. 7.7 Variation in potential energy along the diffusion path

and one can show that the probability that an atom can have sufficient vibrational energy to overcome the potential barrier of ΔH_m is given by

$$\exp(-\Delta H_m/RT) \tag{7.15}$$

This is called the Maxwell Boltzmann distribution. The interstitial diffusivity of many materials is found to obey Arrhenius equation, according to which diffusivity is expressed as,

$$D = D_0 \exp(-\Delta H_m/RT) \tag{7.16}$$

where D_0 is called pre-exponential factor. The diffusivity also depends upon the probability of finding a neighbouring site vacant where the impurity can jump. In case of interstitial diffusion, this probability is nearly unity since most of the interstitial sites are vacant. Thus, the activation energy is same as enthalpy of motion in this case.

In case of substitutional diffusion, impurity atoms diffuse through vacancy mechanism. One can define unit step of diffusion as the jump of an atom to its neighbouring vacant site. For such a process the vacancies are to be created i.e. we have to take into account the probability of finding the neighbouring site to be vacant. The number of vacant sites n in thermal equilibrium can be expressed as

$$n/N = \exp(-\Delta H_f/RT) \tag{7.17}$$

where ΔH_f is the enthalpy of formation of vacancies. The expression n/N also represents the probability that a given atomic site is vacant. Thus substitutional diffusion depends on the following factors:

- The probability that a substitutional atom will have enough energy to jump to the neighbouring vacant site (if available) by crossing kinetic barrier.

- The probability that a neighbouring vacant site is available for the vibrating substitutional atom to jump into it.

So the substitution diffusion coefficient is proportional to the product of these two probabilities,

$$D = D_o \exp(-(\Delta H_m + \Delta H_f)/RT) \tag{7.18}$$

where one can define activation energy, Q for this process as $(\Delta H_m + \Delta H_f)$ i.e. the sum of enthalpy of motion and that of formation of vacancy. The activation energy Q can be defined as the energy required for the atom to cross the potential barrier while moving from one site to the next one. All these energies are expressed in the unit of joules per mole. Usually the interstitial diffusion process is much faster than the substitution diffusion process since the activation energy is much less in the former case. In other words, the probability of finding neighbouring vacant sites to diffuse is nearly one for interstitial diffusion process. For example, diffusion coefficient of carbon in α-iron (bcc) is 3.625×10^{-9} m^2/ s at 900°C whereas iron in α-iron (bcc) is 3.66×10^{-15} m^2/s at 900°C.

Calculation of D_o

Diffusion can be explained with the help of atomic processes. This method helps us in estimating D_o in terms of atomic parameters. During a diffusion process the diffusing atoms jump from one site to another with a certain frequency say ν, which is related to ν_o frequency of vibration of atoms ($\approx 10^{13}$/s) in solid and is given by,

$$\nu = \nu_o \exp(-\Delta H_m/RT), \tag{7.19}$$

and,

$$\Delta G = \Delta H_m - T\Delta S_m \tag{7.20}$$

where ΔG is the free energy of motion of activated complex. It is also the work done against the elastic strain set up in the lattice during the motion of diffusing atom from one site to another and can be approximated as $\mu\varepsilon_o^2$, μ being appropriate elastic modulus for the solvent and ε_o, representative strain for the matrix when the diffusing atom is at the saddle point.

Out of ν_o vibrations of an atom about its equilibrium position per second only ν vibrations give rise to successful jumps which are associated with energy $\geq \Delta H_m$ and are responsible for diffusion. If n jumps are made in randomly distributed directions in time t, the total distance R_n through which diffusion has taken place is given by

$$\overline{R_n^2} = nd^2 \propto na^2 \tag{7.21}$$

and

$$Dt \propto \overline{R_n^2}$$

so,

$$Dt = \gamma na^2 \tag{7.22}$$

where d is the average distance through which jumps takes place and is equal to interplanar distance. 'a' is the lattice parameter, γ is the constant of proportionality and $\nu=n/t$. Therefore, diffusion constant D can be expressed as

$$D = \gamma a^2 \nu, \tag{7.23}$$

The diffusion of an atom can take place by jumping into an adjacent vacant site, if one of the adjacent sites is vacant. Thus, we must include a factor P for the probability that an adjacent site will be vacant, in the expression of D i.e.

$$D = \gamma a^2 \nu P, \tag{7.24}$$

where P (= n/N) is given by Eq. (7.17).

$$P = n/N = \exp-(\Delta H_f)/RT$$

Substituting Eq (7.19)

$$D = \gamma a^2 \nu_o \exp-(\Delta H_m + \Delta H_f)/RT \tag{7.25}$$

Experimentally obtained D_o values vary from calculated values since the effect of increase in entropy, while atom is jumping, has not been considered. During the diffusion of an atom, the surrounding atoms are under elastic strain, which causes increase in entropy ΔS_m. ΔS_m is called the entropy of activation or entropy associated with motion due to diffusion. Hence, diffusion coefficient can be expressed as

$$D = \gamma a^2 \nu_o \exp-(\Delta H_m + \Delta H_f - T\Delta S_m)/RT \tag{7.26}$$

Comparing Eqs. (7.19) and (7.26), we get

$$D_o = \gamma a_o^2 \nu_o \exp(\Delta S_m / R) \tag{7.27}$$

When ΔG_o is equal to ΔG at $0°K$, one can define

$$\Delta S_m = -\Delta G_o \frac{d(\Delta G/\Delta G_o)}{dT} \approx -\Delta H_o \frac{d(\mu/\mu_o)}{dT} \tag{7.28}$$

Since, $\Delta F_o \equiv \Delta H_o \approx \mu_o \varepsilon_o^2$. The final expression of ΔS_m is given by

$$\Delta S_m \approx \beta \frac{\Delta H_m}{T_m}, \tag{7.29}$$

where $\beta = d(\mu/\mu_o)/d(T/T_m)$ and T_m is the melting point of the solvent. The value of β varies between 0.25 and 0.45 for most metals. For diffusion by vacancy mechanism, one can express more appropriately,

$$D_o = \gamma a_o^2 \nu_o \exp((\Delta S_f + \Delta S_m)/R) \tag{7.30}$$

where $\Delta S = \Delta S_f + \Delta S_m$ can be estimated using Eq. (7.29), considering $\Delta H = \Delta H_f + \Delta H_m$ in place of ΔH_m. Diffusivity data are given in Table 7.2 for selected cases.

7.6 HARTLEY-KIRKENDALL EFFECT

A diffusion couple can be formed by application of suitable amount of heat or pressure after placing the blocks of two materials in intimate contact. Electroplating of one metal on the other so that the atoms of each material can diffuse into other can also produce a diffusion couple e.g. Cu and Ni, Cu and Zn etc. This process is called inter-diffusion and is used for diffusion welding.

The rate of diffusion of atoms of material A in material B is not usually equal to that of material B in material A in a diffusion couple. This phenomenon gives rise to Hartley-Kirkendall effect. Consider that a diffusion couple of materials A and B is such that the atoms of A diffuse in B faster than that of B into A. As a result, porosity is created in A side. Now if inert markers are placed at the interface of the two materials, more number of A atoms diffuse through inert markers toward B than the number of B atoms moving toward A. As a result, the inert markers move towards A. This shift of inert markers is known as Kirkendall shift.

The marker shift depends on the following factors:
(i) The shift increases with temperature for a constant interval of time.
(ii) The increase in difference in the composition between two materials A and B also results in the increase of shift.
(iii) The shift is proportional to the square root of the time during which the temperature is kept constant.

Out of the two materials, the one with lower melting point diffuses faster than the other. The position of the inert markers after Kirkendall shift is called Kirkendall interface and the original interface of the two materials is called Matano interface.

Table 7.2 Diffusivity data

Solute	Solvent	D_o 10^{-4} m²/s	Q kJ/mole	Solute	Solvent	D_o 10^{-4} m²/s	Q kJ/mole
C	fcc Fe	0.25	144.21	Al	fcc Cu	0.38	179.74
C	bcc Fe	0.20	84.02	Cu	fcc Cu	0.2	196.88
N	bcc Fe	0.14	73.99	Cu	fcc Ag	1.23	192.70
H	bcc Fe	0.001	13.00	Zn	fcc Cu	0.34	190.61
Fe	bcc Fe	118.00	280.89	Ag	fcc Cu	0.63	194.37
Fe	fcc Fe	0.58	283.82	Au	fcc Cu	0.69	207.75
Fe	fcc Cu	1.40	216.52	Au	fcc Ag	0.26	190.19
Ni	fcc Fe	0.50	275.88	Ag	fcc Au	0.024	154.66
Mn	fcc Fe	0.35	282.15	Ag	fcc Ag (lattice)	0.72	188.10
C	hcp Ti	5.06	181.83	Ag	fcc Ag(G boundary)	0.14	89.87
Ni	fcc Cu	2.70	236.17	Co	fcc Cu	1.93	226.14
Cu	fcc Al	0.08	136.27				

The total diffusion coefficient for the above system is given by Darken as,

$$D = X_B D_A + X_A D_B \quad (7.31)$$

where X_A, X_B are mole fractions of A and B respectively and D_A and D_B are the respective diffusion coefficients. Hartley (1946) observed this effect while experimenting with a diffusion couple of cellulose acetate and acetone and Kirkendall observed this shift for a couple of α-brass and copper. In case of α-brass and copper couple, voids are observed in brass due to net loss of Zn atoms from brass to copper. So the inert marker shifts towards brass end. Similarly, inert markers shift toward Au end in case of Au-Cu or Au-Ni couples because gold atoms diffuse faster in copper (nickel) than copper (nickel) atoms in gold.

7.7 DIFFUSION IN COMPOUNDS

Diffusion processes in ionic compounds become faster in the presence of Schottky and Frenkel defects. In case of Frenkel defects, some cations occupy interstitial sites leaving cation vacancies. These cation interstitials diffuse through ionic compounds to contribute to diffusion flux. When there are Schottky defects, the cation vacancies carry diffusion flux i.e. a cation vacancy cannot attract anions, which are its nearest neighbours since they have wrong charges. Hence the diffusion of cations takes place from second neighbouring sites thereby causing cation vacancies to move. This type of diffusion requires ions to move from second neighbouring sites across a relatively high energy barrier. Similar process takes place when the anion vacancies diffuse through the ionic compound.

The diffusion coefficients are different for cations and anions due to difference in size, charge and structure. For example, diffusion coefficient of Na^+ ion in NaCl is five times as large as that of Cl^- ion at 1000°K. This happens because size of Na^+ ion is 0.98 Å where as that of Cl^- ion is 1.81 Å. It has been observed that the diffusion coefficient of O^{2-} is 10^7 times larger than that of U^{4+} in uranium oxide UO_{2-x}. This can be explained due to the following reasons:

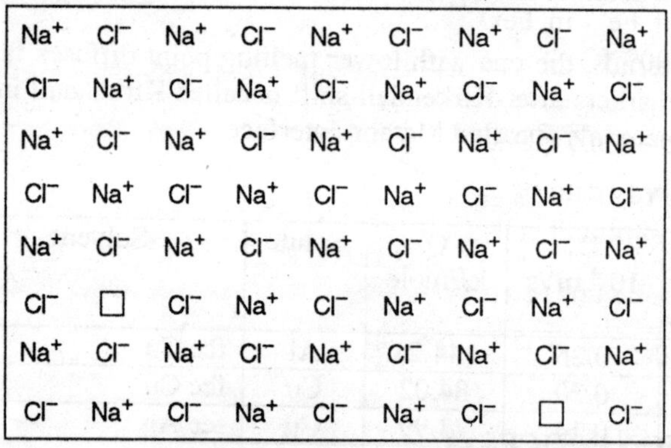

Na^+	Cl^-	Na^+	Cl^-	Na^+	Cl^-	Na^+	Cl^-	Na^+
Cl^-	Na^+	Cl^-	Na^+	Cl^-	Na^+	Cl^-	Na^+	Cl^-
Na^+	Cl^-	Na^+	Cl^-	Na^+	Cl^-	Na^+	Cl^-	Na^+
Cl^-	Na^+	Cl^-	Na^+	Cl^-	Na^+	Cl^-	Na^+	Cl^-
Na^+	Cl^-	Na^+	Cl^-	Na^+	Cl^-	Na^+	Cl^-	Na^+
Cl^-	□	Cl^-	Na^+	Cl^-	Na^+	Cl^-	Na^+	Cl^-
Na^+	Cl^-	Na^+	Cl^-	Na^+	Cl^-	Na^+	Cl^-	Na^+
Cl^-	Na^+	Cl^-	Na^+	Cl^-	Na^+	Cl^-	□	Cl^-

Fig.7.8 Diffusion in NaCl

(i) The energy requirement is more to activate U^{4+} as compared to O^{2-} in uranium oxide since U^{4+} has higher charge.

(ii) UO_{2-x} is a nonstoichiometric ionic compound with anion vacancies, which contribute to higher O^{2-} mobility.

The formation of ion vacancies due to the deviation from stoichiometry has already been discussed in Chapter 6. Excess of Zn^{2+} ions in the interstitial sites of nonstoichiometric ionic compound $Zn_{1+x}O$ increase the mobility of Zn^{2+} ions. Hence the diffusion coefficients of Zn^{2+} in the above compound increases.

If $Fe_{1-x}O$ is dissolved in MgO under reducing conditions to form solid solution, the mobility of Mg^{2+} ions increases as compared to those in MgO. This is due to the fact that $Fe_{1-x}O$ introduces divalent cation vacancies in the host matrix of MgO. Under more oxidizing conditions, cation vacancy appears for every two Fe^{3+} ions formed hence the diffusivity of Mg^{2+} ions increases. When Cr_2O_3 is dissolved in MgO to form solid solution, for every two Cr^{3+} ions occupying cation sites of Mg^{2+} require the omission of three Mg^{2+} ions to preserve the charge neutrality. Consequently one cation vacancy is created assuming all anion sites are occupied. The diffusion coefficient of Mg^{2+} increases due to increase in cation vacancies.

The electrical conductivity for ionic crystals can arise due to the conduction of electrons as well as due to diffusion of ions. At high temperature, the diffusion of ions contributes more to the conductivity than the familiar process of conduction by flow of electrons. Hence the electrical conductivity is called ionic conductivity for ionic crystals. Ionic conductivity σ is expressed as

$$\sigma = nZe\mu \qquad (7.32)$$

where μ is mobility of ions and is given by $CZeD/kT$. Thus, the relation between electrical conductivity σ and diffusivity D is linear and is expressed as

$$\sigma = CnZ^2e^2D/kT = (CnZ^2e^2D_o/kT) \exp(-Q/RT) \qquad (7.33)$$

where the constant C is unity for interstitial ions and slightly greater than one for vacancies, n is the density of defects and Ze is the charge of the defect. Transport number t+ or t– defines the fraction of conductivity due to positive and negative ions respectively and are expressed as,

$$t+ = \sigma+/\sigma$$

and
$$t- = \sigma-/\sigma \qquad (7.34)$$

Example 7.9

Diffusion coefficient of Fe^{2+} in FeO (NaCl type) is 10^{-13} m^2/s at 909°K. Calculate the ionic conductivity of FeO at 909°K.

Solution:

$$\sigma = CnZ^2e^2D/kT$$

Considering C=1, we get,

$$n_{Fe2+} = (4/\text{unit cell})/(4.30 \times 10^{-10}m)^3/\text{unit cell} = 5.031 \times 10^{28}/m^3$$

$$\sigma = \frac{(5.031 \times 10^{28})(2 \times 1.6 \times 10^{-19} C)^2 (10^{-13}m^2/s)}{1.38 \times 10^{-23} \times 909} = 0.041\Omega^{-1}m^{-1}$$

7.8 OTHER DIFFUSION PROCESSES IN CRYSTALS

Diffusion of defects in crystal depends upon the kinetic barrier to the path of diffusing atoms in the host lattice. This is also called lattice diffusion. Lattice diffusion requires higher activation energy compared to that for diffusion through grain boundaries since many bonds are absent along a grain boundary.

The diffusion through cracks or crystal surface requires even less activation energy as compared to that through grain boundary since atoms at the surface have neighbours only inside the surface. Thus the variation of experimentally determined activation energy can be expressed as

$$Q_{lattice} > Q_{gb} > Q_{surface} \qquad (7.35)$$

Similarly, variations of pre-exponential coefficients D_o for various cases can be written as follows:

$$D_{o\ lattice} > D_{ogb} > D_{o\ surface} \qquad (7.36)$$

The diffusion process cannot always occur through grain boundary or crystal surface although activation energy is low for such process. This is because the cross-sectional area through which diffusion occurs in case of grain boundary or crystal surface is usually much smaller compared to that for bulk region of the lattice. At lower temperature the diffusion through surface and grain boundary predominates but at higher temperature the diffusion through bulk region becomes more important. This can also be explained from Eq.(7.16), where pre-exponential factor D_o is dominant at low temperature but exponential part becomes dominant at higher temperature. For example, analysis of diffusion data for single crystal and polycrystalline silver reveals that the coefficient of diffusion through surface or grain boundary is higher for polycrystalline silver than that for single crystal at lower temperature but the diffusivity becomes same for both at higher temperature. Higher diffusivity in polycrystalline substance can be explained due to the presence of grain boundary, Fig. 7.9. At a higher temperature, the diffusion through lattice becomes more and hence the diffusivity becomes same for both the cases.

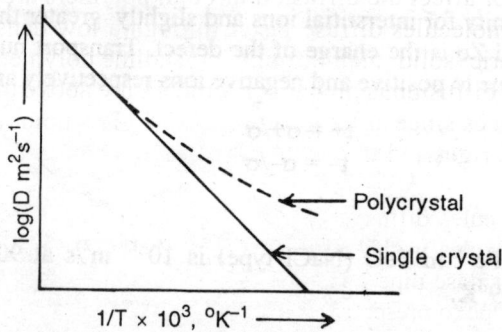

Fig. 7.9 Schematic representation of diffusivity versus temperature

Example 7.10

Compare vacancy diffusion rate with the interstitial diffusion rate in Ag, given that ΔH_f and ΔH_m are 97 kJ mole^{-1} and 80 kJ mole^{-1}, respectively, at temperature 100°C. Assume that the pre-exponential factors to be the same for both the diffusion rates.

Solution:

Diffusion due to interstitial diffusion mechanism / Diffusion due to vacancy mechanism

$$= \frac{\exp(-\Delta H_m/RT)}{\exp[(-\Delta H_m - \Delta H_f)/RT]} = \exp(\Delta H_f/RT) = \exp 97000/8.314 T = 3.8 \times 10^{13}$$

7.9 DIFFUSION IN POLYMERS

The diffusion of polymer molecules through solid polymer is quite less probable due to the fact that the size of polymer molecule is large and sometimes complex. Hence it requires high activation energy. Also polymer molecules are sometimes coiled, kinked or they have three-dimensional framework structures and therefore cannot easily move past other molecules while diffusing through the host matrix. The diffusion process can be facilitated at higher temperatures but in case of polymers the temperature has to be limited to the range 100°K to 1000°K otherwise softening of polymer occurs. Diffusion of small solute molecules takes place easily along the linear molecular chain through polymeric solid because the activation energy for such diffusion process is small due to the presence of weak secondary bonding between linear molecules. Such diffusion processes have lot of importance in case of polymeric film or fibre products where polymer molecules are linear or two-dimensional. These types of diffusion take place in the following cases:

- The diffusion of O_2 through polyethyene bags can oxidize and decompose food material. CO_2 gas generated in the process diffuses out of the bag.

- The gradual diffusion of O_2 and N_2 molecules through inner tube of a tire reduces the pressure in the tyre. That is why tire flattens out with time even when a vehicle is not used.
- Diffusion process is used as a means of dye penetration in the textile industries.
- Many biological processes occur by selective diffusion through membranes.
- Helium plant near a vacuum unit seriously affects the vacuum process since helium atoms easily diffuse through glass chambers.

Factors that influence or affect the diffusion through polymers are the following:
- Small solute molecules diffuse faster through a low density noncrystalline/ amorphous polymeric solid than through crystalline polymer.
- Coefficient of diffusion is higher at a temperature above the glass point for an amorphous substances since at temperature below glass point the structure of polymeric solid becomes rigid and the diffusion process requires higher activation energy.

When the macromolecules diffuse through polymeric solids, the volume is added due to the presence of solute molecules in the polymer structure. This is called swelling. It is a common experience that a rubber hose pipe left outside swells and becomes hardened. This process also makes the pipe brittle. Due to presence of micromolecules in the polymer structure the polymeric solid is plasticized leading to lowering of the glass transition temperature and there is change in the mechanical property. The plasticized polymer can undergo plastic deformation easily. The crystalline or cross-linked polymers cannot be plasticized and hence they are required wherever the structural stability is needed.

7.10 EFFECT OF DIFFUSION ON GRAIN GROWTH

Grain growth results in the increase of grain size. During diffusion process, the atoms move across grain boundary as well as within crystals. If grain boundary is a plane, the movements of atoms across the boundary are equally probable along both the directions into and out of the grain. The curved grain boundary indicates an atom on the surface has more nearest neighbours in case of concave surface than when it is a part of a convex surface. So the surface energy is less for atoms on concave surface as compared to those on convex surface. As a result, the curved boundary itself moves toward its center of curvature and there is an overall tendency to decrease grain boundary area, which will also decrease the surface energy. Normally in a single phase polycrystalline substance the microstructure can be visualized through simple geometry, Fig.7.10. From this figure one can easily observe that at a suitable temperature, the grains with convex surface grow smaller and large grains with concave surface grow larger until the former eventually disappear. Finally a few larger grains are left. The rate of growth of grain can be expressed as

$$dL /dt = k/L^n \tag{7.37}$$

where L is the diameter of the grain and k is a constant. Integrating we obtain,

$$L^{n+1} - L_0^{n+1} = Ct \tag{7.38}$$

where L_0 is the initial diameter of the grain which is smaller than L. If $L_0 << L$, then one can simplify Eq.(7.38) to obtain the average diameter of grain at time t. The average diameter is expressed as

$$L = (Ct)^{1/(n+1)} , n \geq 2 \tag{7.39}$$

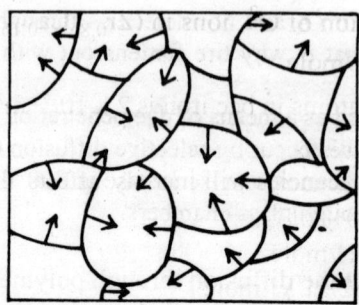

Fig.7.10 Microstructure in polycrystalline substance

The parameter n varies from one material to another and attains values greater than 5 if there is presence of impurity phases which inhibit the grain growth. The grain boundaries become planar where they intersect the external surfaces and hence the grain growth is arrested. The rate of grain growth increases pronouncedly at higher temperature since diffusion increases exponentially as temperature increases. When the material is kept at a temperature greater than 0.5 T_m (T_m is melting temperature), the pattern of variation of L with time can be plotted in log scale and n can be calculated as follows

$$\ln L = \ln C' + (1/(n+1)) \ln t \qquad (7.40)$$

where $C' = C^{(1/(n+1))}$

Example 7.11

Let us consider a sample of brass being annealed at temperature 500°C. After 2.5 minutes of annealing, the size of grain is 0.01 mm and it becomes 4 times after 2500 minutes, find the value of n.

Solution:

$$1/(n+1) = (\ln L_2 - \ln L_1)/\ln (t_2/t_1) = \ln 4/\ln (1000)$$
i.e. $\qquad n = 3.98$

EXERCISES

7.1 What is diffusion? Explain whether it is microscopic or macroscopic in nature. What is the difference between Fick's first and second laws ? Does diffusion coefficient change with concentration? Is it isotropic?

7.2 Compare the diffusion fluxes of C in ã-Fe and Fe in ã-Fe at 900°C, if the concentration gradient of C as well as Fe atoms in the host material is 1.4×10^{26} atoms / m^4.

Ans: 7.11×10^5

7.3 Diffusion coefficient of nickel in iron at 1027°C is 2×10^7 times the diffusion coefficient at 527°C. Calculate the activation energy of diffusion of Ni in γ-Fe. Calculate the diffusion coefficient in m^2/s for diffusion of nickel in γ-iron (fcc) at 927°C, if $D_o = 7.7 \times 10^{-5}$ m^2/s.

Ans: 290.6 kJ/mole, 1.7×10^{-17} m^2/s

7.4 Given that the diffusion coefficient $D = 3.98 \times 10^{-15}$ m^2/s and 10^{-12} m^2/s at T = 560°C and

932°C, respectively, for diffusion of O^{2-} ions in $(Zr_{0.85}Ca_{0.15})O_{1.85}$. Calculate D_o & Q.
Ans: 2.368×10^{-7} m²/s; 124 kJ/mole

7.5   If the diffusivity of iron atoms in bcc iron is 2×10^{-23} m²/s at 400°C and it becomes 2×10^7 times at 800°C, calculate
   i)   the factor by which vacancies will increase at 800°C as compared to that at 400°C, considering $\Delta H_f = 60\%$ of Q,
   ii) activation energy in kJ/mole.
   Ans: (i) 2.4×10^4; (ii) 252.21 kJ/mole

7.6   Find the ratio of diffusion coefficients of Cu in Cu and Zn in Cu at 900°C. Also explain the difference in these values.
   Ans: 0.3

7.7   The diffusion coefficient of Cu in Ni at 1000°C is given as 9.175×10^{-16} m²/s and $D_o = 2.7 \times 10^{-5}$ m²/s. Calculate the value of D at 500°C.
   Ans: 1.55×10^{-22} m²/s

7.8   In diffusion annealing of nickel in fcc iron, if the initial temperature is increased from 900°C to 1090°C, by what factor does the penetration depth increase?
   Ans: 7.2

7.9   Pressure in a cycle tube (surface area 0.35 m² and volume 4.3×10^{-3} m³) is 3 atm which reduces to 2 atm in 20 days. What is the diffusion flux?
   Ans: 1.74×10^{17} molecules/m²s

7.10 Given the following data for Na^+ conductivity with NaCl at 550°C and 727°C:

Temperature	Transport numbers	σ
550°C	0.98	$2 \times 10^{-4} \Omega^{-1} m^{-1}$
727°C	0.80	$2.5 \times 10^{-2} \Omega^{-1} m^{-1}$

   Calculate the diffusion coefficient at 550°C, Q and D_o.
   Ans: 3.78×10^{-15} m²/s; 187 kJ/mole; 2.7×10^{-3} m²/s

7.11 Hydrogen gas can be separated from other common gases using hot palladium membranes in a closed apparatus, since hydrogen gas diffuses through hot palladium metal easily whereas other common gases do not. If the concentration of H_2 is 1kg/m³ on one side of the Pd membrane (of area 0.0012 m² and thickness 10^{-3} m) and that on the other side it is 0.2 kg/m³, calculate the volume (in m³ at STP) of H_2 gas that diffuses through the membrane per hour, where D for H_2 in Pd is 10^{-8} m²/s, and ρ for H_2: 0.0899 kg/m³.
   Ans: 384.43 cm³

7.12 Aluminium is diffused into Si single crystal with Q = 78 kcal /mole and D_o =5.55 x 10^{-4} m²/s. Find the value of diffusion coefficient at 1100°C. If the diffusion of Al atoms into Si wafer takes place with no previous aluminium in it at a temperature of 1100°C for 3 hours, what is the depth below the surface at which the concentration is 10^{22} atoms/m³. The surface concentration of aluminium atoms is 10^{24} atoms/m³.
   Ans: 2.10×10^{-16} m²/s, 5.36 μm

7.13 When boron is diffused into a thick slice of silicon with no previous boron in it at 1100°C for 4 hr, find the concentration of boron at a depth of 1 μm below the surface, if the surface concentration of boron is 10^{24} atoms/m³.(D = 4 x 10^{-17} m²/s at 1100°C)
Ans: 3.26 x 10^{23} atoms/m³

7.14 Heat treatment is done on an alloy of Cu in Al at 550°C. When the alloy is quenched to 27°C the diffusion rate of Cu is increased by a factor of 10^8. Explain the reason for increase in the diffusion rate. If the enthalpy of motion of vacancy in the alloy is 50 kJ/mole, calculate the fraction of vacancies that are created at 550°C, will be retained at 27°C.
Ans: 0.028

7.15 Calculate the diffusion coefficient of O^{2-} ions in $(Zr_{0.85}Ca_{0.15})O_{1.85}$ at 744°C, if the ionic conductivity of the above material is 0.3×10^2 $\Omega^{-1}m^{-1}$ due to diffusion of vacancies. Assume that the ionic conductivity of the above ceramic is almost entirely (t = 0.99) due to anions. Also calculate D_o, if Q = 29 kcal/mole and lattice parameter a = 5.45 Å.
Ans: 1.098×10^{-9} m²/s, 3.396×10^{-8} m²/s

7.16 Consider the ionic compound UO_2 where diffusion takes place due to both cations and anions. If the ratio of diffusion coefficients for U^{4+} and O^{2-} ions is 10^7, calculate the ratio of conductivities considering the structure of ceramic to be fluorite.
Ans: 1.25×10^6

7.17 If the average grain size of a polycrystalline solid is 0.8 mm and the grains are cubic with grain boundaries 3 Å thick, find the ratio of diffusion rates through lattice and through grain boundary at the temperature 500°C. Given the following data:

	D_o(m² s⁻¹)	Q(kJ mole⁻¹)
For lattice diffusion	0.8×10^{-4}	188
For grain boundary diffusion	0.09×10^{-4}	81

Ans: 0.69

7.18 Steel is getting carburized at 910°C, such that 0.8 w/o carbon is found at a depth of 0.2 mm after 1 hr. Find
i) the time required to have the same concentration at the same depth, if carburization is done at 1027°C.
ii) the depth at which same concentration is observed after 2 hr, if carburization is done at 910°C.
Ans: (i) 26 min 9s; (ii) 2.83 mm

7.19 Concentration gradient of copper (fcc) in a luminous bar decreases from 0.4 a/o at the surface to 0.2 a/o one mm below the surface. If the above gradient is maintained, what is the net movement of atoms across a plane of area 5m², one mm below the surface if the bar is at 500°C. Given D_o = 1.5×10^{-5} m²/s; Q = 126kJ/ mole; a = 3.61Å; R = 8.314J/mole°K.
Ans: 3.86×10^{16} atoms/s

7.20 How much should the concentration gradient be for nickel in iron if a flux of 100 nickel atoms/cm²-s is to be realized at 800°C? Given $D_o = 7.7 \times 10^{-5}$ m²/s and Q=280 kJ/mole.
Ans: 5.64×10^{23} atoms/m⁴

7.21 Tick the correct answer. Interstitial diffusion is generally
(a) much faster than,
(b) much slower than,
(c) as fast as
the substitutional diffusion by vacancy mechanism.
Ans: (a)

7.22 Selenium has the following set of experimental data for vacancy diffusion process.

T°C	35	40	46	56
D(m²/s)	7.7×10^{-16}	2.4×10^{-15}	3.2×10^{-14}	3.2×10^{-13}

Calculate the following:
i) Activation energy for the above process.
ii) Enthalpy of motion, if the fraction of vacancies is 10^{-10} in thermal equilibrium at 50°C.
Ans: (i) 182.14 kJ/mole; (ii) 120.2 kJ/mole

7.23 To produce a p-type semiconductor boron is doped in pure silicon. The doping is done in presence of BO vapour phase at 1120 °C for 5 hours. If doping of boron in silicon takes place through diffusion and a constant boron concentration gradient is assumed; then calculate the concentration of boron needed to be maintained at the surface of silicon in order to get the concentration of boron at a depth of 2 μm equal to 10^{23} atoms per m³.
Ans: 1.05×10^{24} atoms/m³

8.1 INTRODUCTION

The state of a material is the physical condition as defined by the quantities of various components present at specific temperature and pressure. Material often exists in a combination of various phases–solid, liquid or gaseous. Phases are chemically homogenous and physically distinct but mechanically separable microstructures present in the material. Thus, phases are different from components. Components have definite chemical formulae whereas phases do not have any specific formula. A range of various weight percent of components can give a specific phase at suitable pressures and temperatures. Material consisting of two components (metals or metallic compounds) is called a binary alloy. The components in the binary alloy can combine in various proportions and a cooling curve can be drawn for each composition. From these cooling curves one can draw equilibrium diagrams, commonly called as the phase diagrams. The weight percent of the components present in an alloy is called the composition of the alloy. As an example, one can think of a material consisting of two components Al_2O_3 and Cr_2O_3. This material can exist in different phases depending upon the pressure, temperature and composition i.e. weight percents of Al_2O_3 and Cr_2O_3. Consider a system consisting of water with floating ice. How many components are present in such a system? Answer to this question is that one component H_2O is present in the system, although the above system consists of two phases water and ice.

A system is in equilibrium, when the properties of the system do not change with time. During a typical reaction, say melting of an alloy, the system exchanges both heat and work with its surroundings. When the system approaches equilibrium the Gibb's free energy of the system decreases and attains minimum value at the equilibrium. The expression for Gibb's free energy is

$$G = U + PV - TS, \tag{8.1}$$

The change in free energy is expressed as

$$dG = dU + PdV + VdP - TdS - SdT \tag{8.2}$$

If the process occurs under constant pressure and temperature, then

$$dG = dU + PdV - TdS \tag{8.3}$$

For a spontaneous reaction, $dG < 0$ i.e.

$$dH - TdS < 0, \tag{8.4}$$

where dH = dU + PdV, H being the enthalpy. Now in case of a solid substance the change in volume is negligible, hence the term PdV= 0 is a good approximation. For a process to occur spontaneously the condition to be satisfied is

$$dU - TdS < 0 \tag{8.5}$$

Consider a system with single component which is undergoing phase change at specific temperature and at standard atmospheric pressure, then dG = 0 i.e. the free energy of the system remains constant during such phase transformation. In case of melting of a metal, the change in enthalpy dH>0, since the free energy of the metal in the crystalline state is lower than that in the liquid state due to bond energy being higher in the solid state. On the other hand, the entropy increases during melting since the atoms are disorganized in the liquid state. At melting temperature T_m,

$$dG = 0,$$

i.e. $$T_m = dH/dS. \tag{8.6}$$

At $T > T_m$, TdS > dH and hence dG < 0.

8.2 PHASE RULE

The condition for thermodynamic equilibrium of a system requires temperature and pressure to be uniform throughout the system. Also each component of the system has same chemical potential in each of the coexisting phases. This gives rise to (P–1) conditions for each component, where P is the number of phases. Gibbs manifests the conditions for equilibrium in the form of phase rule. According to Gibbs, the requirements for equilibrium limit the degrees of freedom for the system. Let us say there is C number of components in the system. Now only the weight percents of (C – 1) components are independent, since the sum of weight percents of all the components should be 100. Since there is P number of phases the number of composition variables is P(C–1). To this we add two more independent variables i.e. pressure and temperature. So the total number of variables N is

$$P(C-1)+ 2.$$

The number of equations is given by

$$n = P - 1 \tag{8.7}$$

which comes from the minimum free energy criterion. Thus the total number of equations arising out of the above constraint is C(P–1). Hence the number of independent variables of the system is

$$F = P(C-1) + 2 - C\,(P-1)$$
i.e. $$F = C - P + 2, \tag{8.8}$$

where F is the number of degrees of freedom. During a phase transformation, sometimes a small quantity of a component of the alloy may transfer from one phase to another coexisting phase. This may disturb the equilibrium of the system. As a result, the compositions of phases change until the free energy of the system is minimized. The total number of variables of a system is greater than or equal to the number of degrees of freedom.

Example 8.1

Find the number of phases coexisting in a system having 3 components, if the degrees of freedom of the system is maximum. What is the total number of variables in this case?

Solution:

We know that the total number of variable F is given by

$$F = C - P + 2$$

substituting C = 3, we get

$$F = 5 - P$$

F will be maximum when P is minimum i.e. when P = 1. The degree of freedom F is 4.

8.3 CLASSIFICATION OF PHASE DIAGRAMS

Phase diagrams represent the combination of phases existing under equilibrium conditions of a material with various compositions at specific pressure and temperature. Depending upon the number of components in the system, phase diagrams can be classified as follows:

    i) Unary phase diagrams
    ii) Binary phase diagrams
    iii) Ternary phase diagrams etc.

8.4 UNARY PHASE DIAGRAMS

Unary phase diagram is an equilibrium phase diagram for a single component system e.g. the phase diagram of water. The curves shown in the phase diagram are univariant as indicated by phase rule since C =1 and P= 2 along the curves.

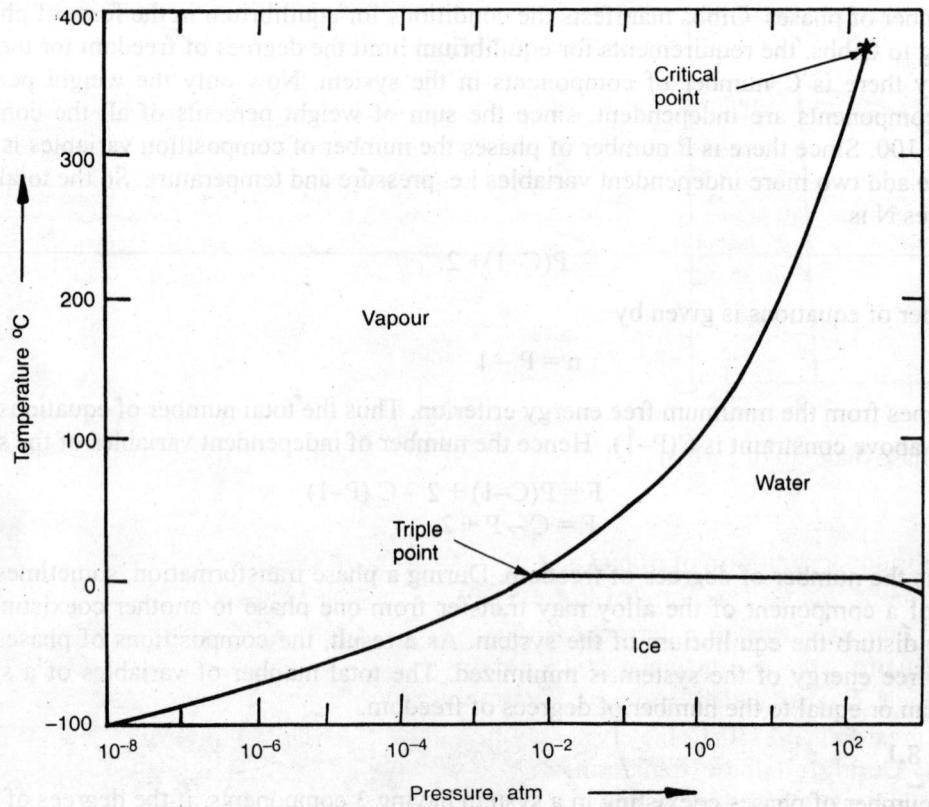

Fig 8.1 Phase Diagram of H_2O.

Hence
$$F = 1 - 2 + 2 = 1 \qquad (8.9)$$

Similarly, the three regions of this phase diagram are bivariant since P = 1 in these regions. The triple point at 0.00785 °C and 4.58 mm pressure is invariant since at this point P = 3; hence the degrees of freedom,

$$F = C - P + 2 = 1 - 3 + 2 = 0 \qquad (8.10)$$

i.e. neither pressure nor temperature can be varied arbitrarily. Another example of unary phase diagram is carbon phase diagram, Fig 8.2 with pressure and temperature as variables. Graphite has hexagonal crystal structure whereas diamond has a typical cubic structure called diamond cubic. Allotropes of carbon and their structures have been described in Chapter 4. The other two phases of carbon are vapour and liquid. All these phases are mechanically separable.

Iron has four allotropes α, γ, δ and ε which are found under different conditions of pressure and temperature. Fig 8.3 shows the various phases of iron that exist at different combinations of pressure and temperature.

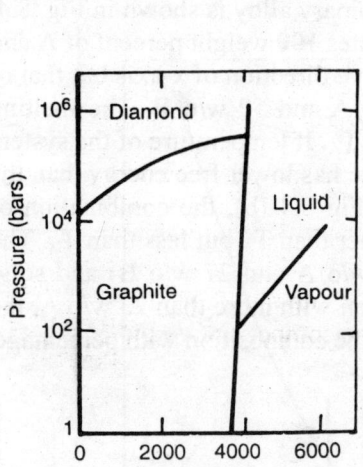

Fig 8.2 Phase diagram of carbon

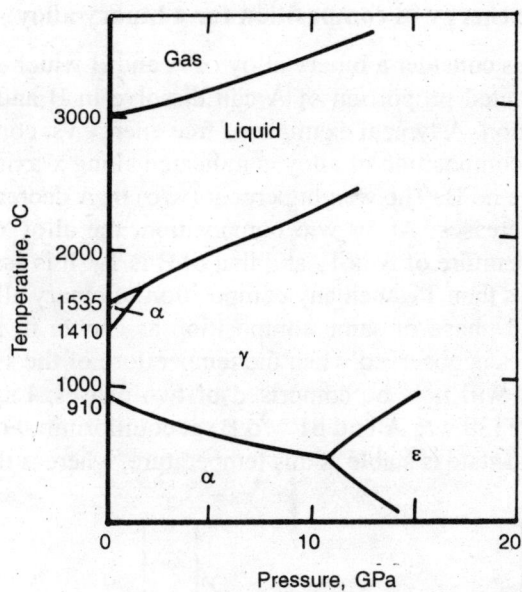

Fig. 8.3 Phase diagram of iron

The equilibrium crystal form of iron at atmospheric pressure and ambient temperature is bcc. It is called α-iron or ferrite. This form of iron is soft and ductile. The tensile strength of α-iron is less than 45,000 psi. This is a ferromagnetic material below 767°C. As the temperature is increased to 910°C, the α-iron undergoes a phase change to form new phase γ-iron with fcc structure which is also called austenite. γ-iron is soft and ductile. It is paramagnetic. At temperature greater than 1410°C, the crystal structure of iron becomes bcc again which is referred to as δ-iron. During all these transformations pressure is atmospheric i.e. 0.101 GPa. At a temperature above 1535 °C there is liquid phase of iron and above 3000°C, the iron is in vapour phase. The variation of phases at different pressures is also clear from the phase diagram. When the pressure is increased to 15 GPa, at a temperature below 910°C α-iron changes to hcp ε-iron. There are two triple points in iron phase diagram: α, ε and γ phases coexist at the first triple point and δ, γ and liquid phases coexist at the second triple point.

8.5 BINARY PHASE DIAGRAMS

Two-component systems have binary phase diagrams in which total number of variables can be four in case P = 2, but the independent variable of this system is 2 only. In multi-component systems, the components can simply form mixture without forming solution or undergoing chemical reactions, e.g. iron reinforced concrete, glass reinforced plastic; while in other cases the components react to form compounds or form solutions. Such types of interactions ultimately reduce the free energy of the system. The two components in a binary phase diagram may or may not be completely soluble in each other. When they are completely soluble in any proportion in solid state as well as in liquid state, it is not possible to detect either of the components in a solution.

Free energy vs composition for a binary alloy

Let us consider a binary alloy of A and B which satisfy Hume-Rothary empirical rules and hence unlimited proportion of A can dissolve in B and vice versa. Such system forms complete solid solution. A typical example of free energy vs. composition for a binary alloy is shown in Fig. 8.4. The composition of alloy is indicated along x-axis where A indicates 100 weight percent of A and hence no B. The weight percent (w/o) of A decreases along positive direction of x-axis but that of B increases. At 50 w/o composition, the alloy contains 50 w/o A and 50 w/o B. The melting temperature of A is T_A and that of B is T_B. It is assumed that $T_B > T_A$. If temperature of the system is less than T_A, then any composition of binary alloy in solid phase has lower free energy than the liquid phase of same composition as shown in Fig. 8.4 (a). In Fig 8.4 (b), the combination of phases is observed when the temperature of the system, T is greater than T_A but less than T_B. The alloy will now be comprised of two phases: Liquid phase (73 w/o A and 27 w/o B) and solid phase (39 w/o A and 61 w/o B) at equilibrium. For the composition with more than 73 w/o A, the liquid state is stable at this temperature, whereas the alloy having the composition with percentage

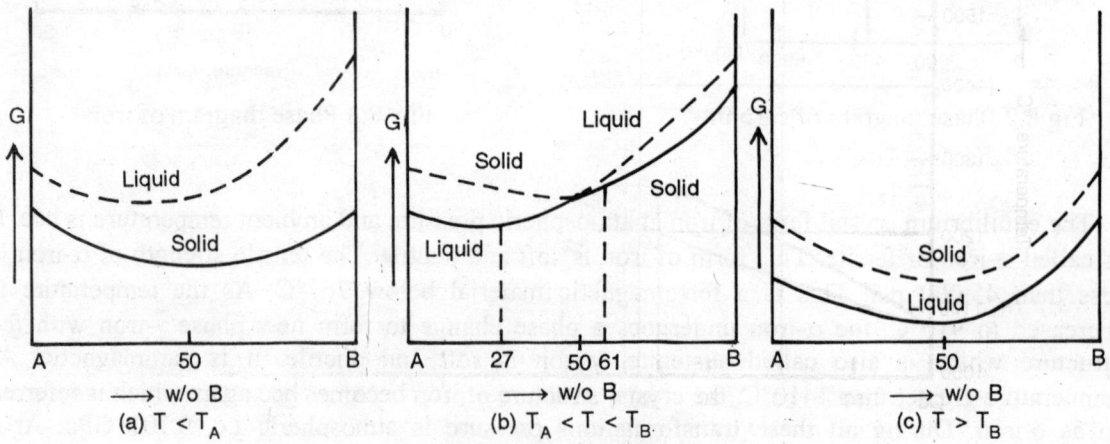

Fig 8.4 Free energy vs composition of a binary alloy

of A lying between 39 w/o and 73 w/o the mixture of solid and liquid phases would provide lower energy than either liquid or solid phase. For compositions with less than 39w/o A, the solid state is the stable one. At a temperature above T_B the liquid phase of the alloy always has free energy lower than that of pure solid phase, Fig 8.4 (c).

The binary phase diagram requires to be represented in a three-dimensional space since in the single phase region of the phase diagram there are three independent variables—pressure, temperature and composition ($C - P + 2 = 3$). In order to represent the phase relationships in a simplified way, binary phase diagrams are drawn at atmospheric pressure such that they can be represented in two dimensions having temperature and composition as ordinate and abscissa, respectively. Moreover pressure has very little effect on binary phase diagrams of solid alloy. So the binary phase diagram for constant pressure will satisfy phase rule

$$F = C - P + 1 \tag{8.11}$$

Isomorphous system

In a binary system when there is complete inter-solubility between components in all phases, the system is isomorphous e.g. Cu-Ni, Al_2O_3-Cr_2O_3, NiO-MgO, MgO-FeO, Nb-W. The phase diagram for Cu-Ni is shown in Fig 8.5, where the composition of the alloy is plotted along x-axis with w/o nickel increasing from left to right starting from 0 to 100 w/o. The percentage of copper varies from 100 to 0 w/o from left to right. Pure copper and pure nickel form left and right ends of the composition axis.

The temperature is plotted along y-axis. The melting point of copper is 1083°C and that of nickel is 1455°C. The phase diagram gives us information regarding the equilibrium states of Cu-Ni alloys at any temperature and composition. Normally, sufficient time should be allowed for the system to reach equilibrium whenever there is any change of variables. In this phase diagram there are two lines connecting the melting points of the pure components Cu and Ni. The upper

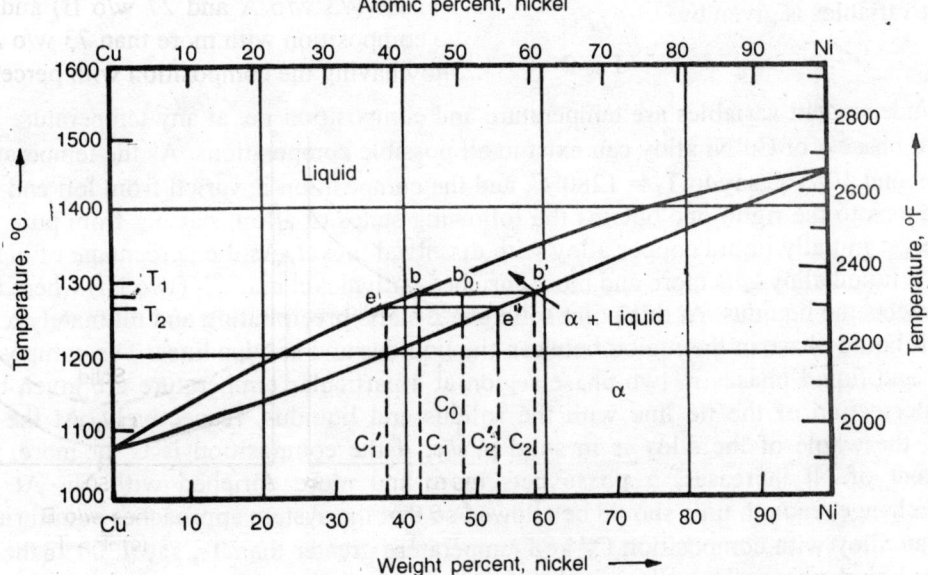

Fig 8.5 Cu-Ni phase diagram.

line is the locus of points at which solidification starts at various compositions. For a specific composition the solidification starts only at a specific temperature. This line is called liquidus. Along this line, liquid phase of alloy and traces of solid phase coexist hence number of independent variables along liquidus is

$$F = C - P + 1 = 1 \qquad (8.12)$$

The lower line in this phase diagram is solidus, which is locus of points at which solidification completes. The region of the phase diagram above liquidus line is liquid alloy and the region below the solidus, the system is entirely solid. The region between the solidus and liquidus lines is a two-phase region where liquid and solid phases coexist. To study the equilibrium states of the system at fixed temperature, a horizontal line is drawn corresponding to that temperature. This is called isothermal line. This line in a two-phase region is called the tie line.

A liquid alloy of specific composition begins to solidify when it is cooled to liquidus temperature depending upon composition. The liquidus temperature of a particular alloy can be determined by drawing vertical line at the overall composition of the alloy. The point of intersection of the vertical line and liquidus gives the liquidus temperature of the alloy. Similarly, solidus temperature of the alloy can be determined from the intersection of the vertical composition line and solidus curve. While using a material in service its service temperature should be less than solidus temperature otherwise the alloy may melt while being used.

Alloys melt and freeze over a range of temperatures, between the solidus and liquidus temperatures, depending upon composition. This is called freezing range of the alloy. The alloy coexists in partially liquid and partially solid phases in the freezing range. An alloy can be cast at liquidus temperature but solidus temperature has to be greater than the service temperature. If the service temperature is known for a specific industrial application, material should be chosen such that its solidus temperature is higher than service temperature. In Cu-Ni phase diagram, all possible compositions of Cu-Ni alloy can be there either in solid phase (α) below 1083°C or in liquid phase (L) above 1455°C. Since in these regions the number of phases is one, the number of independent variables is given by

$$F = 2 - 1 + 1 = 2 \qquad (8.13)$$

The two independent variables are temperature and composition i.e. at any temperature below 1083°C solid phase α of Cu-Ni alloy can exist in all possible compositions. As the temperature is increased beyond 1083°C say to $T_1 = 1280$°C, and the composition is varied from left end of the composition axis to the right, one obtains the following states of alloy, starting from pure liquid copper. We get initially liquid copper alloy with dissolved nickel. As the percentage of nickel is increased, the liquid alloy gets more and more enriched with nickel until C_1 (w/o Ni) where the tie line bb' intersects the liquidus. At this point solid phase starts precipitating and ultimately α phase coexists with liquid phase in the region between the liquidus and solidus lines. The compositions of the solid and liquid phases in two-phase region at a particular temperature are given by the points of intersection of the tie line with the solidus and liquidus, respectively. At the same temperature, the whole of the alloy is in solid phase, if the composition is C_2 or more. If the weight percent of Ni increases, α phase gets more and more enriched with Ni. At every composition change, enough time should be allowed so that the system approaches equilibrium.

Consider an alloy with composition C_1' at a temperature greater than T_1, say 1300°C the alloy would be in liquid phase. The alloy is then gradually cooled to room temperature ensuring equilibrium at every stage. As the temperature decreases from 1300°C, the alloy continues to exist in liquid phase until the liquidus is reached i.e. at temperature T_2. At a temperature below T_2 the solid phase α starts appearing and below solidus line the alloy is only in solid phase α. The cooling should be done very slowly such that the system attains equilibrium at each temperature. At the temperature T_2, the above alloy coexists both in solid and liquid phases. The compositions of the solid phase and liquid phase at this temperature can be found out by drawing the tie line. The tie line intersects liquidus and solidus at points e and e', respectively. The corresponding compositions C_1' and C_2' are the compositions of liquid and solid phases, respectively. If the alloy is cooled below 1200°C, it exists only in α-phase. Thus, by varying temperature the alloy in a single phase region can cross into an equilibrium state in two-phase region, and then into another single phase region. Similarly, an alloy passes from a single phase region to another single phase region through two-phase region by variation of composition. This is called "1–2–1" rule.

Now how to find out the fraction of liquid and solid phases in the alloy at a particular temperature and for a given overall composition (C_o) of the alloy. To find this we have to perform the following steps:

* Draw a tie line at a given temperature T_1, which intersects the liquidus and solidus at points where compositions are C_1 and C_2 w/o of Ni, respectively.
* Draw a vertical line at the overall composition C_o w/o Ni, which intersects the tie line at b_o.
* Let us say that the w/o liquid phase at temperature T_1 is w_L and that of solid phase is w_S in the given alloy i.e. if the sample is 100 g there is w_L g of liquid phase and $w_S = (100 - w_L)$g of solid phase.
* w_L g of liquid phase contains $w_L \times (C_1/100)$ g of nickel. Also w_S gms of solid phase contains $w_S \times (C_2/100)$ g of nickel. The total sample contains.

$$(w_L + w_S) \times C_o/100 \text{ g} \tag{8.14}$$

of nickel. Since the total amount of nickel in the liquid and solid phases should be equal to the amount of nickel in the given alloy, we get

$$(w_L + w_S) C_o/100 = w_L(C_1/100) + w_S(C_2/100) \tag{8.15}$$

$$w_L(C_o - C_1) = w_S(C_2 - C_o) \tag{8.16}$$

Hence

$$\frac{w_L}{w_S} = \frac{C_2 - C_o}{C_o - C_1} \tag{8.17}$$

The weight fraction of liquid phase present in the alloy is

$$\frac{w_L}{w_L + w_S} = \frac{C_2 - C_o}{C_2 - C_1} \tag{8.18}$$

and the weight fraction of the solid phase present in the alloy is

$$\frac{w_S}{w_L + w_S} = \frac{C_o - C_1}{C_2 - C_1} \tag{8.19}$$

In other words the amount of liquid phase present in the alloy is proportional to the length b_0b' of the tie line and that of solid phase is proportional to the length bb_0 of the tie line. This is termed as lever rule.

Lever rule

Lever rule is used to find out the fractions of various phases present in an alloy at a specific temperature. This rule is based on Eq.(8.17). According to this rule, the tie line is considered as lever arm and the intersection of tie line with the vertical line passing through the overall composition is taken as the fulcrum of a simple lever system. The relative lengths of the lever arms multiplied by the corresponding amounts of phase present must balance. This rule is valid for only two-phase region irrespective of the number of components in the system.

Whatever has been discussed so far with respect to the melting point also holds for the boiling point with the only difference is that now we have liquid and vapour phases as single phase regions. In the two phase region both liquid and vapour phases coexist. Fig 8.6 shows a complete representative diagram for a two component isomorphous system.

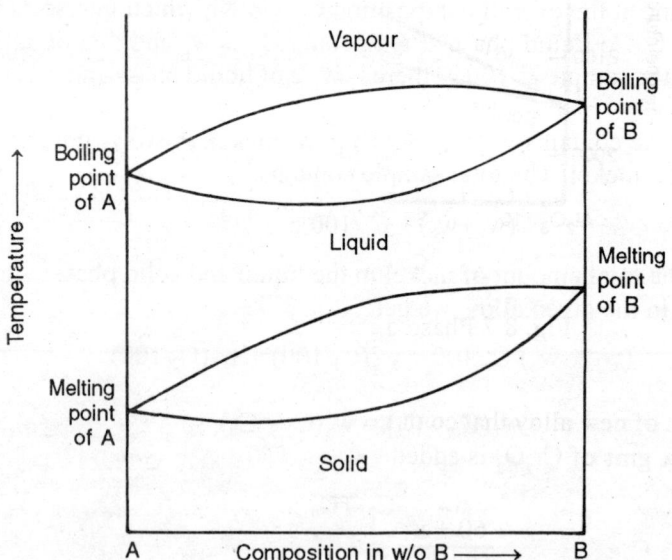

Fig 8.6 Representative phase diagram for isomorphous systems

Example 8.2

Consider the phase diagram of Al_2O_3-Cr_2O_3, Fig. 8.7.
  (a) If the alloy contains 60 g Cr_2O_3 and 40 g Al_2O_3, at temperature 2150°C, find the w/o of the liquid and solid phases present in the sample.
  (b) If we add some amount of one of the components, keeping the temperature same, such that the new alloy so formed contains only solid phase with traces of liquid phase. Find the composition of the new alloy. Also determine the name and the fraction of the component added to the initial alloy.
  (c) If now the temperature of the new alloy is increased such that it contains only liquid phase with traces of solid phase, find the composition of liquid and solid phases at this temperature and also calculate the temperature.

Solution:

(a) From Fig. 8.7 we find that the tie line corresponding to 2150 °C intersects the liquidus at 40 w/o Cr_2O_3 and the solidus at 67 w/o Cr_2O_3. Hence

$$w/o \ \ liquid = \frac{67-60}{67-40} \times 100 = 25.93$$

and

$$w/o \ \ solid = \frac{60-40}{67-40} \times 100 = 74.07$$

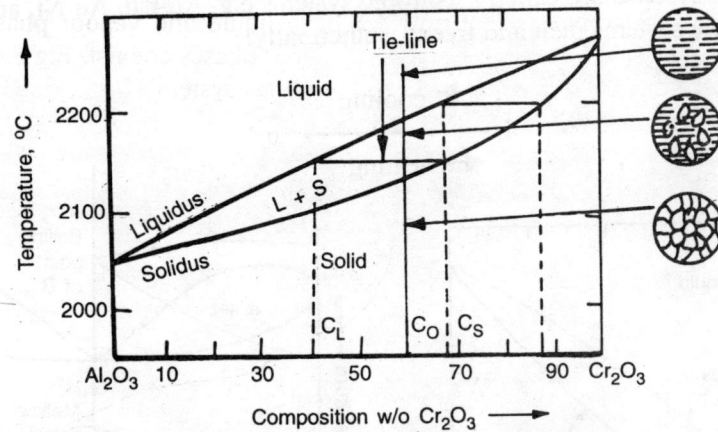

Fig. 8.7 Phase diagram of Al_2O_3-Cr_2O_3 alloy

(b) Composition of new alloy that contains only solid phase is 67 w/o Cr_2O_3 & 33 w/o Al_2O_3. Let us consider x gms of Cr_2O_3 is added to the original alloy such that the total weight becomes $100 + x$

$$\frac{60+x}{100+x} = 0.67,$$

Amount of Cr_2O_3 added = 21.21 g.

(c) From Fig. 8.7 we find that the temperature at which the new alloy has only liquid phase with only traces of solid phase is 2200°C. Also from Fig. 8.7 we find that the composition of the liquid phase is 67 w/o Cr_2O_3 and 33 w/o Al_2O_3 and the composition of the solid phase is 87 w/o Cr_2O_3 and 13 w/o Al_2O_3.

Example 8.3

Consider that NiO-MgO bricks are to be used in furnace. The maximum temperature of the furnace is 2200°C and the brick materials are melted and cast at 2600°C. Find the composition of the alloy. Find its freezing range. (At wt of NiO = 74.71 amu ; At wt of MgO = 40.31 amu)

Solution:

The alloy should be chosen such that it has a liquidus temperature below 2600°C but a solidus temperature above 2200°C. The study of MgO-NiO phase diagram, Fig. 8.8, reveals that the alloy should contain MgO less than 64 mol% (= 49 w/o MgO) so that liquidus temperature is less than 2600°C. The solidus temperature should be above 2200°C i.e. there must be MgO greater than 36 mol%(= 23 w/o MgO). Consequently the MgO-NiO alloy having MgO between 49 w/o and 23 w/o can be used for above purpose.

Azeotropic systems

In some of the isomorphous binary systems, the liquidus touches the solidus tangentially at a minimum temperature, which is lower than the melting temperature of either of the two components. Such systems are called azeotropic systems e.g. Au-Cu, Au-Ni, and Cr-Fe etc. (Fig. 8.9). The azeotropic systems melt and freeze isothermally:

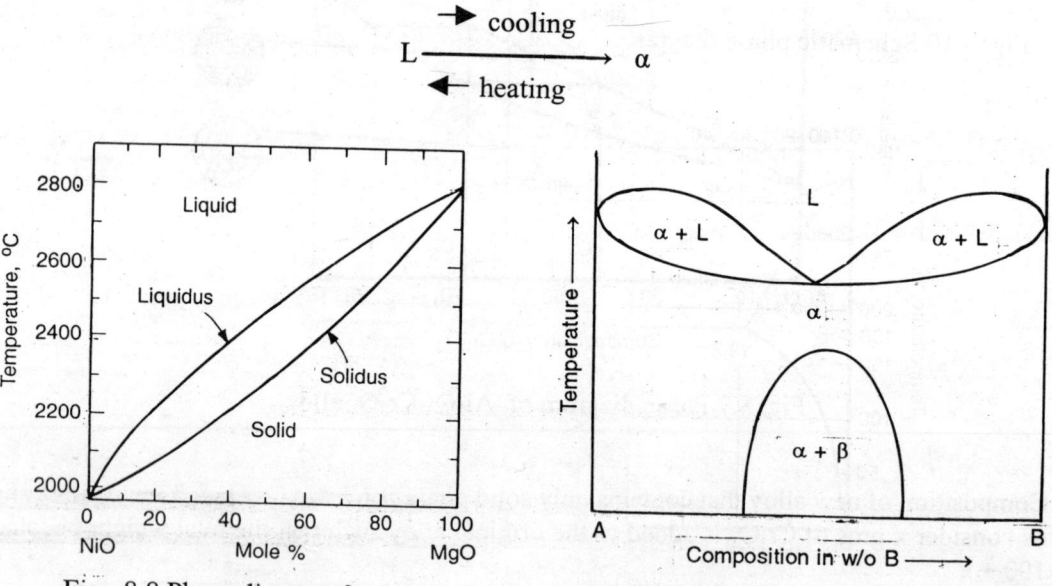

Fig. 8.8 Phase diagram for NiO - MgO system

Fig. 8.9 Phase diagram of a binary azeotropic system, showing a large solubility transformation.

Incomplete solubility in binary systems

In certain binary system, incomplete intersolubility is observed below the transformation temperature and hence we find a solubility gap, Fig. 8.10, for example Cu-Pd system.

Eutectic systems

If Hume-Rothary rules are not obeyed by the components of a binary alloy, intersolubility of the components decreases, solubility gap increases and the upper transformation curves overlap the gap. As a result, we obtain eutectic phase diagram, Fig. 8.11. When the melting points of two components of a binary alloy do not differ much (i.e. differ by 100°C to 300°C) and there is incomplete solubility in solid phases, the phase diagram of such a system is called eutectic. As

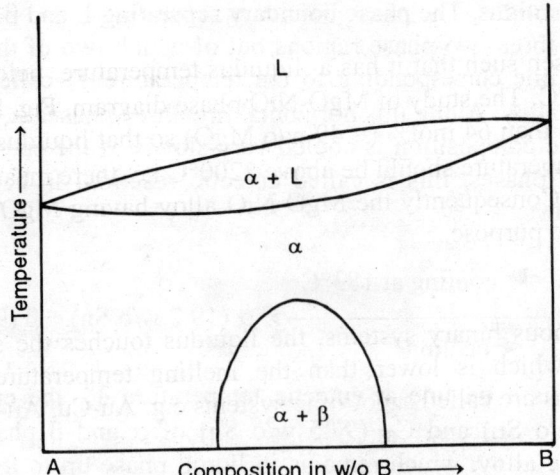

Fig. 8.10 Schematic phase diagram of A - B system showing a small solubility gap.

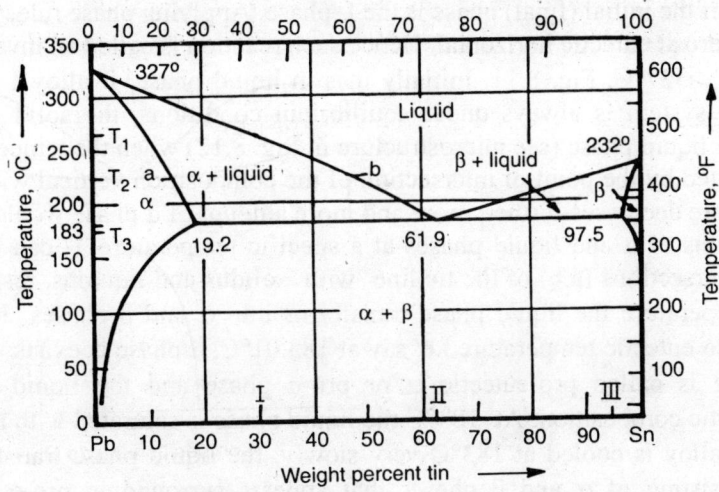

Fig. 8.11 Eutectic phase diagram of Pb-Sn.

there is complete solubility in liquid phase, the liquid phase exists for all compositions above melting temperatures of the components. If we apply Gibb's phase rule in this region degrees of freedom F becomes 2. The solid phase α is lead-rich binary alloy which dissolves only limited amount of tin. Solubility of tin in lead increases with increase in temperature. As quantity of tin is increased in α phase at a fixed temperature, α phase ultimately gets saturated with specific weight percent of tin and any amount of extra tin dissolved in the alloy would result in precipitation of tin rich β phase. Composition of saturated α phase varies with temperature. Thus, the limit of the solubility of tin in α phase is indicated by the phase boundary between α and $\alpha + \beta$, which is called solvus. The solid phase β is tin-rich phase and is shown at the right end of the phase diagram. The phase boundary between β and $\alpha + \beta$, is also called solvus. The phase boundary between α and two-phase region $\alpha + L$ is called solidus. Similarly, the phase boundary separating β and β + L region is also solidus. The phase boundary separating L and β + L (or L and α + L) is called liquidus. There are three two-phase regions out of which two of them are separated from the third by a horizontal line corresponding to the temperature T_e called eutectic temperature which is 183°C for Pb-Sn alloy. Along this horizontal line three phases α, β and L coexist. When the liquid phase of specific composition is cooled very-slowly at the eutectic temperature and it transforms into two solid phases, this is called eutectic reaction. In case of Pb-Sn alloy, the eutectic reaction is as follows:

$$\text{L (61.9 w/o Sn)} \xrightarrow[\text{heating}]{\text{cooling at 183°C}} \alpha \text{ (19.2 w/o Sn)} + \beta \text{ (97.5 w/o Sn)} \qquad (8.20)$$

The eutectic horizontal is a tie line at eutectic temperature T_e, the ends of which give the compositions $C_{\alpha e}$ (19.2 w/o Sn) and $C_{\beta e}$ (97.5 w/o Sn) of α and β phases, respectively. The specific composition of the alloy, which remains in liquid phase up to the eutectic temperature during cooling, is called eutectic composition C_e (61.9 w/o Sn). The composition of the system is hypereutectic if the percentage of tin is more than C_e and is hypoeutectic if the percentage of tin is less than C_e. During the eutectic reaction the heat is evolved (added) from (to) the system at constant temperature if the initial (final) phase is the L phase. Applying phase rule, we get degrees of freedom equal to zero at eutectic horizontal. Hence such reaction is called an invariant reaction. Consider alloy I at T >327°C, Fig. 8.11. Initially it is in liquid phase. If alloy I is cooled very slowly such that the system is always under equilibrium conditions, the solid phase α starts precipitating from the liquid phase (see microstructure in Fig. 8.12) when the temperature reaches the value T_1 as indicated by the point of intersection of the composition vertical with the liquidus line. As the temperature decreases further, more and more amount of α phase would appear in the alloy. The compositions of α and liquid phases at a specific temperature T_2 can be determined from the points of intersections (a,b) of the tie line with solidus and liquidus, respectively, Fig. 8.11. At eutectic temperature the liquid phase transforms into α and β phases. Just before the system cooled down to eutectic temperature i.e. say at 183.01°C, α phase coexists with the liquid phase. This α phase is called pro-eutectic α or pro-α phase and the liquid phase at this temperature has eutectic composition. At 183°C the liquid phase is saturated with respect to both lead and tin. As the alloy is cooled at 183°C very slowly, the liquid phase transforms into the eutectic mixture consisting of α and β phases that appears surrounding pro-α crystals. The microstructure of eutectic mixture is composed of alternate platelets of α and β phases. The α and β phases which appear at the time of eutectic reaction are called eutectic α and eutectic β phases. The compositions of α and β phases can be obtained from the end points of the eutectic horizontal, Eq (8.20).

Let us now consider alloy II of eutectic composition at any temperature (say T_1) above 183°C. It is in liquid phase. When it is cooled slowly at 183°C, the whole alloy becomes eutectic mixture. The microstructure of eutectic mixture can also have other shapes, Fig. 8.13, depending upon the presence of other materials, heat treatment conditions etc. Similarly, when alloy III is cooled to temperature T_2, the alloy is in the two-phase region L + β, the β phase appears in the liquid phase. When alloy III is cooled at 183°C, lamellar eutectic microstructure appears surrounding proeutectic β (pro-β) phase. Fig. 8.12 shows the schematic diagram of microstructures at various stages of cooling. All these structures can be observed provided the cooling is slow such that equilibrium is achieved at every stage of cooling.

Other invariant reactions

There are other invariant reactions like monotectic, eutectoid, peritectic, peritectoid, monotectoid and syntectic which occur in binary systems with incomplete intersolubility. Schematic representation of various reactions along with examples is given in Table 8.1.

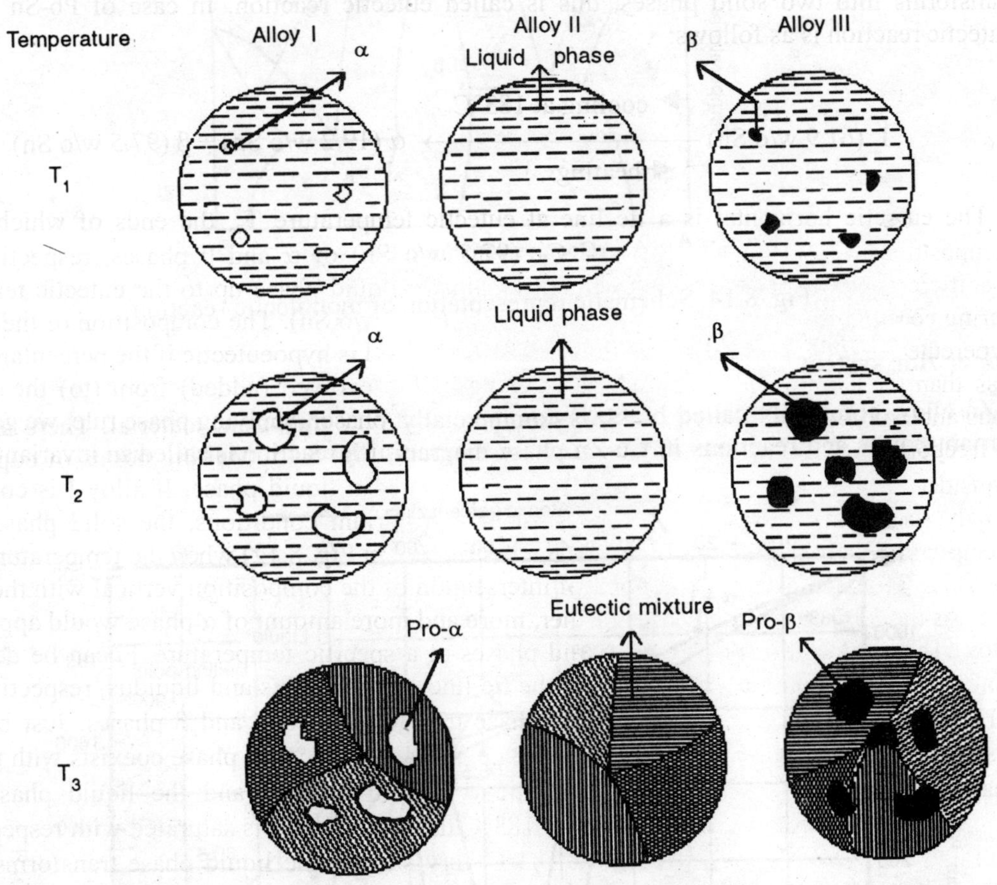

Fig. 8.12 Schematic representation of microstructures of alloy of Pb-Sn at various stages of cooling

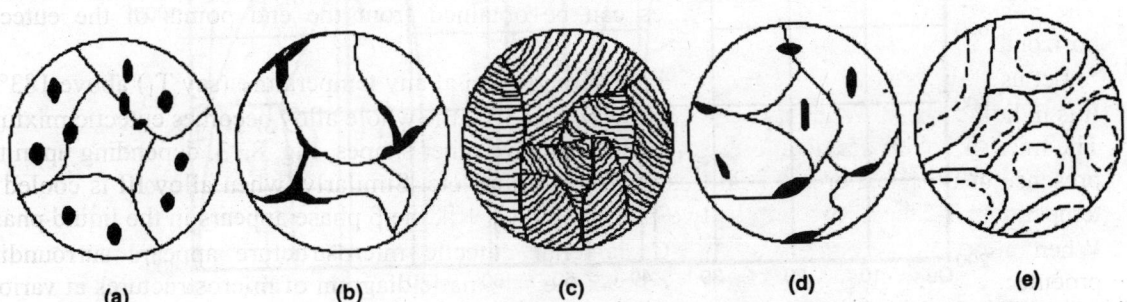

Fig. 8.13 Schematic representation of microstructures of other types of eutectic compositions:
(a) nodular, (b) Chinese script, (c) lamellar, (d) acicular and (e) divorced.

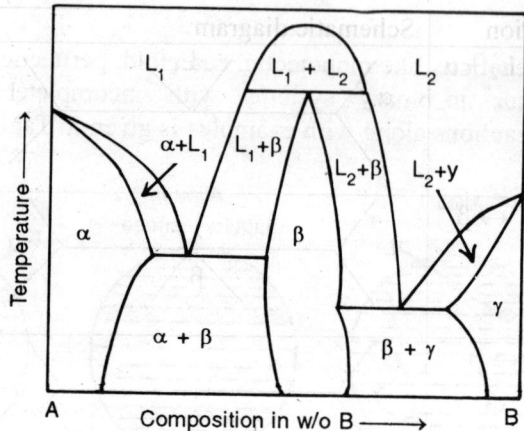

Fig. 8.14. Schematic representation of monotectic reaction

Copper zinc system

Cu-Zn alloy, commonly called brass, is commercially very important material. There are some invariant points and reactions in Cu-Zn phase diagram, Fig. 8.15. Melting points of copper and

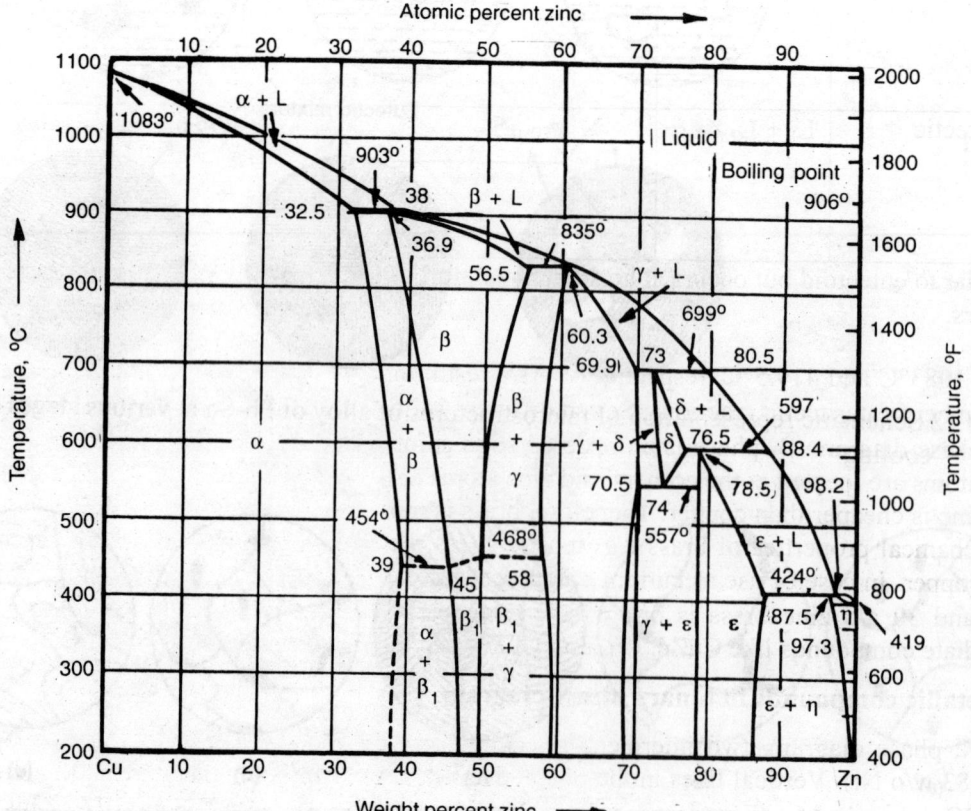

Fig. 8.15 Cu-Zn phase diagram.

Table 8.1. Other invariant reactions in binary systems.

Name	Reaction	Schematic diagram	Examples
Monotectic	$L_1 \leftrightarrow L_2 + \alpha$		$U - Th$ $FeO_2 - SiO_2$
Peritectic	$L + \alpha \leftrightarrow \beta$		$U - Mo$ $Fe - C$
Peritectoid	$\gamma + \alpha \leftrightarrow \beta$		$U - Zr$ $Cu - Al$
Eutectoid	$\gamma \leftrightarrow \alpha + \beta$		$U - Al$ $Fe - C$
Monotectoid*	$\gamma \leftrightarrow \alpha + \beta$		$Th - Zr$
Syntectic	$L_1 + L_2 \leftrightarrow \alpha$		$Na - Zn$

*Similar to eutectoid but occurs at higher temperatures than that at which eutectoid reaction occurs.

zinc are 1083°C and 419.5°C, respectively. As a result of large difference in melting points there are a number of peritectic reactions. Seven different solid phases α, β, β_1, γ, δ, ε and η appear in Cu-Zn phase diagram. β_1 phase is an ordered solid solution (50 a/o Cu and 50 a/o Zn) in which copper atoms are located at the corners and zinc atoms at the body centre of cubic unit cell or vice versa. Zinc is cheaper than copper. Therefore, brass is more economical to use wherever possible. The mechanical properties of brass e.g. tensile stress, percentage elongations are different from that of copper. In case of use as current carrying conductors copper is better than brass. A 70 a/o copper and 30 a/o zinc brass is one of the commercial brasses and is called α-brass. Some intermediate compounds like $CuZn$, Cu_5Zn_8, $CuZn_3$ can also be observed.

Intermetallic compounds in binary phase diagram

In Mg-Ni phase diagram, two intermetallic compounds are found viz. Mg_2Ni (54 w/o Ni) and $MgNi_2$ (83 w/o Ni). Vertical lines in the phase diagram represent these compounds. $MgNi_2$ melts at 1070°C, whereas Mg_2Ni undergoes peritectic decomposition at 760°C, Fig. 8.16. This phase diagram can be visualized as one peritectic diagram sandwiched between two eutectic phase diagrams joined back to back — one for Mg-Mg$_2$Ni system and the other for MgNi$_2$-Ni system. Thus, one can easily find two eutectic horizontals and a peritectic horizontal in this phase diagram.

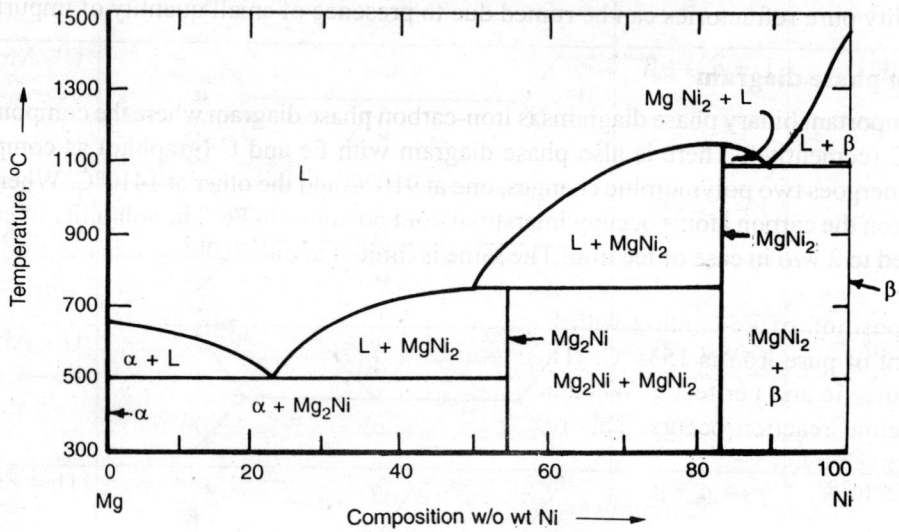

Fig. 8.16 Phase diagram for Mg-Ni

Phase diagram for magnesium oxide-aluminium oxide system

The MgO-Al$_2$O$_3$ phase diagram, Fig. 8.17, can also be visualized as combination of two eutectic diagrams with an intermediate solid solution called spinel (MgO.Al$_2$O$_3$). There is a range of compositions over which spinel is a stable compound. Hence the spinel is represented as a single-

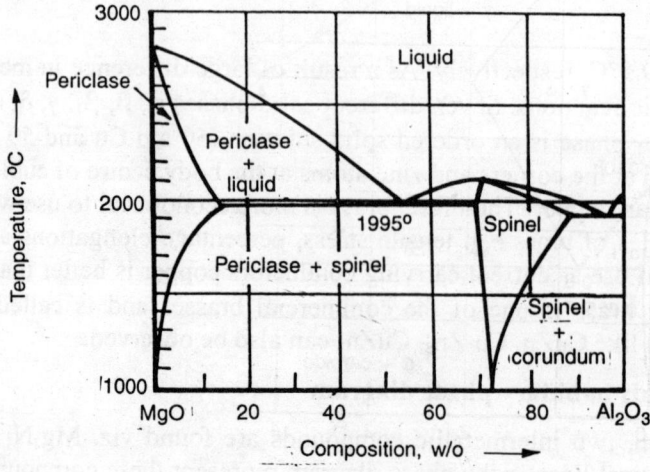

Fig. 8.17 Phase diagram for MgO-Al$_2$O$_3$ system

phase region rather than a vertical line Al_2O_3 is soluble in a limited way in MgO at higher temperatures. Only traces of MgO are soluble in Al_2O_3. MgO-rich alloy is called periclase. Al_2O_3-rich alloy is called corundum. The phase diagram for ceramics is usually more complex than those for metallic alloys. It is quite clear from the phase diagram that melting point of MgO decreases as more and more Al_2O_3 is mixed with MgO, although pure MgO has a very high melting point. So high quality pure refractories can be ruined due to presence of small quantity of impurities.

Iron-carbon phase diagram

One of the important binary phase diagrams is iron-carbon phase diagram where the components are Fe and Fe_3C (cementite). There is also phase diagram with Fe and C (graphite) as components. Pure iron undergoes two polymorphic changes, one at 910°C and the other at 1410°C. When carbon is added to iron the carbon atoms occupy interstitial void positions in Fe. The solubility of carbon in iron is limited to 2 w/o in case of fcc iron. The same is limited to only 0.008 w/o in bcc iron at room temperature.

The composition of Fe-C alloy with 0 to 6.67 w/o C is plotted along x-axis, Fig. 8.18. The melting point of pure iron is 1535°C. There are three invariant reactions in the Fe-Fe$_3$C system: eutectoid, eutectic and peritectic. When austenite i.e. γ-iron (0.8 w/o C) is cooled very slowly at 723°C eutectoid reaction occurs. This reaction results in simultaneous formation of ferrite and carbide, as given by,

$$\gamma\ (0.8\ w/o\ C) \underset{\text{heating}}{\overset{\text{cooling at 723°C}}{\rightleftarrows}} \alpha\ (0.025\ w/o\ C) + Fe_3C\ (6.67\ w/o\ C) \qquad (8.21)$$

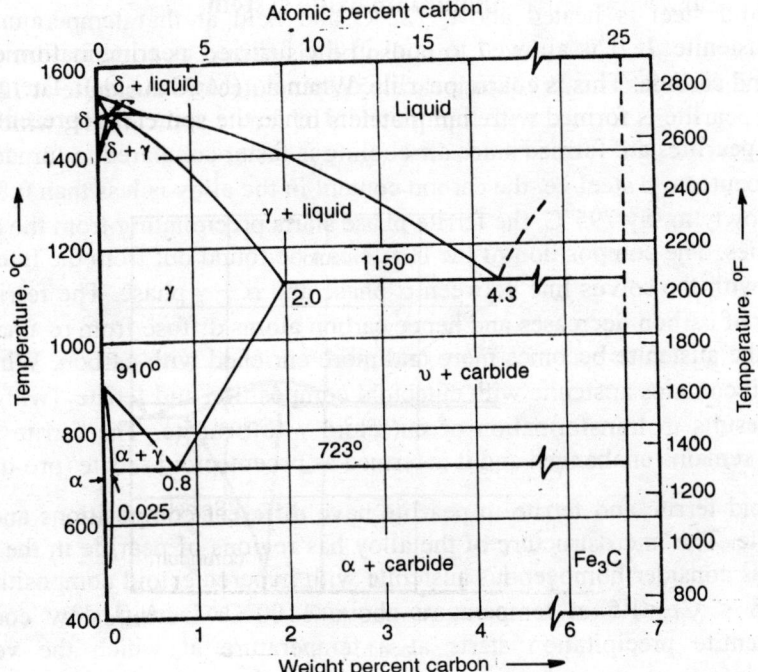

Fig. 8.18 Fe-Fe$_3$C phase diagram

The above composition of austenite is called eutectoid composition. The eutectoid ferrite has 0.025 w/o carbon. The eutectoid mixture of carbide and ferrite is often lamellar if the cooling rate is very slow i.e. there are alternate layers of ferrite and carbide. This microstructure has the appearance of mother of pearl, when viewed through microscope, hence called pearlite. Pearlite is a two-phase lamellar mixture of ferrite and carbide.

Example 8.4

Find the fraction of ferrite and cementite in pearlite.

Solution:

Pearlite is eutectoid mixture containing alternate layers of thin, parallel plates of ferrite (α) and cementite (Fe_3C) which is obtained from austenite with 0.8 w/o C. Using lever rule, one can obtain from Fig. 8.18,

$$f_\alpha = \frac{6.67 - 0.8}{6.67 - 0.025} = 0.88; \qquad f_{cementite} = 0.12$$

As the fraction of iron carbide (cementite) is less than that of ferrite, the layers of iron carbide are thinner than ferrite. The microstructure of eutectic mixture may not be always lamellar, Fig. 8.13. The nuclei of cementite crystal are formed at austenite grain boundaries. Carbon atoms diffuse from austenite to cementite. As the austenite is depleted in carbon, it is gradually transformed to ferrite. Nucleation and growth of alternate layers of cementite and ferrite take place along austenite grain boundaries. This growth of pearlite takes place until the entire eutectic austenite is consumed. The iron carbon alloy with eutectoid composition is called eutectoid steel. When the eutectoid steel is heated above 723°C and held at that temperature, it is totally transformed to austenite. If it is allowed to cool in the furnace, pearlite is formed with thicker layers of ferrite and carbide. This is coarse pearlite. When eutectoid austenite at 723°C is allowed to cool in air, the pearlite is formed with thin platelets of ferrite and carbide, which is called fine pearlite. The fine pearlites are formed since air-cooling is faster compared to furnace cooling.

If we have hypoeutectoid steel i.e. the carbon content in the alloy is less than 0.8 w/o, at 900°C and it is cooled slowly to say 795°C, the ferrite phase starts precipitating from the austenite along the grain boundaries. The composition of the ferrite can be found out from the intersection of the tie line at 795°C with the solvus line between α phase and $\alpha + \gamma$ phase. The ferrite phase being bcc, the solubility of carbon decreases and hence carbon atoms diffuse from α- phase to austenite (γ -phase). Thus the austenite becomes more and more enriched with carbon. While the alloy is cooled to 723°C, it contains austenite with eutectoid composition and ferrite (with 0.025 w/o C). Further cooling results in transformation of eutectoid γ to pearlite. The ferrite formed before eutectoid reaction remains unchanged and it is termed as proeutectoid ferrite (pro-α).

The pro-eutectoid ferrite and ferrite in pearlite have different compositions and orientations, relative to austenite. The microstructure of the alloy has regions of pearlite in the ferrite matrix, Fig. 8.19(a). Let us consider homogenous austenite with hypereutectoid composition, i.e. w/o of carbon is C_1, $0.8 < C_1 < 1.5$ at temperature above 1100°C. During slow cooling process, proeutectoid cementite precipitation starts at a temperature at which the vertical overall composition line intersects the solvus line between γ and $\gamma + Fe_3C$ region. As the cooling continues, the solubility of carbon in austenite decreases and more quantity of cementite is

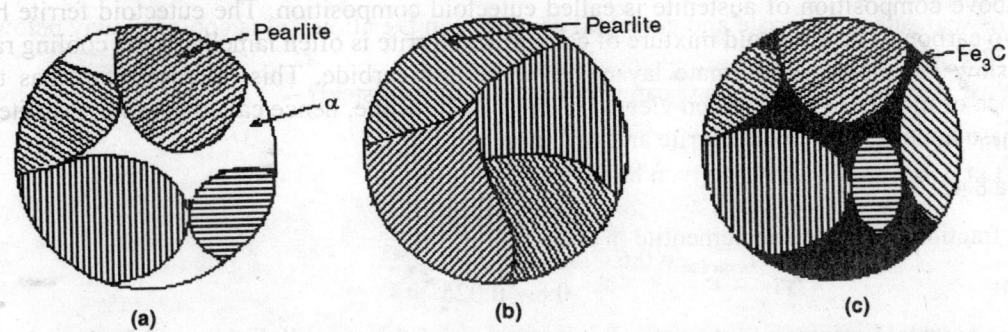

Fig. 8.19 Schematic diagram of microstructure of steel after eutectoid reaction in
(a) Hypoeutectoid steel, (b) Eutectoid steel (c) Hypereutectoid steel.

formed. Thus, the solubility limit follows the solvus line. When the above sample is cooled slowly below 723°C, the remaining eutectoid austenite transforms into pearlite. The microstructure of the alloy has regions of pearlite in the matrix of cementite, Fig. 8.19(c). The proeutectoid cementite has usually thin and needle like structure, whereas proeutectoid ferrite tends to be thick and round. The upper limit of proeutectoid cementite that can be present in hypereutectoid carbon steel is nearly 10 w/o, whereas proeutectoid ferrite in hypoeutectoid steel can be even up to hundred weight percent. For example, consider a configuration of hypereutectoid steel with C_o =1.4 w/o C which contains maximum proeutectoid cementite i.e.

$$\{(1.4 - 0.8)/(6.67 - 0.8)\} \times 100 = 10.22 \text{ w/o}. \tag{8.22a}$$

When hypoeutectoid steel contains 0.025 w/o C, the percentage of proeutectoid ferrite is

$$\{(0.8 - 0.025)/(0.8 - 0.025)\} \times 100 = 100 \text{ w/o}. \tag{8.22b}$$

The other invariant reactions, that are present in Fe-Fe$_3$C phase diagram, are eutectic and peritectic reactions. The eutectic invariant reaction of Fe-Fe$_3$C alloy occurs at 1150°C as given below:

$$\text{L(4.3 w/o C)} \xrightleftharpoons[\text{heating}]{\text{cooling at 1150°C}} \gamma \text{ (2 w/o C)} + \text{Fe}_3\text{C(6.67 w/o C)}. \tag{8.23}$$

A peritectic reaction occurs at 1494°C with the peritectic composition of steel having 0.18 w/o carbon as given below

$$\delta \text{ (0.09 w/o C)} + \text{L(0.53 w/o C)} \xrightleftharpoons[\text{heating}]{\text{cooling at 1493°C}} \gamma \text{ (0.18 w/o C)} \tag{8.24}$$

The Fe-C alloys can be broadly classified as (a) steels and (b) cast irons.

Example 8.5

A hypoeutectoid steel is cooled slowly from 940°C to a temperature just slightly above 723°C. If it contains 60 w/o austenite, calculate the following:

(i) The w/o C present in the alloy,

(ii) The w/o eutectoid ferrite and eutectoid cementite in the alloy if it is cooled slowly below 723°C,

(iii) What percent of increase in C can give rise to eutectoid steel?

Solution:

i) Let w/o C in austenite is given be f,

$$f_{austenite} = 0.6 = \frac{f - 0.025}{0.8 - 0.025}$$

i.e.

$$f = 0.49 \text{ w/o C}$$

ii)

$$f_{euc-\alpha} = 0.6 \times 0.88 = 0.528$$

$$f_{euc-cemenite} = 0.6 \times 0.12 = 0.072$$

iii) 100 g of steel contains 0.49 g of carbon and 99.51 g of iron. If y g of carbon is added to this sample such that carbon w/o becomes 0.8 then

$$0.8 = \frac{0.49 + y}{100 + y} \times 100$$

Hence

$$y = 0.3125 \text{ g}$$

The percentage increase in carbon is $(0.3125/0.49) \times 100 = 63.8$ w/o

(a) Steel

Fe-C alloys containing carbon 0 to 1.4 w/o are called steels. The constituents of carbon steels are ferrite, pearlite and cementite depending upon the percentage of carbon. Low carbon steel contains ferrite which is relatively soft and of low strength. The pearlite has considerably higher strength than ferrite. Steels of higher strength can be obtained by increasing the carbon content so that a high percentage of pearlite is produced. Cementite is stronger than pearlite but it is brittle. The yield strength and tensile strength increases with the increase of carbon content but ductility decreases. The classification of steel can be done as follows:

Table 8.2 Classification of steel

Type of steel	Carbon (w/o)	Characteristics	Used for
Mild steel or soft iron	0 – 0.3	Malleable	Chain links, rails, rivets, car bodies, fridges etc.
Medium carbon steel	0.3 – 0.7	Toughness, wear resistance and hardness	Axles, wheels for trains, rails etc.
High carbon steel	0.8 – 1.4	High wear resistance and hardness	Axes, hammers, knives, blades, scissors, files etc

When the carbon content of the alloy is in excess of 1.2 w/o, hardness of steel increases. To explain this let us discuss again some of the basic properties of the phases of iron carbon alloy. As the weight percent of carbon increases in the alloy, more amount of cementite is formed. The hardness of cementite is nearly 12 times more than that of ferrite, hence the hardness increases with the increase of carbon content.

Hard and brittle particles of cementite resist the motion of the dislocation lines, so the ductility and toughness decreases. Consequently, there is increase in tensile strength and yield point but decrease in percent elongation, impact strength and reduction in the area. After annealing the tensile strength is reduced due to precipitation of secondary cementite along the grain boundaries of the former austenite. During tensile test, high stresses are produced in the material and the cementite fails because of its brittleness. As the carbon content in steel is increased, there is reduction in density, electrical conductivity, thermal conductivity, residual induction and magnetic permeability but there is an increase in coercivity.

Effect of minor constituents on the properties of steel

Carbon steels contain not only iron and carbon but there may be insignificant amounts of other elements like Si, Cr etc. and other alloys.

(i)Effect of sulphur

Sulphur is a harmful impurity in steel. It reacts with iron to form FeS, which is insoluble in iron in solid state, but is soluble in liquid metal. FeS forms a eutectic with Fe. Its melting point is 988°C. When the steel is heated to the rolling or forging temperature at 1000°C or above, the eutectic melts. As a result, cracks are formed in the eutectic when the metal is worked. This is called red-shortness or hot-shortness.

The presence of manganese in steel practically removes the possibility of red-shortness, since Mn easily forms MnS, which has high melting point. The presence of sulphur reduces the impact strength, ductility in the transverse direction to the direction in which the steel is drawn in rolling.

(ii)Effect of silicon and manganese

The percentage of silicon in carbon steel is usually in the range from 0.1– 0.3 w/o which acts as deoxidizer. Silicon forms SiO_2, when it reacts with oxygen present in steel. Silica goes into the slag and the properties of steel are improved. Rimming steel contains traces of silicon. Sometimes silicon content is made 0.5 w/o to increase the strength of steel casting. The presence of silicon impurity increases the yield point of steel. So it reduces the drawing capacity of the steel. Thus the silicon content must be kept low in steel intended for cold working and cold heading. Silicon is added to increase the toughness of steel which are used for making spring steel, chisels etc. Permeability of iron increases with increase in silicon percentage. Nearly 5 w/o silicon is added to steel to have magnetic material which is used in transformers, motors and generators.

The maximum weight percent of manganese in steel may vary from 0.5 to 0.8. It also acts as deoxidizer i.e. it interacts with FeO and forms manganese oxide. Manganese removes undesirable effect of sulphur by forming manganese sulphide. Manganese increases the strength of the steel without reducing its ductility. It reduces brittleness at high temperature. Manganese is usually added in free cutting and structural steels. Tough, wear resistant and nonmagnetic steel can be obtained if approximately 14 w/o manganese is added.

(iii) Effect of phosphorus

In most steels, phosphorous is a detrimental impurity. It forms Fe_3P, if the phosphorous content is more than 15.62 w/o. The presence of this impurity severely distorts the crystal lattice and greatly reduces the ductility and toughness but increases tensile strength and yield point. Even presence of 0.01% of phosphorous raises the cold-shortness threshold by 20°– 25°C.

(iv) Effect of nitrogen, oxygen and hydrogen

Nitrogen and oxygen are usually present in the steels in the form of oxides, nitrides e.g. FeO, SiO_2, Al_2O_3, Fe_4N_2 etc. or as interstitial impurities. These impurities raise the cold-shortness threshold and reduce the resistance to brittle failure, endurance limit and impact strength. Hydrogen dissolved in steel causes severe embrittlement. It promotes the formation of flakes in rolled and forged steel which drastically deteriorates properties of steel. It causes cracks in the welding.

(b) Cast Iron

Fe-C alloys containing 2 w/o or more carbon are called cast iron. In fact alloys having composition range shown by eutectic line in Fig. 8.18 are known as cast iron. Consider cooling of liquid Fe-C alloy with 2.5 w/o carbon. When the temperature reaches such that the alloy is in L + γ region, proeutectic γ crystallizes first. On further cooling slowly at eutectic temperature the liquid alloy of eutectic composition forms a mixture of austenite and cementite. This is also called ledeburite. When it is cooled slowly at eutectoid temperature (723°C), eutectoid austenite forms pearlite. Thus, the structure of this cast iron consists of cementite and pearlite. This is called white cast iron. The white cast iron is hard and brittle due to presence of large amount of cementite.

Sometimes graphite crystallizes in the liquid phase from a single centre and branching out in various directions, takes the shape of flakes. During prolonged heating of cast iron the cementite may decompose into graphite and ferrite at temperatures below 738°C or into graphite and austenite at a temperature above 738°C. On cooling slowly graphite may separate out of the austenite and form eutectoid graphite near eutectoid temperature (723°C –738°C). If silicon is present as alloying element in iron, rapid graphitization of cementite can take place. Thus the cast iron in which flakes of graphite form during the solidification of the casting, is called gray cast iron, since the presence of graphite gives the cast iron gray color. The gray cast iron can be further classified as pearlitic gray cast iron when the microstructure consists of pearlite and graphite flakes and as ferrite gray cast iron when the microstructure contains ferrite and flaked graphite. The graphite flakes with sharp tips increase the stress. The brittleness of gray cast iron increases under tensile loads although graphite is soft. The brittleness can be reduced by suitable heat treatment as a result of which spheroidal graphites (SG) are formed and the cast iron is called SG iron.

When the white cast iron is heat treated at about 900°C for a prolonged period and then cooled very slowly, the malleable cast iron is formed. Representative applications for this material include connecting rods, transmission gears etc., for automotive industry, pipefittings, flanges for rail road and other heavy duty services. If the alloy contains maximum 1 w/o Si, the cementite is formed instead of graphite flakes. After suitable heat treatment of the malleable cast iron, the cementite decomposes into spherical particles of temper carbon. A finer flake size can be produced by inoculating ferrosilicon and calcium silicon in the molten state of the alloy. Graphite nodules are formed if the molten alloy is treated with magnesium or cerium. The silicon content must be 2.5 w/o to facilitate the graphitization. SG iron has superior strength and ductility.

The nodules of graphite would be embedded in the pearlite or ferrite matrix depending on heat treatment. Normally the matrix phase is pearlite, but after a suitable heat treatment for several hours at about 700°C, a ferrite matrix phase is formed. The castings for this type of iron are stronger and much more ductile than gray iron. The ductile iron has tensile strength ranging from 55000 – 70000 psi and ductility (as percentage elongation) varying from 10% to 20%. Typical applications for this material are in the manufacturing of valves; pump bodies, automotive spares and components etc.

Stainless steel

Stainless steel is highly resistant to corrosion or rusting in a variety of environments, especially under normal atmospheric conditions. Chromium is added to steel with a concentration of as high as 11w/o or more to make it corrosion resistant. Corrosion resistance of steel can also be increased by addition of nickel and molybdenum. Addition of molybdenum reduces the possibility of formation of pitting corrosion. The stainless steel can be classified broadly as martensitic, ferritic and austenitic on the basis of phases present in the microstructure. The name of the stainless steel originated from its prime microconstituent. Martensitic stainless steel has martensite as prime microconstituent. Heat treatment of this steel can be done in such a way that its prime microconstituent remains martensite. Austenitic and ferritic stainless steels cannot be heat treated but they can be hardened and strengthened by cold working. The austenitic steels are nonmagnetic and are very high corrosion resistant because of nickel as well as high chromium contents.

8.6 PROPERTIES OF MULTIPHASE SYSTEMS

Distribution of phases is not normally homogeneous in a multiphase system,. Phase distribution is influenced by relative interfacial energies and nonuniform mixing of components at the time of solid state reaction. The multiphase microstructure having lowest total free energy is the most stable configuration. Minimization of interfacial energy leads to the geometric equilibrium. In this process, the atoms of smaller particles go into the solution, diffuse through the matrix and finally precipitate onto a larger particle of same phase. The free energy is higher for the finer particles and as a result they dissolve into the matrix. This may result in the exceeding of the solubility limit of solution near coarse particles and the atoms from solution precipitate onto coarse particles. Hence the coarse particles grow in size at the expense of finer particles. This is termed as agglomeration. The rate of agglomeration, which reduces the surface energy, depends on the rate of diffusion through the solution, the temperature,. and distance between particles, the difference in particle sizes, and the interfacial energy. The equilibrium shapes of particles are usually spherical, which reduces the free energy since for a given volume, surface area becomes minimum for spherical shape.

Density and heat capacity

The density of a multiphase system, ρ_{system} is not sensitive to the geometry of microstructure. It can be determined by suitable weighted average of the densities of each of the individual phases. Thus,

$$\rho_{system} = v_1\rho_1 + v_2\rho_2 + v_3\rho_3 + \ldots \tag{8.25}$$

where v_1, v_2, etc. are volume fractions and ρ_1, ρ_2 etc. are densities for phases 1,2,....respectively. The product $v\rho$ is zero for pores. The heat capacity per unit volume has a similar additive relationship.

Thermal and electrical conductivity

In case of parallel phase combination Fig. 8.20a, the average thermal conductivity of the system is expressed as

$$\kappa_{parallel} = v_\alpha \, \kappa_\alpha + v_\beta \, \kappa_\beta + ... \tag{8.26}$$

where κ_α, κ_β....... are conductivities of phases $\alpha, \beta,.....$, respectively. In case of series phase combinations, Fig. 8.20b, the conductivity of system is expressed as

$$1/\kappa_{series} = v_\alpha/\kappa_\alpha + v_\beta/\kappa_\beta + ... \tag{8.27}$$

for a two series phase microstructure

$$\kappa_{series} = \frac{\kappa_\alpha \ \kappa_\beta}{v_\alpha\kappa_\beta + v_\beta\kappa_\alpha} \tag{8.28}$$

In case of matrix with dispersed phase (α) in a continuous phase (β) the conductivity expression is complex and is given by,

$$\kappa_{system} = \kappa_\beta \frac{1 + 2v_\alpha\{(1 - \kappa_\beta/\kappa_\alpha)/(2\kappa_\beta/\kappa_\alpha+1)\}}{1 - v_\alpha\{(1 - \kappa_\beta/\kappa_\alpha)/(2\kappa_\beta/\kappa_\alpha+1)\}} \tag{8.29a}$$

This can be also written in a simpler form as

$$\kappa_{system} = \kappa_{series} \frac{\kappa_\alpha + 2 \ \kappa_{parallel}}{\kappa_\alpha + 2 \ \kappa_{series}}, \tag{8.29b}$$

where $\kappa_{parallel}$ and κ_{series} are as given by Eqs.(8.26) and (8.28) for two phase systems.

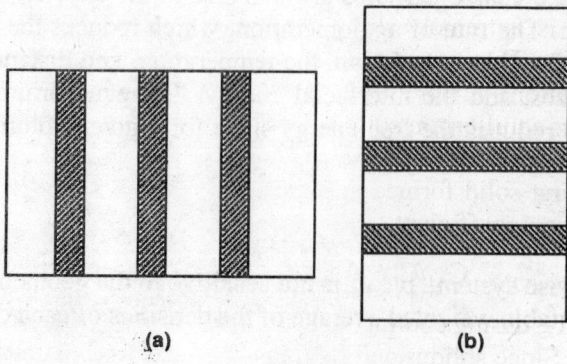

(a)　　　　　　(b)

Fig. 8.20 Phase combination for calculation of conductivity (a) parallel (b) series.

Example 8.6

60 w/o of very fine quartz powder is added to the 40w/o of phenol formaldehyde resin as a filler. Find the density, if ρ_{quartz} = 2650 kg/m³ and ρ_{pf} = 1300 kg/m³. Given that κ_{quartz} = 0.1672 Jm/°Cm²s and κ_{pf} = 12.54 Jm/°Cm²s. Find also the thermal conductivity of the sample if silica is the dispersed phase here.

Solution:

Consider a sample of 100 kg.

$$\text{Volume of 60 kg of quartz} = 60/2650 = 0.0226 \text{ m}^3$$
$$\text{Volume of 40 kg of pf} = 40/1300 = 0.0308 \text{ m}^3$$
$$\text{Total volume} = 0.0534 \text{ m}^3$$

Volume fractions:
$$V_{Quartz} = 0.0226/0.0534 = 0.423$$
$$V_{pf} = 0.0308/0.0534 = 0.577$$
$$\rho_{system} = (0.6)(2650)+(0.4)(1300) = 2110 \text{ kg/m}^3$$

Using Eq.(8.29a) we get:

$$\kappa_{system} = 12.54 \ \frac{1+2 \times 0.423 \left[\dfrac{1-(12.54/0.1672)}{2(12.54/0.1672)+1} \right]}{1 - 0.423 \left[\dfrac{1-(12.54/0.1672)}{2(12.54/0.1672)+1} \right]} = 6.08 \text{ Jm /°C m}^2\text{s.}$$

We obtain

$$\kappa_{parallel} = 0.423 \times 0.1672 + 0.577 \times 12.54 = 7.3063 \text{ Jm /°Cm}^2\text{s.}$$
$$\kappa_{series} = (0.1672 \times 12.54)/(0.423 \times 12.54 + 0.577 \times 0.1672) = 0.3882 \text{ Jm /°Cm}^2\text{s.}$$

Alternatively using Eq. (8.29b) we get:

$$\kappa_{system} = 0.3882(0.1672 +2 \times 7.3063)/(0.1672 + 2 \times 0.3882) = 6.08 \text{ Jm /°Cm}^2\text{s.}$$

8.7 ZONE REFINING

During solidification of alloy it is assumed that the complete equilibrium is maintained between solid and liquid phases if the process is carried out very slowly. This condition can be maintained only when cooling rate is extremely slow, hence most of the time the alloys experience non-equilibrium solidification. This experience can be utilized to purify metals, to even out non-uniformities in concentration.

To understand the non-equilibrium solidification, consider the cooling of a liquid alloy having composition C_{L1} w/o solute, A which is dissolved in the liquid alloy, Fig. 8.21. As the cooling is performed slowly, the first solid formed at temperature T_1 has C_{A1} w/o A and the ratio is called the equilibrium distribution coefficient.

$$C_{A1}/C_{L1} = k_o \tag{8.30}$$

At a lower temperature T_2, during cooling, the compositions of solid and liquid phases are C_{A2} and C_{L2} w/o A, respectively. Since solidus and liquidus lines are nearly straight, the ratio

$$C_{A2}/C_{L2} = k_o \tag{8.31}$$

During solidification process the nth layer of solid phase formed has composition C_{An} w/o A such that

$$C_{An}/C_{Ln} = k_o \qquad (8.32)$$

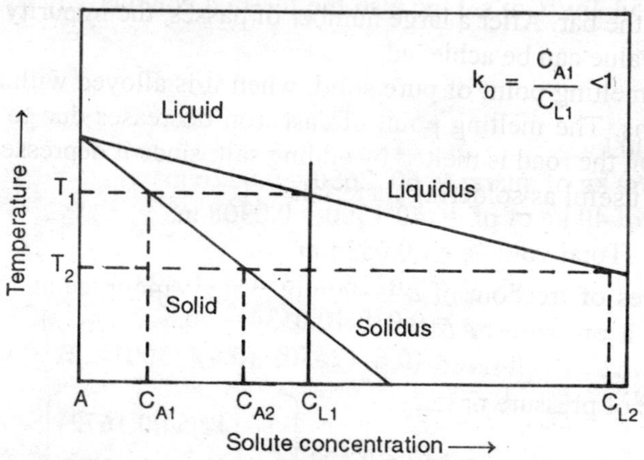

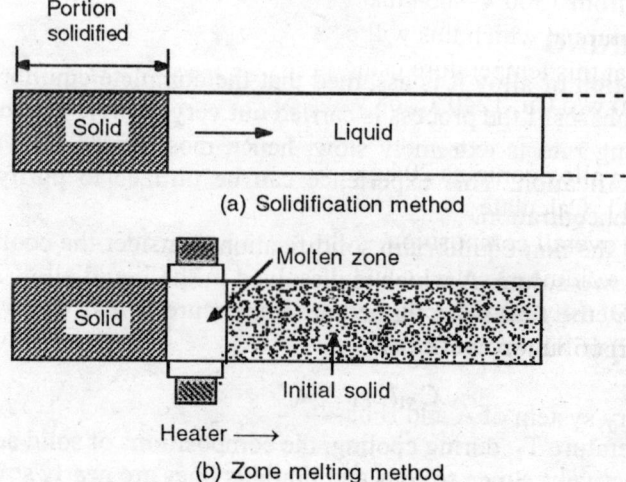

Fig. 8.21 Variation of concentration with temperature during zone refining

The solid alloys formed at various temperatures have nonuniform concentration. If the solidification is being done in the form of bar from the liquid phase, the concentration of the solute increases along the length of the solidifying bar, Fig. 8.22, if $k_o < 1$. In case $k_o > 1$, raw initial solid phase has higher concentration of solute than the liquid from which it is forming, as a result of which the concentration of solute decreases along the length of the bar. The purification of metal can be done using a convenient method, which takes advantage of the difference in solute content of a liquid and the solid formed from it. Consider that the metal A, which is to be purified is in the form of a bar and B is the impurity, which is mixed with A in the form of solute. Initial concentration of B is C_o in the metal bar. An induction coil or any other form of heater is used to melt only a small zone of length l at any time. The molten zone is initially at one end. The heater

Fig. 8.22 Variation of concentration of solute along the length of the specimen.

is slowly moved from one end to the other, continuously solidifying the molten zone and melting the fresh zone. When the melted zone is cooled, purer metal A is separated from the liquid phase.

The molten zone is held in place due to strong surface tension forces. Usually the molten zone is passed through the bar a number of times, each time the material at the starting end becomes purer from the rest of the bar. After a large number of passes, the impurity concentration as low as 10^{-4} times the initial value can be achieved.

The depression of melting point of pure solid, when it is alloyed with another component, has interesting applications. The melting point of cast iron decreases due to the eutectic reaction in Fe-C alloys. The ice on the road is melted by adding salt, since it depresses the freezing point. Pb-Sn eutectic alloys are useful as soldering materials.

EXERCISES

8.1   Find the degrees of freedom of a system of water vapour ice and water in equilibrium at 0.01°C and at a pressure of 613Pa. Under what conditions the degrees of freedom are increased by two?
Ans: 0; change of pressure or temperature

8.2   Is it possible for a two-component system to have four phases in equilibrium? Write clearly yes or no with reason.
Ans: Yes; when degree of freedom $F = 0$.

8.3   Consider a three-component system. Calculate the maximum and minimum number of phases possible for the system. What are the total number of variables and degrees of freedom for this system with the maximum number of phases?
Ans: 5, 1; 12, 0

8.4   Consider an alloy of Cu and Ni with the weight ratio of solid to liquid phases as 3:13 at 1300°C where liquid phase contains 53 w/o Cu and the solid phase contains 37 w/o Cu. Calculate the overall composition of the alloy analytically. If the same alloy is cooled very slowly from 1300°C such that only traces of liquid phase are present in the alloy, find the temperature at which this will occur and also calculate the compositions of liquid and solid phases at this temperature using Fig. 8.5.
Ans: 50 w/o Cu; 1240°C, 68 w/o Cu, 50 w/o Cu

8.5   A Pb-Sn alloy contains 20 kg of Sn such that the fraction of α phase is 0.4 at 183.5°C, Fig. 8.11. Calculate
i)   the overall composition,
ii) the volume of eutectic and pro-eutectic constituents at 182.5°C.
Data: Density of β is 7300 kg/m³ and that of α is 10300 kg/m³.
Ans: (i) 66.18 w/o Sn; (ii) $V_{eut\,\alpha}$ = 1174 cm³, $V_{eut\,\beta}$ = 1986 cm³, $V_{pro\,\beta}$ = 497.7 cm³

8.6   A binary system of A and B has the following invariant reaction occurring at 780°C

$$L(30 \text{ w/oB}) \xrightarrow[\text{Heating}]{\text{Cooling at 780°C}} \alpha\,(10 \text{ w/o B}) + \beta\,(90 \text{ w/o B})$$

This alloy with an unknown overall composition has 80 w/o of α phase at 781°C. Find the following:
i) Overall composition of the alloy.
ii) The amount of á phase in eutectic mixture in 100 g of the alloy at 779°C.
iii) The amount of B to be added to 100 g of the alloy in order to get 80 w/o of â phase at 779°C.
Ans: (i) 26 w/o B; (ii) 75 g; (iii) 169.23 g of B

8.7 (a) A hypoeutectoid plain carbon steel is annealed at 940°C and cooled to 723.5°C very slowly. It is found that the alloy contains 50 w/o α phase. Calculate the overall composition of the alloy.
(b) If the w/o carbon is now increased by 2.5 times and the alloy is brought to equilibrium at 722.5 °C, calculate the w/o and composition of proeutectoid phase.
Ans: (a) 0.41 w/o C; (b) 4 w/o, pro-eutectoid cementite with 6.67 w/o C

8.8 10 kg iron-carbon steel of 99.4 w/o Fe and 0.6 w/o C is cooled from 900°C.
a) Find the fraction and amount of pearlite, ferrite, proeutectoid ferrite and cementite at 722°C.
b) Find the amount of carbon needed to convert this to eutectoid steel.
Ans: (a) 0.742, 7.42 kg; 0.91, 9.1 kg; 0.258, 2.58 kg; 0.09,0.9 kg; (b) 0.02 kg

8.9 Consider an alloy containing 25 mole % MgO and 75 mole % FeO. Calculate the composition of the ceramic in weight percent.
Ans: 16.09 w/o MgO

8.10 Consider the phase diagram of Ge-P system. In the liquid alloy containing 62 a/o phosphorus, Ge, GeP are in equilibrium at 723°C, whereas in the liquid alloy having 94.5 a/o of phosphorus, GeP and P are in equilibrium at 577°C. The melting point of Ge and P are 937°C and 593°C, respectively. What are the invariant reactions in this diagram? Name them.

8.11 Find the liquidus and solidus temperatures and the freezing range for the following MgO-FeO ceramic compositions: (a) MgO – 40 w/o FeO, (b) MgO – 60 w/o FeO. (Fig. 8.23)
Ans: (a) 2400°C, 2000 °C, 400 °C; (b) 2100 °C, 1700 °C, 400 °C

8.12 Estimate the carbon content of the steel containing 4 v/o of pearlite and rest is proeutectoid ferrite when cooled slowly through the eutectoid temperature. The densities of α and pearlite are 7870 kg/m³ and 7820 kg/m³, respectively.
Ans: 0.37 w/o C

8.13 In nodular cast iron, graphite is present as spheroids. Estimate the density of this alloy (97 w/o Fe-3 w/o C) if the densities of the two phases are 7800 and 200 kg/m³, respectively. The alloy contains approximately 12 v/o graphite.
Ans: 6888 kg/m³

8.14 Determine the amount of various phases in 200 kg of steel casting which contains 0.4 w/o C as it is cooled to (a) 871°C, (b) 724°C, and (c) 722°C.
Ans: (a) Traces amount of α and rest is γ phase; (b) α phase: 103.22 kg, γ phase: 96.78 kg; (c) Fe₃C phase: 11.29 kg, α phase: 188.71 kg.

8.15 In a binary system of A (m.p.=400°C) and B (m.p.=300°C), the α-phase, liquid phase and the β-phase are in equilibrium at 250°C with their respective compositions 20 w/o, 65 w/o and 90 w/o B. At room temperature, the maximum solubility of B in a-phase is 10 w/o and that of A in β-phase is 5w/o.

(i) Sketch the phase diagram showing the phases in each area.

(ii) What is the reaction taking place at 250°C?

(iii) Just below 250°C, find the compositions at which the proeutectic phase is 1.5 times the eutectic mixture.

Ans: (ii) Eutectic; (iii) 38 w/o B, 80 w/o B

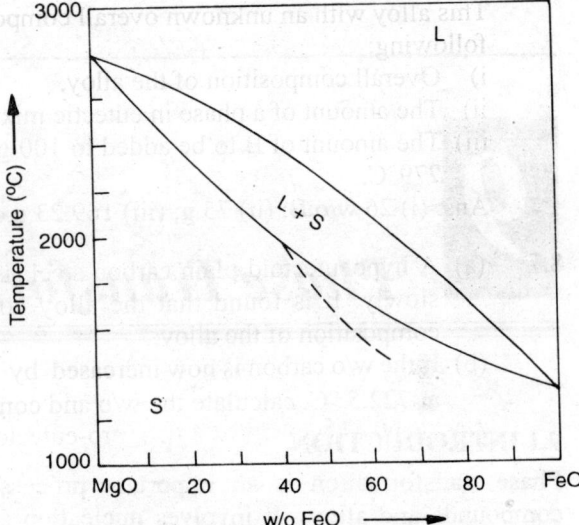

Fig. 8.23 Phase diagram for the MgO-FeO system

8.16 In the Fe-Fe₃C phase diagram, the reactions for 0.2% C-steel are:

At 1493°C: δ (0.1% C) + L (0.5% C) — Cooling → γ (0.18% C)
At 723°C: γ (0.8% C) — Cooling → α (0.02 w/o C) + Fe₃C (6.67% C)

Name the phases and their fractions

(i) just above 1493°C, and
(ii) just below 723°C.

Ans: (i) δ phase: 0.75, L phase: 0.25;
(ii) α phase: 0.97, Fe₃C phase: 0.03

8.17 A Co-50 w/o percent W alloy, Fig. 8.24, is heated to 1500°C. Find out the following:

(a) The composition of the solid and liquid phases in atomic percent.

(b) The amount of phases present in atomic percent at 1400°C.

(c) If the density of solid is 16.05 mg./m³ and of liquid is 13.91 mg/m³, determine the amount of phases present in volume percent.

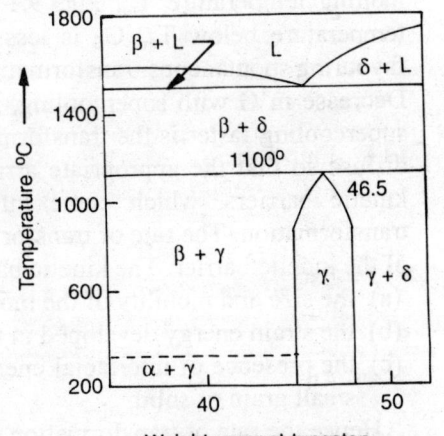

Fig. 8.24 Phase diagram of Co-W system.

9

Phase Transformations

9.1 INTRODUCTION

Phase transformation is an important process in the production and treatment of metallic compounds and alloys. It involves nucleation and growth. Phase transformation occurs due to change of pressure, temperature or composition of the alloy. The rate of phase transformation plays a crucial role in deciding the microstructure of the new phase formed. Deliberately inducing or suppressing a phase transformation can control the microstructure of a phase.

9.2 RATE OF PHASE CHANGE Vs RATE OF COOLING

Time taken to complete a phase transformation may vary from microseconds to years. Consider that there is a phase transformation from the liquid to the solid state. The free energy of the liquid phase (G_L) is equal to that of the solid phase (G_S), when the two phases are in equilibrium at the melting temperature T_m, Fig. 9.1. At a temperature above T_m, G_L is less than G_S and at a temperature below T_m, G_S is less than G_L. The change in free energy ΔG becomes negative favouring spontaneous transformation from liquid to solid state as supercooling is done below T_m. Decrease in G with supercooling, makes the rate of transformation faster. Hence, the more the supercooling faster is the transformation. During the solidification process, atoms and molecules diffuse so that the appropriate arrangement of atoms can lead to crystallization. But there are kinetic barriers, which come into play along the path of diffusion and slow down the transformation. The rate of transformation from liquid to crystalline state is decided by the height of the kinetic barrier. The kinetic barrier depends on

(a) the size and mobility of the molecules/atoms taking part in diffusion,
(b) the strain energy developed in the parent phase during the process of transformation,
(c) the presence of interfacial energy in the liquid phase as a result of formation of a relatively small grain of solid.

Hence the rate of transformation may be slow or rapid depending upon the above factors. If the supercooling below the melting temperature is done very fast, then the transformation from liquid to solid also takes place very fast allowing a short interval of time for the atoms or molecules to diffuse and to crystallize with proper atomic arrangement. As a result the liquid may transform into a glassy state in which long range order does not exist.

In some cases, the transformation takes place very fast since the kinetic barrier is low. The crystallization of such material takes place in spite of high cooling rate e.g. in case of metals the crystallization can be prevented only if the cooling rate is greater than or equal to $10^6 \,^\circ Ks^{-1}$. Such high cooling rate results in formation of metallic glass and is called splat cooling.

Cooling at a very fast rate, such that a fast transformation can be stopped, is called quenching. During quenching the inner part of the material remains at higher temperature compared to that of the outer part, which in turn may set up a thermal gradient. Outer part of the material condenses faster while the inner part cools slowly. This gives rise to thermal stress in the material. Vacancies and other types of defects are formed in the material during heat treatment. When the material is cooled slowly, the atoms from the grain boundary or interstitial sites may diffuse to the vacant sites, which decreases the number of defects. If the material is quenched, defects may remain trapped in it. This gives rise to drastic changes in the properties of the material e.g. increase in hardness and brittleness, decrease in ductility.

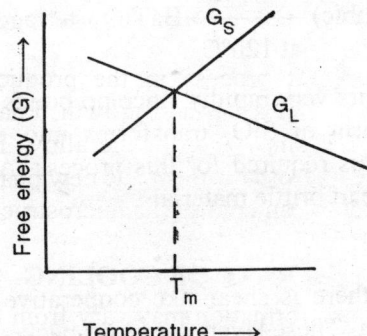

Fig. 9.1 Schematic diagram for variation of the free energy of a material in the liquid (G_L) and crystalline forms (G_S)as a function of temperature.

The transformation that takes place in microseconds is called extremely fast transformation. Such transformation cannot be suppressed. Normal phase transformations take place within time interval of seconds to hours and are usually suppressible. In the other extreme situations, there are virtually impossible transformations, where the transformation takes years to complete e.g. one may have to wait for years to form silicate crystal. In other words, the crystallization of silicate is unachievable even if the cooling rate is fraction of a Kelvin per hour.

9.3 DIFFERENT TYPES OF PHASE TRANSFORMATIONS

An alloy in α phase can spontaneously transform to β phase only if $G_\alpha > G_\beta$. Such processes can take place through distinctly different types of phase transformations. The properties of the final phase depend on thermal and mechanical treatments, which lead to the phase transformation. Major types of phase transformations are the following:

(i) Congruent transformation

In this type of phase transformation, a single component material crystallizes from liquid state with no change in composition e.g. crystallization of copper from melt. The chemical formula of the material remains unchanged under such transformations. Congruent transformation can be two types:

- Reconstructive transformation,
- Displacive transformation.

In **reconstructive phase transformation**, the atoms diffuse across the α-β interface to transform α phase into β phase through a gradual process in which the atomic coordination is altered. The atoms of α-phase rearrange to form not only β-phase but also intermediate phases β', β" etc., if such transformations are more favourable. There are other subsequent reconstructive transformations through which β', β" phases are converted to more stable β phase. This type of transformation requires high activation energy since there would be a significant rearrangement of bonds. Examples of reconstructive transformations are polymorphic transformation from α-Ti (hcp) to β-Ti (bcc), cristobolite to tridymite etc.

A **displacive transformation** arises from the cooperative movements of a large number of neighbouring atoms. The example of such transformation is

$$BaTiO_3 \text{ (cubic)} \xrightarrow[\text{at } 120°C]{\text{cooling}} BaTiO_3 \text{ (tetragonal)} \qquad (9.1)$$

This type of transformation occurs very rapidly since no bonds are broken during the process. For example, low temperature form of SiO_2 transforms into more symmetric form at high temperature. A high driving force is required for this process. Associated volume change may lead to fracture, particularly in case of brittle materials.

(ii) Shear Transformation

During the shear transformation, there is shear like cooperative displacement of one layer of atoms with respect to the next layer. These layers are close-packed planes. The transformation of Co (fcc) to Co (hcp) is an example of this type of transformation.

$$Co(fcc) \xrightarrow[\text{at } 1120°C]{\text{cooling}} Co(hcp) \qquad (9.2)$$

The atoms of (111) planes of fcc Co move through a fraction of interatomic distance to give hcp stacking during this transformation, Fig. 9.2. Consequently, ABCABC...stacking of fcc changes to ABAB...stacking of hcp. The atoms of C-layer in fcc stacking move together such that these atoms are arranged on A-valleys i.e. they form A-layer. The A-layer atoms are displaced in a

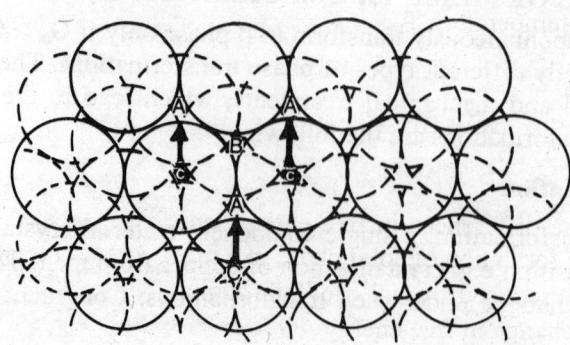

Fig. 9.2 Schematic representation of shear transformation Co (fcc) → Co(hcp)

cooperative manner such that they form B-layer and B-layer atoms form A-layer etc. This is also called martensitic transformation. Glide motion of dislocations contribute to this shear transformation. Supercooling and mechanical stress facilitate the shear transformation. The free energy change ΔG is the usual source of driving force that causes such transformation.

Stability

To understand stability during the phase transition let us examine the free energy versus configuration curves:

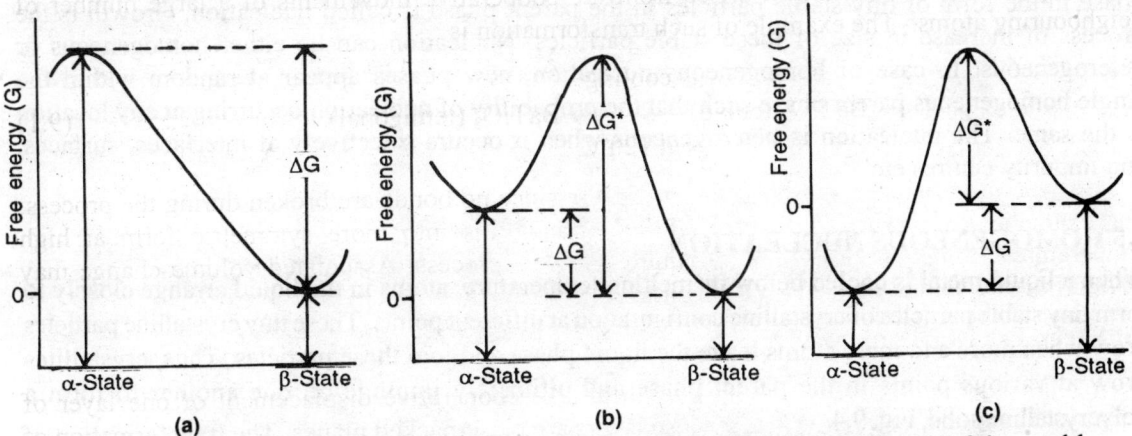

Fig. 9.3 Schematic variation of free energy with configuration when (a) β is a stable phase and α is unstable (b) when α is a metastable state (c) when β is unstable.

If $G_\alpha > G_\beta$ and there is no energy barrier in the path of transition from α to β phase, Fig.9.3 (a) there is spontaneous transformation from α to β phase without any requirement of extra energy. In Fig.9.3 (b) although $G_\alpha > G_\beta$ but an energy barrier exists in the path of transitions hence α is a metastable phase relative to stable β phase. If $G_\alpha < G_\beta$ and also there exists an energy barrier, Fig.9.3(c), the instability refers to the metastable phase β only since unstable phases do not exist.

9.4 MECHANISMS OF PHASE TRANSFORMATION

In this section, the basic steps of phase transformation from the liquid phase to the solid phase are discussed. At the melting temperature T_m, the free energy of the liquid phase and that of the solid phase are equal i.e.

$$\Delta G = \Delta H - T_m \Delta S = 0 \tag{9.3}$$

Hence
$$\Delta S = \Delta H/T_m, \tag{9.4}$$

where ΔG, ΔH, ΔS are the differences in free energy, enthalpy and entropy of the two phases, respectively. Consider the formation of solid phase in a liquid metal at a temperature $T < T_m$, Fig. 9.1. As the temperature is reduced from T_m to T through undercooling, such that $\Delta T = T_m - T$ is small, the corresponding change in free energy ΔG is given by

$$\Delta G = \Delta H - T \Delta S \approx \Delta H - T \Delta H/T_m = \Delta H (T_m - T)/T_m = \Delta H (\Delta T)/T_m \tag{9.5}$$

If ΔH is negative, ΔG becomes negative and such a transformation is called a spontaneous transformation. The above relation is also valid for gas to liquid phase transformation. It is observed that ΔG is directly proportional to ΔT. Normally solidification does not start just below the temperature T_m, since ΔG is very small in that case. The rate of transformation is dependent on ΔG value. An appreciable supercooling i.e. high value of ΔT and hence of ΔG, results in fast rate of phase transformation. Consider a liquid metal at a temperature much above T_m. When the cooling is done just below T_m, the crystallization may not start immediately, since a certain amount of supercooling is required before the solid phase starts growing.

The solidification process involves two steps: nucleation and growth. The appearance of a new phase in the form of tiny stable particles in the parent phase is called nucleation. Growth is the process of increase in size of these stable particles. Nucleation can be either homogeneous or heterogeneous. In case of homogeneous nucleation, new phases appear at random within the single homogeneous parent phase such that the probability of nucleation occurring at any location is the same. The nucleation is heterogeneous when it occurs selectively at interfaces, surfaces, and impurity centres etc.

9.5 HOMOGENEOUS NUCLEATION

When a liquid metal is cooled below its melting temperature, atoms in the liquid arrange closely to form tiny stable particles of crystalline configuration at different points. These tiny crystalline particles grow when more and more atoms leave the liquid phase and join these particles. Thus, crystallites grow at various points in the parent phase and ultimately impinge on one another to form a polycrystalline solid, Fig. 9.4.

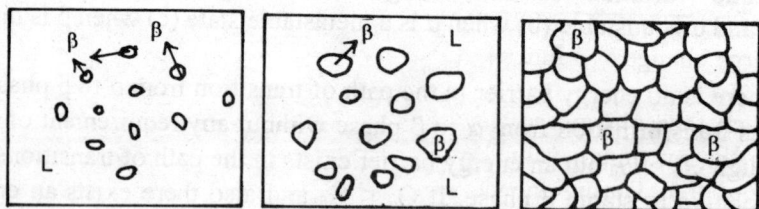

Fig. 9.4 Formation of polycrystalline materials.

Let us consider $\Delta G = G_S - G_L$ is the energy released when a solid particle is formed. The free energy change per unit volume of the solid phase created from the parent phase is Δg_v, which is expressed using Eq.(9.5) as

$$\Delta g_v = \Delta h(\Delta T)/T_m \tag{9.6}$$

where Δh is the enthalpy change per unit volume of the product phase i.e. latent heat of fusion which is assumed to be independent of temperature, ΔT is the degree of supercooling. When the solid phase is formed from the liquid phase, γ is the surface energy per unit area of the solid-liquid interface created. Hence the total change in free energy per new solid particle (taken to be spherical in shape) is given by

$$\Delta G = 4\pi r^2 \gamma + (4\pi/3)\, r^3\, \Delta g_v \tag{9.7}$$

The first term in Eq.(9.7) is the surface energy, which is positive. This term is proportional to surface area. The second term is the volume energy released by the condensing phase and it is negative. This term is proportional to volume. Initially the particles of solid phase are small in size. For a spherical particle of radius r, the ratio of surface area to volume is 3/r. Hence for small values of r, surface energy predominates over the volume energy. So the initial formation of solid phase requires supply of energy. As a result, the free energy of the system may tend to increase and the formation of tiny particles of solid phase is hindered. The positive surface energy of the particle is responsible for instability in its formation. When the size of the solid particle increases such that for small change in r, change in total free energy is zero then the particle is said to have nucleated. The size of the particle is termed as critical (r*). When r increases further, the contribution of the volume energy becomes higher than that of the surface energy.

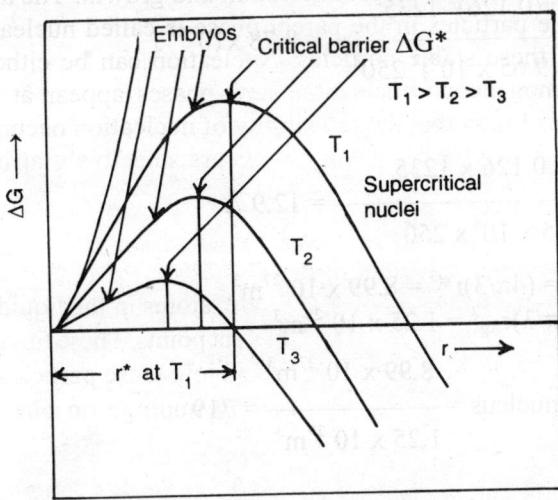

Fig. 9.5 Free energy change as a function of size (r) of the particle at different temperatures.

Thus, initially when the solid phase starts precipitating in the parent phase, ΔG increases. Finally, ΔG reaches the maximum value with increase of r and then it starts decreasing as r increases further, Fig. 9.5. During the decrease of ΔG, the contribution of volume energy becomes larger compared to surface energy term. The critical size r*, at which ΔG becomes maximum can be calculated by putting first order derivative of ΔG with respect to r equal to zero. We obtain r* as

$$r* = - (2\gamma /\Delta g_v) \qquad (9.8)$$

The corresponding maximum value, $\Delta G*$ is

$$\Delta G* = 16\pi\gamma^3/(3(\Delta g_v)^2) \qquad (9.9)$$

$\Delta G*$ is called critical free energy of nucleation. Using Eq.(9.6) and Eq.(9.9) we get

$$\Delta G* = 16\pi\gamma^3 \, T_m^2/3(\Delta h)^2 \, (\Delta T)^2 \qquad (9.10)$$

$\Delta G*$ is expressed in kJ per mole when it refers to change of free energy per mole of nuclei. Similarly,

$$r* = 2\gamma T_m/(\Delta h \times \Delta T) \qquad (9.11)$$

Example 9.1

Calculate the critical free energy of nucleation for Ag with supercooling of 250°C. Calculate the number of atoms in critical nucleus when solid Ag is formed. Given that $\gamma = 0.126$ Jm^{-2}, latent heat of fusion $\Delta h = 9.65 \times 10^8$ Jm^{-3} and the radius of silver atom is 1.44 Å.

Solution:

$$\Delta T = 250°K,$$
$$T_m = 962°C = 1235°K$$
$$\Delta G^* = \frac{16\pi(0.126)^3\ 1235^2}{3\ (9.65 \times 10^8)^2\ 250^2} = 8.78 \times 10^{-19}\ J$$

Critical radius at 712°C

$$r^* = \frac{2 \times 0.126 \times 1235}{9.65 \times 10^8 \times 250} = 12.9\ Å$$

Volume of critical nucleus = $(4\pi/3)r^{*3} = 8.99 \times 10^{-27} m^3$.
Volume of silver atom = $(4\pi/3)r_{Ag}^3 = 1.25 \times 10^{-29} m^3$

No. of Ag atoms in critical nucleus $= \dfrac{8.99 \times 10^{-27} m^3}{1.25 \times 10^{-29} m^3} = 719$

Example 9.2

Calculate the critical radius and critical free energy of nucleation of lead at (a) 327°C, (b) 322°C and (c) 247°C. Given that the enthalpy of fusion of lead is 2.37×10^8 Jm^{-3}, the surface energy is 0.033 Jm^{-2} and the melting point of lead is 327°C.

Solution:

(a) At 327°C, there is no supercooling i.e. $\Delta T=0$. So $\Delta G^* = \infty$ and there is no possibility of nucleation.

(b) At 322°C the degree of supercooling is 5°C i.e. $\Delta T = 5°K$

$$\Delta G^* = \frac{16 \times 3.142 \times (0.033)^3 \times (600)^2}{3 \times (2.37 \times 10^8)^2\ (5)^2} = 1.54 \times 10^{-16}\ J$$

$$r^* = \frac{2 \times 0.033 \times 600}{2.37 \times 10^8 \times 5} = 3.34 \times 10^{-8} = 334\ Å$$

(c) At 247°C the degree of supercooling is 16 times larger than at 322°C i.e. $\Delta T = 80°K$. So ΔG^* is smaller by a factor of 256 and r^* is smaller by a factor of 16.

Hence $\Delta G^* = 6.016 \times 10^{-19}\ J$
$$r^* = 20.89Å$$

The free energy change at a fixed temperature during the process of nucleation is a function of r, Fig. 9.5. Once the solid particle attains the size equal to or greater than the critical value r*, it tends to grow spontaneously. Such particles are called nuclei. Particles of radii r* are called critical nuclei. Smaller particles (r < r*) are unstable and tend to decrease in size. These are called embryos. At a lower temperature, i.e. for a large value of supercooling ΔT, Δg_v becomes more negative and hence the values of ΔG^* and r* become less (Eqs. (9.8) and (9.9)). The maximum of ΔG versus r curve shifts to the left with increase in supercooling. The expression of r* in Eq. (9.8) has a physical meaning only when Δg_v is negative. The value of r* approaches infinity at T = T_m i.e. when $\Delta g_v = 0$, so no nucleation is possible at melting temperature. Thus, nucleation is possible only with cooling below T_m. More amount of supercooling (ΔT) leads to smaller critical nuclei and faster rate of nucleation. Since atomic movements are restricted at very low temperature, so the rate of nucleation becomes slow. The rate of nucleation is a function of supercooling temperature and attains a maximum value at a specific supercooling temperature. During phase transformation, there will be particles of new phase having whole range of sizes. Using Maxwell-Boltzmann statistics, one can show that the number of critical size nuclei per unit volume is given by

$$n^* = n \exp(-\Delta G^*/RT) \tag{9.12}$$

where n is the number of solid particles per unit volume of the liquid phase, R is the gas constant and ΔG^* is expressed in the units of J/mole. The formation of stable nucleus occurs when the critical sized nucleus becomes just supercritical. This process can be identified as diffusion of one atom from the parent phase to the critical sized nucleus. Any further addition of atoms to the supercritical nucleus will result in the growth. Thus the rate of formation of stable nuclei is directly proportional to

- the number of atoms (m*) at the interface of the liquid phase and the critical nucleus;
- the number of critical nuclei (n*) per unit volume.
- the frequency (v') with which any of the m* atoms in the parent phase will cross the boundary and make the critical nucleus, supercritical.
- the probability (p) that the atom in the parent phase is vibrating towards the nucleus and not bouncing back due to elastic collision. Thus the rate of nucleation

$$I = n^* m^* p v' \tag{9.13}$$

where
$$v' = v_o \exp(-\Delta H_d/RT) \tag{9.14}$$

v_o being the frequency of vibration of atom in the parent phase, ΔH_d is the activation energy required for a mole of atoms for jumping from the liquid onto the solid nucleus. Using Eqs. (9.12) and (9.14) in Eq. (9.13) we get:

$$I = nm^* p v_o \exp[-(\Delta G^* + \Delta H_d)/RT] \tag{9.15}$$

Thus,
$$I = I_o \exp[-(\Delta G^* + \Delta H_d)/RT] \tag{9.16}$$

where

$$I_o = n \, m^* \, p\nu_o \qquad (9.17)$$

is quite large and is of the order of 10^{42} m^{-3}s^{-1}. ΔG^* is the critical free energy of nucleation per mole.

Example 9.3

Find the total number of solid lead particles per meter cube of the parent phase in Example 9.2 if the number of critical size particles n* is 5×10^{20}/mole at 247°C.

Solution:

$$n^* = n \exp(-\Delta G^*/RT) = 5 \times 10^{20}$$

where

$$T = 247°C = 520°K$$
$$R = 8.314 \text{ J/mole } °K$$
$$\Delta G^* = 6.016 \times 10^{-19} \text{ J per critical sized particle}$$
$$V_{molar} = 1.83 \times 10^{-5} \text{ m}^3$$
$$\Delta G^* = 6.016 \times 10^{-19} \times 5 \times 10^{20} = 300.8 \text{ J/mole}$$
$$n = 5 \times 10^{20} \exp[\, 300.8/ \, (8.314 \times 520)] = 5.36 \times 10^{20}/\text{mole} \ = 2.93 \times 10^{25} \text{ particles/ m}^3$$

Example 9.4

Calculate the value of ΔH_d at 247°C in Example 9.2 if nucleation rate is 10^6/ m^{3}s.

Solution:

Nucleation rate
$$I = I_o \exp [-(\Delta G^* + \Delta H_d)/RT] = 10^6/ \text{ m}^3\text{s}$$
$$I_o = 10^{42} \text{ m}^{-3}\text{s}^{-1}, \ T = 520°K$$
$$\Delta G^* \ = 300.8 \text{ J/mole}$$
$$\Delta G^* + \Delta H_d = [\, (42-6) \times 2.303 \times 8.314 \times 520] = 358.43 \text{ kJ/mole}$$
$$\Delta H_d = 358.13 \text{ kJ/mole.}$$

9.6 HETEROGENEOUS NUCLEATION

Nucleation processes are accelerated by heterogeneity of the parent phase. Such a nucleation process is called heterogeneous nucleation. Heterogeneous nucleation occurs in the presence of insoluble impurities in the parent phase, the surface of the container where there are irregularities in crystal structure. Irregularities in crystal structure like dislocations, vacancies, and impurities possess strain energy and provide energy for nucleation. The nucleation process is facilitated if it reduces the strain energy. The critical energy for heterogeneous nucleation is lower than that for homogeneous case. Heterogeneous nucleation is preferred in the industrial processes since it requires less amount of supercooling. Heterogeneous nucleation is facilitated by an external surface, a grain boundary or a stacking fault. Consider the nucleation of α phase from a gaseous phase on the external surface of a solid material (δ phase). Consider a particle formed on the surface of δ phase having the shape of a plano-convex lens. Under the equilibrium conditions, the surface tension forces at O satisfy,

$$\gamma_{g\delta} = \gamma_{g\alpha} \cos\theta + \gamma_{\alpha\delta}, \qquad (9.18)$$

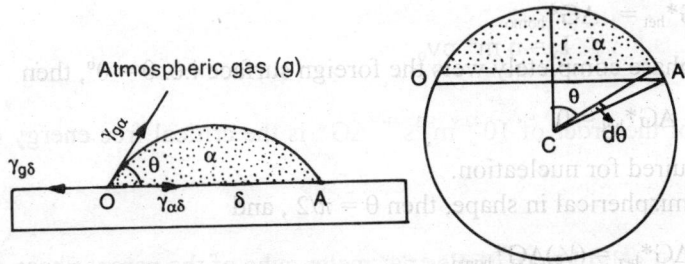

Fig. 9.6 Heterogeneous nucleation of α phase from gaseous phase g on the surface of δ, which is a foreign inclusion.

where θ is the angle of contact. The expression for ΔG can be expressed in terms of the volume and surface energies of the interfaces. A new α-δ interface is created in place of g-δ interface of same area. Also a new α-g interface is created. Thus,

$$\Delta G_{het} = V_\alpha \Delta g_v + S_{g\alpha}\gamma_{g\alpha} + S_{\alpha\delta}(\gamma_{\alpha\delta} - \gamma_{g\delta}) \qquad (9.19)$$

where $S_{g\alpha}$, $\gamma_{g\alpha}$ are the area and surface energy of the interface between gaseous and α phases respectively, and $S_{\alpha\delta}$, $\gamma_{\alpha\delta}$ are the area and surface energy of the interface between α and δ phases, respectively. Similarly, $\gamma_{g\delta}$ is the surface energy of interface between g and δ phases and V_α is volume of the α phase.

$$\Delta G_{het} = \int_0^\theta (\pi r^2 \sin^3\theta' \, rd\theta')\Delta g_v + \int_0^\theta 2\pi r \sin\theta' \, rd\theta' \, \gamma_{g\alpha} + \pi r^2 \sin^2\theta \, (\gamma_{\alpha\delta} - \gamma_{g\delta})$$

Using Eq.(9.18), we obtain,
$$\Delta G_{het} = (\pi r^3/3) [2-3\cos\theta + \cos^3\theta] \Delta g_v + 2\pi r^2 (1-\cos\theta) \gamma_{g\alpha} - \pi r^2 \sin^2\theta \, \gamma_{g\alpha} \cos\theta$$
$$= \pi r^2 [(r/3) \Delta g_v + \gamma_{g\alpha}] [2-3\cos\theta + \cos^3\theta] \qquad (9.20)$$

$$(\partial(\Delta G_{het})/\partial r)\big|_{r^*} = [\pi r^{*2} \Delta g_v + 2\pi r^* \gamma_{g\alpha}] [2-3\cos\theta + \cos^3\theta] = 0$$

So
$$r^* = -2\gamma_{g\alpha}/\Delta g_v \qquad (9.21)$$

where r* corresponds to the critical radius of nucleation. Putting this in Eq. (9.20) we get the critical value of free energy of nucleation as

$$\Delta G^*_{het} = \frac{4\pi\gamma_{g\alpha}^3}{3(\Delta g_v)^2} [2-3\cos\theta + \cos^3\theta] \qquad (9.22)$$

Using Eq.(9.9) we get:
$$\Delta G^*_{het} \approx \frac{1}{4} \Delta G^*_{homo} [2 - 3\cos\theta + \cos^3\theta] \qquad (9.23)$$

So the critical value of the free energy of nucleation depends on the angle of contact θ and interface energies.

(i) If the particle of α phase makes only point contact with the foreign surface, then θ = 180°,

$$\Delta G^*{}_{het} = \Delta G^*{}_{homo}$$

(ii) If the particle of α phase completely wets the foreign surface i.e. $\theta = 0°$, then

$$\Delta G^*{}_{het} = 0$$

i.e. no energy is required for nucleation.

(iii) If the particle is hemispherical in shape, then $\theta = \pi/2$, and

$$\Delta G^*{}_{het} = (\frac{1}{2})\Delta G^*{}_{homo}.$$

If the solidifying liquid metal wets the surface of solid nucleating agent, the angle of contact becomes zero and this facilitates the heterogeneous nucleation. The nucleating agent can be chosen suitably such that θ can be minimized. In Eq. (9.18), $\gamma_{g\alpha}$ is fixed, but $\cos\theta$ depends on $\gamma_{\alpha\delta}$ and $\gamma_{g\delta}$ i.e.

$$\cos\theta = \frac{\gamma_{g\delta} - \gamma_{\alpha\delta}}{\gamma_{g\alpha}} \qquad (9.24)$$

If the crystal structures of nucleating agent δ and the product phase α are similar, and the lattice parameters are nearly equal then $\gamma_{g\delta} \approx \gamma_{g\alpha}$ and $\gamma_{\alpha\delta}$ is very small. This leads to $\cos\theta \approx 1$ i.e. $\theta \approx 0$. Examples of suitable nucleating agents for seeding rain-bearing clouds are NaCl, AgI etc. and that for the production of artificial diamonds from graphite is nickel. AgI and NaCl (fcc) have atomic planes (111) match with (0001) planes of hexagonal ice crystal. If the nucleation rate of ice crystals is increased, the size of ice crystal would be smaller, which minimizes damages due to hailstorm. This can be achieved by injecting above nucleating agent. Nickel and diamond both have fcc structure and nearly the same lattice parameters, so nickel is used for the production of diamond.

Example 9.5

The surface energy of water(L) γ_{LG} is 0.07 J/m^2, that for glass(S) and atmospheric gas(G) interface is $\gamma_{SG} = 0.58$ J/m^2 and for water and glass interface $\gamma_{LS} = 0.54$ J/m^2 .
   (a) Find the angle of contact.
   (b) What should be the value of γ_{SG} so that there is complete wetting by the liquid phase? Assume γ_{LG}, γ_{LS} are unchanged.
   (c) If γ_{SG} and γ_{LG} are unchanged, is it possible for the liquid phase to have point contact?

Solution:

   (a) $\cos \theta = (\gamma_{SG} - \gamma_{LS})/\gamma_{LG} = 4/7$ i.e. $\theta = 55°9'$
   (b) For complete wetting $\theta = 0°$ which gives $\gamma_{SG} = 0.61$ J/m^2.
   (c) Point contact implies $\theta = 180°$ which requires $\gamma_{LS} = 0.65$ J/m^2.

9.7 GROWTH

Consider a phase transformation in which α phase changes to β phase. After nucleation, the product phase starts growing at different nucleation sites. The growth rate depends on the extent of supercooling ΔT. During growth, atoms jump from the parent phase α to the product phase β. The activation energy required for such a transformation is the same as the activation energy for diffusion ΔH_d across the interface. Since α and β phases have the same free energy at equilibrium temperature T_m, there is equal probability of an atom jumping from α to β phase and vice versa. Hence, the nucleation as well as growth is not possible at T_m. The rate of growth of β phase depends on the rate at which atoms jump from α phase to β phase and also on the rate at which atoms can be added into the crystal structure in β phase. The crystallization also depends on the nature of sites on the interface where atoms are added and removed.

When a liquid metal is supercooled below T_m, the transformation of liquid to solid phase takes place faster near the surface of the container than that in the bulk liquid. That is why the nuclei are formed near the wall of the container faster than those in the bulk region. They grow rapidly into the grains of nearly spherical shape. During this process, large latent heat of fusion is liberated. The heat of liquid metal and latent heat of fusion dissipate through the wall and a layer of solidified metal is formed on the surface. As a result, the rate of the growth of the crystal is reduced. A small value of ΔT helps in the advancement of the grain growth but nucleation of new grains is hindered. The product phase formed from the liquid phase may have variety of forms, depending upon the types of crystallographic growth, temperature distribution and solute redistribution between liquid and solid phases during crystallization. Two common types of grain structures, Fig. 9.7, are

a) Equiaxed grains
b) Columnar grains

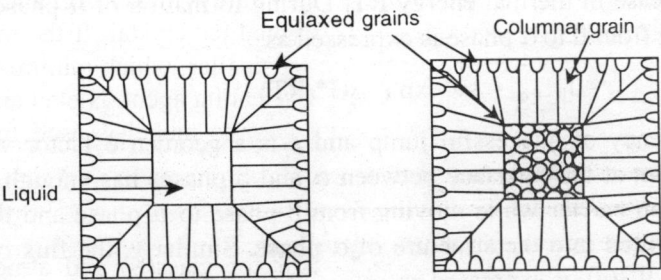

Fig. 9.7 Main types of grain structures

Equiaxed grains

Equiaxed grains are those, which grow equally in all directions. If the nucleation and growth conditions are suitable in the liquid metal, which is cooling slowly, one obtains equiaxed grains. Equiaxed grains are normally grown near a cold wall, Fig.9.7. Large amount of supercooling results in the formation of a large number of nuclei.

Columnar grains

When a liquid metal solidifies relatively slowly in the presence of a steep temperature gradient, columnar grains are formed. Columnar grains are long, thin, and coarse grains. These grains grow perpendicular to the mould walls, since large thermal gradient is present in this direction. Diffusion is an important mechanism for this type of growth. It is thermally activated growth and can occur in various ways. It can occur through the process of recrystallization if it decreases strain energy in the crystal. Recrystallization can also occur due to a difference in surface energies of neighbouring grains. During recrystallization

- larger grains grow at the expense of smaller grains using diffusion mechanism so that the surface energy decreases
- the phase with lower free energy grows at the expense of the phase with higher free energy.

In this precipitation process, growth occurs due to release of chemical energy. During growth, the curved surface of a grain moves toward its centre of curvature, Fig.7.10 i.e. the grain size increases with the increase of curvature. The rate of grain growth can be expressed as

$$g = dr/dt = k/r \qquad (9.25)$$

where r is the radius of grain and k, the constant of proportionality, is a function of temperature and surface energy. Thus,

$$r^2 - r_o^2 = kt \qquad (9.26)$$

where $r = r_o$ at time $t = 0$. The grain growth in a polycrystalline material can be expressed in the form of an empirical equation as

$$r^{1/n} - r_o^{1/n} = kt \qquad (9.27)$$

where n depends on the composition and temperature.

The growth rate increases initially with the increasing degree of supercooling and reaches a maximum for a particular value of supercooling. With more supercooling, the growth rate decreases due to decrease in thermal energy RT. During formation of α phase from β phase the flux of atoms jumping from β to α phase is expressed as

$$J_{\beta \to \alpha} = sv' \exp(-\Delta G^*/RT) \qquad (9.28)$$

where v' is the frequency of successful jump and s is a geometric factor which includes the probability that an atom at the interface between α and β phases has enough thermal energy to overcome the activation barrier while moving from β phase to α phase and the probability that such an atom can be fitted into the structure of α phase. Similarly, the flux of atoms that jump from α to β phase, Fig.9.3(c), is expressed as

$$J_{\alpha \to \beta} = sv' \exp-(\Delta G^* - \Delta G)/RT, \qquad \Delta G < 0 \qquad (9.29)$$

Thus, the net flux of atoms from β to α phase is given by

$$J = J_{\beta \to \alpha} - J_{\alpha \to \beta} = sv' \exp[-\Delta G^*/RT] (1 - \exp(\Delta G/RT)) \qquad (9.30)$$

i.e. the growth rate

$$g = (dr/dt) = J V_\alpha$$
$$= sv'V_\alpha \exp[-\Delta G^*/RT] (1 - \exp(\Delta G/RT))$$
$$= KD (1 - \exp(\Delta G/RT)) \qquad (9.31)$$

where V_α is the volume of an atom of the product phase, $D = D_o \exp[-\Delta H^*/RT]$ is the diffusion coefficient for atoms jumping from β phase to α phase and K is a constant. The growth rate $g = 0$ at equilibrium temperature T_m, since ΔG^* is infinity for $\Delta T = 0$. The overall transformation rate (dx/dt) is given as function of g and I i.e.,

$$dx/dt = f(g, I)$$

If the rate of nucleation is constant, nucleation occurs randomly in β phase and the particles of α phase grow at a constant rate, then the overall transformation rate becomes

$$dx/dt = (1-x)I(4\pi/3)(gt)^3 \qquad (9.32)$$

where x is the fraction of β phase transformed to α phase, gt is the approximate radius of the grain at time t, and I is the nucleation rate. Thus, x versus t is a sigmoidal function, Fig. 9.8.

$$x = 1 - \exp[-(\pi/3)Ig^3t^4]. \qquad (9.33)$$

The rates I, g and (dx/dt) are plotted against supercooling ΔT, Fig. 9.9.

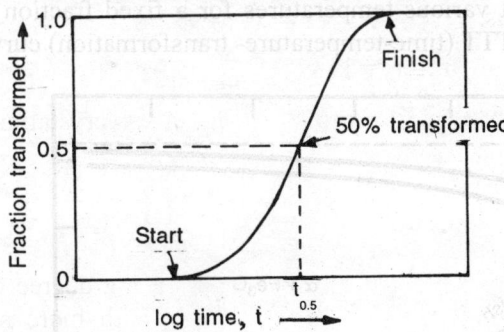

Fig. 9.8 Fractional transformation versus time.

Growth of single crystals

Growth of single crystal is very important for industrial applications. Single crystals can be grown through controlled solidification. A widely used technique is that of crystal pulling. First, the material is melted in a crucible and held at a temperature just below the melting point. A

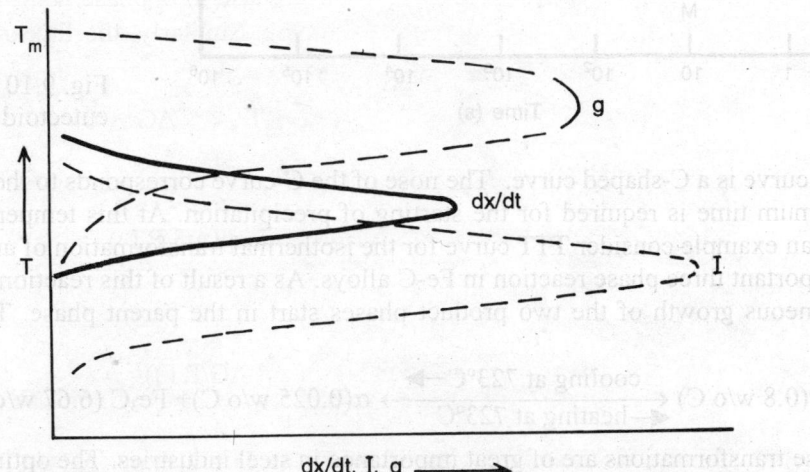

Fig.9.9 Rates (dx/dt), I, and g are plotted against supercooling ΔT.

small single crystal seed is brought into contact with melt carefully to avoid formation of other nuclei. A crystal puller does the job of slowly withdrawing and rotating the oriented crystal as it grows. Size of the single crystal is controlled by varying the temperature of the melt and the rate of pulling. The anisotropy property of single crystals is useful for applications in electronic technology. Single crystals of silicon and gallium arsenide, grown by this technique, are commonly used for semiconductor and optoelectronic devices. Ruby and garnet crystals are used for lasers. Super alloy used in turbine blades of jet engines is also grown by this method.

9.8 TIME-TEMPERATURE-TRANSFORMATION CURVES

The phase transformation in alloys depends on the rate of cooling, the extent of supercooling and the time elapsed after the cooling. If an alloy is quenched and held at a temperature below the solubility limit, the precipitation of the new phase starts after some time. This time interval is longer, if supercooling is less. The complete precipitation of the new phase requires long time. The plot of time taken at various temperatures for a fixed fraction of precipitation is given in Fig.9.10 in the form of a TTT (time-temperature- transformation) curve.

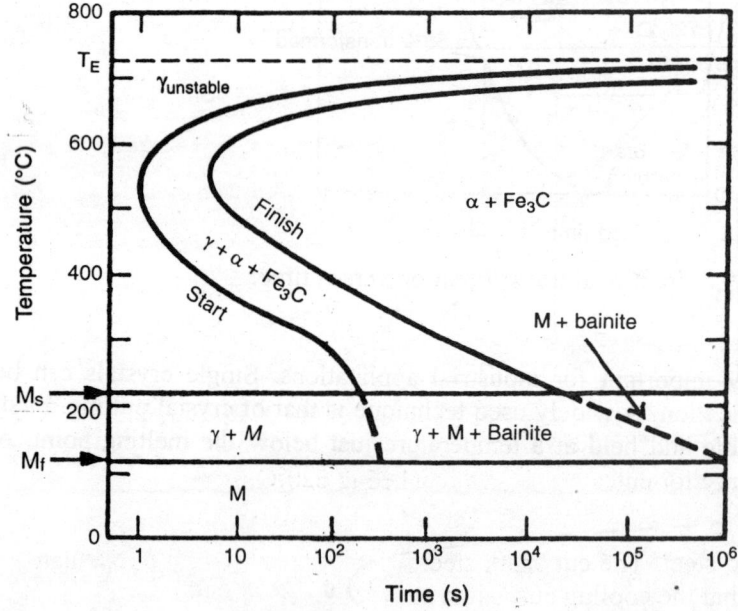

Fig. 9.10 TTT curves for eutectoid steel.

The TTT curve is a C-shaped curve. The nose of the C-curve corresponds to the temperature at which minimum time is required for the starting of precipitation. At this temperature (dx/dt) is highest. As an example consider TTT curve for the isothermal transformation of austenite. This is the most important three phase reaction in Fe-C alloys. As a result of this reaction, the nucleation and simultaneous growth of the two product phases start in the parent phase. The reactions is given as,

$$\gamma(0.8 \text{ w/o C}) \underset{\xleftarrow{\text{heating at } 723°C}}{\xrightarrow{\text{cooling at } 723°C}} \alpha(0.025 \text{ w/o C}) + Fe_3C \ (6.67 \text{ w/o C}) \qquad (9.34)$$

Such phase transformations are of great importance in steel industries. The optimum properties of steel depend on the heat treatment mechanism and kinetics of phase transformation. The isothermal transformation is studied for steel with eutectoid composition where there are no proeutectoid phases. The result of such study is TTT curve, Fig. 9.10. Start of the reaction is indicated by appearance of precipitation of a two-phase mixture of α and Fe_3C. The study of microstructure reveals that the intermixture of two product phases arises as a consequence of the cooperative growth of these two phases during eutectoid reaction. Similarly, the completion of transformation of austenite into two phase mixture of α and Fe_3C indicates the completion of reaction.

Eutectoid steel

Consider a sample of eutectoid steel (0.8 w/o C) heated to a temperature, a few degrees above eutectoid temperature 723°C, and held there for a long time such that its structure becomes homogeneous austenite. This heat treatment is called austenizing. When austenite is initially quenched into a hot bath to a temperature 620°C and held there for varying time intervals for the sample to decompose before it is finally quenched to room temperature. The unstable austenite, after the initial quench, transforms isothermally, to ferrite (α) and Fe_3C along horizontal path at constant temperature 620°C, Fig. 9.11.

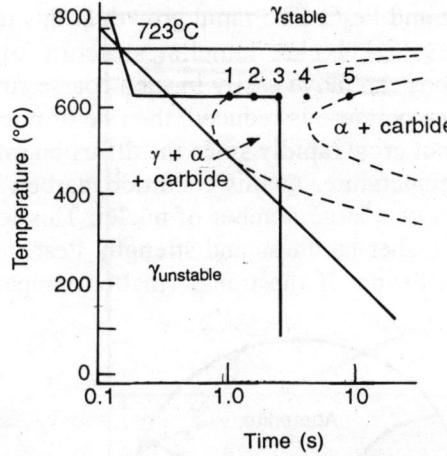

Fig. 9.11 TTT Curve for eutectoid steel quenched to 620°C.

Example 9.6

Consider the TTT curve of eutectoid steel. The eutectoid steel is austenized at 780°C. Calculate the rate of continuous cooling such that the cooling curve just misses the knee.

Solution:

To calculate dT/dt, one must note that the abscissa is in log scale.

$$\frac{dT}{dt} = \frac{dT}{d(\log t)} \cdot \frac{d(\log t)}{dt} = \frac{dT}{d(\log t)} \cdot \frac{1}{2.30\, t}$$

A straight line drawn starting at 780°C is just missing the knee of TTT curve at t = 0.8 s and temperature 510 °C. Hence

$$\frac{dT}{dt} = \frac{1}{2.30 \times 0.8} \cdot \frac{510 - 780}{\log 0.8 - \log 0.1} = -162.5°C/s$$

The precipitation of intermixed phases of α and Fe_3C starts at point 1 i.e. after one second of isothermal cooling. Mixed phase of γ, α and Fe_3C are present at points 2,3,4 and 5 of the path.

The study of microstructure indicates the existence of pearlite and γ along the horizontal path from point 1 to 5 with increasing amount of pearlite as time lapses. The transformation from γ to α and Fe_3C proceeds slowly at first and then it proceeds more rapidly. Finally, it stops gradually near the completion of the reaction in the sigmoidal fashion. Once the path of cooling enters α + carbide region, where there is no γ phase left, the microstructure of steel does not change on further cooling. Finally, fine pearlite (thin lamellar two phase mixture of α and Fe_3C) is obtained. The same sort of isothermal transformation occurs at any other temperature between 250°C to 723°C, but with some differences e.g.

- times for start and end of transformation are different at different temperatures,
- the microstructure and hardness of pearlite depend on the transformation temperature.

At low supercooling, only a few pearlite patches nucleate which grow rapidly, leading to a coarse pearlite, with thicker plates of α and Fe_3C. The rapid growth at this temperature is due to higher diffusion rate of carbon, Fig. 9.12. Thicker lamellar structure of pearlite results in a soft and low strength steel. Dislocations can move easily in such coarse structure with less number of barriers. As the transformation temperature is reduced, the rate of nucleation becomes faster and many nuclei that are formed cannot grow rapidly since the diffusion rate of carbon decreases with the decrease in transformation temperature. At this condition, carbon atoms need to diffuse over shorter distance due to formation of a large number of nuclei. This results in formation of finer pearlitic microstructures having higher hardness and strength. Pearlite nucleates at the austenite grain boundaries and grows by diffusion. If the transformation temperature is much lower than

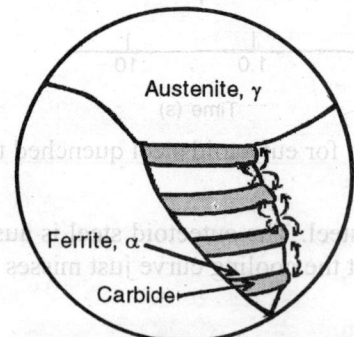

Fig.9.12 Diffusional transformation during pearlite formation.

the nose of C-curve (say 400°C), austenite isothermally decomposes into bainite. Bainite is formed by a combination of shear and diffusion process. Though bainite contains ferrite and Fe_3C, its microstructure is nonlamellar. Bainite microstructure has the cementite p...ies in the form of short needles embedded in the plates of ferrite. The finer bainites that are produced below 350°C, are harder than the bainites formed between 350°C and 450°C. Finer bainite contains fine particles of carbides dispersed throughout the ferrite plates, giving rise to a hard and tough structure.

Example 9.7

Consider a sample of eutectoid steel undergoing the cooling processes as shown in Fig. 9.13. Find out the microstructures at the end of each cooling process.

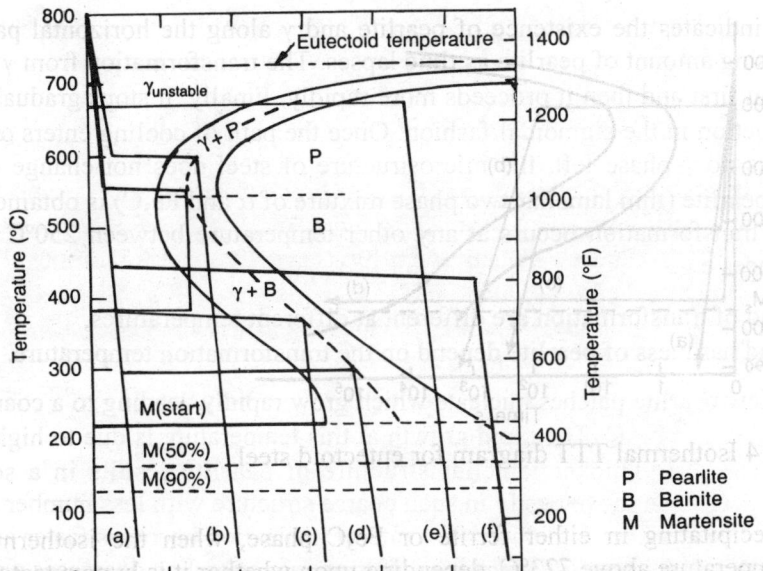

Fig. 9.13 Various paths of cooling shown in TTT curves for eutectoid steel

Solution:

Path	Microstructure	Path	Microstructure
a	100% martensite	d	50% lower bainite & 50%martensite
b	50% coarse pearlite & 50% martensite	e	100% coarse pearlite
c	50% fine pearlite, 25% upper bainite & 25% martensite	f	50% upper bainite & 50% martensite

Example 9.8

Consider a thin sample of eutectoid steel austenized at 750°C. The sample is given the following heat treatments, Fig.9.14. Find out the microstructure of the sample after each heat treatment:
(a) Quenched to room temperature
(b) Quenched to 620°C and held for 90 s and then water quenched
(c) Quenched to 390°C and held for 90 s and then water quenched
(d) Quenched to 250°C and held for 10 hours and then water quenched.

Solution:

The cooling paths are indicated in Fig. 9.14 and the microstructures are given below:
(a) 100% martensite
(b) 100% coarse pearlite
(c) Approximately 50% upper bainite and 50% martensite
(d) 100% lower bainite

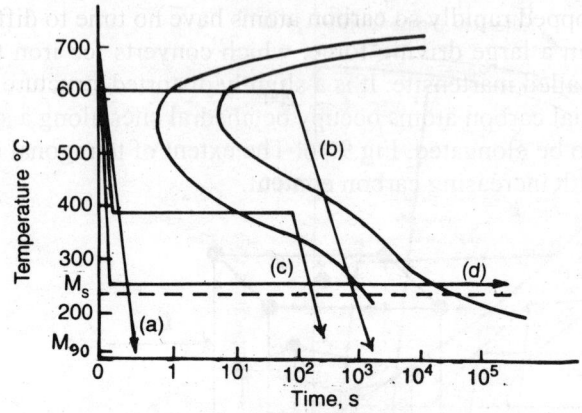

Fig. 9.14 Isothermal TTT diagram for eutectoid steel.

Noneutectoid steel

Noneutectoid steel starts precipitating in either ferrite or Fe_3C phase, when the isothermal transformation occurs at a temperature above 723°C, depending upon whether it is hypoeutectoid or hypereutectoid. The ferrite so formed is called proeutectoid ferrite and the cementite is called proeutectoid cementite. The isothermal transformation below eutectoid temperature, converts hypoeutectoid (hypereutectoid) steel to $\gamma + \alpha$ ($\gamma + Fe_3C$) first and then γ transformed to pearlite, Fig.9.15.

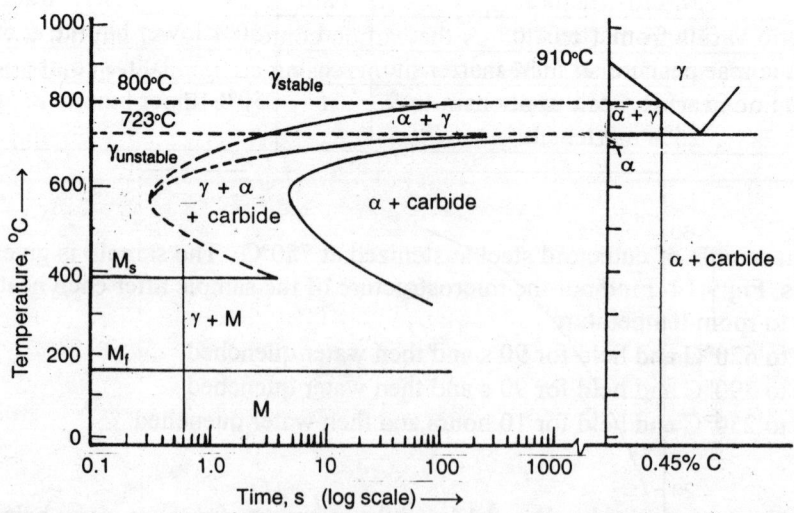

Fig. 9.15 TTT curve for noneutectoid plain carbon steel (containing 0.45 w/o C)

Transformation of austenite to martensite

If during the isothermal transformation of eutectoid steel to pearlite, it is quenched to a temperature below the bainite range (say 250°C) before completion of isothermal transformation, then the diffusion of carbon required for formation of pearlite becomes negligible due to lack of thermal energy and the metastable austenite attempts to reject carbon atoms present in eutectoid steel (0.8 w/o C). This is because the solubility limit of C is only 0.01 w/o in stable α iron. Since the temperature is dropped rapidly so carbon atoms have no time to diffuse. Thus, rapid and large supercooling results in a large driving force, which converts fcc iron to bct iron, supersaturated with carbon. This is called martensite. It is a slightly distorted structure, as compared to fcc phase of iron, since interstitial carbon atoms occupy octahedral sites along a set of parallel edges (say c-axis) causing c-axis to be elongated, Fig.9.16. The extent of tetragonal distortion, as measured by c/a ratio, increases with increasing carbon content.

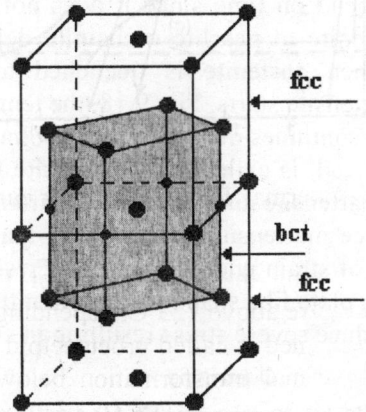

Fig. 9.16 Transformation from austenite to martensite

The c/a ratio varies from 1 to 1.08 as the carbon content varies from 0.0 w/o to 1.2 w/o. The hardness of the martensite also increases with increasing carbon content and attains more or less constant value of 65 on R_c scale at about 0.6w/o carbon, Fig.9.17.

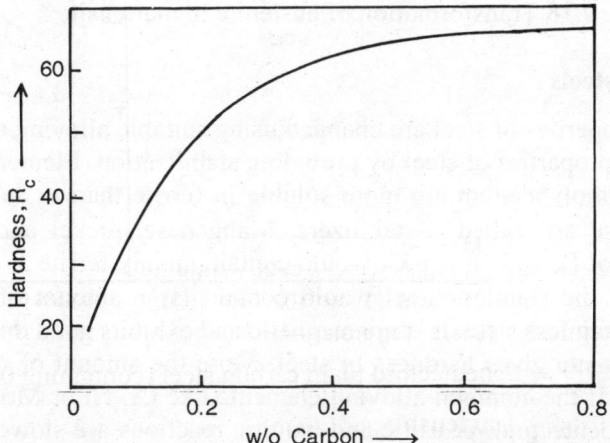

Fig. 9.17 Hardness of martensite on the R_c scale versus w/o carbon in steel.

The density of dislocation increases during martensitic transformation. The martensitic steel has high strength and low ductility. The tensile strength increases with increasing hardness. At a value of 65 on R_c scale, the steel becomes brittle, resulting in a low value of tensile strength.

Thus, the characteristics of martensitic steel make it misfit for structural applications, in the as-quenched condition. Martensite is only a metastable phase and its brittleness can be reduced by reheating (tempering). The hardness of tool steel used for cutting glass is over R_c 65 whereas the steel for a chisel or a punch requires a hardness of R_c 45–55 and also some amount of shock resistance. The formation of martensite from austenite is due to shear transformation. Martensite is not found in the equilibrium phase diagram since it is a metastable phase and its free energy is higher than that of the phases found in Fe-Fe$_3$C phase diagram at the corresponding temperatures. The amount of martensite that forms is a function of temperature to which austenite is cooled. This transformation does not depend on time since it does not involve diffusion. On the other hand, the transformation of austenite to pearlite or bainite depends on time since it involves diffusion of carbon atoms. When austenite is quenched to a temperature at which the transformation of austenite to martensite starts, Fig. 9.18, the temperature is called M_s (Martensite start). Martensitic transformation continues during further cooling. The temperature, at which the transformation is virtually completed, is called M_f (Martensite finish). Between M_s and M_f, the sample material is a mixture of martensite and austenite. Martensitic transformation changes the shape of transforming region since martensite is less dense than austenite. This transformation results in a considerable amount of strain energy due to the resistance from the matrix phase to the shape change. Martensite has plate like shape in an austenitic matrix. If the volume occupied by this phase is large it may introduce severe stress resulting in cracks.

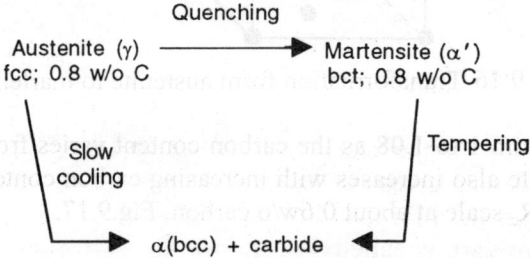

Fig. 9.18 Transformation of austenite to martensite

Transformation in alloy steels

The microstructure and properties of steel are changed using suitable alloying elements. Alloying elements can improve the properties of steel by providing stabilization. Elements such as silicon, chromium, vanadium and molybdenum are more soluble in ferrite than in austenite and favour formation of ferrite. These are called α-stabilizers. Manganese, nickel and cobalt act as γ-stabilizers. A binary alloy of Fe and 18 w/o Cr would contain mainly ferrite at room temperature, but if 8 w/o Ni is added, the stainless steel would contain large amount of γ–phase at room temperature. This is 18/8 stainless steel. It is nonmagnetic and exhibits good ductility.

The formation of martensite gives hardness to steel. More the amount of martensite formed, harder would be the steel. If the common alloying elements like Cr, Ni or Mo are present in the alloy steel, the rates of proeutectoid, pearlitic and bainitic reactions are slowed down, Fig.9.19. Hence, martensite can be formed with less severe quenching since the formation of pearlite or bainite would take longer time. Thus, alloy steels have better hardenability than plain carbon steels without resorting to severe quenching. These alloying elements change the tempering characteristics of steel. Precipitation and coarsening of carbides are affected due to presence of alloying elements. Silicon slows down the precipitation and coarsening of cementite, which results in decreasing of hardness. Alloying elements Cr, Mo, Ti, V and W form carbides, which are more stable than iron carbide. Carbides of alloying elements are formed at tempering temperature in the range 450°C–600°C.

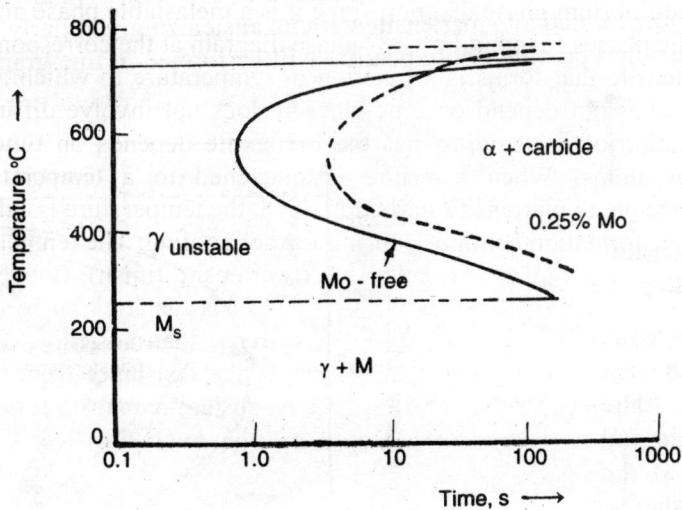

Fig. 9.19 TTT curve for alloy steel.

Particles of alloy carbide in the steel increase hardness, which is termed as secondary hardening. Secondary hardening gives rise to steel of high strength and hardness but with better ductility. The steel, which is used for making tools for cutting and dies for forming or shaping a material into component or part, is called tool steel or high-speed steel. Tools made out of this steel retain their strength at elevated temperatures. Such steel usually contains very fine tungsten carbide particles.

Continuous cooling transformation

TTT diagrams are important for finding the cooling rate required to produce a steel of desired microstructure depending upon its application. These diagrams are obtained by quenching a sample of austenite, keeping it for isothermal transformation for various intervals of time. Finally, it is again quenched to room temperature. The continuous cooling transformation curves are obtained for different rates of continuous cooling. In continuous cooling transformation, C-curves are displaced to the right as compared to the isothermal transformation curve. During continuous cooling, pearlite and bainite transformations occur at lower temperature as compared to that observed in the isothermal TTT diagram. In this section, the progress of transformation, with different rates of cooling, is being considered.

A severe quench below the temperature 250°C results in the partial transformation of austenite to martensite and no pearlite is formed. Slow cooling permits pearlite to form but the precipitate appears after a longer duration as compared to that in case of isothermal transformation, since during continuous cooling, some time is spent at higher temperatures where reaction rates are slower. If the cooling rate is such that it just misses the nose of the transformation curve, then there will be no bainite or pearlite and only martensite is formed. This is called minimum rate of cooling (C_M) for 100% martensite formation, Fig.9.20. As the cooling rate decreases, both pearlite and martensite are formed. If the cooling rate is such that the cooling curve touches the nose of the inner C-curve, it produces 100% pearlite. This is called maximum cooling rate (C_P). Thus, continuous cooling of plain carbon steel produces only pearlite if the cooling rate is equal to or less than C_P, only martensite if the cooling rate is equal to or greater than C_M and a mixture of the two if the cooling rate is in between C_M and C_P. Once pearlite is formed, it cannot be transformed into martensite without austenizing. Similarly, martensite cannot be directly transformed into pearlite. More pearlite will be formed in fine-grained steel than in a coarse grained steel.

Effect of Austenite Grain Size on Transformation

Transformation of austenite to pearlite takes place through diffusion process. The nucleation of carbide and ferrite takes place at the austenite grain boundaries. Steel with smaller austenite grains provides more grain boundary area per unit volume at a given temperature and hence, more nuclei will be produced per unit volume. Thus, a coarse-grained austenite produces more martensite as compared to a fine-grained austenite, at the same cooling rate.

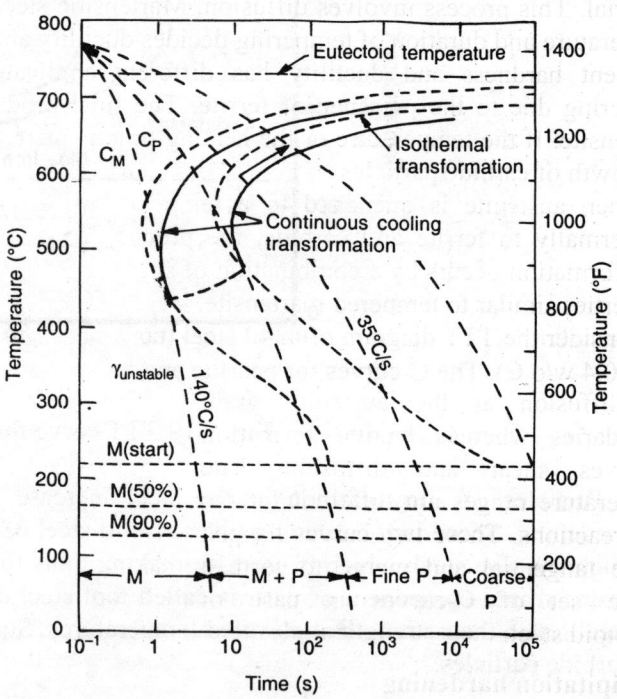

Fig. 9.20 Continuous-cooling transformation of eutectoid steel.

Heat treatment in transformation processes

A slow cooling from the austenitic temperature range is called annealing. During annealing, coarse pearlite is grown from austenite. Austenite is transformed to martensite without forming ferrite and carbide if quenching rate exceeds C_M. When austenite is transformed to martensite or carbide and ferrite, there is always decrease in density. Hence the centre of steel expands due to formation of martensite after the surface has already transformed to martensite and as a result, cracking is observed in high carbon-steel during martensitic transformation. To avoid such cracking, austenite is quenched rapidly to a temperature just above M_s, avoiding the nose of the C-curve so that no ferrite and carbide are formed and then cooling is continued at a slow rate till room temperature is attained. This process ensures the transformations of surface and the center to take place simultaneously avoiding quench cracking. This process of heat treatment is called martempering or marquenching or interrupted quench.

Sometimes austenite is plastically deformed during the quench interruption and then it is again quenched to the ambient temperature, which results in the formation of martensite. Strain hardening is added in this process. The martensitic plates formed are smaller in size, yielding greater ductility for the same strength after tempering. This is called ausforming.

Since martensite is brittle, it has to be tempered before using for engineering applications. During reheating at a temperature between 200°C and 400°C, for a long interval of time, this metastable phase is decomposed to stable phases: ferrite and very fine particles of carbide. Some carbide particles form within the metastable martensite, which gives hardness and strength to the material. This process involves diffusion. Martensitic steel becomes ductile after tempering. The temperature and duration of tempering decides ductility and hardness of the final product. Steel of different hardness and ductility has different applications. Toughness also increases after tempering due to the presence of ferrite. The final product after tempering is called tempered martensite. If the temperature or the heating time is increased, the martensite becomes softer due to growth of carbide particles.

When austenite is quenched to a temperature T, above M_s and is allowed to transform isothermally to ferrite and carbide, the process is called austempering. During this process, transformation occurs by a combination of shear and diffusion to produce bainite. Bainite has the properties similar to tempered martensite.

Consider the TTT diagram of 4340 steel (containing 1.65-2 w/o Ni, 0.4-0.9 w/o Cr, 0.2-0.3 w/o Mo. 0.4 w/o C). The C-curves for pearlite and bainite are separate in Fig. 9.21. The pearlite grows by diffusion at the austenite grain boundaries whereas bainite formation involves shear and diffusion. The temperature ranges are different for the two reactions. These two curves happen to be tangential and merge to form a single set of C-curves in case of eutectoid steel.

Precipitation hardening

Precipitation hardening is one of the most common processes of strengthening of materials where the precipitated phase in a solid solution contributes to the increase of yield strength, tensile strength and hardness with some loss of ductility. Consider a solid solution in equilibrium at a temperature greater than the room temperature. It becomes unstable and supersaturated as the temperature is decreased.

As the solute-rich phase starts precipitating, the solution becomes poorer in solute. The solute-rich precipitates nucleate and grow due to diffusion of solute atoms from the solution to the

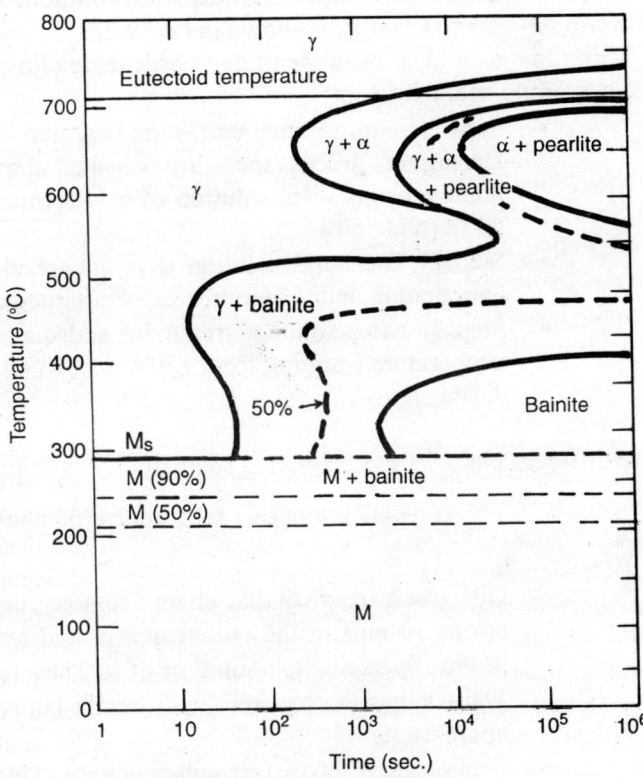

Fig. 9.21 TTT curves for 4340 steel.

precipitate phase. The rejection of the solute-rich phase in the form of the precipitate is favoured since the energy of the system decreases in this process. The concentration of solute in the solution decreases at lower temperature and ultimately the solution will have equilibrium solubility corresponding to that temperature. In certain temperature range near room temperature, the diffusion rate is slow. As a result very fine particles of precipitated phase gets dispersed throughout the medium. This is called dispersion hardening. The increase in strength of the material depends on the following factors:

- Volume fraction of the dispersed phase
- Size of the particles
- Interparticle spacing

The solution is strengthened by this method since the movement of dislocations is restricted due to dispersed precipitate particles. This process is also called aging. There are a wide variety of alloys, which exhibit the precipitation hardening characteristics. The precipitated phases exhibit decreasing solubility with decreasing temperature. There are a large number of alloys which are precipitation hardened to high strength and are being used for commercial purposes e.g. copper based precipitation hardened alloys containing Be are used for nonsparking tools (like wrenches, chisels, hammer etc.). Aluminium-copper alloys are strengthened by precipitation hardening. Precipitation hardening is also observed in certain ceramics like $MgO-Fe_2O_3$ system. Precipitation hardening is also called as age hardening. Hardening occurs at early stages when precipitation starts from a supersaturated solid solution. The process of precipitation hardening of 95.5 w/o Al-4.5 w/o Cu alloy is shown in Fig. 9.22.

This alloy is an ideal example of age-hardenable alloy. The following are the three steps in the age hardening heat treatment:

- Step 1: Solution treatment is the first step in the process of precipitation hardening. During this process the alloy is heated above the solvus temperature and held until homogeneous solid solution of α is formed. The alloy is solution treated between 500°C and 550°C.
- Step 2: The solid solution α is quenched to a lower temperature, usually room temperature and it becomes supersaturated α.
- Step 3: After solution treatment and quenching, the alloy is heated to the aging temperature (ranging from 130°C to 190°C). The alloy is precipitation strengthened during aging .

Five different phases are precipitated sequentially as follows:

$$\alpha \rightarrow GP1 \text{ zones} \rightarrow GP2 \text{ zones } (\theta'' \text{ phase}) \rightarrow \theta' \rightarrow \theta \ (CuAl_2)$$

where

- GP1 zones are small disc-shaped regions (thickness 0.4 nm to 0.6 nm and diameter 8 nm to 10 nm) in the parent matrix and are created by copper atoms segregating in the supersaturated solution of α. They form on the {100} planes of the matrix. These zones are coherent with matrix lattice and can be observed through electron microscope.
- GP2 zones (θ'' phase) are coherent with {100} planes of the alloy. They have smaller thickness as compared to GP1 zones.

- θ' phase nucleates on defects especially on dislocations and is incoherent with the matrix since it has a crystal structure different from that of the matrix and also it consists of disk-shaped regions.
- θ phase is expressed by chemical formula $CuAl_2$. This phase has bct structure and is incoherent with the matrix of the alloy.

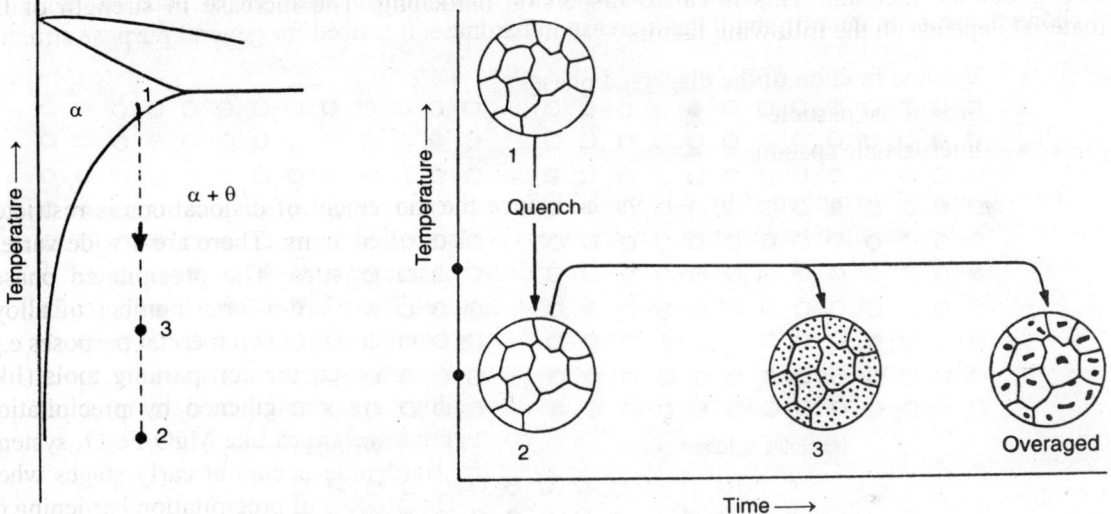

Fig. 9.22 Precipitation hardening of 95.5 w/o Al-4.5 w/o Cu alloy.

Aging process

Aging process at room temperature is called natural aging. Aging at elevated temperatures is called artificial aging. Aging can be reinitiated by heating the solid solution to a temperature where the process will be faster and can be controlled easily. If the rate is too fast, the precipitate agglomerates and the material becomes overaged and soft. The initial precipitate particles provide much of the hardening and strengthening. Just after quenching the solute atoms are randomly distributed at the initial stage, Fig. 9.23. As aging progresses, the precipitate atoms cluster but still maintain lattice coherency with the surrounding crystal. In this case the boundary energy for the particle is low because of coherency but the lattice surrounding the cluster gets distorted and strained. As a result, anchor points are created in the path of dislocation motion. These anchor points lock the dislocation movement. The region around a cluster is strained further. The strained volume surrounding the cluster is sometimes greater than the actual volume of the cluster. As the precipitation progresses with time, clusters become larger and fewer in number. The precipitates become noncoherent. The spacing between the particles also increases which allows the slip to occur more easily and softening of the material occurs. This is called overaging. The phenomena of overaging in iron due to carbide and nitride precipitation are shown in Fig. 9.24.

The overaging at an elevated temperature occurs faster than that at a lower temperature. The combined effect of age-hardening and cold-working gives more strength to the material as compared to that which can be obtained by either process alone.

Some aluminium-copper alloys are used extensively in various structural applications where high strength to weight ratio is needed. Consider the duralumin alloy at temperature 560°C. It exists as a single phase and solid solution. A temperature between 100°C and 200°C is the optimum aging temperature for this alloy. Normally this alloy is aged to the maximum hardness at such temperature and then it would not overage during service at room temperature. Duralumin alloy 2024 has a tensile strength five times larger than that of pure aluminium (about 100 MPa). The heat treated alloy 7075 (5.6w/o Zn, 2.5w/o Mg, 1.6w/o Cu, and 0.23w/o Cr) has a tensile strength of 504 MPa and is mainly used for aircraft. Alloy 6061 (1w/o Mg,0.6w/o Si, 0.27w/o Cu and 0.2w/o Cr) has a tensile strength three times higher than that of pure aluminium and shows decrease in ductility with the increase in hardness. It is used for general purpose structures.

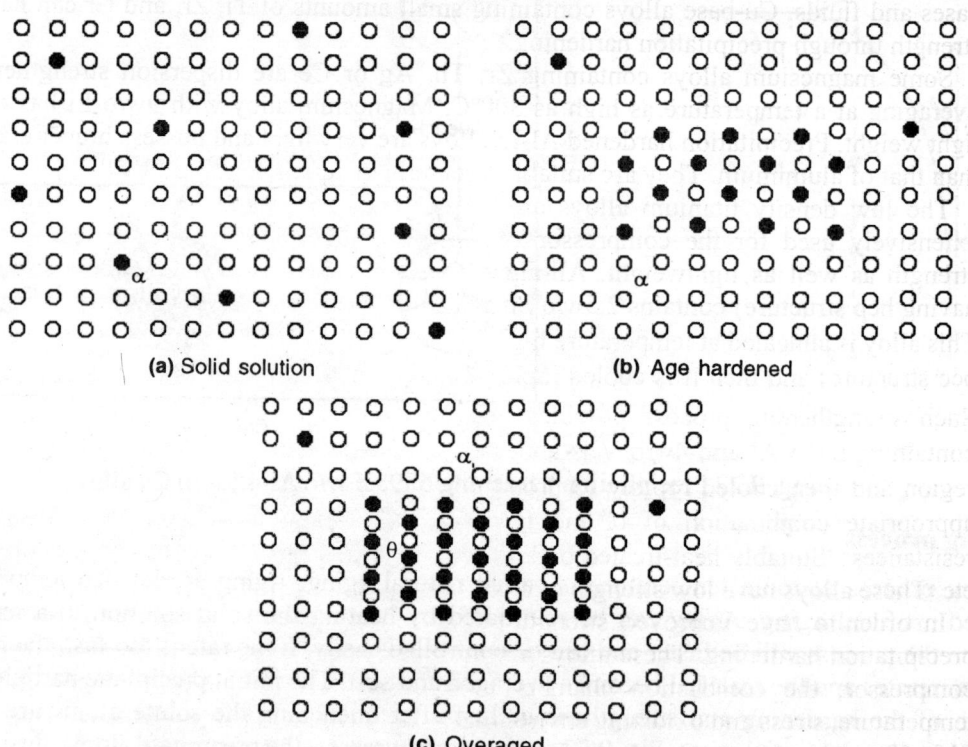

Fig. 9.23 Schematic representation of aging process.

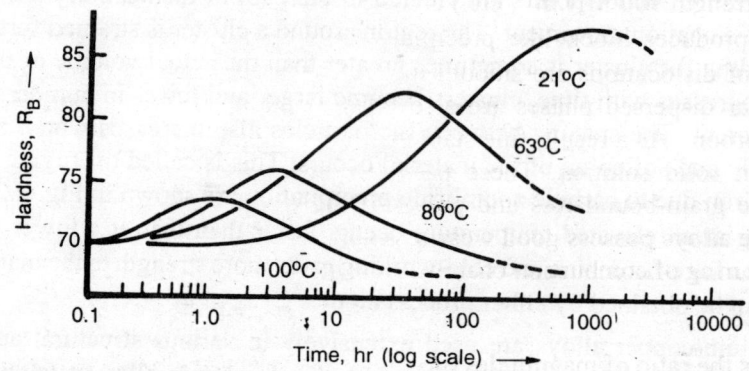

Fig. 9.24 Overaging in iron due to carbide and nitride precipitation

Although precipitation treatment can strengthen an alloy but sometimes quenching can also strengthen the material. Quenching helps in retaining supersaturated solid solution. 5w/o Mg-Al alloy can attain 4 times yield stress through above process as compared to annealed Al. The size and distribution of the precipitate phase obtained by controlled nucleation and growth can be optimized to provide maximum strength. Copper alloy containing 2w/o Be can be hardened to such an extent that its tensile strength becomes 1.38 GNm^{-2} which is three times higher than that of very heavily cold worked pure copper. This alloy is used for spring contacts. Beryllium-copper alloys have nonsparking qualities, which make them useful for tools to be used near flammable gases and fluids. Cu-base alloys containing small amounts of Ti, Zr, and Cr can have improved strength through precipitation hardening.

Some magnesium alloys containing Zr, Th, Ag or Ce are dispersion strengthened to resist overaging at a temperature as high as 300ºC. Magnesium alloy with 9w/o Li has exceptionally light weight. Precipitation hardened Al-Li alloys are very light and possess higher elastic modulus than that of aluminium. They are suitable structural materials for aircraft.

The low density titanium alloys are strengthened through precipitation hardening and are extensively used for the compressor of aircraft engine since they meet the requirements of strength as well as lightweight. All-alpha alloy of α titanium (an allotropic form of titanium having hcp structure) contains 2.5w/o Sn and 5w/o Al which provide solution strengthening to it. This alloy is annealed at temperature higher than 882ºC i.e. in the β phase region (titanium having bcc structure) and then it is cooled rapidly to α region to produce acicular grain structures of α.

Such strengthening process provides good resistance to fatigue and creep. Titanium alloy containing 6w/o Al and 4w/o V is annealed at a temperature corresponding to the $\alpha + \beta$ phase region and then cooled rapidly to produce acicular α grains. Suitable heat treatments produce appropriate combination of α and β phases and such alloys have low fatigue and creep resistances. Suitably heat-treated α-β alloys are used for making airframes, rockets, jet engines etc. These alloys have low strength at high temperatures.

In order to have improved strength at high temperature, super alloy should be used after precipitation hardening. These alloys are used for making the components for tail portion of the compressor, the combustion chamber and the turbine where materials have to face high temperature, stress and oxidizing environment. Certain Ni and Co-based super alloys (e.g. Inconel 718: 53w/o Ni, 19w/o Cr, 19w/o Fe, 5w/o Nb, 3w/o Mo, 1w/o Ti having yield strength 1262 $MN.m^{-2}$; Stellite 6B: 60w/o Co, 30w/o Cr and 4.5w/oW having yield strength 710 $MN.m^{-2}$ and tensile strength 1220 $MN.m^{-2}$; Hastelloy B-2: Ni-28w/o Mo having yield strength 415 $MN.m^{-2}$ and tensile strength 900 $MN.m^{-2}$, etc.) are useful for such applications. Presence of aluminium in such alloys produces block like precipitates of γ' phase $(NiAl_3)$ distributed in the γ phase. Movements of dislocations are smooth in γ phase but are severely hindered due to dispersed γ' phase. Similar dispersed phases are also observed for $NiTi_3$. All these alloys contain a small amount of carbon. As a result, fine stable particles of carbides of W, Cr, Mo etc. are formed and distributed in solid solution. These particles hinder the movement of dislocations. Carbides collect at the grain-boundaries and prevent the deformation of the material under mechanical stress. These alloys possess good wear resistance at high temperatures and are suitable materials for manufacturing of combustion chamber, turbine vanes etc.

EXERCISES

9.1 What is the ratio of magnitudes of volume energy to total free energy at critical condition?

   Ans: 2:1

9.2   In a phase transformation process, the rate of nucleation is $10^6 \, \text{m}^{-3}\text{s}^{-1}$ at 27°C, when the critical nucleation energy is 2.07 x 10^{-19}J. Calculate the value of $(\Delta G^* + \Delta H_d)$ in kJ/mole. Calculate the enthalpy of activation for diffusion. By what factor ΔG^* should be changed to increase the nucleation rate to $10^{10} \, \text{m}^{-3}\text{s}^{-1}$ at the above temperature? If the nucleation rate is increased by a factor 10^4 due to change in the interfacial energy, calculate the percentage change in the interface energy.

Ans: 207 kJ/mole, 82.32 kJ/mole, decrease by a factor 0.814, −6.63%

9.3   For the given TTT diagram of eutectoid steel, Fig.9.25, label all the phases in regions (1) to (7) if these phases appear after quenching from eutectoid temperature and then cooling isothermally. Indicate whether the phases are stable, unstable or metastable.

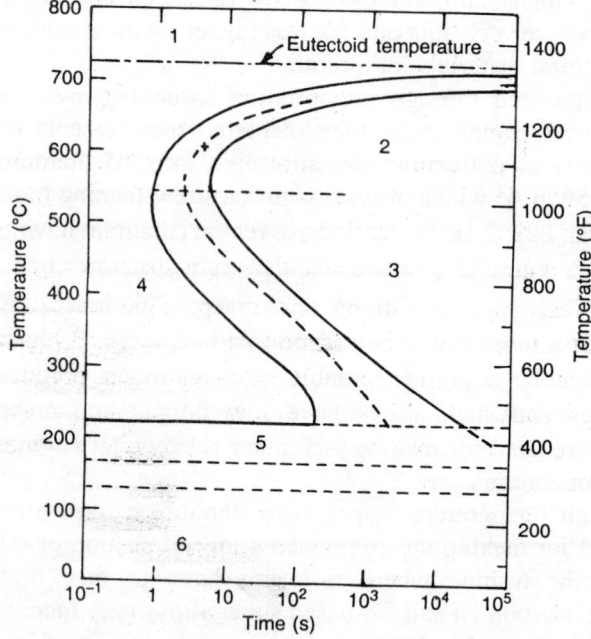

Fig. 9.25 TTT diagram for steel showing regions of stable, unstable and metastable phases.

Ans: (1) γ : stable ; (2) α +Fe₃C : stable; (3) α +Fe₃C : stable; (4) γ : unstable ;
    (5) γ : unstable and martensite : metastable ; (6) martensite : metastable;
    (7) γ : unstable and α + Fe₃C : stable

9.4   Consider the continuous cooling transformation of eutectoid steel, Fig.9.20. State the product(s) that would be obtained at the end of each of the following processes carried out on fully annealed steel at 730°C, cooled at the rate of
(i) 1°C/min for 1 hr, and then quenched to 20°C.
(ii) 2.5°C/min for 1 hr and then quenched to 20°C.
(iii) 20°C/s up to a temperature of 600°C and then quickly quenched to 0°C.
(iv) 150°C/s for 5s, then maintained at a constant temperature for 1 min; then reheated to 300°C for very long time.
Ans: (i) Pearlite; (ii) same as (i); (iii) Martensite with traces of α, Fe₃C and γ;
    (iv) tempered martensite with ferrite and fine particles of carbide are formed within it.

9.5 Referring to Fig. 9.10, mention the simple process steps required for converting
(i) pure bainite into a mixture containing nearly 50w/o bainite and 50w/o martensite,
(ii) coarse pearlite into fine pearlite.
Ans: (i) Austenize at 750°C, quenched to 390°C, kept for 100 s and then water quenched;
(ii) Austenize at 750°C, quenched to 640°C, kept for 5 min and then water quenched

9.6 Consider Al-Cu phase diagram. Draw the microstructure that may develop during cooling of an Al-4w/o Cu alloy from 660°C to room temperature. At room temperature the composition of α is 0.02 w/o Cu. The composition of α after quenching is still 4 w/o Cu.

9.7 Compare the composition of α solid solution in the Al- 4w/o Cu alloy at room temperature when the alloy cools under equilibrium conditions with that when the alloy is quenched.

9.8 For an Fe-0.45w/o carbon alloy, Fig.9.15 determine the
(a) temperature of the start of transformation of austenite on cooling,
(b) composition and amount of each phase present at 724°C,
(c) composition and amount of each phase present at 722°C.
Ans: (a) 800°C; (b) 45w/o α (0.025w/oC) and 55w/o γ (0.8 w/o C);
(c) 93.6w/o α (0.025w/oC) and 6.4 w/o Fe$_3$C (6.67 w/o C);

9.9 Consider the phase transformation when solidification of copper takes place from liquid copper at 980°C. The enthalpy of fusion of copper is 1.88 GJ/m^3. Find the liquid-crystal interfacial energy if the critical radius of the nucleus is 20 Å. Calculate the critical free energy of nucleation. Calculate the supercooling required, so that the critical radius is twice. If there is no supercooling, what should be the critical radius and critical free energy of nucleation?
Ans: 0.14 J/m^2 , 9.02 x 10^{-18}J, −51.5°C, zero critical radius, ∞.

9.10 If the rate of nucleation from liquid phase is 10^9 m^{-3} s^{-1} at room temperature, calculate the value of $(\Delta G^* + \Delta H_d)$ in kJ/mole. When the free energy of critical nucleus is 9.6 × 10^{-20} J, calculate the enthalpy of activation for diffusion ΔH_d. If the supercooling is increased by 27°, what should be the free energy of critical nucleus and the rate of nucleation, where $\Delta h = 0.42 \times 10^9$ J/m^3 and $\gamma = 0.055$ J/m^2, $T_m = 505$ °K?
Ans: 189.6 kJ/mole, 131.9 kJ/mole, 45.1 kJ/mole, 1.36 × 10^8/m^{3}s

9.11 Calculate the angle of contact θ, Fig. 9.6, if $\gamma_{\alpha g} = 0.55$ J/m^2, $\gamma_{\alpha \delta} = 0.05$ J/m^2 and $\gamma_{g \delta} = 0.55$ J/m^2. Hence calculate the ratio $\Delta G^*_{het}/\Delta G^*_{homo}$. At what value of θ, this ratio is one. Find a suitable $\gamma_{\alpha \delta}$, so that the above ratio can be obtained.
Ans: 24.6°, 6.022 × 10^{-3}, 180°, 0

9.12 Consider a strip of eutectoid steel heated for 1 hour at 850°C and then subjected to the following heat treatments as successive steps. Find the phases present at each step.
(a) Quenched to 550°C and held there for 2 s
(b) Quenched to 200°C and held there for 10 s
(c) Quenched to 30°C and held there for one hour.
Ans: (a) $\gamma_{unstable}$, α and Fe$_3$C; (b) $\gamma_{unstable}$, α, Fe$_3$C and Martensite; (c) α, Fe$_3$C and Martensite.

9.13 After austenitizing eutectoid steel at 850°C, it is subjected to the following heat treatments as listed below. Using the isothermal transformation diagram of Fig.9.10, determine the microstructure of the sample after each heat treatment.

(a) Quenched to 150°C.

(b) Quenched to 690°C and held there for 2 hr.

(c) Quenched to 610°C and held there for 3 min.

(d) Quenched to 300° and held there for 30 min.

(e) Quenched to 300° and held there for 5 hr.

Ans : (a) $\gamma_{unstable}$ and Martensite; (b)Coarse pearlite consisting of α and Fe_3C; (c) Coarse pearlite consisting of α and Fe_3C; (d) $\gamma_{unstable}$ and bainite consisting of α and Fe_3C; (e) Bainite consisting of α and Fe_3C.

9.14 Consider a piece of wire of 1045 steel received one of the following heat treatments. Indicate the phases and its microstructure at the end of each sequence (Fig. 9.15).

(a) Heated to 825°C and kept there for 2 hr and then quenched to 100°C.

(b) Heated to 750°C and kept there for 2 hr and then quenched to 100°C.

(c) **Heated to 825°C and kept there for 2 hr and then quenched to 580°C, held there for 3 s, quenched to 250°C.**

(d) Heated to 825°C and kept there for 2 hr and then quenched to 560°C, held there for 20s, heated 750°C and held there for 2 hr.

Ans: (a) Martensite; (b) α and Martensite;(c) $\gamma_{unstable}$, α, Fe_3C and Martensite; (d) α and γ

9.15 A small piece of steel containing 0.45w/o of carbon is heated to 850°C, quenched to –60°C, reheated to 700°C and held there for 5 s. Then the sample is quenched to 20°C. What are the phases present?

9.16 Sketch TTT diagram for 0.2w/o C steel. M_s = 450°C, M_f = 290°C; α and γ are stable between 725°C and 550°C. At 550°C, 1w/o transformation is complete in 0.2 s and 99w/o transformation is finished in 3 s.

9.17 The steel mentioned in Ex. 9.14 is heated to 870°C until equilibrium is attained and quenched in the ways indicated below. Indicate the phase(s) at the end of each sequence.

(i) Quenched to 280°C.

(ii) Quenched to 660°C, held for a long time and then quenched to 350°C.

(iii) Quenched to 600°C, held for a long time, heated to 800°C and then held for a long time.

Ans: (i) $\gamma_{unstable}$ and Martensite; (ii) Coarse pearlite consisting of α and Fe_3C; (iii) First pearlite is formed and then γ is formed.

9.18 The surface energy of a liquid/gas interface is 0.07 J/m² and that for a glass/gas surface is 0.58 J/m². What should the liquid/glass interface energy be for complete wetting?
Ans: 0.51 J/m

9.19 Consider the transformation

$$\alpha \xrightarrow{\text{cooling at } 1025°C} \beta,$$

where the interfacial energy between α and β is 0.5 J/m^2 and the value of Δg_v for α to β is -4×10^8 J/m^3 at 1000°C. Determine the critical nuclear radius and critical nucleation energy at 1000°C.

Ans: 25Å, 1.309 x 10^{-17} J

9.20 Indicate whether the following statements are true or false and justify your answer:
i)   Presence of alloying elements in steel shifts the nose of C-curve towards right.
ii) Austempering is a process used for the formation of pearlite.
Ans: i) True: ii) False (Bainite is formed)

9.21 In a phase transformation, the critical nucleation energy is 2.5×10^{-19} J, when an appreciable rate of nucleation 10^6 m^{-3} s^{-1} occurs at room temperature (27°C). The energy of interface between the product and the parent phase is 0.1 Jm^{-2}. Calculate the nucleation rate, if the interfacial energy is 5% larger than this.

Ans: There is approximately no change in nucleation rate

9.22 In some heterogeneous nucleation the product phase completely wets the surface if
(a) $\Delta G^*_{het} = 0$
(b) $\Delta G^*_{het} = \infty$
(c) $\Delta G^*_{het} = \Delta G^*_{homo}$
a)   None of the above.
Ans: a

9.23 Consider α-phase undergoes a homogeneous transformation to β phase. The free energy difference between β and α phases, when both are unstrained, is Δg_v per unit volume and γ is the surface energy per unit area between two phases. Deduce the expressions for the critical radius and critical free energy if β phase has strain energy ε per unit volume and α phase is unstrained. Assume the nucleus of β to be spherical in shape.

Elastic Behaviour of Materials

10.1 ELASTIC BEHAVIOUR AND RELATION BETWEEN ELASTIC CONSTANTS

When an external force is applied on a material body, the body undergoes changes in shape and size. It may come back to its original shape and size after removal of the applied force. Such behaviour of the material is called elastic behaviour. This behaviour is observed as long as the applied force per unit area is less than or equal to a critical value which corresponds to the elastic limit. For example, a spring shows elastic behaviour when relatively small force is applied to elongate or compress it. However, if the applied tensile force is large enough then initially the spring is extended and the length of the spring remains longer than the original length after the removal of the external force. Therefore we say that the spring has undergone permanent deformation. This is also called plastic deformation. When an external force **F** is applied on a body, an equal and opposite reaction force **F** comes into play within the body. If this reaction force is acting on an area A_o, then the stress is defined as

$$\sigma = F/A_o \qquad (10.1)$$

When the external force is applied along the length of a wire, the tensile stress is developed and extension or contraction of length occurs depending upon the direction of the force. The tensile strain is defined as change in length per unit length of the wire and is given by

$$\varepsilon = \frac{\Delta l}{l_o} \qquad (10.2)$$

where Δl is the change in length and l_o is the initial length of the wire. If the strain is reversible when the stress is removed i.e. if recovery of the material is 100% after removal of stress, the material is said to obey Hooke's Law. According to Hooke's Law strain is proportional to the stress, when the stress is within elastic limit i.e. $\varepsilon \propto \sigma$ or

$$Y = \sigma/\varepsilon \qquad (10.3)$$

where Y is constant of proportionality and is called Young's modulus, Fig. 1.2. If a rod is stretched along the length elastically by force F, then

$$F = k\Delta l, \qquad (10.4)$$

where k is constant of proportionality and is called the tensile stiffness constant. This is an important property of engineering material. One can express Young's Modulus of a rod as:

$$Y = \frac{F/A_o}{\Delta l/l_o} = \frac{k\Delta l/A_o}{\Delta l/l_o} = \frac{k\, l_o}{A_o} \qquad (10.5)$$

Bulk modulus

In case of hydrostatic compression, volume strain or bulk strain $\Delta V/V_o$ is developed, where ΔV is the change in volume and V_o is the initial volume. The bulk modulus κ is defined as,

$$\kappa = \frac{F/A_o}{\Delta V/V_o} = \frac{F\, V_o}{A_o \Delta V} \qquad (10.6)$$

or

$$\kappa = \frac{F\, l_o}{\Delta V} \qquad (10.7)$$

for a bar-shaped object. **Compressibility** is reciprocal of bulk modulus.

Rigidity modulus

The shear stress is developed when tangential force **F'** acts on the surface of area A_o of the body and as a result of which shape of the body is changed (Fig. 10.1). The change in the shape of the body is measured by an angle θ, which is the change in angle between the two lines, which were initially perpendicular. The shear strain is given by

$$\gamma = \tan\theta \approx \theta, \qquad (10.8)$$

when θ is very small. The shear stress τ is given by F'/A_o. Hence the rigidity modulus μ is expressed as,

$$\mu = \tau/\gamma \qquad (10.9)$$

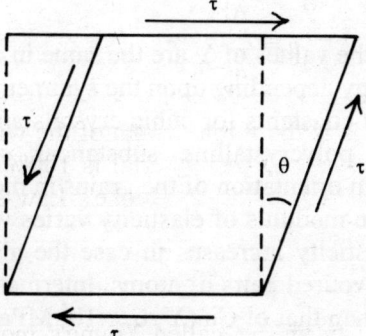

Fig. 10.1 Representation of shear strain $\gamma = \tan\theta$

When tensile stress is present in a body the shear stress is also observed in it. For example, consider a long cylindrical rod of cross sectional area A_o being acted upon by a tensile force **F** parallel to its axis, Fig. 10.2.

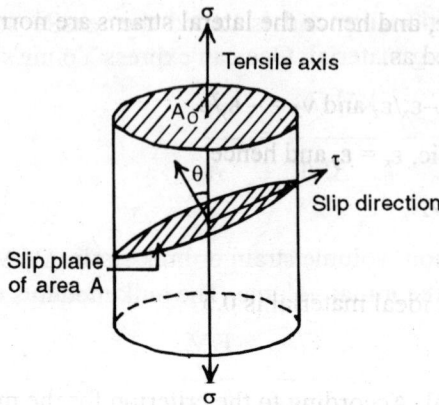

Fig. 10.2 Representation of shear stress and tensile stress in a body acted upon by tensile force

Consider the stresses acting on a slip plane of area A whose normal makes an angle θ with the axis of the rod. Let us resolve the force **F** along the normal and parallel to the plane, which are given by $F_n = F \cos\theta$ and $F_t = F \sin\theta$ respectively. Hence

$$\sigma = \frac{F \cos\theta}{A} = \frac{F \cos^2\theta}{A_o} \qquad (10.10)$$

$$\tau = \frac{F \sin\theta}{A} = \frac{F (\sin\theta \cos\theta)}{A_o} = \frac{F \sin2\theta}{2A_o} \qquad (10.11)$$

where σ is tensile stress and τ is shear stress. Hence,

$$Y = \sigma/\varepsilon = (\sigma/\Delta l)\, l \qquad (10.12)$$

and

$$\mu = \frac{\tau}{\gamma} \approx \frac{\tau}{\theta} \qquad (10.13)$$

A material is elastically isotropic if the values of Y are the same in all directions of the crystal. Usually a single crystal shows anisotropy depending upon the symmetry of the crystal system. For example, there are at least three elastic constants for cubic crystals and 21 constants for triclinic crystals. However quite often the polycrystalline substances show elastically isotropic characteristics because of nearly random orientation of the grains in most polycrystalline solids.

In case of an ideal solid solution, the modulus of elasticity varies with the atomic fractions of elements present. The modulus of elasticity increases in case the alloy is ordered, since more energy is required to break bonds of favoured pairs of atoms. Intermetallic compound Cu_3Au has Y value 7.6×10^4 MPa which is higher than that of Cu (Y=6.9×10^4 MPa) and of Au (Y= 4.14×10^4 MPa).

Poisson's ratio

Under the tensile force, a material in the form of rod, wire etc., extends mainly longitudinally and this results in a decrease in the lateral dimensions. If the longitudinal strain ε_z is positive, then the lateral strains ε_x and ε_y are negative, and hence the lateral strains are normally expressed as $-\varepsilon_x$, $-\varepsilon_y$. The Poisson's ratios are defined as,

$$v_1 = -\varepsilon_x/\varepsilon_z \text{ and } v_2 = -\varepsilon_y/\varepsilon_z \qquad (10.14)$$

If the material is elastically isotropic, $\varepsilon_x = \varepsilon_y$ and hence

$$v_1 = v_2 \qquad (10.15)$$

Example 10.1

Prove that the Poisson's ratio of an ideal material is 0.5

Solution:

Consider a cube of an ideal material. According to the criterion for the material to be ideal,

$$V_o = V_f$$

where V_o is the initial volume of the cube before application of tensile force along z-direction and V_f is the final volume. Hence

$$V_o = l_o^3$$

$$V_f = (l_o + \Delta l_z)(l_o + \Delta l_x)(l_o + \Delta l_y)$$

where Δl_z, Δl_y, Δl_x are the extensions of the sides of the cube in z, y and x directions, respectively. Now assuming $\Delta l_x = \Delta l_y$ we get:

$$V_f = l_o^3 + 2 l_o^2 \Delta l_x + l_o^2 \Delta l_z + \text{terms containing higher powers of } \Delta l_x \text{ and } \Delta l_z.$$

Using $V_o = V_f$ we get:

$$l_o^3 \approx l_o^3 + 2 l_o^2 \Delta l_x + l_o^2 \Delta l_z$$

Neglecting terms containing higher powers of Δl_z, Δl_x etc., we finally obtain,

$$0 = 2 (\Delta l_x/l_o) + (\Delta l_z/l_o)$$

i.e. $\qquad\qquad 2\varepsilon_x = -\varepsilon_z$

Hence $\qquad\qquad v_1 = -\varepsilon_x/\varepsilon_z = 0.5$

Thus, four elastic constants Y, κ, μ and v are defined but out of these, only two are independent for an isotropic material since there are two relations which express μ and κ in terms of other two constants as given below,

$$\mu = \frac{Y}{2(1+v)} \qquad (10.16)$$

and $\qquad\qquad \kappa = \frac{Y}{3(1-2v)} \qquad (10.17)$

10.2 ELASTIC ENERGY DENSITY

When a force **F** is applied parallel to the axis of a rod, taken to be x-axis, the rod is extended in length by x (say). Then the work done by the force **F** is expressed as.

$$W = \int_0^x \mathbf{F} \cdot dx = -\frac{1}{2} kx^2 = -\frac{1}{2} Y(A_0/l_0) x^2 \qquad (10.18)$$

which is the elastic potential energy stored in the stretched rod. The magnitude of elastic energy density U is given by

$$U = \frac{W}{V_0} = \frac{W}{A_0 l_0} = \frac{\frac{1}{2} Y(A_0/l_0) x^2}{A_0 l_0} = \frac{1}{2} Y(x/l_0)^2 = \frac{1}{2} Y \varepsilon^2 = \sigma^2/2Y = \frac{1}{2}\sigma\varepsilon \qquad (10.19)$$

10.3 ELASTIC MODULI AND ATOMIC BONDING

The elastic moduli are directly dependent on the type of bonding and bond energy. Materials with deep and narrow energy troughs have high elastic modulus, Fig. 10.3(a) whereas that with shallow and nonsymmetrical energy trough has low elastic modulus, Fig. 10.3(b). The potential energy of bonding atoms is a function of interatomic distance (r) and is given by

$$W = -\frac{A}{r^n} + \frac{B}{r^m} \qquad (10.20)$$

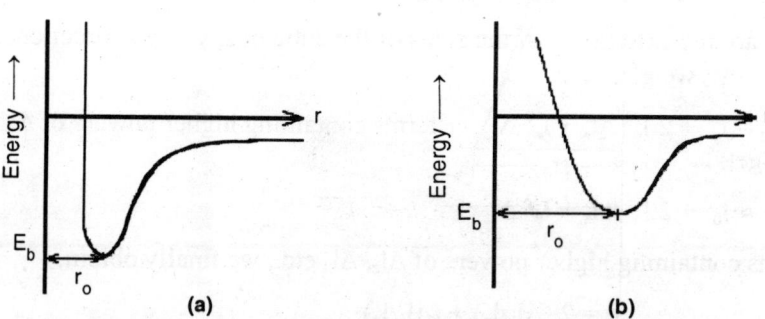

Fig. 10.3 Potential energy versus interatomic distance

where A, B, n and m are constants for a specific material and m >n. We know that the interatomic force is given by

$$F = -\frac{dW}{dr} = -\frac{nA}{r^{n+1}} + \frac{mB}{r^{m+1}} \qquad (10.21)$$

The equilibrium state is the state with minimum potential energy i.e. zero interatomic force. At equilibrium, interatomic distance r_0 is calculated by equating dW/dr to zero. Thus one can find out

$$r_0 = (mB /nA)^{(1/(m-n))} \qquad (10.22)$$

When a compressive force **F'** is applied, the interatomic distance is decreased to r'; whereas the tensile force **F"** causes increase in interatomic distance from r_0 to r". Young's modulus is directly proportional to the slope of the curve at r_0, Fig 10.4. To derive this relation, consider dF as the tensile force applied on an elemental area da of the body at equilibrium. Assuming $da \approx r_0^2$ and dr equal to increase in interatomic distance, we obtain

$$Y = - \left. \frac{dF/r_0^2}{dr/r_0} \right|_{r=r_0} = - \left. \frac{1}{r_0} \frac{dF}{dr} \right|_{r=r_0} = \left. \frac{1}{r_0} \frac{d^2W}{dr^2} \right|_{r=r_0} \qquad (10.23)$$

Using Eq.(10.21)

$$Y = - \frac{1}{r_0^3} \left[\frac{n(n+1)A}{r_0^n} - \frac{m(m+1)B}{r_0^m} \right] \qquad (10.24)$$

Material with strong atomic bonding has narrow and deep potential well, which gives rise to large values of elastic moduli. Similarly, shallow potential well results in small values for elastic moduli. Table 10.1 shows the values of elastic moduli and type of bonding of various materials.

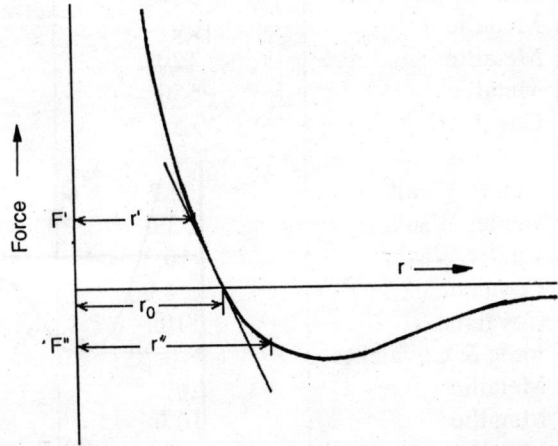

Fig. 10.4 Interatomic force versus interatomic distance

Example 10.2

Estimate Young's modulus of a material, which has bonding characteristics given by Eq.(10.20), where n=1, m=8, A=7.7×10^{-29} J m and r_0=2 Å.

Solution:

Using Eq.(10.22)

$$B = \frac{A \times r_0^7}{8} = 1.23 \times 10^{-97} \text{ J m}^8$$

Using Eq.(10.24)

$$Y = - \frac{1}{r_0^3} \left[\frac{2A}{r_0} - \frac{72B}{r_0^8} \right] = \frac{7A}{r_0^4} = 337 \text{ GNm}^{-2}$$

Table 10.1 Elastic moduli and type of bonding of certain materials

Metal/Alloy/ Compound	Bonding	Young's modulus GN/m^2	Shear Modulus GN/m^2	Poisson's ratio
Alumina (Al$_2$O$_3$)	Ionic & Covalent	400	--	0.23
Aluminium	Metallic	69	26	0.33
Aluminium alloy 2014	Metallic	72	26	0.33
α Iron	Metallic	207	83	0.2972
Brass	Metallic	101	37	0.35
Bronze	Metallic	101	--	--
Cast iron(gray)	Metallic	83–170	--	--
Cast iron (white)	Metallic	207	--	--
Carbon (diamond)	Covalent	1140	--	--
Copper	Metallic	110	46	0.35
Magnesium	Metallic	45	17	0.29
Magnesia (MgO)	Ionic & Covalent	310	--	0.19
Nickel	Metallic	207	76	0.31
Niobium	Metallic	120	--	--
Osmium	Metallic	550	--	--
Phenol formaldehyde (Bakelite)	Covalent	5	--	0.3
Polyethylene	vander Waals	0.2	--	0.4
Polystyrene	vander Waals	3.0	--	0.33
Rubber (Natural)	vander Waals	10.3	--	0.49
Rubber (Hard)	Covalent	4.0	--	0.39
Silicon	Covalent	110	--	--
Silica Glass	Ionic & Covalent	0.70	--	0.2
Steel	Metallic	207	83	0.27
Titanium	Metallic	107	45	0.36
Tungsten	Metallic	407	160	0.28

10.4 TENSION TEST

The tension test is one of the most common mechanical stress-strain tests. In this test, a standard tensile specimen is deformed gradually by increasing uniaxial longitudinal force perpendicular to its cross-section, which may be circular or rectangular, Fig. 10.5. This longitudinal force causes elongation of the specimen until it fractures. The tensile testing machine elongates the specimen at a constant rate and measures the instantaneous applied load and elongation simultaneously. Stress-strain test is destructive since in this test the specimen is usually fractured. Measurements are recorded on a strip chart. Elongation is dependent on load as well as on the cross-sectional area. If the cross-sectional area is doubled, the load has to be twice to achieve the same elongation and that is why the stress (not load) is plotted against elongation. Similarly, longer the specimen more is the elongation under the same stress condition.

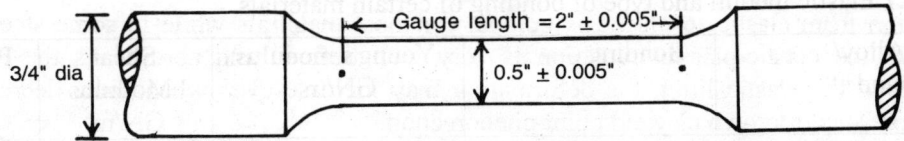

Fig. 10.5 A specimen for a standard tensile test

10.5 COMPRESSION AND SHEAR TESTS

Compression stress-strain tests may be conducted by a similar method except that the force is compressive and the specimen is compressed along the direction of stress. During this process shear stress and strain may also develop. Compressive stress is considered as negative and the corresponding strain is also negative since the length of the specimen decreases under compressive stress. The test for shear stress and strain is conducted by application of tangential force F' along faces of the specimen as shown in Fig. 10.1. The shear stress can be measured as

$$\tau = \frac{F'}{A_o} \qquad (10.25)$$

and shear strain is tangent of the strain angle θ.

10.6 NONLINEAR ELASTIC PROPERTIES

Most of the metallic substances show elastic behaviour up to elastic limit. When stress increases beyond elastic limit, the linear relationship between stress and strain is no longer valid. The value of elastic limit varies from one material to another and is also called yield point or yielding stress, Y_S. When stress exceeds Y_S, the material undergoes a nonreversible or permanent plastic deformation. It is difficult to find the exact elastic limit, so a convention has been adopted according to which a straight line can be drawn parallel to the elastic portion of the stress-strain curve at some specified strain offset (see Fig. 1.2), usually 0.2%. The stress corresponding to the point of intersection of the straight line and stress-strain curve is normally known as yielding strength. In other words, the stress, which causes a finite permanent strain 0.2%, is called yield strength. The yield strength is a measure of resistance offered by a material to plastic deformation.

When the stress is increased, after yielding has occurred, the plastic deformation increases until the stress reaches a maximum value σ_M. Beyond this point, the stress decreases while plastic deformation increases until fracture occurs. The maximum stress that can be tolerated by a material structure in tension is called tensile strength. If this stress is maintained, the fracture results. Fracture is the separation of a specimen under stress into two or more parts. Certain materials may undergo extensive plastic deformation before fracture since crack propagation is slow in such materials. This type of fracture is called ductile fracture. In contrast there are other materials (some metals, oxides and alloys) where cracks propagate rapidly and fractures occur with very little plastic deformation. Such type of fracture is called brittle fracture. Brittle fracture proceeds along specific crystallographic planes, called cleavage planes, when stress is perpendicular to these plane e.g. high stress normal to (0001) plane of zinc single crystal (hcp) causes brittle fracture. Similarly, many bcc crystals (e.g. Mo, α-Fe etc.) fracture in a brittle manner at low temperature under high strain rates. The stress at which fracture occurs is called breaking strength at fracture. This is also called fracture or rupture strength.

The transition from elastic to plastic is gradual for most materials while in some steels and other materials the elastic-plastic transition is very well defined and abrupt. As the load is increased beyond the elastic limit, the deformation may increase even when some decrease in loading is there. This is known as yield point phenomenon.

In some materials a kink is observed in the stress-strain curve, Fig. 10.6. For such materials, there are two yield points: an upper yield point, which is a point in the stress-strain curve, where plastic deformation starts with an actual decrease in stress and a lower yield point, from where the deformation continues at a constant stress. The yield strength for these materials is taken as the stress corresponding to lower yield point.

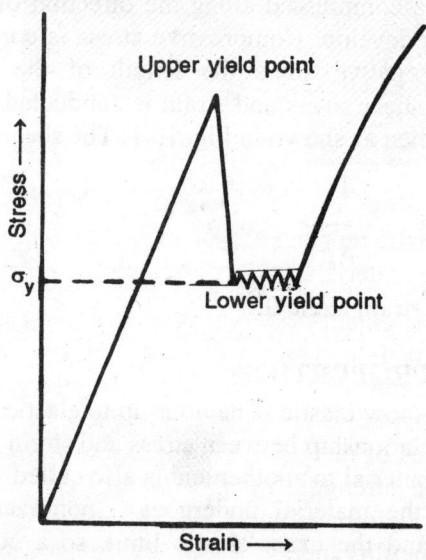

Fig. 10.6 Schematic representation of yield point phenomena in stress-strain curve.

Ductility and malleability

Most engineering materials undergo plastic deformation in service. Ductility is the measure of the degree of plastic deformation the material has undergone by stress in tension before fracture. A material that experiences very little or no permanent deformation before fracture is called brittle. A brittle material will have a very low value of ductility as compared to the ductile material. The fracture can occur at a very low value of strain in brittle material whereas in a ductile material fracture occurs when the strain is substantially high, Fig. 10.7. Ductility is expressed quantitatively as either percentage elongation or reduction in the original area A_o when the specimen is under tension i.e.

$$\text{Ductility} = (\Delta A / A_o) \times 100\%$$ (10.26)

where reduction in area ΔA is calculated from the minimum diameter of the neck, Fig. 10.5. For thicker and flat rectangular specimen under tension, the area of after fracture is approximated as

$$A = h(d_c + 2d_e)/3$$ (10.27)

where h is the width of the specimen, d_c the thickness at the center of the specimen and d_e thickness at the end of the specimen. Similarly, percentage elongation is given by $(\Delta l/l_0) \times 100\%$.

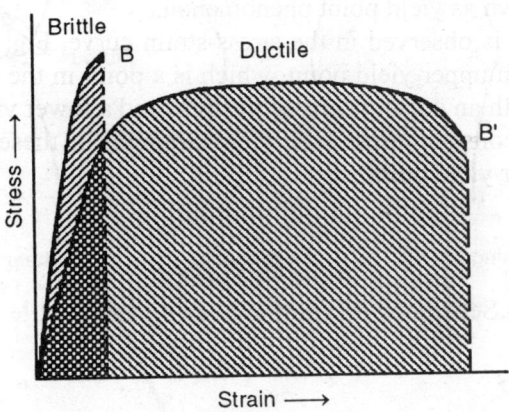

Fig. 10.7 Schematic representation of stress-strain curves for brittle and ductile materials loaded to fracture.

Malleability is defined as the extent of permanent deformation of material when subjected to rolling or hammering. Ductility is distinguished from malleability since while the former is a tensile property, the later is a compressive property. A brittle material cannot be mechanically worked to various sizes and shapes without fracture or breaking. Thus, we conclude that several important mechanical properties of metals may be derived from stress-strain test in tension. The properties like ductility, yield strength and tensile strength are sensitive to the changes of temperature, existence of prior deformation and the presence of impurities. The yield strength and tensile strength decrease but ductility increases with the increase of temperature.

When the applied stress exceeds the tensile strength of the material, elongation may not remain uniform along the direction of tension in the specimen. The percentage elongation is more at certain portions of the specimen as compared to that in the other regions. The former portion is called neck, Fig.10.8. When a specimen forms a neck, cavities are formed within the necked region. These cavities coalesce to form a crack at the center of necked region. The crack propagates to the surface in the direction normal to the direction of the stress and results in a cup and cone fracture.

Resilience

A material absorbs energy during elastic deformation and gets rid of it when unloaded. This capability of material is called resilience. The modulus of resilience is the strain energy per unit volume required to stress the original material up to the yield point i.e. as per Eq. (10.19) the modulus of resilience is given by

$$U_R = \frac{1}{2}\,\sigma_o\varepsilon_o = \sigma_o^2/2Y \qquad\qquad (10.28)$$

where σ_o, ε_o are the stress and strain at yield point, respectively. For example an ideal mechanical spring should have high yield stress and a low modulus of elasticity so that it can have high value of modulus of resilience.

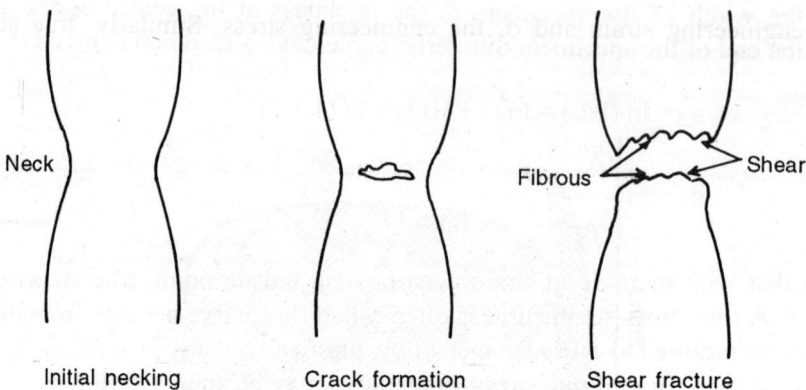

Fig. 10.8 Schematic diagram of necking and ductile fracture.

Toughness

Toughness is one of the important properties of a material that determines its suitability for specific applications in which the material has to face a lot of wear and tear. It is the ability of a material to absorb energy up to the point of fracture. It is usually represented by the area under the stress-strain curve up to the point of fracture and quantitatively it is the energy absorbed per unit volume of the material up to the fracture point. In SI units toughness is expressed as J/m^3 (or Pascal). Tough materials show both strength and ductility and usually ductile materials are tougher than brittle ones which is demonstrated in Fig.10.7. But the brittle materials usually have high yield and tensile strengths.

10.7 TRUE STRESS-STRAIN BEHAVIOUR

The engineering stress-strain curve shows there is a maximum value of stress beyond which the deformation increases but stress decreases i.e. the material becomes weaker. But this is not the true representation of the strength of material. In fact the cross-sectional area of the specimen under tensile stress decreases rapidly as the length of the specimen increases. This reduces the load bearing capacity of the material when the stress is calculated; the area of cross-section is normally taken as the original area. While calculating stress, if the load F is divided by the instantaneous cross-sectional area A_i we get the value of true stress as,

$$\sigma_T = F/A_i \tag{10.29}$$

To calculate true strain (ε_T), we first find the instantaneous strain $d\varepsilon = dl/l$ and hence

$$\varepsilon_T = \int_{l_o}^{l_i} dl/l = \ln (l_i/l_o) \tag{10.30}$$

where l_i is the instantaneous value of length and l_o is the original length. Assuming the material to be ideal one (whose volume does not change during application of stress) we get,

$$A_i\, l_i = A_o\, l_o$$

$$\sigma_T = F/A_i = (F/A_o)\,(l_i/l_o) = \sigma\,(l_o+\Delta l)/l_o = \sigma\,(1 + \varepsilon) \tag{10.31}$$

where ε is the engineering strain and σ, the engineering stress. Similarly, true strain can be expressed as

$$\varepsilon_T = \ln (l_i/l_o) = \ln (l_o + \Delta l)/l_o = \ln (1 + \varepsilon) \qquad (10.32a)$$

also

$$\varepsilon_T = \ln \frac{A_o}{A_i} \qquad (10.32b)$$

Thus we find that with increase in strain even beyond tensile point, true stress continues to increase, Fig. 10.9. A true stress-strain curve is often called flow curve because from this curve one can find out the stress required to cause the metal flow plastically to any give strain. The behaviour of the material in plastic region of true stress-strain curve may be approximated by

$$\sigma_T = k\varepsilon_T{}^n \qquad (10.33)$$

where the parameters n and k vary from material to material and n is called strain-hardening exponent.

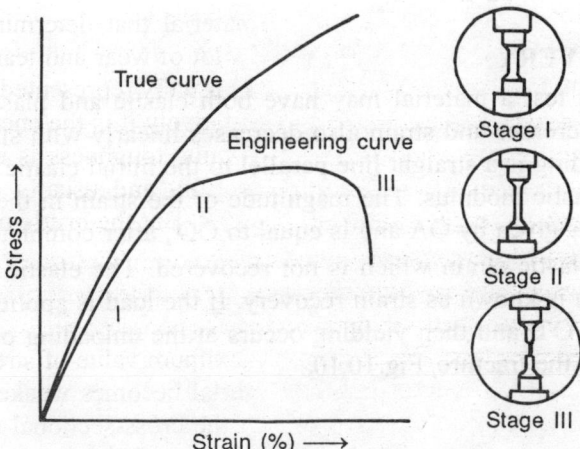

Fig. 10.9 True stress-strain curve in comparison with engineering stress-strain curve.

Example 10.3

Consider an alloy in which stress of 375 MPa produces a strain of 0.105. Find the value of the strain-hardening exponent if k value for the material is 1035 MPa.

Solution:

Using Eq. (10.33) one can express

$$n = \frac{\ln \sigma_T - \ln k}{\ln \varepsilon_T}$$

where $\sigma_T = \sigma(1+\varepsilon) = 375 (1 + 0.105) = 414 \text{MPa}$ and $\varepsilon_T = \ln (1 + \varepsilon) = \ln (1 + 0.105) = 0.1$.
Hence

$$n = \ln(414/1035)/\ln 0.1 = \ln 0.4/\ln 0.1 = 0.4$$

Example 10.4

When a cylindrical specimen of steel having original cross-sectional area as 1.29 cm^2 is tensile tested to fracture, it is found to have σ_f (fracture strength) equal to 460 MPa. If the ductility of the material is 30%, calculate the true stress at fracture.

Solution:

Using Eq. (10.26) we get:

$$\text{Ductility} = \Delta A / A_o \times 100\% = 30\%$$

i.e. $\quad\quad \Delta A = 0.3\, A_o = 0.3 \times 1.29 = 0.39 \text{ cm}^2$

and $\quad\quad A_f = A_o - \Delta A = 0.9 \text{ cm}^2$

Fracture strength F is given as:

$$F = \sigma_f\, A_o = 460 \times 10^6 \text{ Pa} \times 1.29 \times 10^{-4} \text{ m}^2 = 5.93 \times 10^4$$

True stress at fracture

$$\sigma_T = \frac{F}{A_f} = \frac{5.93 \times 10^4}{0.9 \times 10^{-4}} = 659 \text{ MPa}$$

10.8 ELASTIC RECOVERY

During the stress-strain test a material may have both elastic and plastic deformations. While unloading, the stress decreases and strain also decreases linearly with stress i.e. the stress-strain curve BO' during unloading is a straight line parallel to the initial elastic portion of the curve. So the slope is equal to elastic modulus. The magnitude of the strain in the material at the starting point of unloading (B) is given by OA and is equal to OO', after complete removal of stress. The residual strain OO' is plastic strain which is not recovered. The elastic strain recovered during unloading is O'A, which is known as strain recovery. If the load is applied again the stress-strain curve is again given by O'B and then yielding occurs at the unloading point B. There is also an elastic strain recovery at the fracture, Fig. 10.10.

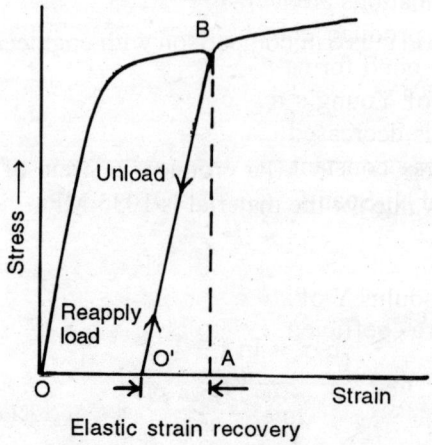

Fig. 10.10 Schematic tensile stress-strain diagram showing elastic recovery during unloading.

10.9 VARIATION IN ELASTIC MODULI WITH CRYSTALLOGRAPHIC DIRECTIONS

The bond length as well as bond energy vary with different crystallographic directions. Hence the Young's modulus varies with crystallographic directions. In case of polycrystalline substances the elastic modulus is a weighted average of anisotropic moduli, Table 10.2. If a material has preferred orientations for example cold rolled copper sheet one should not use average value of moduli. In case of coat hanger (a cold drawn thick steel wire), which has got preferred orientation after cold working, the Young's modulus along the length of the hanger wire is approximately 24.2 x 10^4 MPa. The mean Young's modulus is 20.7 x 10^4 MPa without any preferred orientation.

Table 10.2   Variation of Young's moduli

Material	$Y_{<111>}$ 10^4MPa	$Y_{<100>}$ 10^4MPa	$Y_{polycrystalline}$ 10^4MPa
Al	7.6	6.4	6.9
Au	11.0	4.1	8.3
Cu	19.2	6.7	11.1
Diamond	120.0	6.7	114.0
Fe (bcc)	28.3	13.1	20.7
MgO	33.6	24.5	31.0
NaCl	4.4	3.2	3.7
Nb	15.2	8.1	11.7
Pb	2.8	0.7	1.4
W	41.0	41.0	41.0

10.10 VARIATION OF ELASTIC MODULI WITH TEMPERATURE

Elastic moduli decrease at higher temperatures since thermal energy helps to overcome the bonding forces. Exceptions occur with rubbers, other elastomers and thermoset plastics. In elastomers, kinked conformations are increased at higher temperatures. So extra stress is required to uncoil elastomers and produce elastic strains. Thermoset plastics become harder at higher temperature due to more   bond formation. Hence more stress is required to produce strains. The change in the magnitude of Young's modulus is more or less linear up to 100°C. For example, Young's modulus of iron is decreased by 5.5 GNm^{-2} between 0° to 100°C, Fig. 10.11 whereas Y values remain more or less constant over this temperature range for certain alloys like invar (64 w/o Fe, 36 w/o Ni) and Ni-Span-C which have low temperature coefficient.

Example 10.5

Given that the Young's modulus Y of aluminium to be 6.9 x 10^4 MPa at 20°C, find the value of Y at 100°C, if the temperature coefficient of Y for Al is –30 MPa/°C.

Solution:

$$\Delta T = 80°C$$

Change in Y = 80 x (–3.0 x 10^7) Pa = – 240 x 10^7 Pa = – 2.4 x 10^3 MPa

$$Y_{100°C} = – 2.4 x 10^9 \text{ Pa} + 6.9 x 10^{10} \text{ Pa} = 6.66 x 10^4 \text{ MPa}$$

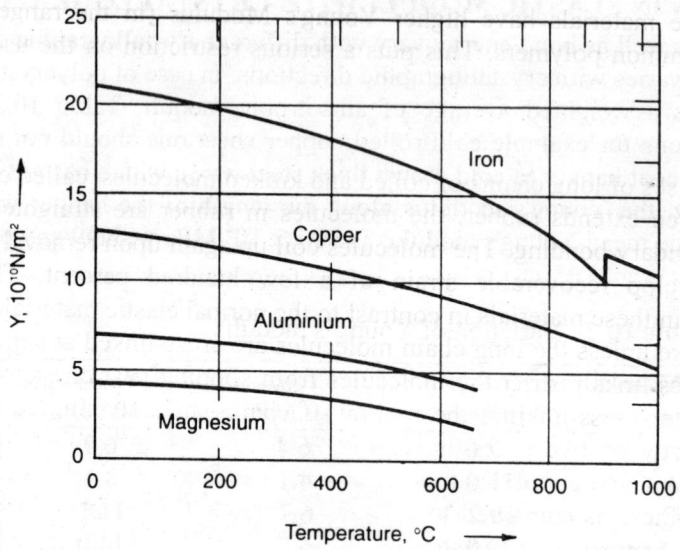

Fig. 10.11 Effect of temperature on the value of Young's modulus for certain metals.

10.11 VARIATION OF ELASTIC MODULI WITH COMPOSITION

There are variations in Young's Modulus if pure metal A is converted to solid solution by alloying it with another metal B. The modulus of elasticity varies linearly with atomic fraction in case of ideal solid solution. Suitable material can be developed by taking proper proportions of components such that Y attains a value suitable for specific applications. When the solid solution of two metals is ideal, ordering can occur, and under these circumstances elastic modulus is increased. This is due to the fact that additional energy is needed to break the bonds of favoured pairs of atoms. Since the intermetallic compounds are usually highly ordered solid solutions, they have higher values of elastic moduli as compared to those of component metals. Young's modulus is not very structure sensitive property but it can be increased by increasing the proportion of higher modulus component.

10.12 ELASTCITY OF POLYMERS

The relationship between bonding and modulus so far described is not strictly followed in polymers because of the existence of mixed bonding. Covalent bonding is present in monomers and also in polymeric chains but the polymeric chains are mostly bonded to each other by secondary type of bonding governed by van der Waals forces or by hydrogen bonding. The values of Young's modulus are quite low for polymers e.g. it is 0.2GPa for unbranched polyethylene. Young's modulus for polymer is dependent on the following factors:

- Nature of secondary bonding
- Presence of bulky side groups
- Branching in chains
- Cross-linking

The elastic response of a three-dimensional covalently bonded network polymers like bakelite or glassy polymers like polystyrene, can be described in terms of bonding since applied stress causes stretching of molecular and atomic bonds. The fully cross-linked rubber also shows similar response. Hence these materials have higher Young's Modulus (in the range of 3–5GPa) as compared to other common polymers. This puts a serious restriction on the use of polymers as structural components.

Rubber elasticity

Natural rubber comprises of long chains of coiled and kinked molecules, called elastomers. When an applied tensile force extends rubber, the molecules in rubber are straightened and aligned. They try to form secondary bonding. The molecules coil up again upon removal of applied force. Thus elastomers develop recoverable strain of a few hundred percent. The stress is not proportional to strain in these materials in contrast to the normal elastic materials. The elastomers are normally liquid like unless the long chain molecules are cross-linked at some points between the chains. These cross-links restrict the molecules from slipping past one another permanently during stretching. After cross-linking, the motion of chains are only limited to the motion of molecular segment between two cross-link points. The response of rubber to stress is almost independent of time to a first approximation. The nonlinear nature of rubber elasticity requires a new definition of stiffness as compared to the elastic moduli for linearly elastic materials. The stiffness is commonly defined as tensile stress to produce 300% percent elongation for rubber like elastic material. The interesting and distinct feature of rubber elasticity in contrast to other elastic material is the increase in elastic modulus with the increase of temperature.

In the rubbery material, the segments between cross-link points are flexible and can be easily straightened out upon application of stress. A large number of distinguishable configurations are possible at a particular temperature in the absence of stress. Hence entropy of the system is large but free energy is low

$$F = U - TS \qquad (10.34)$$

As the stress is applied, the coiled molecules are straightened. The stretching of flexible segments reduces the number of configurations and hence lowers the configurational entropy. Fully stretched natural rubber has nearly zero configurational entropy (stretching does not change bond length or bond angle). If f is the stretching force, which causes the change in length dl in the direction of the force then, the energy absorbed by the polymeric substance is

$$fdl = dU - TdS \qquad (10.35)$$

Let us assume the process to be very slow i.e. isothermal process, dU = 0. Hence there is no change in relative arrangement of atoms or ions. Hence

$$fdl = - TdS \qquad (10.36)$$

$$f = - T(dS/ dl)_T \qquad (10.37)$$

Also the entropy (S) of the system is related to thermodynamical probability (W) as given by Boltzmann's law,

$$S = k \ln W, \qquad (10.38)$$

where k is the Boltzmann's constant. In order to find out the number of configurations that a chain may assume or the thermodynamic probability, one can proceed as follows: Let the end-to-end distance of a coiled elastomer be h. The chain is maximally coiled then h = 0 and h is equal to the hydrodynamic length (L) of the chain, if it is maximally straightened out without disrupting the bond angle and bond length, Fig. 10.12. If the polymeric chain is of large size, then there is no correlation between orientations of mer units, which are remote from one another. Thus, one can consider that the real chain is divided up into a number of independent random sections, say m, each of length A. So L = m A. The number of conformations that a chain can assume, is given by

$$W(h)dh = (3m/(2\pi L^2))^{3/2} \exp((-3m\ h^2)/2L^2)\ 4\pi\ h^2\ dh \qquad (10.39)$$

Using this relation one can calculate

$$\sigma = \rho RT/M_n\ (\lambda - 1/\lambda^2) \qquad (10.40)$$

where $\lambda = L'/L$ = relative elongation, L′ being the length of the deformed polymeric chain.

ρ = polymer density

M_n = molecular mass of segment between cross-link points

and $Y = 3RT\rho/M_n$

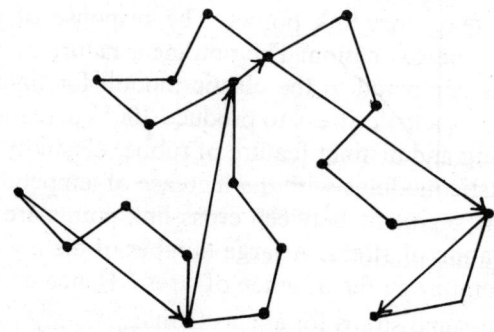

Fig. 10.12 Marking off independent chain segments

The phenomena of rubber-like elasticity of real rubbers are quite complex and can be described as follows, Fig. 10.13:

1. Deformation of real rubber involves a change in its volume i.e. the mean distance between the chains changes. Hence there is change in the entropy and internal energy.

2. Rubber like deformation may be observed only in a certain range of stresses. After application of stress, the flexible molecular chains of elastomers straighten out, such that there is possibility to form bonds and hence, to form crystalline substances. As a result, the flexibilities of molecular chains decrease and the modulus of elasticity increases sharply.

3. If the stress is increased so much that the relative displacement of straightened chains occurs, the relative displacement is called flow.

4. Another important property of real rubber is the nonequilibrium nature of deformation.

10.13 ANELASTIC BEHAVIOUR

When certain material is subjected to a constant stress below the elastic limit, the strain developed initially would increase gradually till it reaches a constant value. This is because it takes some time for atoms or molecules of certain materials to relocate them within the total structure. The reasons for such behaviour may be attributed to size of molecule, bond energy and the structure of material. Although the strain decreases after the removal of applied stress but certain amount of strain remains which slowly decreases with time until the material comes back to its original state. This behaviour of material is called anelastic. The time-dependent nature of strain, during application and removal of stress, is known as the elastic after-effect. The realistic stress-strain relationship for an elastic material can be obtained as

$$\alpha_1 \sigma + \alpha_2 \dot{\sigma} = \beta_1 \varepsilon + \beta_2 \dot{\varepsilon} \qquad (10.41)$$

where α_1, α_2, β_1 and β_2 are constants for the material.

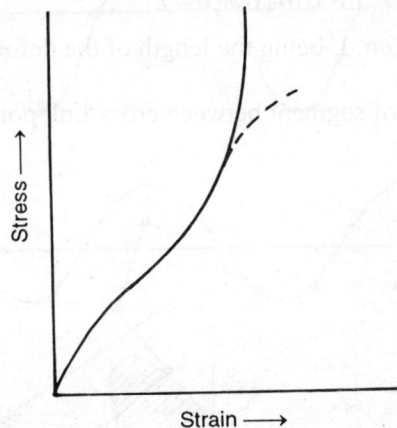

Fig. 10.13 Schematic representation of stress strain curve for rubber–like substance.

The time-dependence of stress and strain for anelastic material is shown in Fig. 10.14. The time independent strain ε_i occurs immediately after the application of stress. The total strain $\varepsilon_{i+}\varepsilon_a$ approaches ε exponentially with time where ε_a is anelastic strain. Similarly, the time independent strain is recovered immediately after the removal of stress but the anelastic strain relaxes exponentially.

The anelastic behaviour of material can be illustrated by considering the response of bcc iron, having carbon as interstitial impurity, to an axial stress along [001] direction. The solubility of carbon atoms in bcc iron is only 0.008 w/o. So only a very few of interstitial sites at the edge centre of iron unit cell are occupied by carbon atoms. As a result of axial tensile stress along [001] direction there are contractions along the other two axes perpendicular to it. The interstitial sites along elongated direction open up whereas, those along [100] and [010] directions become smaller in size. So, carbon atoms occupying the edge centers of the contracted sides (i.e. (0, ½,0) and (½,0,0)) jump into the unoccupied edge centers (0,0, ½) of the elongated sides. These jumps require energy, which is supplied by high strain energy developed around the interstitial atoms. As a result of these jumps (i) there is additional elongation along the direction of the applied stress, (ii) there is a stress-induced ordering process and also (iii) strain energy of the material reduces. Thus this additional elongation is time dependent and explains the time dependent part of strain.

Relaxation time

When the strain is time dependent, then the phenomenon of relaxation can be observed provided the process is neither too fast nor too slow as compared to the time interval during which stress is applied. If the strain develops very slowly as compared to the above time interval it will not be possible to find any measurable strain. On the other hand, if the strain develops very fast, the stress-strain relationship becomes independent of time.

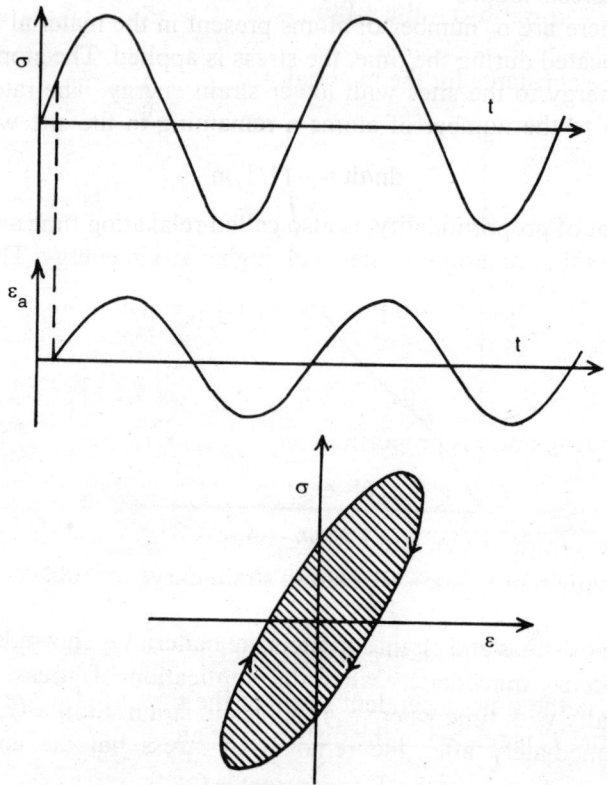

Fig. 10.14 Schematic representation of time dependence of stress and strain for anelastic material.

If the time scale of the process of development of strain is comparable to above time interval, the relaxation time for the process can be measured. Consider the example of carbon atoms as interstitial impurities in fcc iron, as discussed in the last section.

(i) If the time interval during the application of stress is short compared to the relaxation time, the anelasticity is not observed. Since the time interval for loading and unloading is very small compared to the relaxation time, the carbon atoms would not have enough time to jump from the position of higher strain energy to that of lower strain energy during loading and vice versa during unloading. So the sample behaves like an elastic material within the elastic limits.

(ii) If the time interval for application and removal of stress is large compared to the relaxation time, carbon atoms can distribute themselves in such a manner that the strain energy of the system reduces. The stress versus strain curve is straight line again with the slope greater than the case (i), since in this case strain is not only due to stretching of bond but also due to preferential arrangement of carbon atoms along the tensile axis.

(iii) When the time taken for above cycles is comparable to the relaxation time the additional elongation other than bond stretching in the direction of the tensile stress occurs. For example, the strain due to preferential occupation of carbon atoms in the interstitial sites along the tensile axis lags behind the strain due to bond stretching. As the stress increases, strain always lags behind the stress, so that during the cycle of loading and unloading the stress-strain curve becomes a hysteresis loop.

Let us consider there are n_o number of atoms present in the material before the application of stress, which are relocated during the time, the stress is applied. The atoms relocate from the sites with higher strain energy to the sites with lower strain energy. The rate of movement of atoms dn/dt is proportional to the number of atoms n remaining in the site with higher strain energy. Hence,

$$dn/dt = - (1/T_r)n \qquad (10.42)$$

where T_r, the constant of proportionality, is also called relaxation time and negative sign indicates the decrease in the number of atoms at sites with higher strain energy. Thus,

$$n = n_o \exp(-t/T_r) \qquad (10.43)$$

At time $t = T_r$, $n = n_o/e$. The number of atoms (n_r) relocated during time t is given by

$$n_o - n = n_r = n_o (1-\exp(-t/T_r)) \qquad (10.44)$$

Since the time dependent strain is proportional to the number of atoms relocated, hence

$$\varepsilon_t = \varepsilon_o (1-\exp(-t/T_r)) \qquad (10.45)$$

where ϵ_o is the total anelastic strain, that may develop exponentially after a large interval of time, t_o, Fig.10.15. Now when the stress is removed at time t_o, the time-dependent strain ϵ_t present at time t is

$$\varepsilon_t = \varepsilon_o \exp[(t_o-t)/T_r] \qquad (10.46)$$

where we assume ε_o is the time-dependent strain at the time of removal of stress. Now it is clear that the relaxation time T_r is related to the diffusion of atoms and hence one can express

$$\log T_r = \text{constant} + Q/(2.3RT) \qquad (10.47)$$

where Q is the activation energy required, for the specific atomic movements to produce anelastic strain.

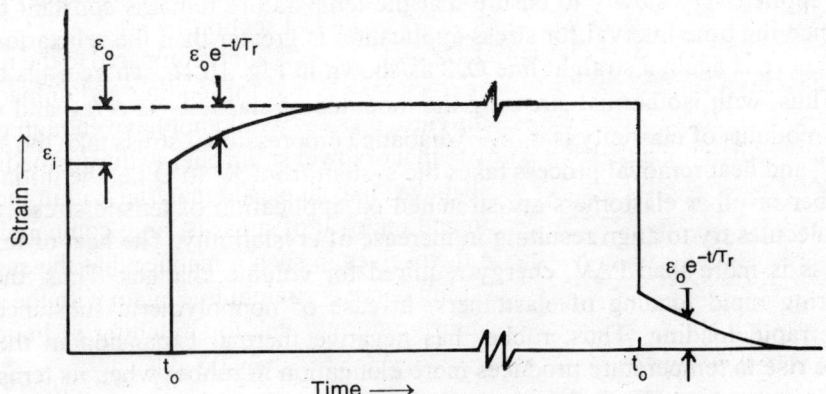

Fig. 10.15 Schematic representation of anelastic strain vs time

10.14 THERMOELASTICITY

Relaxation time in materials varies from 10^{-18}s for atomic vibrations to more than 10^6 s for viscous flow. As discussed in the last section, the relaxation time is depends on temperature. It decreases with the increase of temperature. When a material (nonpolymeric) is adiabatically stressed by applying tensile force, the strain in the material increases from 0 to ε_X. This also results in finite increase in volume since Poisson's ratio is usually less than 0.5 ($\Delta V = 0$ if Poisson's ratio is 0.5). The volume increase of material requires $P\Delta V$ energy, where P is the pressure and ΔV is the volume change in the material and this energy is supplied from the thermal sources. The temperature of the material drops as a consequence of this energy transfer and strain increases from ε_X to $\varepsilon_{X'}$ at constant stress σ_Y. If the sample is heated during the application of stress such that the temperature remains constant, there is increase in strain from 0 to $\varepsilon_{X'}$. The above phenomenon can be explained in terms of relaxation time as follows: When the material is

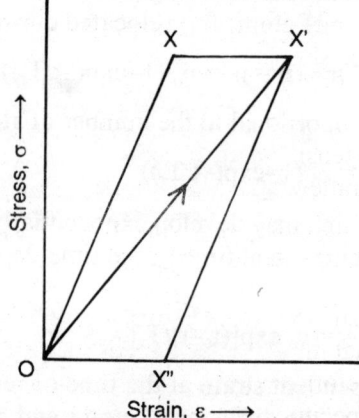

Fig. 10.16 Schematic representation of thermoelastic deformation in stress-strain curve

adiabatically stressed, the time interval during which stress is applied, is much smaller than the relaxation time hence anelastic strain is developed later than the instantaneous elastic strain which also explains the increase of strain from ε_X to $\varepsilon_{X'}$. On the other hand, during isothermal process the stress is applied very slowly to ensure that the temperature remains constant throughout the process. Hence the time interval for stress application is greater than the relaxation time and the stress-strain curve is again a straight line OX' as shown in Fig. 10.16, where $\varepsilon_{X'}$ is the final strain developed. Thus, with isothermal stressing the modulus of elasticity is $\sigma / \varepsilon_{X'}$ and with adiabatic stressing the modulus of elasticity is σ / ε_X. Adiabatic compressional stress take the state of system from X' to X" and heat removal process takes the system from X" to O i.e. the initial state.

When rubber or other elastomers are stretched on application of tensile stress, the elastically elongated molecules try to align resulting in increase of crystallinity. The heat of fusion given off in this process is more than $P\Delta V$, energy required for volume changes. Thus, the temperature increases during rapid loading of elastomers. In case of nonpolymeric substance temperature drops during rapid loading. Thus, rubber has negative thermal expansion in the direction of stressing. The rise in temperature produces more elongation in rubber when its temperature drops to its initial temperature. Thus the stress-strain relation for rubber as well as nonpolymeric substance is similar although reasons are different.

EXERCISES

10.1 A small block of steel having a cross-section 0.5 cm × 1.5 cm is pulled in tension with 30,000 N force producing only elastic deformation. Calculate the resulting strain. Given that $Y = 2.07 \times 10^{11}$ N/m^2.
Ans: 1.93×10^{-3}

10.2 (a) How much will a 3 mm diameter steel wire with a length of 9 cm be strained if it supports a load of 50000 N?
(b) What is the diameter of an aluminium wire 9 cm long, which is strained to the same extent as in (a) using the same load? Given that Y for aluminium = 6.9×10^{10} N/m^2.
Ans: (a) Extension in length = 3.07 mm; (b) 5.2 mm

10.3 If the same energy density is produced by a tensile stress and a shear stress separately on a specimen of copper calculate the ratio of tensile strain to the shear strain. Given that Y for copper is 1.1×10^5 MN/m^2 and the shear modulus for copper is 4.6×10^4 MN/m^2.
Ans: 0.647

10.4 A specimen of copper is under a tensile stress of 15000 psi. What are the dimensions of the unit cell assuming the stress is parallel to the c-axis? Given that Poisson's ratio = 0.35, a = 3.5 Å and 1 psi = 6.9×10^{-3} MPa.
Ans: a = b = 3.4988 Å, c = 3.5033 Å

10.5 For brass, the stress at which plastic deformation begins is 250 MPa.
(a) What is the maximum load that may be applied to a specimen with a cross-sectional area of 150 mm^2 without plastic deformation.
(b) If the length of the original specimen is 15 cm, what is the maximum length to which it could be elongated without plastic deformation. Given Y for Brass = 1.01×10^5 MPa.
Ans: (a) 3.75×10^4 N; (b) 15.037 cm

10.6 A 50cm copper rod, cylindrical in shape, has a yield strength of 300 MPa. It is subjected to a load of 9000 N. What must be the diameter that allows an elongation of 1 mm? Find the maximum load that can be applied to a copper rod without producing plastic deformation.
Ans: 7.2 mm; 12273 N.

10.7 A stress of 475 MPa is applied in the [001] direction of a unit cell of a fcc copper single crystal. Calculate the resolved shear stress on the (111) plane in the following directions
(a) [101] (b) [110] (c) [011].
Ans: (a) 1.9395×10^2 MPa, (b) 0, (c) 1.9395×10^2 MPa

10.8 Experiments conducted on a specimen of brass alloy reveal that it has an yield strength of 300 MPa, a tensile strength of 400 MPa and an elastic modulus of 10^5 MPa. A cylindrical specimen of this alloy 3 cm long and 1.5 cm diameter is bound to have an elongation of 0.6 cm when a stress is applied. Calculate the magnitude of the load (if possible). If not possible, explain why?
Ans: 3.5×10^6 N (equivalent to a stress of 2000 MPa). Such stress is not possible to apply because the specimen breaks.

10.9 Calculate the compressibility and the Poisson's ratio when a bar of a metal X of the shape of a cube with a volume of 1.4560 cm^3 is compressed with a hydrostatic pressure 9.67×10^8 N/m^2 accompanied by a reduction in volume of 0.009 cm^3. Given that $Y = 2.067 \times 10^{11}$ N/m^2.

Ans: 6.39×10^{-12} m^2/N, 0.28

10.10 Compare the behaviour of an elastomer with that of other materials under a tensile load.

10.11 When a stress of 60 MPa is applied to a plastic showing anelastic characteristic, a strain of 0.014 was produced instantly. After 2 minutes, the strain rose to 0.021 and after a long interval of time strain becomes 0.024. Then the stress was removed. After a time t, the strain dropped to 0.002. Calculate relaxation time T_r and t. Assume relaxation time is the same for both elongation and contraction.

Ans: 49.83 s, 80.21 s.

11
Viscoelasticity

11.1 INTRODUCTION

Plastic deformation occurs in a crystalline solid usually by movement of dislocations under the influence of shear stress. In case of noncrystalline substances, the deformation occurs due to viscous flow. When shear stress is applied to a ceramic material, small atoms, molecules or ions move past their neighbours easily causing deformation and also provide viscosity to the medium. The rate of deformation is proportional to the applied stress. The factors, which influence the viscosity of a medium, are

* shape and size of the molecule
* molecular weight
* polar character of the molecules
* cross-linking or framework structure in molecules

A fused silica glass has a rigid framework structure with covalently bonded silicon and oxygen atoms such that each silicon atom is bonded to four oxygen atoms and each oxygen atom is bonded to two silicon atoms. Effectively each SiO_4^{4-} unit is bonded to four such neighbouring units. The shear stress causes breaking and reforming of bonds when molecules move past their neighbours. Thus, in case of fused silica very high stress is required for deformation to occur. Polar molecules form ionic/ hydrogen bonding in contrast to other molecular solids where intermolecular bonding is secondary e.g. van der Waals type. The viscosity of the later is smaller than that of the former. For example pentane, being nonpolar, has lower viscosity than that of benzene and H_2O.

Example 11.1

Young's modulus for diamond is 10^3 GPa and soft wax has viscoelastic modulus 1 GPa. Given that viscoelastic modulus for linear polymer composed of diamond and wax, is 2 GPa. Calculate the fraction of bonds that are primary i.e. bonds of diamond in the polymer.

Solution:

$$2 = [\, p/10^3 + (\, 1 - p)/1\,]^{-1}$$

where p is the fraction of primary bonds present in the linear polymer.

So $p = 0.5$

The molecular weight as well as size of the molecule are also important factors in deciding the viscosity of a material e.g. polyethylene has a high value of viscosity because of its molecular size. The viscosity of a noncrystalline material is a measure of resistance to plastic deformation.

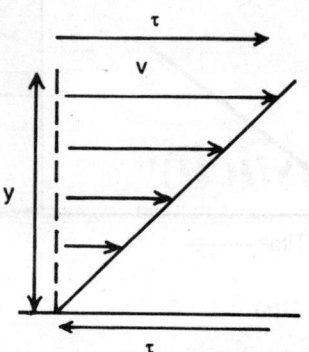

Fig. 11.1 Schematic representations of viscous drag in fluid due to velocity gradient.

The result of application of shear stress (τ) on the surface of a liquid or viscous medium is setting up of velocity gradient in a direction (y-axis) perpendicular to the surface, Fig. 11.1. Viscosity of the medium is defined as

$$\eta = \frac{\tau}{dv/dy} \qquad (11.1)$$

since $v = x/t$ and shear due to viscous flow $\gamma = x/y$

$$\eta = t\,\tau/\gamma \qquad (11.2)$$

Viscosity of a medium is dependent on the temperature. It decreases with the increase of temperature and is given by

$$\eta = \eta_o\, e^{Q/RT}, \qquad (11.3)$$

where η_o is a constant for the medium and Q is activation energy for viscous flow of atoms or molecules moving past their neighbours. The fluidity is reciprocal of viscosity and expressed as

$$f = 1/\eta \qquad (11.4)$$

11.2 DISPLACEMENT MODELS

When an elastic material is stressed, the response is linear within elastic limits. In case of an ideal fluid, the flow displacement γ_f is observed as a function of time when the constant shear stress is applied.

Dashpot model

The behavior of an ideal fluid can be modeled as that of dashpot, Fig. 11.2. A dashpot is a loosely fitting piston in a vessel containing the fluid. When a force is applied to the piston, it moves down gradually and the fluid flows around its edges. The displacement is slower, if the fluid is

The molecular weight as well as size of the molecule are also important factors in deciding the viscosity of a material e.g. polyethylene has a high value of viscosity because of its molecular size. The viscosity of a non-crystalline material is a measure of resistances to plastic deformation.

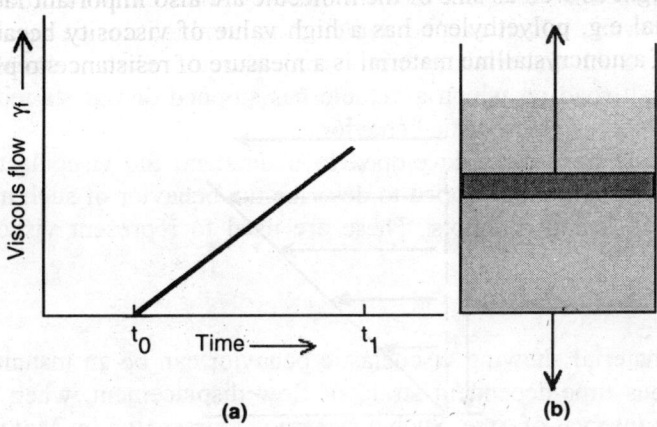

Fig. 11.2 (a) Flow displacement vs time. (b) Dashpot model

more viscous. The dashpot does not go back to its original position after removal of the force. Reversal of dashpot motion occurs when the force is reversed. The stress relation can be expressed in this case as,

$$\tau = \eta \gamma_v / t \qquad (11.5)$$

where the subscript 'v' denotes the properties related to viscous fluid.

Spring model

Strain develops almost instantaneously when a load is applied on an elastic material and it also recovers instantaneously when the load is removed. The behavior of such material can be represented by that of a spring, Fig. 11.3, so elastic modulus (rigidity modulus) is given by

$$\mu = \tau / \gamma_e \qquad (11.6)$$

where subscript 'e' signifies properties of elastic material.

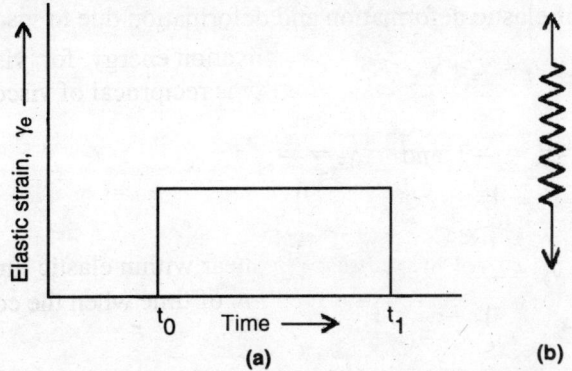

Fig. 11.3 (a) Elastic displacement vs time; (b) Spring model

11.3 VISCOELASTIC BEHAVIOUR OF MATERIALS

The amorphous polymeric solids behave like a glassy substance at low temperature i.e. there is no sharp melting point, a rubbery solid at intermediate temperature and viscous liquid at a higher temperature. At intermediate temperatures the rubbery solid exhibits combination of elastic as well as viscous behavior, which can be depicted as combination of mechanical characters of spring and dashpot. Such materials exhibit viscoelasticity when a stress is applied e.g. the response of an asphalt road on which a vehicle has stopped or has started moving. Silicone, a polymeric substance shows viscoelastic behavior.

Macroscopic models have been developed to understand the viscoelastic behavior of these materials. Suitable models are developed to describe the behavior of such materials with various combinations of springs and dashpots. These are used to represent viscoelastic behaviour of materials:

a) Maxwell model

b) Voigt model.

The response of a material showing viscoelastic behavior can be an instantaneous elastic strain followed by a viscous time-dependent strain or flow displacement, when it is subjected to an applied stress for an interval of time. Such a system is represented by Maxwell model, Fig. 11.4 in which the spring and dashpot are connected in series. When a viscoelastic material is stressed and its response to stress is nonlinear and time-dependent. Suitable model to represent the behavior of such material is Voigt model or combination of Voigt and other models.

Creep test

When a material is under constant stress for an interval of time, the material may develop strain gradually increasing with time. This time-dependent deformation of the material is called viscoelastic creep. This type of creep may be observed at room temperature when the stress does not exceed yield strength of the material. For example, the tyres of a vehicle may become flat at the regions where they are in contact with the ground if the vehicle remains standing for a long period.

a) Maxwell model

When the response of a material under constant stress is an instantaneous elastic strain followed by a viscous time-dependent strain, the viscoelastic behaviour of the material can be studied using Maxwell model, where the elastic and viscous elements are in series. In this case, the total deformation is the sum of elastic deformation and deformation due to viscous flow,

$$\gamma = \gamma_e + \gamma_v, \tag{11.7}$$

where
$$\gamma_e = \frac{\tau}{\mu} \quad \text{and} \quad \gamma_v = \frac{\tau t}{\eta}$$

So,
$$\gamma_v = \frac{\tau}{\eta} \tag{11.8}$$

Here we have considered that the stress τ is constant. After an interval of time, when the stress is removed, the elastic strain recovers immediately but the viscous time-dependent strain remains until the reversal of stress. This is called creep test. Viscoelastic response of constant stress can be studied using creep test.

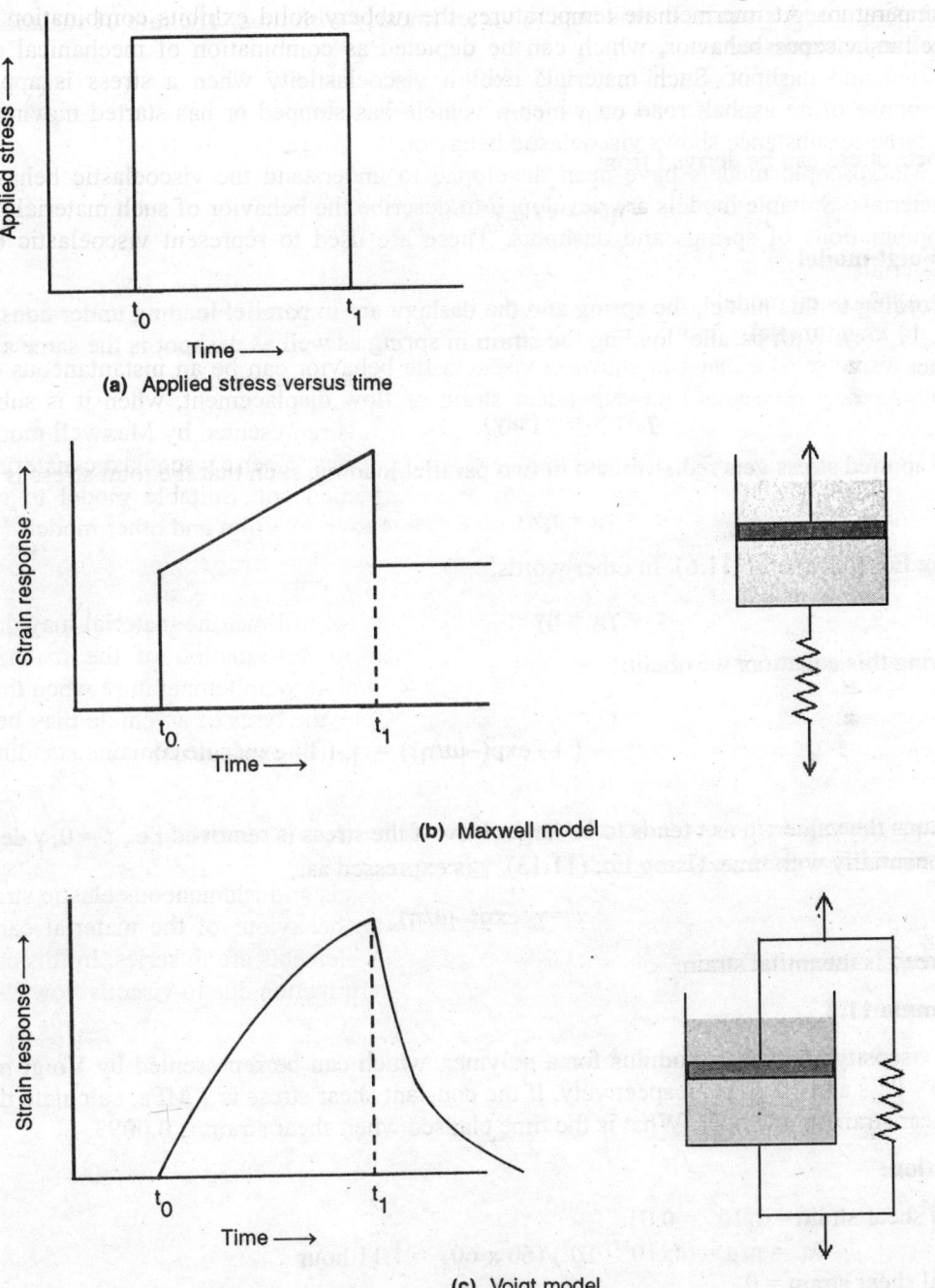

Fig. 11.4 (a) Stress versus time; Strain versus time showing (b) viscoelastic leathery behaviour; (c) viscoelastic rubbery behaviour

When the stress is varying with time, elastic strain in the spring component depends on the instantaneous value of stress, whereas the dashpot responds slowly giving rise to progressively increasing viscous strain. In this case γ_v varies linearly with time. The rate of variation of γ with time can be expressed as

$$\dot{\gamma} = \tau / \mu + \dot{\tau} / \eta \qquad (11.9)$$

Hence, stress can be derived from

$$\dot{\tau} = \eta(\dot{\gamma} - \tau / \mu) \qquad (11.10)$$

b) Voigt model

According to this model, the spring and the dashpot are in parallel loading under constant stress, Fig. 11.4(c). With parallel loading the strain in spring as well as dashpot is the same at any time. Hence,

$$\gamma_e = \gamma_v = \gamma \text{ (say)} \qquad (11.11)$$

The applied stress gets redistributed in two parallel loading, such that the total stress is

$$\tau = \gamma\mu + \eta\gamma/t \qquad (11.12)$$

using Eqs.(11.5) and (11.6). In other words,

$$\tau = \gamma\mu + \eta\dot{\gamma} \qquad (11.13)$$

Solving this equation, we obtain

$$\gamma = \frac{\tau}{\mu} \{ 1 - \exp(-\mu t/\eta) \} = \gamma_o \{ 1 - \exp(-\mu t/\eta) \} \qquad (11.14)$$

γ attains the value τ/μ as t tends to be large. Now if the stress is removed i.e., $\tau = 0$, γ decays exponentially with time. Using Eq. (11.13), γ is expressed as,

$$\gamma = \gamma_o \exp(-\mu t/\eta), \qquad (11.15)$$

where γ_o is the initial strain.

Example 11.2

The viscosity and shear modulus for a polymer, which can be represented by Voigt model, are 4×10^{11} Pa-s and 10^2 MPa, respectively. If the constant shear stress is 1 MPa, calculate the change in shear strain in one hour. What is the time elapsed when shear strain is 0.0093.

Solution:

Final shear strain $= 1/10^2 = 0.01$

$$\lambda_v = \eta/\mu = (4 \times 10^{11} / 10^8)/(60 \times 60) = 1.11 \text{ hour}$$

Initial shear strain $= 0$

Shear strain after an hour is given by

$$\gamma = \tau/\mu \{1 - \exp(-t/\lambda_v)\} = 0.01 \{1 - \exp(-1/1.11)\} = 0.0059$$

The strain increased to 0.0059 in one hour. When $\gamma = 0.0093$, time elapsed is t_1 say and

$$0.0093 = 0.01 \{1 - \exp(-t_1/1.11)\}; \text{ so, } t_1 = 2.95 \text{ hours.}$$

Stress relaxation test

In many engineering applications the system is under constant strain. Viscoelastic behavior of such systems can also be observed through stress relaxation test, in which stress varies in such a way as to keep the strain constant. Under this condition the viscoelastic component of strain (if any) allows the stress to relax, without any change in it, Fig. 11.5.

During stress relaxation test, material is initially strained rapidly to a predetermined value by applying tensile stress. The stress necessary to maintain the fixed value of strain decreases exponentially with time, as more and more plastic strain occurs in place of elastic strain as observed in case of **Maxwell model**, Fig. 11.5(b). Since the deformation remains constant,

$$\frac{d\gamma}{dt} = \frac{d\gamma_e}{dt} + \frac{d\gamma_v}{dt} = 0 \tag{11.16}$$

Using the definition of modulus of rigidity, and using Eq. (11.5)

$$d\gamma_v/dt = \tau/\eta$$

and

$$d\gamma_e/dt = -d\gamma_v/dt = d(\tau/\mu)/dt$$

So

$$\frac{1}{\mu} \frac{d\tau}{dt} = -\frac{\tau}{\eta} \tag{11.17}$$

Hence

$$\tau = \tau_o e^{-(\mu/\eta)t} = \tau_o \exp(-t/\lambda_v) \tag{11.18}$$

where τ_o is the shear stress at the initial time $t = 0$. The above equation shows stress relaxation and λ_v is the relaxation time during which τ_o becomes τ_o/e. We know that resolved shear stress τ is proportional to axial stress σ, hence

$$\sigma = \sigma_o \exp(-t/\lambda_v) \tag{11.19}$$

where σ_o is the initial axial stress at $t = 0$ and λ_v is the relaxation time during which σ_o is reduced to σ_o/e and $\lambda_v = \eta/\mu$.

Certain materials show a constant value of stress during the period a constant strain is applied. This behaviour is represented by **Voigt model**, Fig. 11.5(c).

Example 11.3

The stress decays from 8 MPa to 5.7 MPa in 105 days
   (a) What is the relaxation time?
   (b) How many days would it take to relax to 3 MPa?

Solution:
   (a)

$$\tau_1 = \tau_o \exp(-t_1/\lambda_v) = 8 \text{ MPa.}$$
$$\tau_2 = \tau_o \exp(-t_2/\lambda_v) = 5.7 \text{ MPa.}$$
$$(\tau_1/\tau_2) = \exp{-((t_1 - t_2)/\lambda_v)}$$
$$(8/5.7) = \exp(105/\lambda_v)$$

Hence

$$\lambda_v = 105/\ln(8/5.7) = 309.75 \text{ days} \approx 310 \text{ days}$$

   (b)

$$\tau_3 = 3 \text{ MPa}$$
$$(\tau_3/\tau_1) = \exp{-((t_3 - t_1)/\lambda_v)}$$

$$(3/8) = \exp-((t_3 - t_1)/310)$$

Time required $= (t_3 - t_1) = 304$ days.

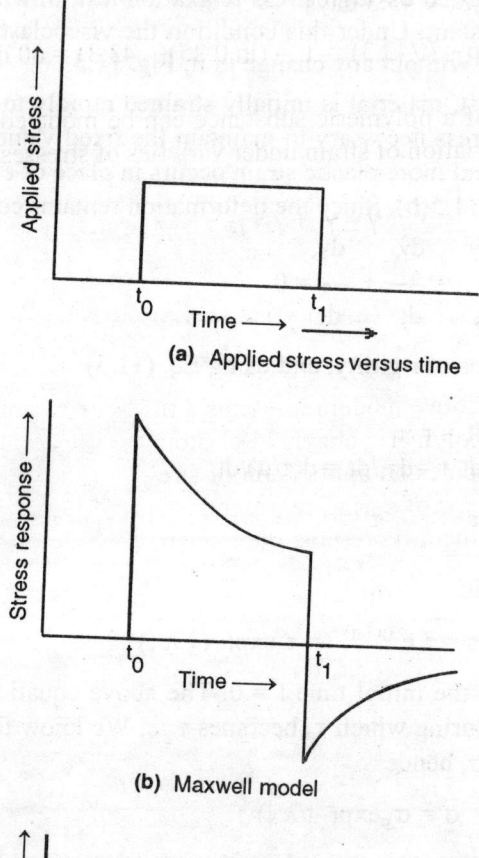

(a) Applied stress versus time

(b) Maxwell model

(c) Voigt model

Fig. 11.5 Stress relaxation under constant strain.

Example 11.4

A tensile stress of 8 MPa is applied to a polymeric substance at constant strain. The stress decays to 50% of its initial value after 30 days at 30°C. Calculate:

(a) The relaxation time constant for this material.
(b) The number of days after which stress is 2 MPa.

Solution:

(a) $\quad \sigma = \sigma_o \exp(-t/\lambda_v)$

$\quad\quad 0.5 = \exp(-30/\lambda_v)$

$\quad\quad \ln 2 = 30/\lambda_v ; \quad \lambda_v = 43.3$ days

(b) $\quad 2\text{MPa} = 8\text{ MPa} \exp - (t/43.3) ; \quad t = (\ln 0.25)(-43.3) = 60$ days.

Sometimes deformation of a polymeric substance can be modeled as represented in Fig. 11.6 such that it can represent variation of strain under varieties of stresses. The total deformation γ, of such system is given by

$$\gamma = \gamma_1 + \gamma_2 + \gamma_3$$

$$\gamma = \frac{\tau}{\mu_1} + \frac{\tau}{\mu_2}(1 - \exp(-\mu_2 t/\eta_2)) + \frac{\tau t}{\eta_3} \quad\quad (11.20)$$

The Voigt element in the above model represents a time-dependent but recoverable strain such as uncoiling and coiling of polymeric chains in elastomers. The time-dependent development of strain like uncoiling of molecules on application of stress is also known as retarded elasticity. In Maxwell element, the response of spring represents stretching of bonds when stress is applied. The dashpot of Maxwell element represents the movement of chain molecules past one another. These movements cause permanent deformation.

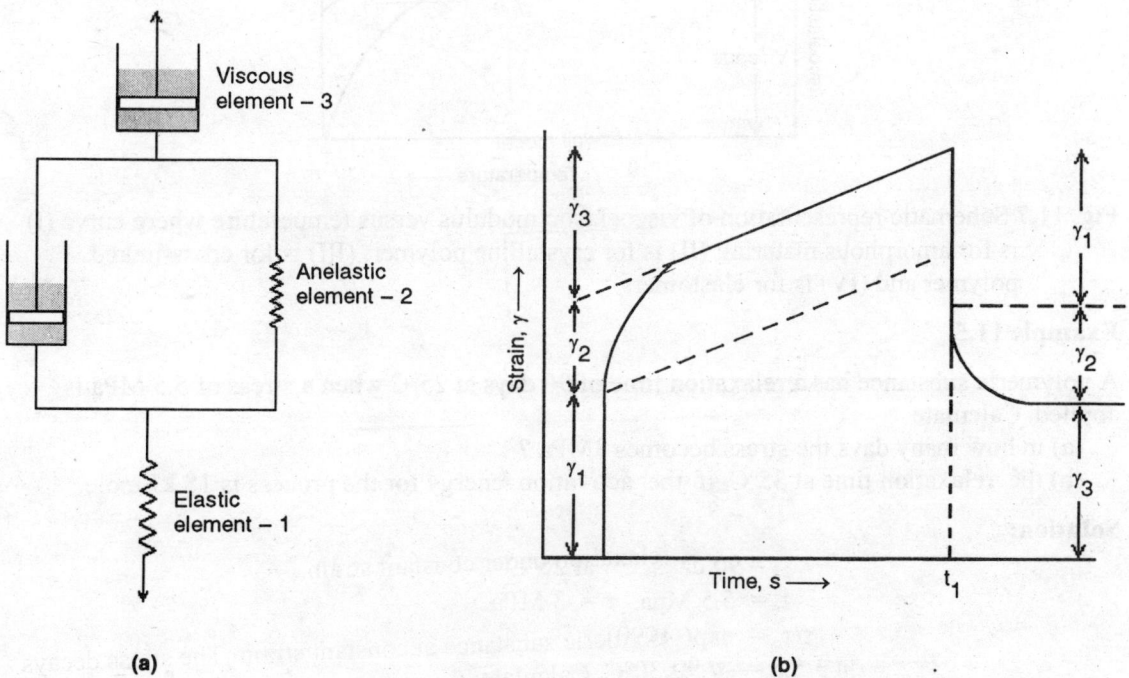

Fig. 11.6 (a) Viscoelastic model, (b) viscoelastic deformation with time

11.4 VISCOELASTIC MODULUS

The viscoelastic modulus E is given by τ/γ. Since τ is a function of time, we can express

$$E = (\tau_0/\gamma) \exp(-t/\lambda_v) = E_0 \exp(-t/\lambda_v) \qquad (11.21)$$

where E_0 is the initial value of viscoelastic modulus. The relaxation time is $\lambda_v = \eta/\mu$, is also a function of temperature since $\eta = \eta_0 e^{Q/RT}$. Thus

$$\lambda_v = (\eta_0/\mu) e^{Q/RT} \qquad (11.22)$$

The dependence of λ_v of temperature also implies E is a function of temperature. Fig. 11.7 shows the variation of E with temperature.

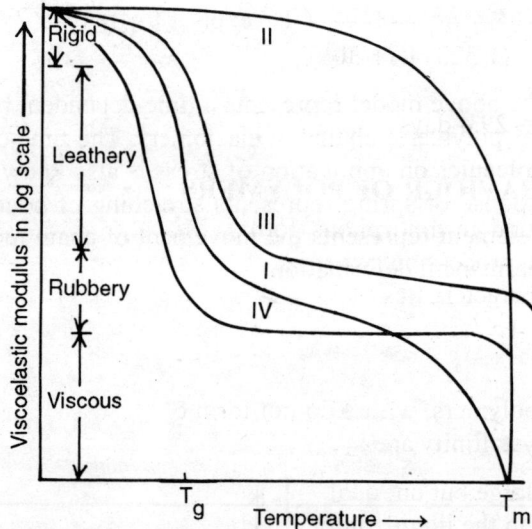

Fig. 11.7 Schematic representation of viscoelastic modulus versus temperature where curve (I)
      is for amorphous material; (II) is for crystalline polymer; (III) is for cross-linked
      polymer and (IV) is for elastomers.

Example 11.5

A polymeric substance has a relaxation time of 90 days at 25°C when a stress of 5.5 MPa is applied. Calculate
   (a) in how many days the stress becomes 3MPa ?
   (b) the relaxation time at 35°C if the activation energy for the process is 18 kJ/mole.

Solution:

$$\lambda_{25} = 90 \text{ days} \qquad \text{at } 25°C$$
$$\tau_0 = 5.5 \text{ Mpa}, \quad \tau = 3 \text{ MPa}.$$
$$\tau/\tau_0 = \exp(-t/90)$$
$$\ln 3/5.5 = -t/90.$$
$$t = 90 \ln(5.5/3) = 54.55 \text{ days} \approx 55 \text{ days}$$
$$\lambda_{25} = \lambda_0 \exp(Q/298R) = 90 \text{ days}$$
$$\lambda_{35} = \lambda_0 \exp(Q/308R) = 71 \text{ days}.$$

Example 11.6

The relaxation time for a polymeric substance is 40 days at 35°C and is 50 days at 25°C. Calculate the relaxation time at 50°C.

Solution:

Using (11.22)
$$\lambda_v = \lambda_o \exp(Q/RT)$$
$$\lambda_{35} = 40 \text{ days} = \lambda_o \exp(Q/308R)$$
$$\lambda_{25} = 50 \text{ days} = \lambda_o \exp(Q/298R)$$
$$40/50 = \exp[(Q/R)\{(1/308) - (1/298)\}]$$
$$\ln(50/40) = (Q/R)\{(1/298) - (1/308)\}$$
$$\ln[\lambda_{50}/40] = (Q/R)\{(1/323) - (1/308)\}$$

$$\frac{\ln(50/40)}{\ln(\lambda_{50}/40)} = \frac{(1/298) - (1/308)}{(1/323) - (1/308)}$$

$$\lambda_{50} = 29.4 \text{ days}$$

11.5 VISCOELASTIC BEHAVIOUR OF POLYMERS

Polymers are commonly found in amorphous form. Moderately crystalline polymers can be obtainable under controlled processing when it is possible to reduce steric hindrance. The behaviour and viscoelastic characteristics of various types of polymers are given below.

a) Amorphous Polymers

There are large varieties of polymers, which do not form crystalline structure after supercooling. The factors governing noncrystallinity are :

* Sizes of molecules are large but unequal.
* They remain entangled in the liquid state.
* The molecules may not align due to large size or due to presence of large side radicals or branching.

That is why supercooling results in amorphous character in polymers. Although polymer molecules are entangled in the liquid state, they try to rearrange in the presence of thermal energy. When temperature $T > T_m$, the melting temperature, the polymer is in liquid state. The thermal agitations of molecules are reduced as the polymeric liquid is cooled. This also reduces the volume of the liquid. The decrease in volume is due to two factors – rearrangement of molecules so that there is better packing and reduction in the amplitude of vibrations of molecules. The volume of the liquid decreases till the temperature is lowered below T_m. The liquid structure is retained for a range of temperature below T_m. This state of polymer is called supercooled liquid state. The flow displacement becomes less with decrease in temperature. At lower temperatures, the movement of molecules is reduced due to lack of space and also due to decrease in mobility of polymer molecules. Below a specific temperature, the rearrangement of atoms or molecules stops i.e. there is no reduction in free volume. This temperature is called glass-transition temperature or glass point T_g, Fig. 11.8. Cooling below T_g can only result in decrease in amplitude of vibrations of molecules and there is discontinuous change in thermal expansion coefficient. Glass point depends on rate of cooling. If cooling rate is slow, molecular rearrangement occurs at lower temperature i.e. T_g is less compared to glass point value at fast cooling rate. Below T_g, material is usually brittle and nondeformable.

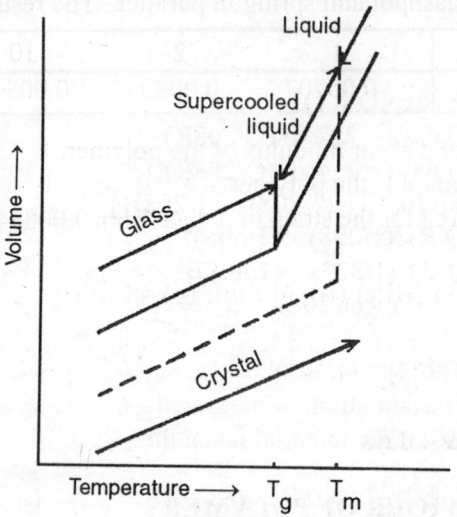

Fig. 11.8 Volume vs temperature curve for glassy material.

At temperature higher than melting point T_m, polymer behaves like a true liquid, which possesses low viscosity. Elastic strain cannot be produced in pure liquid and stress causes viscous flow. Thus γ_3 in Eq. (11.20) predominantly represents the deformation of polymeric materials.

b) Crystalline Polymers

A polymer with high crystallinity does not show rubbery or leathery behavior since the dominant term in deformation given by Eq. (11.20) is γ_1. If plastic product is to bear high stress at temperatures nearing melting point, the crystalline polymers are desirable for such applications.

c) Cross-linked Polymers

Cross-linked polymers are hard and brittle at $T < T_g$. These materials exhibit anelastic behaviour at T, $(T_g < T < T_m)$, when stress is applied. Molecular rearrangements of these polymeric materials take place in this temperature range and strain developed, is much less as compared to that in polymeric substances without cross-linking. Vulcanized polymers belong to this class of material. In framework polymers γ_1 has major contribution to deformation. Phenol formaldehyde is an example of framework polymer.

d) Elastomers

Elastomers have very low moduli of elasticity due to presence of coiled and kinked polymer molecules. These materials have low values of μ_2 and η_2, (Eq.(11.20)). Molecules are unkinked and uncoiled very easily at $T > T_g$. At higher temperatures, thermal energy available for coiling and kinking is greater than that at $T < T_g$. Hence more stress is required to produce anelastic strain at $T < T_g$. At higher temperature, μ_2 increases. Hence modulus of viscoelasticity also increases. At a temperature $T > T_m$, γ_3 has dominant contribution to the deformation.

EXERCISES

11.1 Shear strains are measured at different times for a polymer when a stress of 2 MPa is applied. The polymer behaves as a dashpot and spring in parallel. The results of measurement are

Time (hrs)	1	2	10
Shear strain	0.0070	0.0082	0.00844

(a) Calculate the viscosity and shear modulus of the polymer.
(b) What is the relaxation time for the polymer?
(c) Calculate the time required for the strain to reduce from 0.00844 to 0.0023 after the stress is removed.

Ans: (a) 4.8×10^{11} Pa-s, 237 MPa; (b) 34 min; (c) 44 min

11.2 A viscoelastic material (viscosity = 4×10^{10} Pa-s) is subjected to a strain of 0.4 and the stress required to produce this constant strain is measured as a function of time. Calculate the shear and viscoelastic moduli of the material when the stress drops from 3 MPa to 1 MPa in 30 minutes.

Ans: 24.4 MPa, 2.5 MPa

11.3 An elastomer is subjected to a stress of 10 MPa at a constant strain at 20°C. After 30 minutes the stress is reduced to half.
(a) What is the relaxation time for this material?
(b) What will be the percentage decrease in the stress after another 30 minutes?

Ans: (a) 43 min; (b) 50%.

11.4 In Exercise 11.1 the data were obtained at 293°K. Let the relaxation time be 1000 s at 300°K.
(a) Calculate activation energy for relaxation.
(b) Find the strain after 2000 s at 300°K during loading.
Assume that μ is constant over the temperature range.

Ans: (a) 74.43 kJ/mole; (b) 0.0073

11.5 The viscosity of a liquid above the glass transition temperature (T_g) can be expressed as:

$$\ln [\eta_T / \eta_{Tg}] = [-17.44(T - T_g)] / [51.6 + T - T_g]$$

If the viscosity of an unknown material is 12×10^3 GPa-s at T_g, calculate its viscosity at 50°K and 200°K above T_g.

11.6 A polymeric material has a relaxation time of 40 days at room temperature (27°C). When a stress of 5 MPa is applied, it gets reduced to 3 MPa after 'd' days. Calculate
(a) d,
(b) Relaxation time at 40°C if the activation energy for this process is 10 kJ/mole?

Ans: (a) 20 days; (b) 34 days

11.7 A stress of 7 MPa is required to stretch a 6 inch rubber band to 7.5 inch. The band exerts a stress of 4.83 MPa in the same stretched position after 30 days.
(a) What is the relaxation time?
(b) What would be the stress exerted by the band after another 30 days in the same stretched position?
Ans: (a) 81 days; (b) 3.33 Mpa

11.8 Calculate the relaxation time for glass [Y = 70 GPa and η = 10^{12} GPa-s].
Ans: 453 years

11.9 A polymer has an elastic modulus of 700 MPa, subjected to a stress of 7 MPa. After 300 s, the strain is measured to be 0.009.
(a) Calculate viscosity if the polymer behaves as per Voigt model?
(b) What will be the strain after another half an hour?
Ans: (a) 1993 GPa-s; (b) 0.00478

11.10 A certain polymer 'X' has a modulus of elasticity of 70 MPa and a relaxation time of 30 days. Both measurements are made at the same temperature. After application of the external force, the strain produced is 0.05 at the same temperature.
(a) What is the initial stress?
(b) What is the stress after 1 month and after 1 year, if strain remains constant?

11.11 A spring (S1) is in series with a parallel combination of a spring (S2) and dashpot (η). Sketch and describe the displacement or strain vs time behaviour after a load is hung on the system.

11.12 Derive an expression for the resulting strain of a polymer in the Voigt model. Can you describe a physical situation in which this formula applies?

11.13 Describe the phenomenon of viscoelasticity in your own words.

11.14 When a stress of 7 MPa is applied to an elastomer at room temperature. After 2 weeks the stress is reduced to 4.2 MPa by stress relaxation. It is reduced to 3.5 MPa in 2 weeks when the temperature is raised to 45°C. Calculate the activation energy of this reaction.
Ans: 13.43 kJ

12 Ceramics: Properties and Applications

12.1 INTRODUCTION

Ceramics are complex compounds of metals e.g. oxides, nitrides, carbides or combination of them. They are usually hard and brittle. Ionic and covalent bonds are observed in this type of materials. So, they are good electrical and thermal insulators and resist chemical degradation. Ceramics have low thermal shock resistance because of low thermal expansion coefficient and low thermal conductivity. These properties are responsible for cracking of an ordinary glass, when boiling water is poured into it. The structure of ceramics is discussed in Sec. 4.5. Properties, processing and applications of ceramics are presented in this chapter.

Ceramics can be usually classified as follows:

- Glass
- Clay
- Refractory
- Abrasive
- Cement

12.2 GLASS

The basic component in glass is silica (SiO_2). Other materials, which are often used with silica, are limestone, soda ash, boric oxide, lead oxide, magnesium oxide, aluminium oxide, sodium oxide, calcium oxide etc. to fabricate glasses for various applications. There are thousands of varieties of glasses depending upon the type and composition. Oxides like silica, B_2O_3, GeO_2, V_2O_5 act as glass formers, whereas PbO, Al_2O_3, BeO etc. do not form glass by themselves but are incorporated into the network structure of glass formers. These are intermediate oxides. The thermoplastic property of commercial glasses can be improved upon by adding metallic oxides like Na_2O, CaO etc. These are added to provide cations to the structure and are called network modifiers. They help to shape the glass in the desired form. Addition of Na_2O to silica introduces nonbridging oxygen ions and twice the number of Na^+ ions, Fig. 12.1. The Na^+ ions introduce holes in the network structure and do not join the network. The extra O^{2-} ions that enter the network make the O: Si ion ratio high such that the network structure is disrupted. The silica tetrahedra form chains, rings or compounds. If O: Si ratio is greater than 2.5, silica glasses are difficult to form. Adding CaO further modifies these glasses.

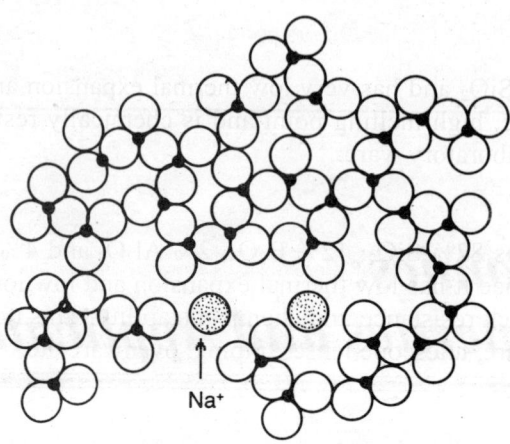

Fig. 12.1 Na₂O modified glass

Example12.1

Calculate the O: Si ratio when 20% Na₂O is added to SiO₂. Explain whether this material will provide good glass-forming tendencies.

Solution

Mol.wt. of SiO_2 = 28.09+32 =60.09 amu.
Consider the sample to be one mole of SiO_2 i.e. 60.09 g.
Na₂O added = 20% of 60.09 g = 12.018 g = 0.194 moles
So the sample now contains one mol of Si and 2.194 moles of oxygen. So O:Si = 2.194, which is much lower than 2.5. Hence the material will have good glass forming tendencies.

Example 12.2

A glass sample contains 70 w/o SiO₂, 20 w/o Na₂O and 10 w/o CaO, what fraction of oxygen is nonbridging?

Solution:

Let the sample be 100g. This contains 70 g of SiO₂, 20 g of Na₂O and 10 g of CaO

$$70 \text{ g of } SiO_2 = 70/60.09 \text{ moles} = 1.165 \text{ moles}$$
$$20 \text{ g of } Na_2O = 20/62 \text{ moles} = 0.322 \text{ moles}$$
$$10 \text{ g of } CaO = 10/56.08 \text{ moles} = 0.178 \text{ moles}$$

1.165 moles of SiO₂ contain 1.165 moles of Si and 2.330 moles of O
0.322 moles of Na₂O contain 0.644 moles of Na and 0.322 moles of O
0.178 moles of CaO contain 0.178 moles of Ca and 0.178 moles of O

There is one nonbridging oxygen for each Na⁺ and two nonbridging O for each CaO molecule.

$$\text{So fraction of nonbridging oxygen} = \frac{0.644 + 2 \times 0.178}{2.330 + 0.322 + 0.178} = \frac{1}{2.830} = 0.353$$

Different types of glasses are fabricated for different applications. Some examples are given below:

Fused silica glass

This type of glass contains 99% SiO_2 and has very low thermal expansion and high viscosity. It has high thermal shock resistance, high melting point and is chemically resistant. Hence it is a suitable material for fabricating laboratory ware.

Pyrex or borosilicate glass

The composition of borosilicate is 82% SiO_2, 12% B_2O_3, 2% Al_2O_3 and 4% Na_2O. It is used for high temperature applications, since it has low thermal expansion and low ion-exchange property. Pyrex glass has good thermal shock resistance and chemical stability. This type of glass is used in laboratory equipment, medical ware, telescope lenses, piping, ovenware etc.

Window glass

Window glass contains 73% SiO_2, 14% Na_2O, 10% CaO, 1% Al_2O_3 and 2% MgO. It has wide applications as a building material since it is quite durable. It is also used for electric bulbs, containers etc.

Fibreglass or aluminosilicate glass

Fibreglass is composed of 54% SiO_2, 15% Al_2O_3, 16% CaO, 11% B_2O_3 and 4% MgO. It is also called E (electrical) glass since it is a good electrical insulator. Modulus of elasticity of E glass is 72.3 GPa and tensile strength is 3.44 GPa. S glass is another type of fibreglass, which has higher magnesia content (10% MgO) and has greater strength per unit mass as compared to E glass. Its tensile strength is 4.48 GPa. It is very important material for telecommunication and networking applications. They are also used for developing fibre-reinforced composites. S glass is used specifically for aerospace applications because of its higher strength. Fibreglass has not only high strength but also has high thermal shock resistant characteristics.

Lead or flint glass

Lead glass contains 56% SiO_2, 4% Na_2O, 2% Al_2O_3, 9% K_2O and 29% PbO. It has improved refractive indices and low melting point. This type of glass is suitable for optical applications. High lead glasses find application for radiation windows, television bulbs and solder sealing glasses.

Porosity in ceramics

Ceramics retain some residual porosity during fabrication process. The pores are usually small and nearly spherical in shape. Pores at the surface can be considered as surface defects. The pores in a ceramic may be either interconnected or closed. The apparent porosity is defined as the volume fraction of the interconnected pores. It determines the ease with which gas or fluid can seep through a ceramic component. The apparent porosity of the material is given by

$$P_a = \frac{M_w - M_d}{M_w - M_s} \times 100\% \tag{12.1}$$

where M_d is the weight of the dry ceramic, M_s is the weight of the ceramic when suspended in water and M_w is the weight of the ceramic when removed from water. The true porosity or volume fraction porosity is defined as

$$P = \frac{\rho - \rho_B}{\rho} \times 100\% = \frac{V_p}{V} \times 100\% \qquad (12.2)$$

where ρ is the true density or specific gravity of the ceramic and ρ_B ($= M_d/(M_w - M_s)$) is the bulk density of the material i.e. the weight of the ceramic divided by its volume. V_p is the volume of the pores and V is the volume of the ceramic.

Adsorption of gas molecules onto the surface or entry of foreign ions in the surface reduces the surface energy and alters the property of ceramic products. Porosity is damaging to mechanical integrity of a ceramic piece. The elastic modulus Y decreases with P. An empirical formula relating porosity and Y is given by

$$Y = Y_o (1 - 1.9 P + 0.9 P^2), \qquad (12.3)$$

where Y_o is the Young's modulus of the nonporous material.

Example 12.3

A mixture of sand (diameter ~ 1.0 mm) and quartz powder (diameter ~ 0.01 mm) is prepared to get maximum density. The sand has density 1600 kg/m^3 and density of quartz powder is 1400 kg/m^3. What fraction of the quartz powder is present in the final mixture? Calculate the maximum density of mixture. Quartz has true density 2650 kg/m^3. Calculate the percentage change in elastic modulus due to increase in density.

Solution:

Consider 1 m^3 of sand. It weighs 1600 kg. Sand is nothing but quartz, which has true density 2650 kg/m^3. So

Volume of sand weighing 1600 kg = 1600/2650 m^3 = 0.604 m^3

True porosity = 0.396

The pores of sand are filled with quartz powder in case the density of the mixture is maximum. The weight of the quartz powder required for filling up pores is 0.396 x 1400 = 554.4 kg.

Fraction of the quartz powder = 554.4/(554.4 +1600) = 0.257

Density of the mixture is 2154.4 kg/m^3.

Percentage change in elastic modulus = $[(Y_o - Y)/Y] \times 100\%$

$$= [(1.9 P - 0.9 P^2) / (1 - 1.9 P + 0.9 P^2)] \times 100\%$$

$$= 157 \%$$

Example 12.4

The specific gravity of Al_2O_3 is 3.96 Mg /m^3. A ceramic component is produced by sintering alumina powder. The weight of the component is 75g. It is 87g when soaked in water and 50g when suspended in water. Calculate the apparent porosity, true porosity and fraction of closed pores.

Solution:

The apparent porosity $P_a = [(87-75)/ (87-50)] \times 100\% = 32.4\%$

Bulk density $\rho_B = 75/(87-50) = 2.027$ Mg/m^3;

$\rho = 3.96$ Mg/m^3 ;

Hence $P = [(3.96 - 2.027)/3.96] \times 100\% = 48.8\%$

Fraction of closed pores = (48.8 - 32.4)/48.8 = 0.336

Glass transition temperature

The major content of glass is silica. At high temperatures, glass behaves as viscous liquid. Normally a liquid has a freezing or melting temperature (T_m) below which the solid phase is more stable than than the liquid phase. During the transformation of liquid to solid phase, activation energy is necessary for structural rearrangement. When a liquid glass is cooled, the heat of fusion released in this process is very low whereas the requirement of energy for crystallization is very high. Hence the process of crystallization is slow at freezing temperature. So, a molten glass requires to be cooled at a very slow rate to achieve crystallization at freezing temperature. However, if the cooling rate is high, then movements of atoms or molecules in glass are frozen, before they are arranged orderly. This is called glassy state.

When the temperature of the liquid is lowered, the volume contraction takes place due to decrease in free volume as well as due to reduction in the amplitude of vibration of atoms. Glass remains in supercooled liquid state, when the molten glass is cooled below the temperature T_m. On further cooling, the atoms or molecules do not rearrange any longer due to lack of sufficient activation energy. This occurs at a specific temperature, called as glass transition temperature or glass point T_g, Fig.11.8. There is no evolution of latent heat at T_g. If the temperature is lowered beyond glass point, decrease in volume is only due to decrease in the vibrational amplitude. Thus, the coefficient of thermal expansion is lower for glassy state as compared to that of supercooled liquid or of liquid state. Above the glass point, supercooled liquid is highly viscous and semi-rigid but as temperature is increased, its viscosity decreases and the true liquid state is obtained above T_m. The change in slope at T_g is not as abrupt as it happens in case of crystalline materials at temperature T_m.

The glassy state consists of incomplete and short-range network of SiO_2 and other oxides. In glass, short ranged network of SiO_2, consisting of silicate islands, rings or chains move past one another due to application of stress. This causes deformation. However, the attraction between these groups of atoms hinders the movement and resists the applied shear stress. This gives rise to viscosity in glass. The viscosity of glass is a function of temperature and is given by the Arrhenius formula:

$$\eta = \eta_o \exp \frac{Q_\eta}{RT} \qquad (12.4)$$

where Q_η is the activation energy related to the movement of atomic groups past one another. η_o is the pre-exponential coefficient. These parameters depend on the bonding and structure of the material. Viscosity of glass decreases as the temperature increases and glass becomes easily deformable. Modifiers are added to reduce the activation energy and hence the viscosity. The glass is easily deformable at a particular range of temperature if its viscosity becomes 10^3 Pa-s. This temperature is called working point. The lower limit of viscosity at which glass blowing can be done is 3000 Pa-s. A typical value of viscosity at blowing point is 4×10^6 Pa-s. The glass becomes soft enough for hot working process at the working point. The glass flows at a moderate rate due to gravity, at a specific temperature, called softening point. Viscous flow is an important property of the glass. The processing of glass requires optimal value of activation energy for viscous flow. Inorganic glasses become readily formable by glass blowing at elevated temperatures at which necking does not occur. For example when a glass tube is heated uniformly in a burner, it can be pulled into a long capillary of uniform cross-section if pulling is done carefully.

The temperature, at which any residual stress can be removed through movement of atoms within a short period of time, is called annealing point. This temperature is much lower than T_m and viscosity of the glass at this temperature is approximately 10^{12} Pa-s. When the viscosity becomes $10^{13.5}$ Pa-s at a temperature lower than the annealing point, it is called strain point. At strain point, the glass becomes rigid enough such that it can be handled without generation of residual stresses. When the temperature is lower than the strain point, fracture occurs before the onset of plastic deformation of the glass. The temperature difference between the annealing point and strain point is the annealing range of a glass.

Example 12.5

The minimum temperature at which glass blowing can be done is 1050°C. If a sample of glass has strain point of 500°C, calculate the softening point.

Solution:

We have the data:

	η(Pa-s)	Temp(°K)
Strain point	$10^{13.5}$	773
Blowing point	4×10^6	1323
Softening point	1000	T_{so}

Using Arrhenius formula we get:

$$10^{13.5} = \eta_0 \exp(Q/(R \times 773)), \text{ so } 31.08 = \ln \eta_0 + Q/(R \times 773)$$
$$4 \times 10^6 = \eta_0 \exp(Q/(R \times 1323)), \text{ so } 15.20 = \ln \eta_0 + Q/(R \times 1323)$$
$$1000 = \eta_0 \exp(Q/(R \times T_{so})), \quad \text{so} \quad 6.91 = \ln \eta_0 + Q/(R \times T_{so})$$

or
$$15.88 = Q((1/773)-(1/1323)) /R$$
$$8.29 = Q((1/1323)-(1/ T_{so})) /R$$

or
$$(15.88/8.29) = ((1/773)-(1/1323))/ ((1/1323)-(1/ T_{so}))$$

or
$$T_{so} = 2105°K = 1832°C$$

Example 12.6

Silica glass has a viscosity of 10^{12} Pa-s at its annealing point 950°C. Find the viscosity of this glass at the softening point 1500°C, if the activation energy for the viscous flow of this glass in this temperature range is 382 kJ/mole.

Solution:

$$\eta_{an} = 10^{12} \text{ Pa-s}$$
$$T_{an} = 950°C = 1223° \text{ K}, \ T_{so} = 1500°C = 1773°K$$

Using Arrhenius formula,

$$\eta_{an} = \eta_0 \exp (Q/RT_{an}) \text{ and } \eta_{so} = \eta_0 \exp (Q/RT_{so})$$

$$\frac{\eta_{so}}{\eta_{an}} = \exp \left(\frac{382 \times 10^3}{8.314} \right) \left[\frac{1}{1773} - \frac{1}{1223} \right], \eta_{so} = 8.68 \times 10^6 \text{ Pa-s}$$

Heat treatment of glasses

When a glass is heated at an elevated temperature and then cooled, usually thermal stresses are developed due to thermal gradient between the surface and interior regions. This happens because the cooling rate and thermal contraction are different at the surface and in the interior. The thermal stresses may weaken the material and cause fracture to the ceramic. This is called **thermal shock**. If the cooling rate is slow, then such thermal shock can be avoided. Heating the material to the annealing point and then cooling slowly to the room temperature can reduce the thermal stresses.

Silicates, borates, phosphates etc. have open structures due to high cation-cation repulsion. In silicates, the tetrahedra are joined at the corners having certain amount of freedom of position for tetrahedral units surrounding a central unit. The heat of fusion Δh is small for such a system. Consequently, the critical energy for nucleation is high since it is inversely proportional to the square of Δh. The enthalpy for motion ΔH_m, which activates the transfer of atoms from the parent phase to the nucleating phase and helps in the growth, increases with the increase of viscosity. Higher ΔH_m value makes the diffusion rate slower. Viscosity of silicates is very high and is of the order of 1000 Pa-s at the equilibrium freezing temperature. Both ΔH_m and Δh values for the above material are unfavourable for nucleation and growth. So there are little chances for such materials to crystallize. Suitable metallic oxides are added to silica to provide heterogeneous nucleating agent and promote crystallization. The formed glass articles are then subjected to special heat treatment to produce better mechanical and thermal properties. A typical example of heat treatment process is described below.

Heterogeneous nucleating agent such as TiO_2 is first dissolved in the molten silicate and mixed thoroughly for homogenizing the mixture. After shaping the material, it is annealed. It is then brought to a temperature corresponding to which nucleation rate is maximum (Fig.12.2) and kept there for a period of time such that very fine particles of TiO_2 are precipitated. Then it is heated to

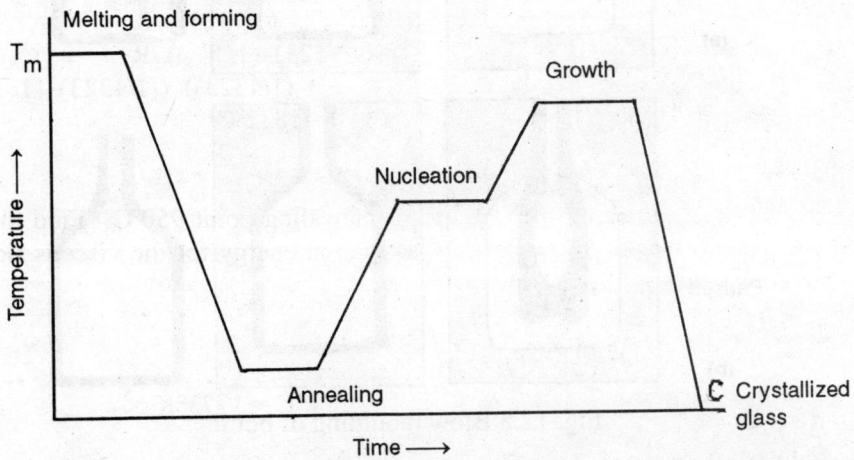

Fig. 12.2 Heat treatment processes for producing a glass-ceramic

a temperature corresponding to maximum growth rate and held there for a period of time, till the stable nuclei grow to large sizes. This polycrystalline glass is called glass-ceramic. The average grain size of glass-ceramic is about 0.1 μm as compared to a fine grain size of 10μm in metals. This process is called crystallization or devitrification of glass. Devitrified glass is polycrystalline and hence is not transparent. Glass-ceramics have good mechanical and thermal shock resistance and have high thermal conductivity. These materials are used to fabricate pore-free cookware, tableware etc. using conventional glass forming techniques. These are corning cookwares, The trade names of glass-ceramics are Pyroceram, Cer-Vit and Hercuvit. These materials are also used as substrate for printed circuit boards. A typical glass ceramic has 70 w/o SiO_2, 18w/o Al_2O_3, 4.5w/o TiO_2 and 2.5 w/o Li_2O, which is used for fabrication of ovenware etc. Glass-ceramics are used for fabrication of guided missile radomes (lenses for high-frequency radio waves), photosensitive glasses capable of storing photosensitive three-dimensional images.

Glass forming

Glass is manufactured by heating the raw materials to an elevated temperature near melting point. The complete melting and thorough mixing of constituent materials give a homogeneous final product. Efforts are made to remove porosity during the process of manufacturing of good quality glasses. Glass forming can be done by various methods like pressing, blowing, drawing etc.

Pressing and blow moulding

In the pressing method of fabrication the raw constituents are pressed in graphite coated cast iron mould of desired shape to produce glass items of required shapes and sizes. This process fabricates thick pieces like plates and dishes. The glass blowing is a very common and automated technique to produce glass jars, bottles and bulbs, Fig. 12.3.

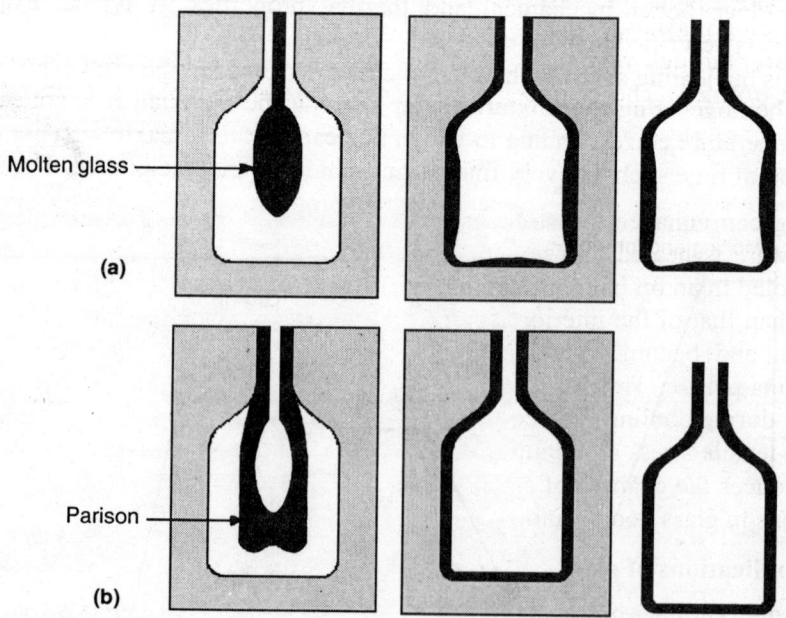

Fig. 12.3 Blow moulding of bottles

The glass gob is inserted into a blow mould and pressing is done to give a rough shape. Finishing is done by air pressure, which is forced into the mould from outside. This process is sometimes done in two stages to achieve uniform thickness of the wall of the bottle. In the first stage a partially blown bottle is produced which is called parison. This parison is so designed that when it is blown in a second (finishing) mould a bottle is produced with a uniform wall thickness. The temperature and rate of the process must be controlled appropriately so that the viscosity is within working range (10^3 Pa-s to 10^6Pa-s). After the bottle is ready to be taken out of finishing mould, cooling should be done below T_g such that the outer surface becomes relatively rigid while inner surface is still at higher temperature. Sometimes during cooling process thermal stresses are set up due to the outer rigid surface, which restrict the rearrangement of the molecules in the interior, and this leads to the premature failure during production. These stresses may be removed by annealing heat treatment in which glass item is heated to the annealing point and then cooled very slowly to the room temperature.

Glass drawing

Long glass pieces like sheet, rod, fibres etc. can be produced by glass drawing process. These products are of constant cross-section. Sheets can also be fabricated by hot rolling. The surface finish of the sheet can be improved by float moulding (e.g. Pilkington float moulding process). In a process for the continuous drawing of glass sheet the raw materials are fed into one end of a large horizontal furnace. The glass is melted and refined here. The glass is fed into a pair of cooled rollers. At the other end, the solid glass sheet emerging from the rollers is then annealed.

In float moulding process soft glass sheet is floated on a bath of molten tin at an elevated temperature. This process smoothens out the surface of sheet in contact with the molten tin. The upper surface of the glass develops by gravity. The sheet is subsequently cooled and annealed. The drawback of this process is that tin may be introduced in the glass.

The fibreglasses are among the recent developments in materials. The fibres (A- and E-types) are used for reinforcing plastics, cements etc. Fibreglasses are produced by forcing molten glass under steam pressure, through very small holes at the chamber base. When it is cooled, network of glass fibres are formed. The chamber and hole temperatures control the desired value of viscosity.

Glass tempering

Thermal tempering can enhance the strength of glass, which is called glass tempering. In this process, the glassware is heated to a temperature in the glass transition range below softening point. It is then cooled in an oil bath or in a jet of air at room temperature. The cooling rate of the surface is higher than that of the interior region. The surface cools more rapidly as compared to the interior region and becomes rigid at strain point. The interior region, being at higher temperature, remains plastic. This results in contraction of the interior region more than that of rigid outer surface during cooling. The cooling at room temperature results in compressive stress on the surface and tensile stress at the interior region. Such distribution of stresses gives strength to the glass and reduces the chances of creation and propagation of cracks. Tempered glasses are used for applications in glass doors, automobile windshields etc.

Properties and applications of glass

Glass is an important engineering material. The mechanical processing like rolling, drawing, blowing, pressing etc.can be done on the glass during fabrication of various items. Glass is a rigid amorphous solid. It has refractive index of about 1.5, which can be varied by changing chemical constituents. The conductivity and heat capacity of glass increase with temperature in the low temperature range and then reach a nearly constant value for temperatures of about a few hundred degrees centigrade. The glass is brittle but it can be made ductile at an elevated temperature. It has greater wear and abrasion resistance than steel. The tensile strength of glass varies with constituents as well as with size. For example, glass fibre of micron diameter has tensile strength 50 times larger than a glass rod of diameter 10 mm. The glass transition temperature varies from one type of glass to another.

Glasses are extensively used in packaging and processing of food materials. Glasses are also used in building construction. Glass coating on ceramics and metal surfaces saves the same from corrosion and increases the beauty. Safety glass is a specially manufactured glass, used in automobiles due to its nontransparency to ultraviolet light and resistance to shattering. Safety glass is made by sandwiching tough transparent material say plastic sheet between two transparent glass sheet and then rolling the sandwich into a single unit.

12.3 CLAY

Clay is a naturally available ceramic raw material. It is widely used for fabrication of large varieties of items, starting from building material like bricks, tiles, and sewer pipes to ceramic goods like pottery, tableware, china and sanitary ware. The constituents of clay vary considerably depending upon the place from where the clay is obtained. Sometimes nonplastic or other ingredients are mixed to increase strength and better finishing. English China contains 90% kaolinite (Al_2O_3. $2SiO_2$. $2H_2O$) whereas ordinary brick may contain only 30% kaolinite. Other constituents of bricks are montmorillonite typified by pyrophyllite ($Al_2(Si_2O_5)_2(OH)_2$ and illite. Clay is a popular raw material because it is easily available and cheap. The forming of clay product is quite easy. When clay, other ingredients and water are mixed in proper proportions, the mixed material becomes plastic hence it can be given any shape and size very easily. After forming, it is dried and then fired at an elevated temperature, which gives mechanical strength to the final product. The clay products are called whitewares when the products become white after firing at elevated temperature. Clay contains varieties of phases but the most common out of these is kaolinite. When water is added to clay, the water molecules form a thin film around clay particles.

Clays are fine-grained hydrated aluminium silicates having sheet like structure or flake form. These sheets can absorb water on the surface between the sheets. The water molecules act as plasticizer for clay. Hence wet clays are malleable and are said to be hydroplastic. The raw materials for whiteware products are clay and triaxial porcelain composition (clay-feldspar-flint). Clay forms a viscous liquid during firing and feldspar acts as a flux and helps in vitrification. Flint is inactive filler to clay at low temperature. The clay with this filler forms a viscous liquid at high temperatures. The final product is quite heterogeneous and is finished by applying a glaze. Fine powder of clay, frit and water mixture is applied on the surface of the final product by spraying or dipping. The whole body is then fired to produce thin glazed coating. Appropriate colorants can be added to the glaze for finishing purposes. High strength clay based products can be made by reducing porosity. Firing temperature and size of the particles are deciding factors for porosity. On the other hand, low porosity products fail under application of stresses.

Vitrification in clay

Ceramic products like porcelain, structural clay products have glass phase. When these types of ceramic materials are fired, the glass phase existing in the material liquefies and fills up the pore spaces in the material. This process is called vitrification. During cooling, the liquid phase solidifies to form glassy matrix, which bonds other unmelted particles. Any vitrified material should contain suitable proportion of glass so that the final product should have high strength. Excessive glass content makes the final product brittle.

12.4 REFRACTORY

The refractory ceramics are quite important in modern industries because of the capacity to withstand high temperatures without melting or decomposing or reacting with the environment, Table 12.1. These materials are inert to many fluxing situations. They also possess reversible thermal expansion and resistance to thermal shocks. The common refractory ceramics are silica-alumina systems, silica-iron oxide systems, magnesia-alumina systems etc. These materials are used for making furnaces, which perform various industrial operations like refining metals, generating power, melting, fabrication of materials, heat treating etc. Requirement of specific refractories for furnaces depends on service conditions, which are quite often combination of corrosive environment, high mechanical stress and high temperature.

Table 12.1 Refractory materials

S.No.	Material	Composition	Maximum working Temperature(°C)
1.	High-heat-duty fireclay brick	Al_2O_3 35-42%, SiO_2 52-60%	1743
2.	High Alumina	Al_2O_3 50 -80%	1866
3.	Kaolin	Al_2O_3 44-45%, SiO_2 51-53%	1760
4.	Mullite	Al_2O_3 60-75%, SiO_2 18.0-34%	1838
5.	Zircon	$ZrSiO_4$	2015
6.	Silica	SiO_2 95-96%	1743
7.	Magnesite	MgO 92-98%, Fe_2O_3 0.1-2%	2199
8.	Zirconia	ZrO_2	2400
9.	Thoria	ThO_2	2700
10.	Dolomite	CaO-MgO	1800
11.	Quartz	SiO_2	1700

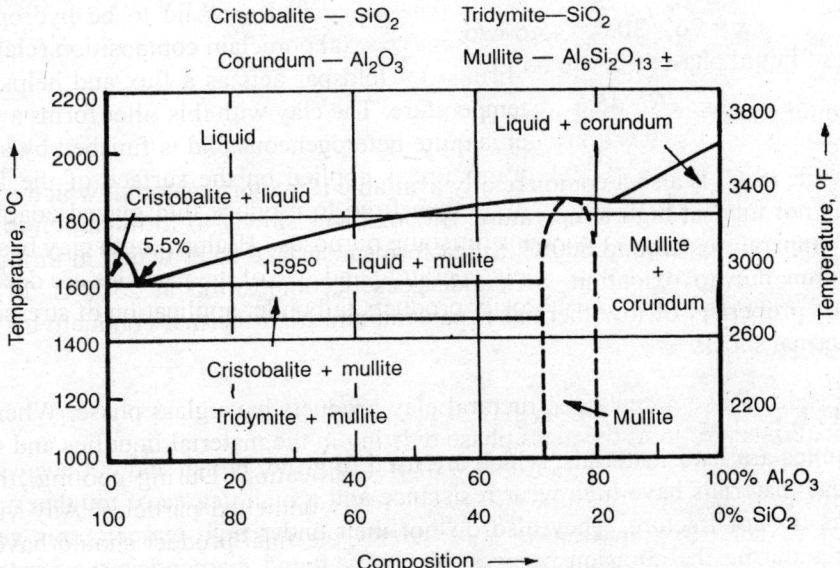

Cristobalite — SiO_2 Tridymite—SiO_2

Corundum — Al_2O_3 Mullite — $Al_6Si_2O_{13}$ ±

Fig. 12.4 Phase diagram for silica-alumina system.

The phase diagram, Fig. 12.4, for silica-alumina system gives the suitable composition for refractories. The composition with 3 to 8 percent of Al_2O_3 should not be used for refractory purposes, since these materials start melting between 1600°C to 1700°C. As the percentage of Al_2O_3 increases the melting temperature of the alloy increases. This also results in the increase of temperature range over which the material softens. The material containing 20 to 50% of Al_2O_3, finds its use in fireclay refractory bricks. Increase in Al_2O_3 improves the performance of the refractory material in service conditions. To increase alumina content, calcined bauxite is used as ingredient, which results in the final product, called bauxite refractories. If Al_2O_3 is increased to more than 72%, then the material can be used even up to 1800°C; the operation temperature is reduced to 1650°C if the material is under stress. Alloys with more than 80 w/o Al_2O_3 are made from fused alumina bonded by fireclay and are better than bauxite refractories.

Example 12.7

A fire clay ($Al_{14}Si_4O_{27}(OH)_4$) is used to produce refractory bricks. It undergoes the dissociation reaction:

$$Al_6Si_4O_{15}(OH)_4 = Al_6Si_2O_{13} + 2SiO_2 + 2H_2O$$

When a fire clay is heated at 1500°C to form refractory bricks, it contains the equilibrium phases cristobalite (SiO_2) and mullite $(Al_6Si_2O_{13})$ of 40% and 60%, respectively.
(a) Calculate the overall composition of the brick.
(b) The brick breaks into mullite and liquid phases at 1591°C. Calculate the composition and weight percent of liquid present at this temperature.

Solution:

Let us consider that the sample contains x w/o Al_2O_3.

$$\text{Hence fraction of cristobalite} = \frac{40}{100} = \frac{76-x}{76-0}$$

Thus $\qquad$ x = 76 − 30.4 = 45.6 w/o
Composition of liquid phase: 5.5 w/o Al_2O_3.

$$\text{Weight of liquid phase} = \frac{76-45.6}{76-5.5} = 0.431$$

Silicon carbide (SiC) is also a commercially available refractory material, which is harder than Al_2O_3 but it is not inert at high temperature. SiC oxidizes slowly to form carbon monoxide gas and silica at temperatures around 800°C. Silica forms a surface layer which in turn protects SiC from degradation due to oxidation. SiC undergoes decomposition at 2300°C. It has a good combination of properties of low thermal expansion and high thermal conductivity so that it is resistant to thermal shock.

12.5 ABRASIVES

Abrasive ceramics are hard materials, which are used to grind, polish and cut away other softer materials. These materials have high wear resistance and a high degree of toughness so that the abrasive grains do not fracture. They also do not melt under high temperatures generated by frictional forces during the abrasion process. Naturally found diamond is the hardest material having hardness number 10 on Mohs scale. Both natural and synthetic diamonds are used as abrasives. The other common abrasives are silicon carbide, tungsten carbide, alumina (hardness number 9), Topaz ($SiAl_2F_2O_4$, hardness number 8), quartz (hardness number 7). Al_2O_3 is a well-known component of grinding wheels and is also called emery.

The friability or relative toughness of an abrasive depends on proportions of various components present in the ceramic. Solid solution of TiO_2 and Fe_2O_3 in Al_2O_3 is much tougher and more resistant to wear and fracture as compared to pure corundum. In case of the later material, the abrasive grains are extremely friable and fracture very easily. This type of material is useful for grinding hard materials. SiC has a good abrasion resistance and is used for ball and roll bearings, rocket nozzles etc. Abrasives are used in the following ways:

i) Bonded to grinding wheels

Organic resins or glossy ceramics are used to bind the abrasive particles to the wheel. The surface of the wheel should have some pores so that flow of air or liquid within the pores that surround refractory grains reduce the temperature, arises due to frictional forces.

ii) Coated Abrasives

The abrasive powder is coated on some type of paper e.g. emery paper, sand paper. The cleaning of surface, polishing etc. is done on metal, ceramics, wood and plastics by using coated abrasives.

iii) Loose grains

Loose grains of abrasives are often employed in some type of oil-based vehicles with the help of grinding or polishing wheels.

12.6 CEMENT

Cement, plaster of paris and lime belong to a class of ceramics, which form paste or a plastic material with water. This paste subsequently sets and hardens. Certain materials of these types chemically bind particulate aggregates into a lump of cohesive material, where cementitious bonds are formed at room temperature. Portland cement is one of the most economic and commonly used cement, which requires lime (CaO), silica (SiO_2), and ferric oxide (Fe_2O_3) as raw materials. High alumina cement is used for refractory purposes. The process of manufacturing Portland cement is shown below.

Step 1: Appropriate proportions of limestone and clay are mixed and ground.
Step 2: The mixture is fed into a rotary kiln where temperature is maintained. at 1400–1600°C. This is called clinkering or calcination.
Step 3: Cool and pulverize.
Step 4: A few percent of gypsum powder ($CaSO_4.2H_2O$) is mixed with the clinker product and ground to fine powder. The final product is Portland cement.

The major constituents of Portland cement are: dicalcium silicate ($2CaO.SiO_2$: 28%), tricalcium silicate ($3CaO.SiO_2$: 46%), tricalcium aluminate ($3CaO.Al_2O_3$: 11%), gypsum ($CaSO_4$: 3%), magnesium oxide (MgO: 3%) and glass.

The amount and composition of the phases in the final product depends on the raw materials used and on the conditions of processing. With higher alumina content in the raw material, luminous cement is obtained. During firing process, some fraction of the charge converts into a liquid phase. This happens also in case of Portland cement but the amount of liquid phase formed during firing is less in case of Portland cement as compared to aluminous cement, which is also called, melted cement.

When water is added to the cement, a plastic paste is obtained. The initial reaction of hydration of tricalcium aluminate proceeds very fast as follows:

$$3CaO.Al_2O_3 + 6H_2O \longrightarrow 3\ CaO.Al_2O_3\ .6H_2O \tag{12.5}$$

This reaction produces heat of hydration. The hydration reaction is reversible at high temperatures and causes dissociation of $3CaO.Al_2O_3.6H_2O$ to $3CaO.Al_2O_3$. This reverse reaction does not affect much of the bulk material since hydration, Eq.(12.5) is only a surface reaction for the cement. For certain period of time there is no change in the state of the paste material this period is called induction period. After a period of time, the material becomes unworkable due to initiation of setting; this is known as initial set. The setting period starts after this. The paste becomes stiffer during this period until the material can be regarded as rigid solid i.e. final set, which is the final concrete. It is formed without enough strength. To hinder this reaction gypsum is added which forms calcium sulphoaluminate. This coats the grains of $3CaO.\ Al_2O_3$ and hence retards the above reaction. The other reactions, which follow in succession with much slower pace, are given below:

$$2(3CaO.\ SiO_2) + 6H_2O \rightarrow 3CaO.\ 2SiO_2\ .3H_2O + 3\ Ca(OH)_2 \tag{12.6}$$

$$2(2CaO.SiO_2) + 4H_2O \rightarrow 3CaO.2SiO_2\ .3H_2O + Ca(OH)_2 \tag{12.7}$$

The first reaction proceeds with a rate such that 70% of the reaction is complete in 30 days, whereas the second reaction proceeds very slowly. Thus it takes years for the concrete to get hardened. The important point to be noted here is that the cement is hardened not by drying but by hydration. Portland cement is sometimes termed as hydraulic cement. During hydration the volume of the concrete is increased by more than a factor of two as compared to anhydrous cement.

The appropriate proportion of water is to be added to the cement to achieve the maximum strength of concrete. The dependence of compressive strength (σ_c) on water to cement ratio (w/c) can be given by empirical formula,

$$\sigma_c = (A/B)^{1.5}\ (w/c) \tag{12.8}$$

A is a constant of value 100 MN/m^2 and the constant B varies with the types of cement.

12.7 OTHER CERAMIC MATERIALS

Alumina

Alumina is widely used as electrical insulators. Alumina can withstand high voltage as well as temperature fluctuations. So it is suitable for use in sparkplug insulators. It is also used as lining of high temperature furnaces since its melting point is very high. Also it is used for tool tips and grinding tools since it has high compressive strength and is wear resistant.

Silicon nitride and sialon

Silicon nitride (Si_3N_4) is brittle and can react with atmosphere. Since it has good thermal shock resistance, high thermal conductivity with a low thermal expansion and high strength, it is used in heat exchanger, furnace components, and crucibles. Sialon is a technical ceramic, which is obtained from Si_3N_4. The silicon nitride is formed with high porosity content, which is quite undesirable. Moreover there are problems with production of Si_3N_4. The structure of β-Si_3N_4 consisting of SiN_4 tetrahedra has the capability of accommodating the simultaneous substitution of aluminium atoms for silicon and oxygen for nitrogen. As a result, a new ceramic β-sialons $Si_{6-z}Al_zO_zN_{8-z}$ is obtained where z denotes the level of substitution. By adding suitable proportion of oxides such as magnesia, alumina some amount of liquid is formed at low temperature at the time of processing which helps in the formation of denser glassy phase without any hot pressing. Silicon mixed with Y_2O_3 after suitable heat treatment gives better strength at high temperature. The material formed by this process is sialon with yttrium-aluminium-garnet at the grain boundaries. This material retains high strength up to temperatures of the order of 1400°C.

Zirconia

Zirconia, ZrO_2 is also a technical ceramic, which can exist in three crystalline forms, cubic, tetragonal and monoclinic. As zirconia is cooled slowly from temperature 2680°C, the cubic phase of the material transforms to tetragonal phase and at a still lower temperature the conversion of tetragonal phase to monoclinic phase takes place through the martensitic transformation. As a result of this martensitic transformation, the volume of the material is increased by 3%. Increase in volume causes cracking in the material. By adding Y_2O_3 in excess of 9 mole percent of zirconia we obtain the stabilized cubic phase at room temperature. This material has application in oxygen sensors, fuel cells etc.

12.8 APPLICATION OF CERAMICS ON THE BASIS OF THEIR PROPERTIES

Electrical, mechanical and thermal properties of ceramics decide the usefulness of these materials for specific applications.

Electrical ceramics

Electrical conductivity of ceramics is quite low. Ceramics are often ionic compounds where cations and anions are strongly bonded to each other. Large ions have low mobility due to their large masses. Low mobility also arises due to hindrance of ionic movements in crystalline solid or highly viscous glasses. The electrical conductivity is expressed as

$$\sigma_i = nq\mu_i \qquad (12.9)$$

where $$\mu_i = qD_i/kT$$

using Einstein relationship. Here q is charge and n is density of the carrier. D_i is ionic diffusion coefficient. Thus,

$$\sigma_i = (nq^2/kT)D_o \exp(-Q/kT) \qquad (12.10)$$

where Q is the activation energy for diffusion and D_o is the preexponential factor which depends on diffusing atoms as well as diffusing medium. These relationships clearly explain the ionic conductivity property for ceramics. It is known that defect structures provide easy path for ionic conduction, so the temperature of stoichiometric compound is raised to introduce defects as well as to increase thermal energy, which in turn accelerates ionic movements. Sometimes, suitable network modifier oxides like Na_2O, K_2O, CaO etc. are used in glass to improve ionic conductivity. Increase in σ value is more when Na_2O is used as modifier as compared to that when CaO is used, because of the following reasons:

- More cations are added per oxygen ion in the former case.
- Na^+ ion is held less strongly than Ca^{2+} ion.
- Na^+ ion has smaller diameter than that of Ca^{2+} ion.

Na_2O and K_2O provide Na^+ and K^+ ions, respectively, which do not enter the network structure and retain metallic ion behaviour. The glass transition temperature also plays an important role in deciding conductivity, since network structure of ceramic can accommodate cation movements and can increase conductivity, if the temperature of the medium is raised above T_g. Semiconducting ceramics show increase in electrical conductivity with increase in temperature, since at higher temperature, there is greater ionic diffusion and more electrons are able to enter the conduction band.

Example 12.8

The electrical conductivity due to Na^+ ions in NaCl is 5.6×10^{-4} ohm^{-1}m^{-1} and 3.3×10^{-3} ohm^{-1}m^{-1} at 858°K and 925°K, respectively. Find the value of activation energy of Na^+ diffusion in NaCl.

Solution:

$$Q_{Na+}/R = [\ln (5.6 \times 10^{-4} \times 858) - \ln (3.3 \times 10^{-3} \times 925)]/((1/925) - (1/858))$$
$$Q_{Na+} = 182 \text{ kJ/mole}$$

Ceramic semiconductors

Ceramic materials which have low conductivities are usually semiconductors and are used for making (i) constant voltage drops i.e. fixed resistors; (ii) heating elements; (iii) thermistors.

Fixed resistors

Ceramic insulation materials are used for making fixed resistors, which can be film type or composition type. As an example, film type of fixed resistor consists of a glass rod with thin conductive layer of semiconducting oxide like SnO. In the composition type resistors, there is a mixture of carbon and insulating powders within a plastic or porcelain jacket which gives low but constant conductivity if the temperature coefficient of resistance is low.

Heating elements

Commonly used nonmetallic heating elements are silicon carbide and graphite. These materials are semiconducting, and are capable of rapidly converting electrical energy into thermal, when suitable voltage is applied. They are refractory in character and hence provide rapid transfer of heat by convection and radiation mechanisms while used in industrial furnace applications. When silicon carbide is used for furnace applications, protective SiO_2 glaze is developed on it. Other ceramic oxides and carbides are also used as high temperature resistors like zirconia.

Thermistors

Thermistors are thermal resistors, which are used as temperature compensators or as temperature sensors. Such devices utilize materials with negative temperature coefficient of resistance. Ceramic semiconductors, which are widely used for thermistors are magnetite (Fe_3O_4), solid solution like $Li_x^+Mn_x^{3+}Mn_{1-2x}^{2+}O$. There are ceramics, which show nonlinear relationship between current and voltage and are utilized in rectifiers and variable resistors. Cu_2O has wide applications in rectifiers. SiC is used both as a variable resistor and a rectifier.

Ceramic insulators

Ceramic insulators have high resistivity, high dielectric strength and low loss factor. The dielectric strength is the voltage gradient, volts/m, which causes electrical breakdown. The loss factor is the energy lost per cycle with ac voltage.

In ceramic materials, metallic elements donate electrons to the nonmetallic atoms, which retain them. Thus, so called conduction electrons of metallic elements are immobilized and ceramic materials behave as good insulators e.g. various compositions of $MgO\text{-}Al_2O_3\text{-}SiO_2$ form very good electrical insulators. Common electrical porcelain, is made out of talc $(Mg_3Si_4O_{10}(OH)_2)$ and clay $(Al_3Si_2O_5(OH)_4)$ using suitable heat treatments.

Piezoelectrics

Piezoelectric ceramics are a special class of materials, which show change in voltage drop between the two electrodes at the ends of the piezoelectric sample under applied pressure, Fig. 12.5. The sample is shown as collection of dipoles arranged in such a way that they produce a charge difference between the two ends. When pressure is applied along the length of the dipoles, dipole moment is decreased due to decrease in the dipole length. This results in voltage change between the two electrodes and a flow of current if the two electrodes are connected. The change in voltage can be amplified in an electric circuit. Such materials can therefore be used in pressure gauge as well as in mechanical-electrical transducers. The reverse effect is the production of mechanical vibrations using alternating voltage. This principle is used to generate ultrasonic vibrations. Examples of such materials are $BaTiO_3$, $PbTiO_3$ etc.

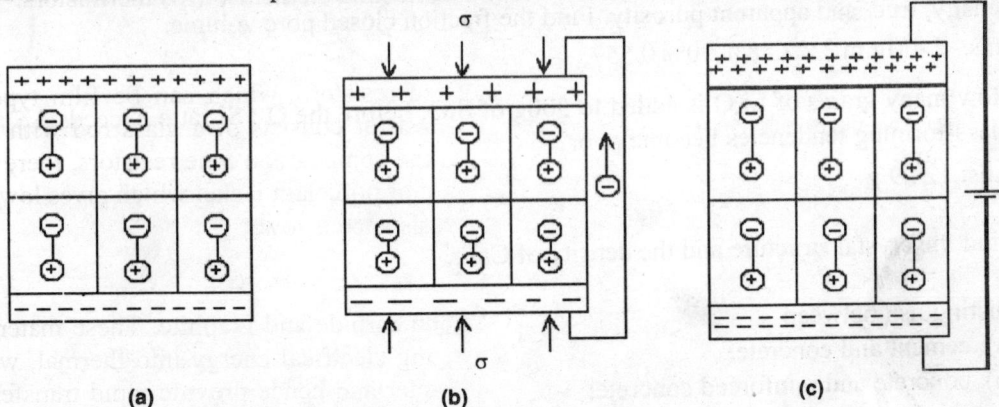

Fig. 12.5 Schematic representation of piezoelectric effect.

Structural ceramics

Cement, concretes, vitrified products like bricks, tiles, whitewares etc. come under the broad class of structural ceramics.

EXERCISES

12.1   Using the $SiO_2\text{-}Al_2O_3$ phase diagram, find the following:
   (a) Compositions of the phases involved in the eutectic reaction at 1595°C.
   (b) Fraction of cristobalite present in the alloy having 4.5 w/o Al_2O_3 and 95.5 w/o SiO_2 at 1600°C.
   Ans: (a) Liquid phase (5.5 w/o Al_2O_3 and 94.5 w/o SiO_2), cristobalite (100 w/o SiO_2) and mullite $(Al_6Si_2O_{13\pm})$; (b) 0.18

12.2 What are bauxite refractories? If Al_2O_3 is increased to more than 80%, what happens to the properties of refractories? What are the advantages and disadvantages of using SiC as refractories?

12.3 What are abrasives? Describe the uses of abrasives. How do you distinguish between abrasives and refractories?

12.4 A 60 kg charge of glass contains 49.6 kg of SiO_2, 5 kg of Na_2O, and 5.4 kg of CaO.
(a) Find the molar percentage of each oxide.
(b) What is the empirical formula for this glass?
Ans: (a) Mole % of SiO_2 = 82.34, mole % of Na_2O = 8.04, mole % of CaO = 9.62;
(b) $(SiO_2)_{82} (Na_2O)_8 (CaO)_{10}$

12.5 Describe the structure of $BaTiO_3$. What are the lattice parameters of $BaTiO_3$, if its density is 6020 kg/m^3.
Ans: 4Å (Assuming $BaTiO_3$ unit cell to be a perfect cube)

12.6 Silicon carbide (SiC) has a specific gravity of 3.1 Mg.m^{-3}. A sintered SiC part of volume 300cm^3 weight 720g. The part weighs 750g when it is soaked in water. Calculate the bulk density, true and apparent porosity. Find the fraction closed pore volume.
Ans. 2.4Mg.m^{-3}, 22.58%,10%,0.557.

12.7 How many grams of Li_2O is added to 500g of SiO_2 before the O : Si ratio exceeds 2.5 and glass-forming tendencies become poor?
Ans: 124.9 g

12.8 Find the crystal structure and the density of CaF_2.

12.9 Distinguish between
(a) cement and concrete;
(b) concrete and reinforced concrete;
(c) clay and glass.

12.10 Cement is hardened not by drying but by hydration. Explain.

12.11 5 g of $YBa_2Cu_3O_{7-x}$, high temperature superconductor (See Chapter 17 for details) is obtained by solid-state reaction of Y_2O_3, Cu_2O and $BaCO_3$. If x = 0.1, a = 3.82 Å, b = 3.88 Å, c = 11.6 Å
(a) How many grams of each constituent must be added.
(b) What should the value of x? Calculate its density.
Ans: (a) 0.849g Y_2O_3, 1.615g Cu_2O, 2.969g $BaCO_3$; (b) x varies from 0.35 to 0.1, 6416 kg/m^3

12.12 What is vitrification? In what type of ceramic materials does vitrification take place? Explain with an example.

12.13 Calculate the density of UO_2 that has CaF_2 structure ($r_{U4+} = 0.105$ nm and $r_{02-} = 0.132$ nm).

Ans: 2734 kg/m^3

12.14 Calculate the ionic packing factor for $SrZrO_3$, which has the perovskite structure. Ionic radii are 0.127 nm, 0.087 nm and 0.132 nm for Sr^{2+}, Zr^{4+} and O^{2-}, respectively.

$(a = r_{Zr4+} + r_{02-})$

12.15 What are pressing and blow moulding? What is parison?

12.16 The ionic conductivity of NiO, which arises due to the movement of cations only, is $1\Omega^{-1}cm^{-1}$ at a temperature of 1150°C. Calculate the diffusion coefficient for Ni^{2+} ion in NiO. Given that NiO has NaCl structure and ionic radii are 0.132 nm and 0.078 nm for O^{2-} and Ni^{2+}, respectively.

Ans: $3.55 \times 10^{-10}\,m^2\text{-s}$

Note: See Chapter 4 for the structural details of ceramic compounds.

13 Composites: Structure and Properties

13.1 INTRODUCTION

With the advent of new technologies, the need for tailor-made materials has been felt more and more for specific applications. For example, the material for aircraft should have the properties like low density, high strength, stiff, abrasion and impact resistant and noncorrosive. This is rather a peculiar combination of properties. Since a strong material is usually dense and stiffness or strength makes the material less resistant to impact. Thus, the need for development of materials with unusual combination of properties led to the development of composites. A composite is a multiphase material, which exhibits combination of properties that makes it suitable for a specific application. The constituent phases are separated by distinct phase boundaries. Multiphase metal alloys, ceramics, and polymers are examples of composites. Fibreglass, which consists of glass fibres embedded in polymer, is a fibre-reinforced composite. Example of a naturally occurring composite is wood consisting of strong and flexible cellulose fibres held together by lignin. As an example of microscopic composites, cell wall of a plant is made up of cellulose fibrils arranged in alternate layers lying at different angles to each other. These layers are interlaced by lignin, apparently acting as cement. The cell wall, a sac like envelope of the plant cell, has tremendous tensile strength exceeding even that of the highest quality steel.

In this Chapter, the discussion is centered on artificially made composites. A common example of composite is the reinforced concrete where steel rods are embedded in concrete. Concrete itself is a composite of cement, sand, stone-chips and water. Glass fibre-reinforced plastics are also composites, which are widely used in military and aerospace applications.

13.2 CLASSIFICATION OF COMPOSITES

Composites can be broadly classified as macroscopic and microscopic composites. Both of these are discussed below:

Macroscopic composites

Macroscopic composites are mainly used for structure e.g. structural laminates, concrete, reinforced concrete. In structural laminates, panels of plywood or fibre-reinforce plastic are placed one above another and then cemented together in such a way that the orientation of high strength direction varies with each successive layer, Fig. 13.1 (a). Hence a laminated composite may have high strength in a number of directions. The lamination sheet can be constructed by embedding cotton, fibres, paper etc. in plastic.

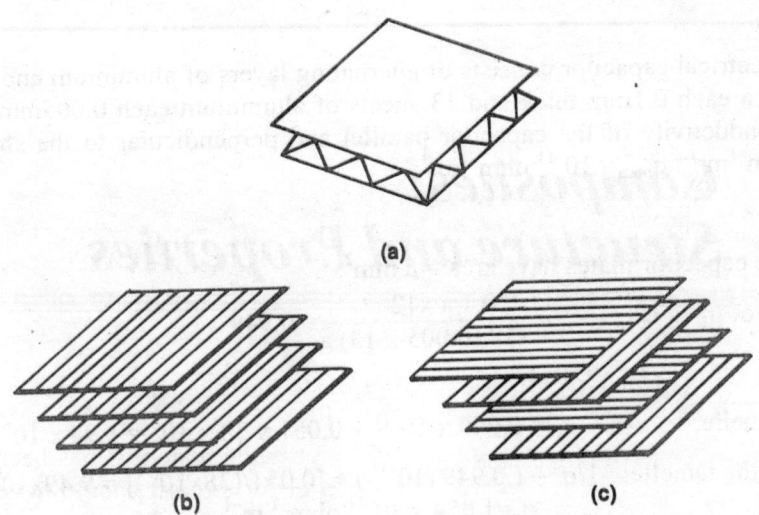

Fig.13.1 (a) Fibre-reinforced layer for structural laminate; (b) and (c) Cardboard with corrugated board in between the cardboard sheets.

When the fibre-reinforced layers are oriented in such a manner that all the directions of high strength fibres are parallel then the laminated structure is stronger in one direction compared to all other directions, Fig. 13.1(b). Ski is an example of a laminated structure. Plywood is made by gluing together thin sheets of wood with their grain directions perpendicular to each other, Fig. 13.1(c). The directions of cellulose fibres in the wood sheet are termed as grain directions. The cladding of aluminium-copper alloy with aluminium gives a material with a better corrosion resistance than that of the alloy. Coating of zinc on steel gives galvanized steel, which has a better corrosion resistance.

Corrugated cardboard has paper corrugations sandwiched between layers of paper. The sandwich structure with layer of foam or corrugation in between two layers of papers gives a light but relatively strong and stiff structure.

Properties of laminar composites

Properties of composites depend on the orientation and the properties of components of the composites. Density ρ_c of the laminar composite is given as $\Sigma_i\, f_i\rho_i$ and other properties of laminar composite are given by

Property	Parallel to the lamellae	Perpendicular to the lamellae
Electrical conductivity	$\sigma_c = \Sigma_i\, f_i\sigma_i$	$1/\sigma_c = \Sigma_i\, f_i/\sigma_i$
Thermal conductivity	$\kappa_c = \Sigma_i\, f_i\kappa_i$	$1/\kappa_c = \Sigma_i\, f_i/\kappa_i$
Elastic modulus	$Y_c = \Sigma_i\, f_i Y_i$	$1/Y_c = \Sigma_i\, f_i/Y_i$

where 'i' subscript indicates the property of the ith component of the composite and f_i is the volume fraction of the i-th component of the composite. However, other properties like corrosion or wear resistance depend on the properties of individual component.

Example 13.1

A multi-layer electrical capacitor consists of alternating layers of aluminium and mica. There are 12 sheets of mica each 0.1mm thick and 13 sheets of aluminium each 0.005mm thick, calculate the electrical conductivity of the capacitor parallel and perpendicular to the sheets. Given that $\sigma_{Al} = 38 \times 10^6$ ohm^{-1}m^{-1}, $\sigma_{mica} = 10^{-11}$ ohm^{-1}m^{-1}

Solution:

Consider that the capacitor plates have area = a mm^2.

Volume fraction of mica $f_{mica} = \dfrac{0.1 \text{ a} \times 12}{(0.1 \times 12 + 0.005 \times 13) \text{ a}} = 0.949$

Hence $\qquad\qquad f_{Al} = 0.051,$

Parallel to the lamelle: $\qquad \sigma_c = 0.949 \times 10^{-11} + 0.051 \times 38 \times 10^6 = 1.94 \times 10^6$ ohm^{-1}.m^{-1}

Perpendicular to the lamelle: $1/\sigma_c = (0.949/10^{-11}) + [0.051/(38 \times 10^6)] = 9.49 \times 10^{10}$
or $\qquad\qquad\qquad\qquad \sigma_c = 1.054 \times 10^{-11}$ ohm^{-1}.m^{-1}

Thus, the capacitor is highly conducting parallel to plates but behaves as an insulator perpendicular to the plates.

Concrete

Concrete is a type of macroscopic composite in which one ceramic material is dispersed in another ceramic material. In concrete, stone chips are the particles of dispersed phase that are bound together in the binding medium of cement, which is the matrix phase. The pavement is usually built by using asphalt concrete. In Rajasthan, the boundary walls are sometimes built out of concrete where the stones pieces from rocks are embedded in a special variety of mud. Portland cement concrete is a composite in which ingredients are sand, gravels, Portland cement and water. The gravels act as filler material which reduce the over all cost of the concrete since this is a cheap constituent.

Example 13.2

To prepare a unit mix of concrete 50 kg of cement, 120 kg of sand and 220 kg of aggregates are added in the mixer. Calculate the cubic meter of concrete that can be made out of it, if 20 liters of water is added. Given the density of cement = 1750 kg/m^3, density of sand = 2560 kg/m^3 and density of aggregate = 2720 kg/m^3.

Solution:

	Weight (kg)	ρ (kg/m^3)	Volume(m^3)
Cement	50	1750	0.029
Sand	120	2560	0.047
Aggregate	220	2720	0.081
Water	20	1000	0.020

The total volume of the concrete = 0.177 m^3

The quality of the concrete depends on its strength and the strength depends on the proportion of various constituents. If the aggregate contains sand and gravels two third of the volume of concrete, it becomes densely packed. The quantity of cement-water mixture should be enough to coat all the aggregate particles such that the complete bonding between cement and aggregate particles is achieved after thorough mixing. Inadequate amount of water leads to incomplete bonding and porosity. As a result, the strength of the concrete becomes below the desired optimal value. The Portland cement concrete is not always appropriate as a structural material. It is weak and brittle in many situations. Its tensile strength is approximately 10 to 15 times smaller than its compressive strength. Weather also affects the concrete structure and as a result of which its strength is lowered. When water penetrates into the pores of the wall, it expands on freezing in cold weather and causes severe cracking. Reinforcement can improve the performance of concrete structure. Materials that serve as reinforcement materials, are described in the next paragraph.

a) Steel

It is one of the most suitable reinforcement materials since the coefficient of expansion of steel is nearly the same as that of concrete and it is not easily corroded in the cement matrix. Introducing contours in the steel increases adhesion between concrete and steel. This results in increase of strength. Normally steel rods are used in reinforced concrete in roofing. Wire meshes are used for making water tanks.

b) Fibres

Fibrous materials made up of glass, steel, carbon, asbestos, nylon and other polymeric substances can be mixed with Portland cement to increase the elastic modulus of the reinforced concrete. Now-a-days agricultural fibres are also used in building materials on experimental basis. The commonly used fibres are usually extracted from various plants like sisal, coconut etc.

Asphalt and asphalt mix

Asphalt is a mixture, in which bitumen is associated with substantial proportion of inert material, obtained during petroleum refining process. It can be extracted from bitumen bearing rock i.e. rock asphalt or from surface deposits in lake, called lake asphalt. Asphalt is a hydrocarbon, which also contains small amounts of sulphur, oxygen, nitrogen, metals and other impurities. Chemical constituents of asphalt are 80–85% carbon, 10% hydrogen, 2–8% oxygen, 1–7% sulphur and small quantities of nitrogen and other impurities. The composition of asphalt varies quite a lot. Crude oil contains asphalt up to a maximum of 60%. The products of asphalt consist of chain, ring and condensed ring hydrocarbons. Asphalt mix is a composite of aggregates and asphalt and is mainly used for road paving. The aggregates are obtained from granite, basalt, quartzite, limestone, sand stone etc. The percentage of asphalt in asphalt mix varies from 3 to 8% such that it is enough to coat the aggregate particles in the densely packed angular aggregates. Asphalt acts as binder for aggregates. If the asphalt content is on the higher side, it concentrates on the road in hot weather and increases the possibility of skidding whereas insufficient amount of asphalt may lead to improper binding of aggregates and leads to uneven road surfaces. Asphalt is also used for roofing and construction.

Strengthening of roof

For the purpose of waterproofing and strengthening of roofs, special type of surface dressing is done. Initially, the surface of the roof is cleaned and then a layer of bituminous primer is flooded on the surface. A layer of hot refined mineral asphalt (1.5 kg/m^2) is then spread over the surface. The surface is then dressed by covering it with polymeric felt (sandwiched composite with asphaltic layer of 1mm thick between two thin plastic sheets). Another layer of hot asphalt followed by a polymeric felt layer gives very good waterproofing. Third layer of hot mineral asphalt is then flooded over the second layer of polymeric felt. Polished ceramic tiles (6 to 8 mm thick) are fixed in the third layer grouted with cement mortar and on verticals with aluminium foil of 0.05mm thickness with butt joints. The butt joints are sealed with strip of felt laid in hot asphalt. On vertical sides 2 layers of Hessian felt (composite consisting of cloth reinforced asphalt with mica grains on the surfaces) are laid in hot asphalt over a coat of bituminous primer. Combination of above materials form a special type of composites, which gives protection to roof. This type of waterproofing has been done on the roof of the Birla Institute of Technology and Science, Pilani, India.

Guniting

Guniting is a process of treating the damaged surface. In this process, mortar (cement and sand in 1:3 ratio) is sprayed, along with water, into the damaged surface (e.g. cracked roof) with pressure. This technique ensures spreading of mortar mixture through the surface material and fills up the cracks. As a result, the surface is strengthened and dressed.

Wood

Wood is a widely used engineering material. It is an example of naturally occurring composite. It contains polymeric molecules of cellulose, Fig. 13.2, whose degree of polymerization is of the order of 12000. Since cellulose has –OH radicals, it can be crystallized. Wood contains more than 50% of cellulose and 10–30% of lignin, a carbohydrate polymer.

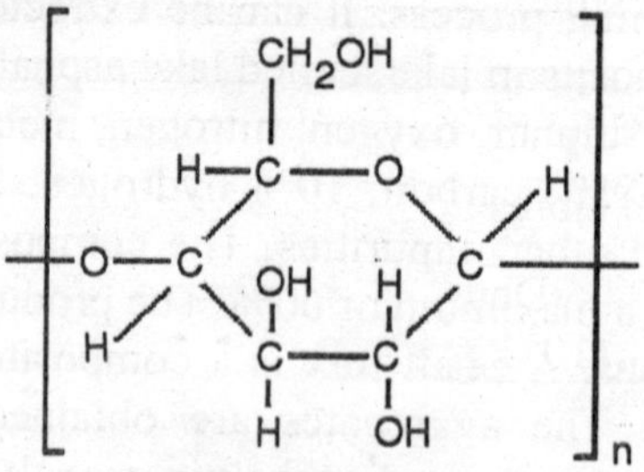

Fig. 13.2 Polymeric molecule of cellulose

Reaction with a mixture of nitric and sulphuric acid converts cellulose into cellulose nitrate where – OH radicals are replaced with – ONO_2. The properties and uses of the product depend upon the extent of nitration. These are the basic molecules for guncotton. Guncotton is used in making smokeless powder. This is also called cellulose trinitrate since there are three nitrate groups per glucose unit. Pyroxylin is also cellulose nitrate, which contains two or three nitrate groups per unit and is used in the manufacture of plastics like celluloid and collodion, in photographic film and in lacquers. The reaction of acetic anhydride, acetic acid and small amount of sulphuric acid with cellulose forms cellulose triacetate where –OH radicals are replaced with acetate radicals ($-OCOCH_3$) in cellulose molecules. Cellulose acetate has replaced the cellulose nitrate in safety type photographic film since the former is less inflammable than the later. Cellulose acetate is used for making acetate rayon.

Example 13.3

A piece of wood contains 20% moisture and weighs 250g. Calculate its weight, when it is completely dried using oven.

Solution:

Amount of moisture in the wood = 250 x 0.2 = 50 g
Weight of dried wood = 250 – 50 = 200 g

Microscopic composites

Microscopic composite are classified into three basic types:
- Fibre-reinforced
- Particle-reinforced
- Dispersion-strengthened

Fibre-reinforced composites

In fibre-reinforced composites, the fibres and matrix provide complimentary properties to the material. Fibres provide high tensile strength and a high tensile modulus, Table 13.1. Fibres are either polycrystalline or amorphous with large length to diameter ratio. They are made up of materials like metals, nonmetals, alloys, polymers, and ceramics. Although the ceramics have high tensile strength, and useful property of low density but they are brittle and any small flaw in the ceramic can reduce the tensile strength drastically. Small diameter i.e. small surface area of the fibre decreases the probability of presence of critical surface flaws, which can cause fracture. The proper choice of matrix material, which is usually ductile, can rectify the disadvantage of using ceramic fibres by
(i) providing binding of fibres,

Table 13.1 Properties of fibres and matrices

Material	Density (10^3 kg/m^3)	Young's modulus (GPa)	Tensile strength (GPa)
Fibres			
Alumina	3.2	170	2.1
Alumina whisker	3.9	1550	20.8
Aramid (Kevlar 29)	1.48	124	3.6
Carbon	1.75	544	2.6
E-glass	2.56	70	3.5
S-glass	2.50	86.9	4.48
Graphite whisker	2.2	704	20.7
Kevlar 49	1.45	125	4.48
Silica	2.2	75	6.0
Matrices			
Epoxy	1.2–1.4	2.8–4.2	0.055–0.13
Polyester	1.10–1.46	2.0–4.4	0.040–0.090

(ii) acting as a medium which transmits and distributes the externally applied stress mostly to the fibres such that only a small fraction of the applied load is sustained by the matrix phase,

(iii) protecting the fibre surfaces from damage and by keeping the fibres apart to resist the crack propagation.

Matrix phase can undergo plastic deformation and ultimately yield under the influence of the external force. This can be avoided by selecting proper combination of matrix and fibre phases such that the capability of stress transmittance from matrix to fibres is enhanced and the adhesive bonding between fibre and matrix is high, which reduces the possibility of pulling out of fibres from matrix phase. Thus, the strong fibres mostly sustain the stress.

Fibres may be continuous i.e. of same length as that of composite or of short length, Fig. 13.3. If these fibres are aligned in the same direction then the composite can have higher strength in that particular direction as compared to other directions. The fibres of short lengths give less strength to the material as compared to the aligned continuous fibres in a composite. The proper combination of fibre and matrix can only give desired characteristics to the composites. The commonly used matrix materials are metals or polymers. The fibres are basically classified into whiskers, fibres and wires.

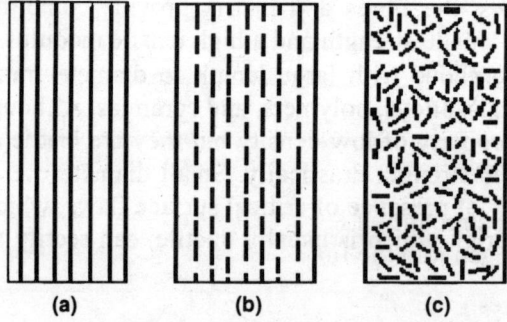

Fig. 13.3 Fibre arrangement in composites

Whiskers are very thin single crystals having large length to diameter ratio and hence of exceptionally high strength. Graphite, aluminium oxide, silicon carbide, silicon nitride etc. are used to form whiskers. Whiskers are not usually used as reinforcement medium since they do not always form strong adhesive bonding with many common matrix materials. Commonly used fibres are made up of glass, nylon, aramids, carbon, alumina, asbestos etc. Example of composites using fibres are glass fibre-reinforced plastics, carbon fibres reinforced epoxy resins. The glass fibres are most commonly used additive materials. There are also composites where glass fibre mats or clothes are used as reinforcing materials.

Carbon fibre-reinforced epoxy resins

Carbon fibre-reinforced epoxy resins have the mixed characteristics of lightweight, very high strength and high stiffness. These materials are ideally suitable for aerospace applications. They are also useful for automobile industry but the cost of these materials are quite high. The carbon fibres are produced from two sources, polyacrylonitrile (PAN) and pitch, which are called precursors. The tensile strength of carbon fibres ranges from about 3 to 4.5 GPa. Its modulus of elasticity varies from 193 to 550 GPa.

Aramid fibres

Aromatic polyamide fibres are called aramid fibres or Kevlar (trade name), Fig. 13.4. Two commercial types of Kevlars are Kevlar 29 and Kevlar 49. Both of these have high strength, low density. Kevlar 29 is used for making ropes, cables and ballistic protection. Fibres made up of Kevlar 49 are used to fabricate the composites for marine, aerospace, automotive and other industrial applications. Polymeric chains of aromatic polyamides are bonded by hydrogen bonding in the transverse direction. The polymeric chains are rigid due to the presence of aromatic rings and this results in rod-like structure of polymeric chains. Kevlar-epoxy composites are used for fabrication of various parts of space shuttles since these materials have high damage resistance, resistance to fatigue and stress rupture.

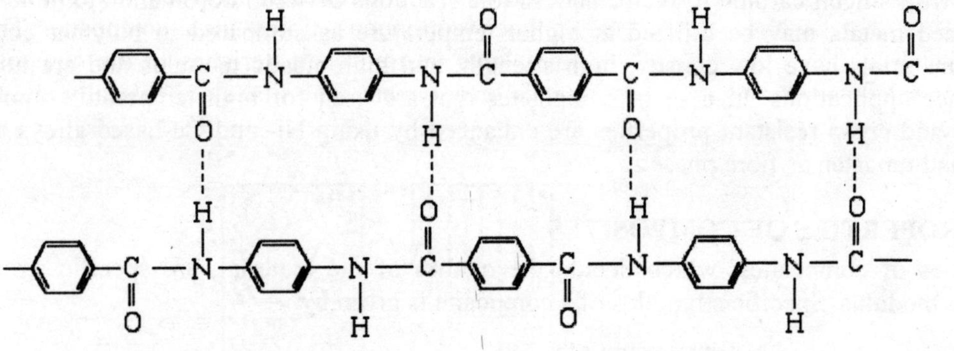

Fig. 13.4 Chemical structure of Kevlar fibres: Secondary bonds between oxygen
and hydrogen join the fibres

Example 13.4

Kevlar 49 fibres are arranged parallely in a fibre-epoxy composite, which contains 70% volume of fibres. The density of Kevlar is 1.48 Mg/m^3 and that of epoxy resin is 1.2 Mg/m^3. Calculate the average density of the composite and weight fraction of fibres present in it.

Solution:

Consider 1 m^3 of composite, which has 0.70 m^3 of Kevlar and 0.30 m^3 of epoxy resin.

Mass of Kevlar = 1.48 x 0.70 Mg = 1.036 Mg

Mass of epoxy resin = 1.2 x 0.30 Mg = 0.36 Mg

Total Mass = 1.396 Mg

Weight fraction of fibre = 1.036/ 1.396 = 0.742

Average density of the composite = 1.396 Mg/m^3

Fibre glass-reinforced polyester resins

The strength of these materials depends on glass content as well as arrangement of glass fibres in the epoxy matrix. If glass fibres are arranged parallely in the matrix, the final product will have higher strength in the direction of alignment of glass fibre, as compared to that of the composite where glass fibres are arranged randomly. The later variety will have lower but equal strength in all directions.

Steel radial

Fine wires are also used as reinforcement. These have larger diameter as compared to that of fibres. Applications of wire as reinforcement material are in radial steel reinforcement in tires, in rocket casing, in high-pressure hoses etc.

Metal matrix fibre composite

In these composites metal alloys of aluminium, magnesium, copper and titanium are used as matrix phase; graphite, silicon carbide, boron and borasic fibres are used in fibre phase. These fibres are produced by vapour deposition of a layer of boron on a thin tungsten wire and are then coated with silicon carbide to retard undesirable reactions between boron and aluminium. Fibre-reinforced metals may be utilized at higher temperature as compared to polymer composites. These materials have low density, high strength, and high elastic modulus and are utilized for aerospace applications. In case of composites that are used for high temperature applications, rupture and creep resistant properties are enhanced by using Ni- and Cu-based alloys as matrix phase and tungsten as fibre phase.

13.3 PROPERTIES OF COMPOSITES

Properties of composites, which decide the quality of the material, are specific strength and specific modulus. Specific strength s of a composite is given by

$$s = \frac{\text{Tensile strength}}{\text{Density}} \qquad (13.1)$$

and specific modulus m is defined as

$$m = \frac{\text{Tensile modulus}}{\text{Density}} \qquad (13.2)$$

Composites, with all the fibres aligned in the same direction, exhibit high tensile strength in that direction. Tensile properties show marked variations if the direction of the applied force makes an angle with the fibre direction e.g. tensile strength becomes one sixth at right angles to the fibre direction. Such materials are therefore showing anisotropy.

Fatigue strength of a polymeric substance can be improved after reinforcement with fibres. In case the matrix phase is metallic, fatigue strength is determined by the matrix phase and hardly any change in fatigue property is possible by reinforcing the metallic phase. The reinforcement can increase creep resistance. Reinforcement of ductile matrix phase with fibres decreases impact strength of pure matrix phase. Impact strength of a brittle matrix phase can be improved by reinforcing this material using ductile fibres.

Example 13.5

Calculate the specific modulus and specific strength of alumina whisker if its density is 3900 kg/m^3; tensile strength and tensile modulus are respectively 20.8 GPa and 1550 GPa.

Solution:

$$\text{Specific modulus} = 1550 \times 10^9 / (3900) = 3.97 \times 10^8 \text{ Nm/kg}$$
$$\text{Specific strength} = 20.8 \times 10^9 / (3900) = 5.33 \times 10^6 \text{ Nm/kg}$$

Strength and elastic modulus of fibre-reinforced material

Consider a composite having all its fibres aligned along its axis; this type of composites shows anisotropy. The elastic modulus for such material can be formulated as follows:

Case-I : Longitudinal Stress

Consider the composites with continuous fibres embedded in the matrix. When the stress is acting along the direction of alignment, the longitudinal strain develops along the aligned direction. Assuming the bonding between matrix and fibres to be strong, both matrix and fibres are strained equally. Consider an idealized composite sample with alternate layers of matrix materials and continuous fibres. The bonding between the layers remains intact during loading. This type of loading is called the isostrain loading. Hence the total load, F_c sustained by the composite is equal to sum of the load on the matrix phase, F_m and that on the fibre phase, F_f,

$$F_c = F_m + F_f \tag{13.3}$$

If the total cross-sectional area of the matrix is approximately a_m, that of fibres is a_f and of the composite a_c, one can write

$$\sigma_c a_c = \sigma_m a_m + \sigma_f a_f, \tag{13.4}$$

where σ_c, σ_m, and σ_f are the stresses experienced by the composite, matrix and fibres, respectively. Hence,

$$\sigma_c = \sigma_f (a_f/a_c) + \sigma_m (a_m/a_c)$$

Assuming the fibre, composite and matrix phases are of equal lengths, we can express $a_f/a_c = f_f$ as volume fraction of the fibre phase, and $a_m/a_c = f_m$ as volume fraction of matrix phase. Hence,

$$\sigma_c = f_f \sigma_f + f_m \sigma_m \tag{13.5}$$

is defined as strength of the composite. The value of the strain in the composite, fibre phase and matrix phase are the same within the elastic limit under isostrain loading i.e.

$$\varepsilon_c = \varepsilon_m = \varepsilon_f = \varepsilon, \tag{13.6}$$

Using Eq.(13.5), the elastic modulus of the composite is given by

$$Y_c = f_f Y_f + f_m Y_m \tag{13.7}$$

where Y_f and Y_m are elastic moduli of fibre and matrix phases, respectively. Since $f_f + f_m = 1$, we can write,

$$Y_c = (1 - f_m)Y_f + f_m Y_m \tag{13.8}$$

If there is no fibre phase in the composite, then $f_m = 1$

$$Y_c = Y_m \tag{13.9}$$

The strength of the composite σ_c is not the stress at the failure of the composite i.e. it is not the fracture stress or the tensile strength (σ_c^{Ts}) of the composite. The value of σ_c^{Ts} depends on the mechanical properties of matrix and fibre phase.

Let the tensile failure strain of fibre is ε_f^* and that of matrix is ε_m^*. If $\varepsilon_f^* < \varepsilon_m^*$, Fig. 13.5(a), and the volume fraction of fibre is greater than a minimum value $(f_f)_{min}$, the composite fails if the fibres fail, so the tensile strength σ_c^{Ts} of the composite is given by

$$\sigma_c^{Ts} = \sigma_f^{Ts} f_f + \sigma_m' f_m \tag{13.10}$$

where σ_m' is the tensile stress in the matrix phase at the failure strain of the composite and σ_f^{Ts} is the tensile strength of the fibres. When the volume fraction of the fibre f_f is less than a minimum value $(f_f)_{min}$, then the matrix material carries the stress as the fibres break. In this case the composite has more or less the characteristics of the matrix phase and it fails only when the stress on the matrix is equal to σ_m^{Ts}. There may be multiple fractures in fibres but the composite does not fail till its strength reaches the tensile strength of the matrix. Thus, the tensile strength of the composite is expressed as

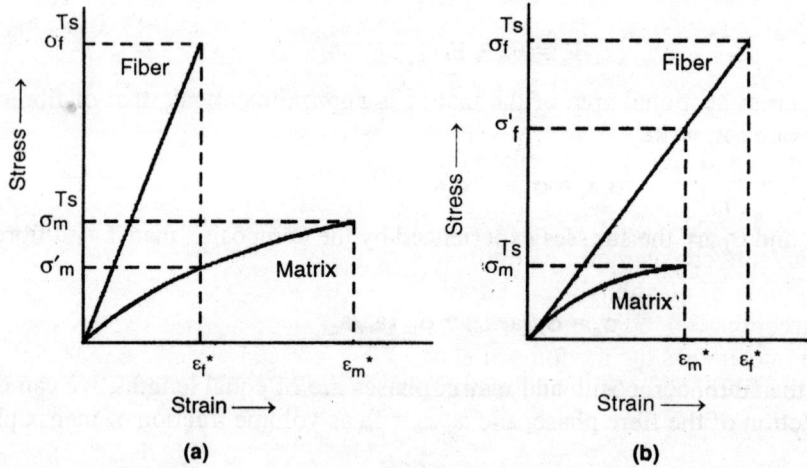

Fig.13.5 Stress-strain curve for two different type of composites (a) $\varepsilon_f^* < \varepsilon_m^*$ (b) $\varepsilon_f^* > \varepsilon_m^*$

$$\sigma_c^{Ts} = \sigma_m^{Ts} f_m \tag{13.11}$$

Equating the right hand side of the Eqs. (13.10) and (13.11), we obtain

$$\sigma_c^{Ts} = \sigma_f^{Ts} (f_f)_{min} + \sigma_m' f_m = \sigma_m^{Ts} f_m \tag{13.12}$$

when $f_f = (f_f)_{min}$. Since $(f_f)_{min} + f_m = 1$

$$(f_f)_{min} = \frac{\sigma_m^{Ts} - \sigma_m'}{\sigma_f^{Ts} + \sigma_m^{Ts} - \sigma_m'} \tag{13.13}$$

When $f_f < (f_f)_{min}$, multiple fractures occur in fibres before the failure of the composite. If $f_f > (f_f)_{min}$, the composite fails when fibres show failure. Composite is strengthened when $f_f > f_c$, a critical value. If $\sigma_c^{Ts} = \sigma_m^{Ts}$ and $f_c = f_f = 1 - f_m$, one can express critical value f_c using Eq.(13.10) as

$$f_c = \frac{\sigma_m^{Ts} - \sigma_m'}{\sigma_f^{Ts} - \sigma_m'} \approx \sigma_m^{Ts}/\sigma_f^{Ts}, \qquad (13.14)$$

since $\sigma_m^{Ts} >> \sigma_m'$ and $\sigma_f^{Ts} >> \sigma_m'$.

When we consider the case of fibre-matrix combination as shown in Fig.13.5 (b),i.e. $\varepsilon_f^* > \varepsilon_m^*$, the tensile strength of the composite is

$$\sigma_c^{Ts} = \sigma_m^{Ts} f_m + \sigma_f' f_f \qquad (13.15)$$

where σ_f' is the stress in the fibres at the failure strain of matrix phase. If f_f is large, the fibre phase is capable of carrying the transferred load when the matrix fractures. The failure of the composite takes place when fractures occur in fibre phase. When f_f is large the Eq. (13.10) reduces to

$$\sigma_c^{Ts} = f_f \, \sigma_f^{Ts} \qquad (13.16)$$

The fracture mechanism changes at $(f_f)_{min}$. Using (13.15) and (13.16)

$$(f_f)_{min} = \frac{\sigma_m^{TS}}{\sigma_f^{TS} - \sigma_f' + \sigma_m^{TS}} \qquad (13.17)$$

where $f_f = (f_f)_{min}$ and $f_m = 1 - (f_f)_{min}$. Since the fractures in fibres are not uniformly distributed, strength also varies in different parts of the material.

Example 13.6

Consider a composite consisting of a continuous glass fibre reinforced epoxy resin where E-glass fibres have modulus of elasticity, 72.3 GPa and epoxy resin has 3.1 GPa. The fibres and epoxy have tensile strengths 2.4 GPa and 0.062 GPa, respectively. Calculate the volume fraction of fibres, if the composite has the modulus of elasticity 51.54 GPa. Also calculate the tensile strength of the composite and the fraction load carried by the fibres.

Solution:
$$Y_c = (1 - f_m)Y_f + f_m Y_m$$
$$51.54 \text{ GPa} = [(1 - f_m)72.3 + 3.1 \, f_m] \text{ GPa}$$
So
$$f_m = 0.3 \, , \ f_f = 0.7$$

Tensile strength of composite = $[0.3 \times 0.062 + 0.7 \times 2.4] = 1.7$ GPa

Fraction of load carried by fibre = $\dfrac{F_f}{F_c} = \dfrac{Y_f f_f}{Y_c} = \dfrac{50.61}{51.54} = 0.98$
(Under isostrain condition)

Case-II: Transverse stress

When the applied stress (σ_c) is perpendicular to the direction of fibre alignment, then

$$\sigma_c = \sigma_m = \sigma_f = \sigma \qquad (13.18)$$

In this situation the system is in isostress state. Now the strain in the composite is partially due to strain (ε_m) in matrix phase and partially due to that (ε_f) in fibre phase and is given by,

$$\varepsilon_c = f_m \varepsilon_m + f_f \varepsilon_f; \quad f_m + f_f = 1 \qquad (13.19)$$

Since $Y = \sigma/\varepsilon$, $\qquad\qquad \sigma_c/Y_c = f_m \sigma_m/Y_m + f_f \sigma_f/Y_f$ (13.20)

Using (13.18), we get $\qquad\qquad 1/Y_c = f_m/Y_m + f_f/Y_f$

i.e. $\qquad\qquad Y_c = Y_m Y_f/(f_m Y_f + f_f Y_m) = Y_m Y_f/[(1-f_f)Y_f + f_f Y_m]$ (13.21)

When the volume fraction of fibre phase $f_f = 0$, $Y_c = Y_m$ (13.22)

Example 13.7

If the elastic modulus of the glass fibre-reinforced epoxy resin is found to be 8.4 GPa when stressed under isostress conditions, where moduli of elasticity of glass fibre and epoxy are 72.3 GPa and 3.1GPa, respectively, calculate the volume percent of continuous glass fibre.

Solution:

$$Y_c = Y_m Y_f/(f_m Y_f + f_f Y_m)$$
$$8.4 = (72.3 \times 3.1)/((1-f_f) \times 3.1 + f_f \times 72.3)$$
So $\qquad\qquad f_f = 0.34.$ i.e. Volume percent of glass fibre = 34%

Case -III: Longitudinal stress transfer to discontinuous fibres

The matrix material in the fibre-reinforced composite has the task of transferring the stress to the fibre through interfacial surfaces by shear mechanism, Fig. 13.6. The extent to which this can be

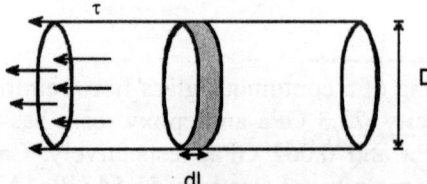

Fig. 13.6 Interfacial surfaces between fibre and matrix

achieved depends on the surface area of a fibre and this is an important factor, which decides the strength of the material. The stress transfer from matrix to fibre of different lengths is shown in Fig. 13.7. The maximum stress (tensile strength of the fibre) is transferred to the fibre if its length L is greater than some critical value L_c. If $L < L_c$, the stress does not attain the maximum value. If τ is the interfacial shear stress, the tangential force F acting on the element dl of the fibre (of diameter D) is given by

$$F = \tau \pi D dl$$ (13.23)

As a result of this applied force F, the stress transferred to the element dl is $d\sigma_f$ (say). So, one can express

$$F = \tau \pi D dl = d\sigma_f(\pi D^2/4)$$ (13.24)

$$\frac{d\sigma_f}{dl} = \frac{4\tau}{D}$$ (13.25)

Integrating we obtain $\qquad\qquad \sigma_f = 4\tau l/D + C$ (13.26)

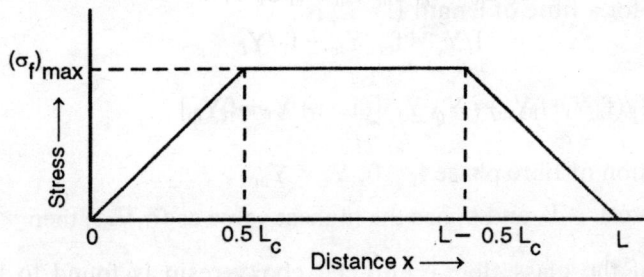

(a)

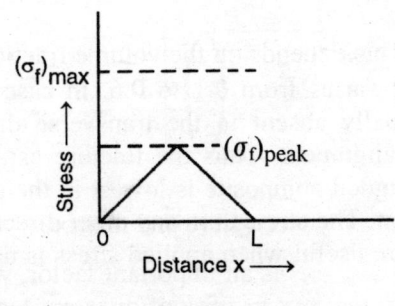

(b)

(c)

Fig. 13.7 Schematic diagram of stress transfer from matrix to fibre for a) $L>L_c$ b) $L=L_c$ c)$L<L_c$

The stress is zero at $l = 0$. Using this boundary condition, constant $C = 0$. Hence

$$\sigma_f = \frac{4\tau l}{D} \qquad (13.27)$$

and σ_f is maximum for a specific value $l = \frac{1}{2}L_c$. So, maximum tensile strength

$$(\sigma_f)_{max} = 4\tau(\frac{1}{2}L_c)/D = 2\tau L_c/D \qquad (13.28)$$

Thus critical length to diameter ratio $L_c/D = (\sigma_f)_{max}/2\tau$. If the length of the fibre $L > L_C$ average stress is given by

$$\overline{\sigma_f} = (\sigma_f)_{max} (1 - (L_c/2L)) \qquad (13.29)$$

and the average stress for a fibre of length $L = L_c$ is

$$\bar{\sigma_f} = (\sigma_f)_{max} \frac{L_c}{2L_c} = \frac{1}{2}(\sigma_f)_{max} \tag{13.30}$$

If the length of the fibre $L < L_c$ and σ_f has the highest value as $(\sigma_f)_{peak}$, then

$$\bar{\sigma_f} = \frac{1}{2}(\sigma_f)_{peak}. \tag{13.31}$$

The strength of the composite is defined as

$$\bar{\sigma_c} = \bar{\sigma_f} f_f + \sigma_m f_m \tag{13.32}$$

Case IV: Elastic modulus of a composite with randomly oriented short fibres

When the short fibres are oriented randomly, the empirical expression for elastic modulus can be expressed as

$$Y_c = \kappa Y_f f_f + Y_m f_m \tag{13.33}$$

where κ is the fibre efficiency parameter. This depends on the volume fraction of fibre phase as well as on the alignment of fibre phase. κ varies from 0.1 to 0.6. In case of aligned fibrous composites, the fibre-reinforcement is virtually absent in the transverse direction i.e. in the direction perpendicular to the direction of alignment. Thus the fracture can occur at very low loading conditions. So the strength of the aligned composite is lowest in the transverse direction and is maximum in the longitudinal direction. The strength in any other direction varies between these two limits. Aligned fibre composites are useful when applied stress is uniaxial and parallel to the direction of alignment.

Example 13.8

Consider a glass fibre-reinforced polyester whose maximum tensile strength is 1.2 GPa and the shear stress τ is 0.025 GPa. Calculate the critical length L_c, if the glass fibre has diameter 6μm. If the tensile strength of the polyester is 0.040 GPa and the length of the fibre is 4 mm, calculate the average strength of the composite, which contains 60% fibre. What should be the strength of the composite, if fibres are of critical length?

Solution:

We know that, $(\sigma_f)_{max} = 2\tau L_c/D$
Thus critical length to diameter ratio $L_c/D = (\sigma_f)_{max}/2\tau = 1.2/0.05 = 24$
So the critical length of the fibre is $24 \times 6 \times 10^{-6}$ m $= 0.144$ mm
To calculate the strength of the composite, we use

$$\bar{\sigma_f} = (\sigma_f)_{max}(1-(L_c/2L)); \sigma = 1.2(1-0.144/(2\times4)) \times 0.6 + 0.040 \times 0.4 = 0.723 \text{ GPa}$$

If $L = L_c$, $\bar{\sigma_f} = (\sigma_f)_{max}/2;$ $\sigma = 1.2(1-0.5) \times 0.6 + 0.040 \times 0.4 = 0.376$ GPa

Particle-reinforced composites

Sometimes filler materials are added to the polymers to improve tensile and compressive strength, abrasion resistance, toughness, etc. This type of combination is often called particle-reinforced composite. The filler material is usually less expensive and hence reduces the material cost. The size of the particle in the filler material is usually of the order of $1\,\mu m$. Presence of these particles in composites resists the movement of matrix phase in the neighbourhood of the particles. The degree of particle reinforcement, and resulting improved mechanical behaviour, depends on the bond strength between the particles and matrix. The particulate composites are designed to produce unusual combination of properties for example metal matrix composites. In these composites large volume fractions (nearly 50%) of particle phases like oxides (e.g. Al_2O_3, ThO_2), carbides (e.g. SiC, BC), boride, AlN, graphite etc. are incorporated in metal matrix to make the resulting composites stiffer, stronger, lighter and more resistant to wear. The particles can have a variety of geometry but they are usually equiaxed. The elastic modulus varies from values

$$Y_c = Y_m\, f_m + Y_p\, f_p \tag{13.34}$$

to
$$Y_c = Y_m\, Y_p / (f_m\, Y_p + f_p\, Y_m) \tag{13.35}$$

where 'p' suffix indicates the properties for particle phase. The particle-reinforced materials are combinations of metals, polymers and ceramics.

Cermets

Cemented carbides or cermets are a group of ceramic-metal composites. These are extremely hard material where refractory carbides like TiC, WC etc are embedded in the metal matrix such as nickel, cobalt. The cermet comprising of tungsten carbide and cobalt can be made first by mixing the carbide powder with cobalt powder and then heating the compacted powder to a temperature greater than melting point of metals. After solidification, the material becomes strong, hard and tough so that it can be used to make tools. This composite is even tougher than carbide since crack propagation through material is hindered. Other examples of this type of composites are glass beads, silica flour and rubber particle reinforced composites. The high impact polystyrene (HIPS) is polystyrene toughened by adding tiny rubber particles. ABS (acrylonitrile-butadiene-styrene) is a terpolymer, obtained by toughening styrene-acrylonitrile copolymer with polybutadiene (occupying 30% of the volume of the composite). This material is tougher than the polymer.

Example 13.9

Calculate the density of a cermet based on cobalt matrix, which contains 65 w/o WC, 15 w/o TiC and 8 w/o TaC. Given that $\rho_{WC} = 15.8$ Mg/m^3, $\rho_{TiC} = 4.9$ Mg/m^3, $\rho_{TaC} = 14.5$ Mg/m^3, $\rho_{Co} = 8.9$ Mg/m^3

Solution:

The density of a composite can be expressed as $\Sigma_i\, f_i \rho_i$ where 'i' subscript is for the i-th component of the composite and f_i is the volume fraction of the i-th component of the composite. Consider the weight of cermet to be 100 Mg. The weights and volumes of different components are :

Component	Weight (Mg)	Density (Mg/m³)	Volume (m³)	Fractional volume
WC	65	15.8	4.114	0.453
TiC	15	4.9	3.061	0.337
TaC	8	14.5	0.552	0.061
Co	12	8.9	1.348	0.149
Total	100		9.075	1.000

$$\rho_c = f_{WC}\, \rho_{WC} + f_{TiC}\, \rho_{TiC} + f_{TaC}\, \rho_{TaC} + f_{Co}\, \rho_{Co}$$
$$= 0.453 \times 15.8 + 0.337 \times 4.9 + 0.061 \times 14.5 + 0.149 \times 8.9 = 11.019 \text{ Mg/m}^3$$

Foams

Foams are also particle-reinforced composites in which particles are bubbles of a gas. Foams are used in cushions, energy absorbent packaging for thermal insulation, for buoyancy and as filler material in the sandwiched panels of wood or cardboard sheets. The characteristics of foams are decided by the type of cellular structure and also by the ratio (R) of the bulk density of the foam and that of matrix material. The cellular structure in the foam can be open type or closed type or mixture of the two. In a closed-cell structure the gas bubbles in the foam-matrix are discrete and not interconnected, whereas in the open-cell structure these are interconnected. The low density foams are specially developed for packaging small sophisticated instrument, which have a low R value e.g. 0:01. Higher R-value foams are used for packing of heavier components. The elastic modulus of foam is given by

$$Y_f = f_m\, Y_m \tag{13.36}$$

where Y_m is the elastic modulus of the matrix material in solid form . We know that,

$$f_m = \frac{V_m}{V_m + V_g} \tag{13.37}$$

where V_m, is the volume of the matrix material which is solid and V_g is that of the gas . The bulk density of the foam

$$\rho_f = \frac{M_m + M_g}{V_m + V_g} \tag{13.38}$$

M_m being mass of the solid and M_g, that of the gas. Since $M_g \approx 0$,

$$\rho_f \approx \frac{M_m}{V_m + V_g} = \rho_m\, f_m \tag{13.39}$$

Hence
$$Y_f = (\rho_f / \rho_m)\, Y_m \tag{13.40}$$

Empirical formula as found out from experimental observation is given by

$$Y_f = (\rho_f / \rho_m)^n Y_m \tag{13.41}$$

The values of n may vary from 1 to 2. The foam composite is very flexible since the elastic modulus of a material decreases when foamed.

Dispersion-strengthened materials

In the dispersion-strengthened composites, the material is treated or fabricated in such a way that the fine particles of hard and inert material are dispersed uniformly in metals or metal alloys. This is a method of strengthening the material. The dispersed phase may be metallic or nonmetallic. The strengthening can be achieved by precipitation hardening e.g. in case of an aluminium-copper alloy precipitation hardening can cause a fine dispersion of an aluminium copper compound throughout the alloy. This results in higher tensile strength since the movement of dislocations is hindered due to the presence of small-dispersed particles. So, the plastic deformation is reduced and yield strength of the material also improves.

Another method of introducing dispersed phase throughout the metal is by sintering. This process involves first powdering the metal, then compacting in a die and at the end heating it to a temperature such that sintering occurs. In this method, metal oxide particles form dispersed phase in the metal e.g. aluminium oxide is formed at surface of aluminium due to presence of oxygen. When aluminium is processed as above, fine powder of alumina is dispersed throughout the metal. This is termed as sintered aluminium powder (SAP). This increases the strength of the metal. Similarly, thoria (ThO_2) dispersed (TD) nickel has enhanced high temperature strength.

13.4 BONDING IN COMPOSITES

One of the most important requirements of composites is that the components should be strongly bonded to one another. Types of bonding may be classified as:
i) Chemical bonding
ii) Sintering

i) Chemical bonding

When chemical bonding occurs between the two components of a composite, the greatest coherency is achieved for example in glass-coated metals, the complex chemical bonds are formed between glass and metals. In welding, the solidification across the joint forms chemical bonding.

ii) Sintering

When powders are agglomerated during sintering a new interface called grain boundary is formed by replacing two previous surfaces of the powder particles. This process reduces surface energy. Thus, sintering occurs naturally with a rate limited by the rate of diffusion of the atoms or of slowest moving ion in an ionic compound. The mechanism of joining of two grains can be explained as follows. The atoms from one grain, at the surface of contact between two grains, move to the pores of the surface of adjacent grain. Since grain boundary diffusion is faster than diffusion through crystal, the pores at the surfaces of the grains are removed faster than the pores within the grains. Care should be taken that the pore removal should take place earlier than the grain growth, otherwise trapped pores make the material less dense and weak.

EXERCISES

13.1 Distinguish between matrix and dispersed phases in a composite material. Also compare and contrast their characteristics (mechanical aspects). Give examples.

13.2 A specimen of Kevlar 49 composite contains 66% by volume of Kevlar immersed in 34% by volume of epoxy resin. Given that the density of Kevlar 49 is 1.48 Mg/m^3 and that of epoxy resin is 1.20 Mg/m^3.

(a) Calculate the weight percent of Kevlar 49 and epoxy resin in the composite.

(b) What is the average density of the composite?

Ans: (a) Kelvar 49 = 70.5 w/o; epoxy = 29.5 w/o; (b) 1.38 Mg/m^3

13.3 Explain how the strength of fibre glass reinforced plastics is affected by the amount and the arrangement of glass fibres.

13.4 Discuss in brief the advantages and disadvantages of concrete as a composite material.

13.5 Calculate the elastic modulus of a continuous and aligned glass reinforced composite containing 35% by volume of glass fibres and 65% by volume of resin. Given that

$Y_{glass} = 7 \times 10^4$ MPa, $Y_{resin} = 3.4 \times 10^3$ MPa

Ans: $Y_c = 2.67 \times 10^4$ Mpa

13.6 In Ex.13.5 if the cross section area is 300 mm^2 and a stress of 50 MPa is applied in the same direction, calculate the magnitude of the load carried by each of the fibre and matrix phases. Also determine the longitudinal tensile strength of this composite if tensile strengths of fibre And matrix phases are 2.4 and 0.062 GPa, respectively.

Ans: $W_f = 1.376 \times 10^4$ N, $W_m = 1.24 \times 10^3$ N, $\sigma_c^{Ts} = 0.88$ GPa.

13.7 The efficiency of reinforcement (η) depends on the fibre length 'l' according to the expression $\eta = [l - 2x]/l$ where x is the length of the fibre at each end that does not contribute to the load transfer.

(a) If x = 2 mm, plot n versus l for a maximum length of 100 mm.

(b) If the reinforcement efficiency is to be 0.97, calculate the length required.

Ans: (b)133.3 mm

13.8 If the tensile strengths of Kevlar 49 and epoxy resin are 3.45 GPa and 0.069 GPa respectively, calculate the volume percent of the components to obtain the strength of the composite material to be 2.07 GPa. Also find the fraction of load F_f that is carried by the Kevlar $Y_{kelvar} = 130$ GPa and $Y_{resin} = 4.2$ GPa

Ans: $f_f = 0.59$, $f_m = 0.41$, $F_f = 0.98$

13.9 In a continuous and aligned fibre-reinforced composite, the elastic module in the longitudinal and transverse directions are 5×10^4 MPa and 4×10^3 MPa respectively. Calculate the elastic moduli of fibre and matrix phases if the fibres occupy 25% volume.

Ans: 1 GPa and 66.67 GPa or 191 GPa and 3 GPa.

13.10 Consider a load of 2267 kg supported by eight rods each with an area of 50 mm^2 and 10 cm long. Four rods are made of brass and other rods of stainless steel. If Y (brass) = 1.1×10^2 GPa and Y(stainless steel) = 2.07×10^2 GPa,

(a) Calculate fraction of the load that is carried by each brass rod at room temperature (27°C)

(b) Suppose each rod has to bear 12.5% of load, calculate the temperature to which the rods must be subjected to achieve this.

13.11 A composite contains 40 v/o glass fibre in a matrix of epoxy resin. Calculate Y for the composite. Given that $Y_g = 70$ GN/m² for glass; $Y_e = 3$ GN/m² for epoxy resin.

Ans: 29.8 GN/m².

13.12 If $Y_{kevlar} = 125$ GPa and $Y_{resin} = 3$GPa. Calculate the value of Y/ρ for the composite given in Ex. 13.2. If Kevlar fibres were to be substituted by another material x with Y = 1550 GPa and density 3.9 Mg/m³ calculate the fraction (by volume) of continuously aligned x which gives the same Y/ρ value.

Ans: 60.52 GPa-m³/Mg, 0.05

13.13 A hybrid composite is one which consists of two different types of fibres oriented in the same direction in the matrix consisting of resin.
   (a) Write an expression for the elastic modulus of a hybrid composite.
   (b) Calculate the longitudinal elastic modulus of a hybrid composite containing brass and glass fibres in fractions of 0.2 and 0.5 respectively by volume in a resin matrix having $Y_{resin} = 4 \times 10^3$ MPa., $Y_{glass} = 69$ GPa, $Y_{brass} = 110$ GPa.
   (c) Can you think of any advantage of using hybrid composites over normal fibre composites?

Ans: (a) $Y_C = f_{resin} Y_{resin} + f_1 Y_1 + f_2 Y_2$; (b) 57.7 GPa

13.14 Briefly explain the difference between cement and concrete as a structural material and hence explain why cement is preferred over concrete.

13.15 To design an aligned, continuous glass fibre reinforced composite with a ratio of $Y_c/\rho = 30$ MPa-m³/kg. Calculate
   (a) the volume fraction of fibre required;
   (b) the fraction of load carried by fibres,
   where $Y_{glass} = 7.6 \times 10^4$ MPa; $Y_{matrix} = 3.4 \times 10^3$ MPa; $\rho_{glass} = 2200$ kg/m³ and $\rho_{matrix} = 1200$ kg/m³.
   Ans: (a) 0.765; (b) 0.986

13.16 An electrical contact material is produced by introducing copper(density 8940 kg/m³) into a porous tungsten carbide (WC) compact (density 15770 kg/m³). The density of the final product is 14000 kg/m³. If all the pores are filled with copper, calculate the
   a) volume fraction of copper in the composite,
   b) density of WC compact before introduction of copper.
   Ans: (a) 0.26; (b) 11670 kg/m³

13.17 Consider a composite having yttria (Y_2O_3) particles (80 nm in diameter) introduced in tungsten by internal oxidation. If the weight percent of Y initially present in the alloy is 2.5, calculate the number of particles per mm³. Also find the volume percent of yttria in the composite. The density of yttria is 5010 kg/m³ and that of tungsten is 19250 kg/m³.
   Ans: 2.35×10^9 particles/mm³, 11.1 v/o

13.18 A grinding wheel of 250 mm diameter and 25 mm thick weighs 3.2 kg. The wheel contains SiC (density 3170 kg/m^3) in the form of 0.5 mm cubes, bonded by silica (density 2500 kg/m^3). If the material has 5 v/o porosity, find

    a) the number of SiC particles present in the wheel,

    b) the weight fraction of SiC lost, when the wheel is worn to a diameter of 210 mm.

    Ans: (a) 3.403 x 10^6 ; (b) 0.294

14 *Plastic Deformation and Strengthening of Materials*

14.1 INTRODUCTION

Movement of dislocation lines causes plastic deformations in materials. The dislocation theory explains the process of plastic deformation and accompanying structural changes. Plastic deformation can be observed macroscopically and can be measured as permanent strain. This deformation can also be explained in atomic scale in terms of slip of dislocation lines or shearing of crystal lattice on crystal plane. In this chapter we initially concentrate on the characteristics of stress-strain curves under plastic deformation and then explain the deformation mechanisms. Various mechanisms of strengthening of material are also discussed in this chapter.

14.2 DISLOCATIONS AND STRESS STRAIN CURVE

The stress-strain curve for a single crystal, as obtained from tensile tests, is shown in Fig. 14.1. The OA part of the curve explains the elastic behaviour of the crystal when applied stress is within the elastic limit. The length of the stage AB depends on many factors such as crystal

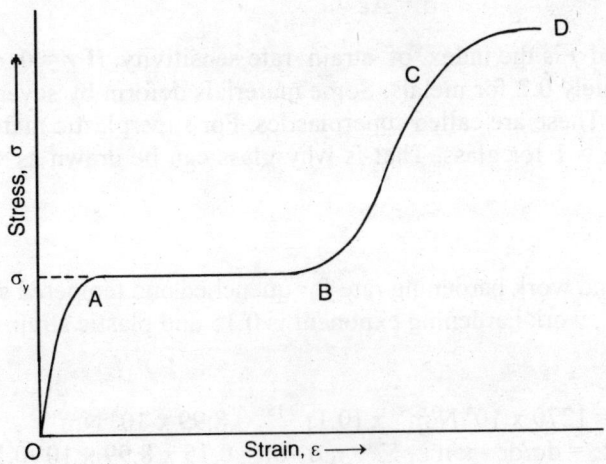

Fig. 14.1 Generalized stress-strain curve.

structure, purity of the specimen, orientation of slip plane etc. The tensile stress required for onset of the plastic deformation is yield stress (σ_y) as indicated at the beginning of the stage AB. The yield stress depends on the angle between the tensile axis and normal to the slip plane and also on the angle between the tensile axis and the slip direction. The stage AB indicates that there is easy slip of dislocation lines i.e. the work hardening rate $d\sigma/d\epsilon$ is very low since the glide motion occurs without much change in stress. The primary slip system becomes active and dislocations do not interfere with one another during the stage AB. Some of these dislocations reach the surface of the crystal and permanent deformation occurs in the crystal. The BC part of the curve indicates that the value of $d\sigma/d\epsilon$ is very high. During this stage the dislocation lines move on from the original slip plane to intersecting slip planes because their motions may be hindered due to presence of impurities or due to entangling of dislocation lines. The movement of dislocation lines becomes more difficult due to increased hindrance arising from increasing strain. Thus, the plastic deformation becomes difficult which is also reflected in high work hardening rate of the BC part. When the stress is increased further, the screw dislocation lines cross-slip to another crystal plane bypassing the obstacles. This process again decreases the work hardening rate since the strain increases with stress in a parabolic fashion in region CD.

A general relationship between the true stress σ_T and true strain ϵ_T can be expressed as,

$$\sigma_T = \kappa \, \epsilon_T{}^n \tag{14.1}$$

where κ is the strength coefficient and n is the exponent of work hardening. σ_T is also called flow stress. The exponent is usually a fraction. The rate of work hardening is given by

$$R_w = d\sigma/d\epsilon = \kappa \, n \, \epsilon_T{}^{n-1} \tag{14.2}$$

In case of metals like copper n = 0.5, a high plastic strain can develop easily, whereas materials with smaller n value develop plastic strain less easily. Sometimes the strain develops gradually after the application of stress as observed in case of material with inelastic behaviour. For such material the stress is related to the strain rate, as follows.

$$\sigma = A\dot{\epsilon}\,\gamma \tag{14.3}$$

where A is a constant and γ is the index of strain rate sensitivity. If $\gamma = 0$, σ is independent of strain rate. γ is approximately 0.2 for metals. Some materials deform by several hundred percent of strain without necking. These are called superplastics. For superplastic materials γ varies from 0.4 – 0.9. The exponent $\gamma \approx 1$ for glass. That is why glass can be drawn as very long rod or as tube without necking.

Example 14.1

Calculate the flow stress and work hardening rate for quenched and tempered steel, if the strength coefficient is 1270 MNm^{-2}, work hardening exponent is 0.15 and plastic strain is 0.1.

Solution:

The flow stress $\sigma_T = \kappa\epsilon_T{}^n = 1270 \times 10^6$ N m$^{-2} \times (0.1)^{0.15} = 8.99 \times 10^8$ Nm^{-2}
The work hardening rate $R_w = d\sigma/d\epsilon = \kappa \, n \, \epsilon_T{}^{n-1} = n \, \sigma_T/\epsilon_T = 0.15 \times 8.99 \times 10^8/0.1$
$\qquad\qquad = 1.349 \times 10^9$ Nm^{-2}

14.3 PLASTIC SLIPS IN ELEMENTAL CRYSTALS

Plastic deformation in materials occurs due to two types of processes (a) slip and (b) twinning.

The movement of dislocation lines in crystal planes produces the plastic deformation. It is called slip. Dislocation lines move due to application of shear stress and the slip occurs when dislocation line reaches the surface of the crystal. The steps are formed at the surface of the crystal during slip, but the orientation of all parts of the crystal remains same after slip. During plastic deformation by slip, atoms move through a large distance equivalent to N x a, where a is the interatomic distance and N is the number of lattice points in the direction of slip.

Slip systems

There are always one or more than one families of preferred crystal planes for a crystal system along which the slip occurs. These are called slip planes. There are also preferred families of directions in a slip plane, called slip directions along which slip takes place. The combination of a slip plane and the slip directions parallel to the slip plane forms a slip system. Slip systems are different for different Bravais lattices. For a particular lattice, the slip planes are usually close-packed planes and slip directions are also close-packed directions. Table 14.1 shows slip systems for certain Bravais lattices.

Table 14.1 Slip systems for certain crystal structures

Metals	Structure	Slip plane	Slip direction	No. of combinations
Ag, Al, Cu, Au, Ni, γ-Fe, Pb	fcc	{111}	<$\bar{1}$10>	12
α-Fe, β-brass, Mo, W	bcc	{101}	<11$\bar{1}$>	12
α-Fe, Mo, W, Na	bcc	{211}	<$\bar{1}$11>	12
α-Fe, K	bcc	{312}	<$\bar{1}$11>	12
Be, Cd, Zn, Mg , α-Ti	hcp	{0001}	<11$\bar{2}$0>	3
α-Ti, Mg	hcp	{10$\bar{1}$0}	<11$\bar{2}$0>	6
NaCl, LiF, MgO, MnS, TiC	fcc	{1$\bar{1}$0}	<110>	6
PbS	fcc	{001}	<110>	6
MnSe	fcc	{$\bar{1}$11}	<110>	12
CsCl	sc	{001}	<100>	6

When we compare interplanar separation of sets of parallel planes in a crystal, it is found that interplanar separation is maximum for close-packed planes. That is why it is easy for slip to occur parallel to close-packed plane. The slip systems for ionic crystals are such that the ions of same polarity do not come next to each other during shear. This can be explained from the fact that the ions of same polarity repel each other and as a result, the energy of the crystal increases.

Critical Resolved Shear Stress

When tensile force F is applied along the axis of a crystalline wire or rod such that it is perpendicular to slip plane S, then there is no shearing stress on the plane S. The tensile stress on S is given by F/A, where A is the area of cross-section of the specimen perpendicular to the axis. If the slip plane of this crystal is such that its normal makes an angle ϕ with the tensile axis, the area of the slip plane is A/cosϕ, Fig 14.2.

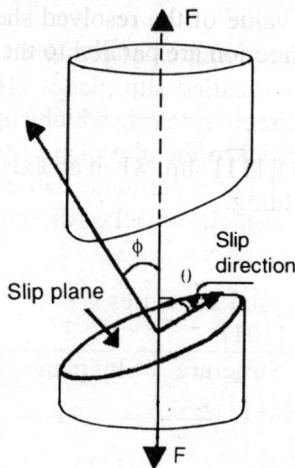

Fig. 14.2 Schematic representation of plastic deformation.

If the slip direction makes an angle θ with the tensile axis, the component of the force F along the slip direction is F cosθ. The resolved shear stress along the slip plane is given by

$$\tau = \frac{F}{A} \cos\phi \cos\theta = \sigma \cos\phi \cos\theta \qquad (14.4)$$

This is known as Schmid's law. The factor (cosϕ cosθ) is called Schmid's factor. The tensile stress on the slip plane is given by

$$\sigma_{slip} = \frac{F \cos\phi}{A/\cos\phi} = \frac{F}{A} \cos^2\phi = \sigma \cos^2\phi \qquad (14.5)$$

The slip can occur easily only when resolved shear stress τ equals or exceeds a threshold value, τ_c that is specific for a crystal. This value is called the critical resolved shear stress. The tensile stress corresponding to τ_c is yield stress. τ_c is also called shear strength. Eq. (14.5) is valid for tensile as well as compressive stress. One can obtain similar equation for twinning but τ_c for twinning is greater than that for slip. Yield stress is different along different crystal directions, which can be obtained for a given single crystal by varying its orientation. All these yield stresses result in the same value of τ_c for a specific crystal.

At lower values of stress, the rate of strain is very slow. When the loading is in the range of τ_c, the material exhibits progressive plastic deformation over an extended period. This is called creep. When the shear stress is greater than τ_c, plastic flow is observed. τ_c depends on the composition, temperature as well as the extent of prior strain in a material. In the presence of certain soluble impurities or alloying materials, movement of dislocation in a material is usually hindered and hence the material does not deform easily. If the material has undergone some deformation, the development of additional strain requires higher value of τ_c. As the temperature increases, dislocation movement becomes easier and as a result τ_c decreases. The value of τ_c is zero at the melting temperature of the metal or solidus temperature of the alloy. If the slip plane is perpendicular to the tensile axis, the value of the resolved shear stress is zero and it is equal to F/A, if the slip plane as well as slip direction are parallel to the tensile axis.

Example 14.2

Calculate critical shear stress for $(101)[1\bar{1}\bar{1}]$ slip, when an axial stress of 1.38 MPa, applied in the [100] direction of a crystal causes yielding.

Solution:

Using Schmid's law

ϕ = Angle between [100] and [101]; $\cos\phi = 1/\sqrt{2}$

θ = Angle between [100] and $[1\bar{1}\bar{1}]$; $\cos\theta = 1/\sqrt{3}$; $\tau_c = (1.38 \text{ MPa}) / \sqrt{6} = 0.56$ MPa

Example 14.3

A shear stress of 3.79 MPa will produce $(1\bar{1}1)[011]$ slip in a copper crystal. What compressive stress in the [111] direction will produce slip of the above orientation?

Solution:

Using Schmid's law

ϕ = Angle between $[1\bar{1}1]$ and [111]; $\quad\cos\phi = 1/3$
θ = Angle between [111] and [011]; $\quad\cos\theta = 2/\sqrt{6}$
$\tau_c = \sigma (2/(3 \times \sqrt{6})) = 3.79$ MPa
Compressive stress, $\quad\sigma = 13.93$ MPa

Shear stress

Shear stress can be estimated by viewing the phenomenon in terms of movement of one crystal plane over another. The crystal planes can be actually visualized as packing of atoms such that each plane perpendicular to the plane of the paper is represented by a row of atoms. When one such plane moves over another plane parallel to it due to shear, this can be depicted as movement of one row of atoms over another row, Fig. 14.3. In Fig. 14.3 (a), no stress is applied and the potential energy is a minimum. The shear stress, required to displace one row of atoms over the other is such that an atom moves from one valley position to the next. The stress required varies with the position of the atom. Initially the atoms in the upper row are in the valley positions of the lower row. During the movement of one plane above another, the shear stress is increased till a specific atom moves through ¼ of the interatomic separation (i.e. a/4). The shear stress then gradually decreases and follows a sinusoidal variation with x as given by, Fig. 14.4,

$$\tau = \tau_{max} \sin (2\pi x /a) \approx \tau_{max} \, 2\pi x/a, \text{ for } 2\pi x/a \ll 1 \qquad (14.6)$$

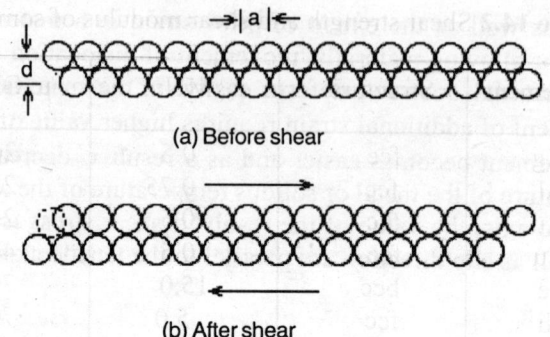

(a) Before shear

(b) After shear

Fig. 14.3 Plastic deformation by shear

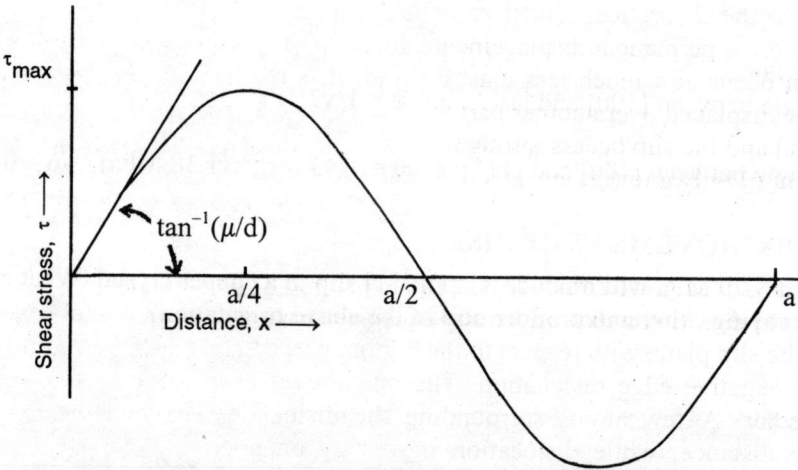

Fig. 14.4 Sinusoidal variation of shear stress τ with distance x.

That is why the curve near x = 0 is linear and obeys Hooke's law. According to Hooke's Law

$$\tau = \mu\gamma = \mu \tan\theta = \mu \,(x/d) \approx \mu\theta \qquad (14.7)$$

where μ is shear modulus and d is the interplanar spacing. For small values of x, one can obtain γ from Eqs. (14.6) and (14.7) as follows

$$\tau = \mu \,(x/d) = \tau_{max}\,(2\pi x/a) \qquad (14.8)$$

Hence
$$\tau_{max} = (\mu a/2\pi d) \qquad (14.9)$$

Taking a = d
$$\tau_{max} = \mu/2\pi \qquad (14.10)$$

Thus, τ for slip is approximately 0.1μ. We know that μ value for copper is 44 x 10^9 Nm^{-2}. Hence shear stress of copper would be of the order of 10^{10} Nm^{-2} and that of steel is 6.8 x 10^9 Nm^{-2}. In actual crystals, the initial shear stress for deformation is much less than the theoretically predicted value. The deformation in single crystal of copper and iron takes place at an initial stress of approximately $10^6 Nm^{-2}$. Values of τ_c and shear modulus of some crystals are given in Table 14.2.

Table 14.2 Shear strength and shear modulus of some crystals

Element	Structure	τ_c, MNm^{-2}	μ, GN m^{-2}
Ag	fcc	0.5	26
Al	fcc	0.75	25
Au	fcc	0.5	26
Cu	fcc	0.5	44
Fe	bcc	15.0	70
Ni	fcc	5.0	70
Zn	hcp	0.3	33

According to the theoretical model explained above, an elastic strain of about 25% would be required before a permanent displacement can occur. The experimental investigation reveals that the slip can occur at a much less elastic strain. It is not always necessary that one part of the crystal to be displaced over another part for the slip to occur. The dislocations are already present in the crystal and the slip occurs through the movement of the dislocation line, which involves the displacement of only a few atoms at a time. Hence, the stress required is only a fraction of τ_{max}.

14.4 SLIP BY MOVEMENT OF DISLOCATIONS

When a shear stress greater than τ_c is applied parallel to a slip plane in a crystal the dislocation line starts moving. Movement of dislocation line causes a displacement of the top portion of the crystal on the slip plane with respect to the bottom part or vice versa depending upon whether it is positive or negative edge dislocation. The magnitude of one step of displacement is equal to Burgers vector. A few atoms surrounding the dislocation line move only a fraction of the interatomic distance, while dislocation moves by one step. Hence less stress is required to produce such slip. The movement of a dislocation line can be compared with the movement of a wrinkle in the carpet, Fig. 14.5. The bodily displacement of the carpet requires much more force than that required by movement of a wrinkle from one end of the carpet to the other end. The presence of dislocations in a crystal ensures easy production of plastic deformations. The single crystal filaments of whisker can be grown without any edge dislocation. So it is very difficult to produce any plastic deformation in such crystals since they resist shear strain.

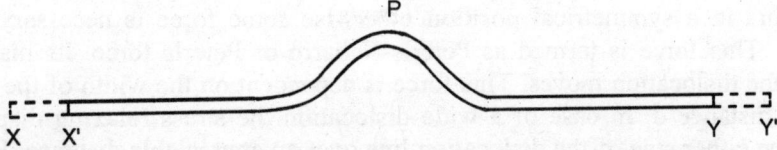

Fig. 14.5 Wrinkle is worked along the carpet to move it through a distance

The dislocation lines are normally surrounded by the atoms, which are displaced from their initial position. The extent of lattice distortion due to presence of dislocation is called width of dislocation. As a result of lattice distortion, the stress field is created in the vicinity of dislocation line, Fig. 14.6. The extra half-plane 1 is shown in Fig. 14.6 (a), which results in the displacement of atoms to the left or to the right, in its neighbourhood. The atoms in the planes 2 and 3 which are parallel to the extra half plane 1, are displaced to the right and those in the planes 2'and 3' are displaced to the left. Now the external force required displacing the extra half plane and hence the dislocation line, to the right or to the left is very small due to the presence of opposing forces on the atoms in the stress field surrounding the dislocation line.

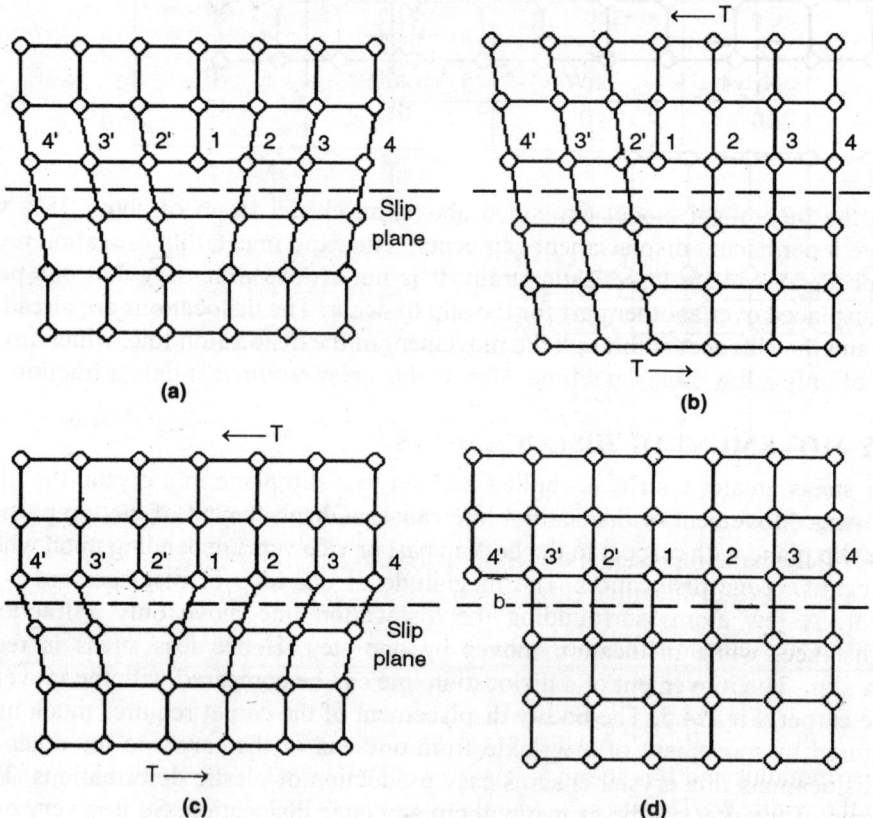

Fig. 14.6 Slip due to movements of dislocation

Cottrell justified that the force required for moving the dislocation line is zero only if the dislocation occurs in a symmetrical position otherwise some force is necessary to displace a dislocation line. This force is termed as Peierls-Nabarro or Peierls force. Its magnitude varies periodically as the dislocation moves. This force is dependent on the width of the dislocation W and interplanar distance d. In case of a wide dislocation the stress relaxing displacements are distributed on the either side of the dislocation line over an appreciable distance, Fig 14.7. Thus, the movement of a wide dislocation requires the changes in the bond lengths distributed over a number of bonds in the region so that the change per bond length is very small. Movement of a narrow dislocation involves larger change per bond length in the region of stress relaxing displacements on the both sides of dislocation. Thus, the narrow dislocation requires more stress to move as compared to that for the wide dislocation.

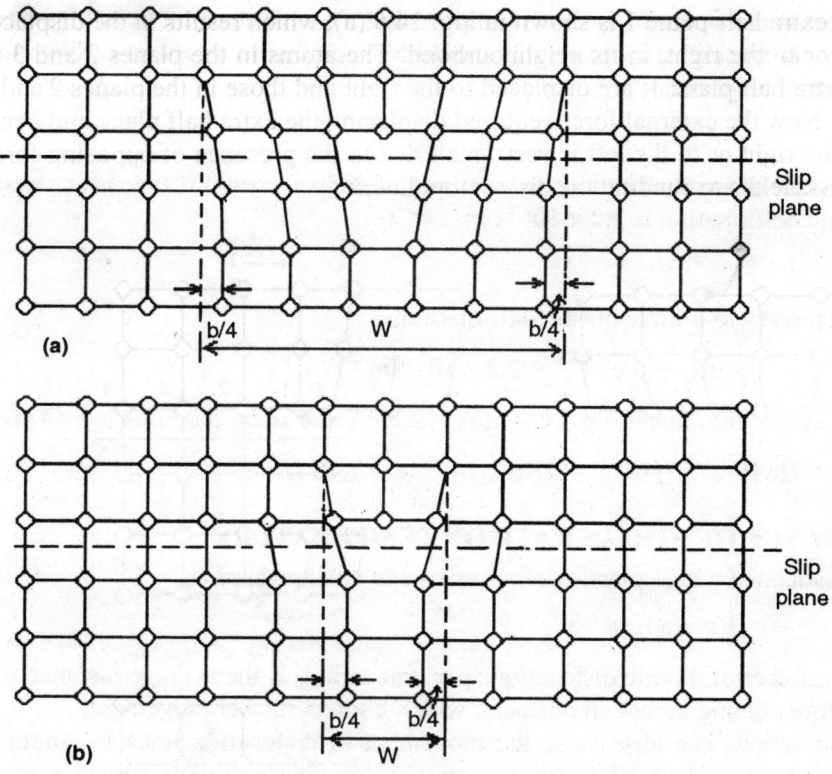

Fig. 14.7 Schematic representation of the width of a dislocation (a) a wide
dislocation, (b) a narrow dislocation

Peierls and Nabarro expressed the stress necessary to move a dislocation i.e. to initiate plastic
deformation as

$$\tau_{PN} \approx \mu \, e^{-2\pi W/b} \qquad (14.11)$$

where μ is the shear modulus and b is the magnitude of the Burgers vector. The Peierls stress is
highly sensitive to the width W. The stress is a maximum when W is zero and it decreases as W
increases or when interplanar spacing increases. Interplanar spacing is a maximum in case of
close-packed planes. The width of the dislocation line becomes large when the bonding forces are
spherically distributed which happens in case of close-packed structures. Thus W is large in case
of fcc and hcp structures and Peierls stress becomes low. The bonding forces are directional in
nature in case of bcc and covalent crystals, which results in the narrow width of dislocation, and
hence large Peierls stress. The magnitude of Peierls stress depends on the variation of interatomic
force with displacement and hence exact estimation of stress is quite difficult.

Metals like copper, silver, aluminium etc., have close-packed structure with no directional
bond. Dislocation width is large for such metals and hence low Peierls stress is required to
produce plastic deformation. These materials are said to be ductile. For example, the copper wire
can be cold drawn such that the length of the wire can become hundred times its original length
without fracture. Materials with covalent character of bonding like iron, the dislocation width is
narrower and hence such material is less ductile. These types of materials are harder than Cu, Al,
Ag etc. The ionic compounds also have nondirectional bonding of moderate strength. So the

plastic deformation can occur in ionic crystal provided the surface of the crystal is free of cracks. The Burgers vector in case of ionic crystal should be such that two cations or two anions should not become nearest neighbours.

Example 14.4

Calculate the width of the dislocations in a sample of copper crystal if the shear stress necessary to initiate plastic deformation is 2.2×10^2 N m^{-2}.

Solution:

Shear stress necessary to initiate plastic deformation is

$$\tau_{PN} = \mu \, e^{-2\pi W/b} = 2.2 \times 10^2 \text{ Nm}^{-2}$$

where Burgers vector for copper $b = 3.61/\sqrt{2}$ Å $= 2.55$ Å and shear modulus $\mu = 44$ GNm^{-2}

Therefore, $W = -(\ln [\tau_{PN} / \mu]) \, b/2\pi = 7.76 \times 10^{-10}$ m $= 7.76$ Å

14.5 DEPENDENCE OF STRAIN RATE ON TEMPERATURE

The strain ε, associated with a plastic deformation can be expressed as,

$$\varepsilon = n \, b \, l_d \tag{14.12}$$

where n is the number of mobile dislocations per unit area, l_d is the average distance traveled by a dislocation before coming across an obstacle, which hinders further movement.

Thermal fluctuations can also cause the movement of dislocation lines in random directions even in the absence of applied stress. Internal stress τ_{PN} is the stress necessary to move dislocation line in the absence of thermal energy. In the presence of thermal energy, less amount of stress, τ_A is required to move the dislocation line. When the dislocations move, they interact with the already existing dislocations present in the slip plane. A dislocation line may cut through the forests of dislocation lines while gliding in a slip plane. As a result, jogs are formed which further hinder the glide motion. The thermal energy can help the movement of the dislocation lines. The volume over which thermal energy has to be concentrated such that the plastic deformation can be activated is called activation volume, v. The activation energy Q required for the movement of dislocation line is a function of stress and is given by

$$Q = Q_o - v \, (\tau_A - \tau_{PN}) \tag{14.13}$$

where Q_o is the activation energy when $\tau_A = \tau_{PN}$. The activation volume may differ depending upon the types of obstacles present in the path of moving dislocation lines. These obstacles may be forests of dislocations, lattice friction, impurity atoms etc. The activation volume is much larger in the presence of forests of dislocations than otherwise. The strain rate can be expressed as,

$$\dot{\varepsilon} = n \, v_d \, b + \dot{n} \, l_d \, b \tag{14.14}$$

where v_d is the average velocity of dislocation and $\dot{n}$ is the variation in the dislocation density. The second term is negligible. Using the approach of Seeger, the strain rate is given by

$$\dot{\varepsilon} = n \, l_d \, b \, v_o \exp \, [-Q/kT] \tag{14.15}$$

or $\qquad$ $\ln \dot\varepsilon = \ln (n \, l_d \, b \, v_o) - [Q/kT]$

where v_o is a frequency factor determined by the nature of the obstacles and k is Boltzmann constant. Since Q is a function of τ_A, Eq. (14.13),

$$\ln \dot\varepsilon = A + B\tau_A \qquad\qquad (14.16)$$

Using Eqs. (14.13) and (14.15), we get the expression for flow stress as,

$$\tau_A = \tau_{PN} + (1/v)(Q_o - kT \ln (n \, l_d \, b \, v_o/\dot\varepsilon)) \qquad\qquad (14.17)$$

where τ_{PN} and Q_o are independent of temperature.

Example 14.5

The activation volume for dislocation motion in a crystal is $15b^3$, where b (= 3Å) is the Burgers vector. The τ_{PN} stress for the crystal is 1000 MNm^{-2}. For a specified rate of dislocation motion $Q = Q_o - 30 \, kT$, calculate the stress required for dislocation movement at 0°K, 50°K, 200°K.

Solution:

We have $\tau_{PN} = 1000 \times 10^6$ J m$^{-2} = 10^9$ J m^{-2} ; $\quad b = 3 \times 10^{-10}$ m

T (°K)	$Q - Q_0$ (= − 30 kT) (J)	τ_A {= $\tau_{PN} - (Q - Q_o)/15 \, b^3$} (Jm^{-2})
0	0	10^9
50	− 2.07 x 10^{-20}	10.51 x 10^8
200	− 8.28 x 10^{-20}	12.04 x 10^8

14.6 MULTIPLICATION OF DISLOCATIONS

The dislocation density in annealed metals varies in the range $10^9/m^2 - 10^{12}/m^2$. When the metals undergo plastic deformation, the dislocations move towards the surface of the crystal and ultimately dislocations disappear at the surface. In this method dislocations are removed. The slip offsets in a deformed crystal are clearly visible in microscope. The magnitude of slip step is of the order of 10^{-6} m. The slip step caused by a dislocation line is of the order of Burgers vector (10^{-10} m). So there should be approximately 10^4 dislocations of same sign on each slip plane. Such a large number of dislocation lines is not normally present in a slip plane. Thus additional dislocations are generated during plastic deformation. This is verified by the observations through electron microscope where it has been observed that the number of dislocations increases from $10^8 - 10^9$ per m^2 to $10^{15} - 10^{16}$ per m^2, as the annealed crystal is heavily cold-worked. This proves that there are sources within the crystal, which generate new dislocations during plastic deformation.

Frank-Read source

Frank-Read source is one of the most important sources of generation of dislocations in a crystal. The mechanism of generation of dislocations can be explained as follows: Let us say PQ is a segment of length l of dislocation line, lies on a slip plane. This segment is considered to be pinned at the two ends by impurity atoms or ions or by interaction with other dislocation lines. The average length of a dislocation line between two pinning points is estimated to be reciprocal of the square root of dislocation density. As the shear stress τ is applied, the segment bends like bow with a radius R, where

$$\tau = (\mu b/l) \tag{14.18}$$

The dislocation bows out further as the applied stress increases. As a result, the radius of curvature decreases till the radius R becomes half of the length of segment i.e. l/2. The dislocation vector **d** changes the direction along with the dislocation line since it is pointing along

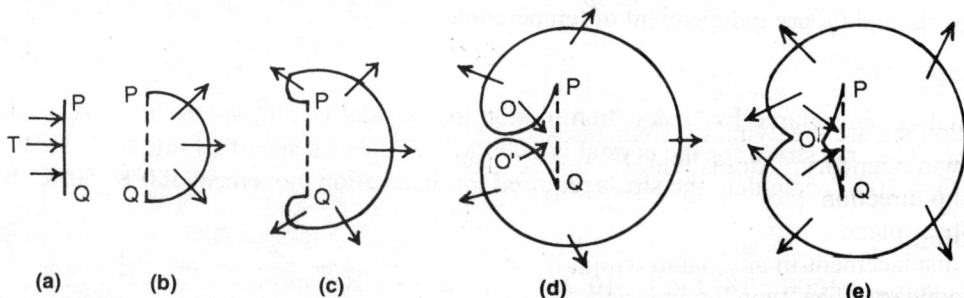

(a)   (b)   (c)   (d)   (e)

Fig. 14.8 Multiplication of dislocation due to Frank-Read Source

the tangent at any point on the dislocation line. When the dislocation line becomes semicircular, it begins to bend around itself. The loop surfaces touch at O and O' and the loop is divided into two parts – dislocation cusp POQ and the loop, Fig. 14.8. After this, cusp POQ and the loop are straightened out. Now finally the straightened segment PQ and a dislocation loop with same Burgers vector are obtained. When the application of stress is continued, the segment PQ bows again to form a second loop while the initial loop is expanded radially with continued application of stress. This source can generate more and more number of dislocation loops, which may move out and disappear at the surface of the crystal. If the loops are piled up against any impurity particle or atoms or grain boundary, the source is eventually shut down by the back stress created by the piled up dislocations.

The longer the segment l or smaller the value of b, less amount of stress is required. The generation of additional Frank-Read sources can be there by screw dislocation through the process of cross slip. A screw dislocation may cross slip twice and ultimately glide on a plane parallel to the initial slip plane.

Example 14.6

Calculate the shear stress required for the slip to occur in a work hardened copper crystal, if the dislocation density is 10^{13} m^{-2}.

Solution:

Shear modulus for copper is 44 GN m^{-2} and b= 3.61/$\sqrt{2}$ Å = 2.55 Å
The average length of a dislocation line, $l = 1/\sqrt{(10^{13})}$ = 3.16 x 10^{-7} m
Shear stress $\tau = \mu b/l = 44 \times 10^9 \times 2.55 \times 10^{-10}/(3.16 \times 10^{-7})$ N/m^2 = 3.55 x 10^7 Nm^{-2}

14.7 TWINNING

During twinning the material undergoes a change in shape. As a result of deformation twinning, a region of crystal undergoes a homogeneous shape deformation in such a way that the resulting structure is same as that of the parents but oriented differently. The twinning occurs as a result of cooperative movement of atoms such that individual atom moves only a fraction of the atomic spacing relative to each other, but the final result is shear. In case of slip, the atoms are displaced through one or multiple of lattice parameter whereas twinning occurs when smaller displacements take place successively in adjacent planes. These displacements occur in shear plane and in a particular slip direction **b**. In a bcc crystal, the shear plane can be (110) and the slip direction can be

$$[\bar{1}11].$$

The following are the twinning elements which characterize a twin:
   (i) Twin boundary: Atoms lying on this boundary are not displaced
   (ii) Slip direction
   (iii) Shear plane

The displacement in any plane within the twin is directly proportional to its distance from the twin boundary. The twin boundary is a planar surface imperfection. The twinning causes a rotation of the certain region of lattice, Fig. 14.9 such that the atomic arrangement in the twin is the mirror image of that of untwined part of the crystal. The twin boundaries occur in pairs, such that the orientation change due to one boundary is restored by the other, Fig.6.24. The layer of lattice sandwiched between two boundaries is called the twinned region. Twins are formed at the time of annealing as well as during cold working.

Twins in crystals are classified as deformation twins and annealing twins. The deformation twins are generated during plastic deformation whereas the annealing twins are formed during prior heat treatment in association with recrystrallization and growth of new grains. The defects in packing sequence (such as stacking faults) may arise in the original grain during the growth of

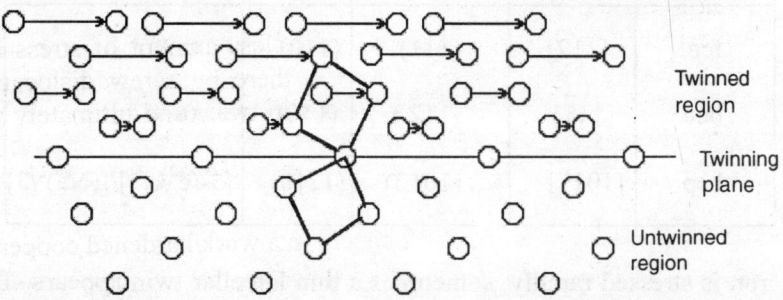

Fig. 14.9 Cooperative movement of atoms in twinning

new grains and as a result, annealing twins are formed. The annealing twins are observed in material, which has low stacking fault energy e.g., brass. Deformation twinning is usually characterized by
   (i) the evolution of energy in the form of sound (cracking of deformed crystal),

(ii) a jump wise variation of the deforming stress in the stress-strain curve.

Distinction between twin and slip

(i) Slip involves translation in multiple of unit displacement (Burgers vector) across a slip plane such that the relative orientation of different regions in slipped cube remains unchanged. Twin occurs due to atomic displacement taking place on all the planes in fractional amounts of Burgers vector within the twin.

(ii) Twinning shear produces mirror image whereas slip occurs along the close-packed direction on the slip plane.

(iii) Twinning strain has specific values, Table 14.3, depending upon the crystal system but slip may continue until failure.

(iv) Slip is heterogeneous and occurs on a fewer planes as compared to twinning. In case of deformation twin, twinned region had undergone a homogeneous shape deformation in such a way that there is reorientation of the lattice as compared to the parent lattice, although the structure of twinned region is identical with that of the untwinned one.

(v) Deformation by twinning requires higher critical shear stress than that is required for slip. At lower temperature, twinning becomes more favourable as compared to slip, since the stress for slip is more sensitive to temperature as compared to that for twin.

(vi) The microscopic observation can easily distinguish between offsets due to slip and deformation twinnning. In case of slip, only straight lines or wavy lines can be observed by microscopic examination of the surface of a prepolished sample. No change in contrast is observed on either side of the slip offset. A change in contrast is observed in case of twin bands.

Table 14.3 Crystallographic characteristics of twinning

Crystal structure	Twin direction	Twinning plane	Normal plane	Shear strain
fcc	$[11\bar{2}]$	(111)	$(1\bar{1}0)$	0.707
bcc	$[11\bar{1}]$	(112)	$(1\bar{1}0)$	0.707
hcp	$[10\bar{1}1]$	$(10\bar{1}2)$	$(1\bar{2}10)$	$[3-(c^2/a^2)]/[(c/a)\sqrt{3}]$

When bcc iron is stressed rapidly, sometimes a thin lamellar twin appears. This type of twins also appears in some alloys of iron. These twins are called Neumann bands. The deformation in close-packed hexagonal crystals can occur easily through twinning.

14.8 WORK HARDENING AND DYNAMIC RECOVERY

When the dislocations glide on the slip plane of a crystal due to application of stress, the crystal may undergo plastic deformation. During the movement of dislocation lines, they interfere with each other's motion and also with the point defects and grain boundaries, which act as barriers to further movement of dislocations. The movement of dislocation lines is hindered, and as a result, the stress required for plastic flow is increased. Thus, as plastic deformation increases, the movement of dislocation lines becomes more difficult due to increase of stress. This phenomenon is called work hardening or strain hardening process. The strain hardening process satisfies the relationship

$$\sigma_{tr} = A (B + \varepsilon_{tr})^n \qquad (14.19)$$

where the subscript 'tr' refers to true stress and true strain. The strain hardening exponent n is always less than one and depends on the structure and properties of a specific material. A, B are constants for the material. The strain hardening is a strengthening mechanism. The stress continues to increase with deformation beyond yield stress until the ultimate tensile stress is attained. At this point neck begins to form in the rod shaped specimen. The work hardening rate decreases with the increase of deformation beyond this point and hence change in stress also becomes smaller. If the material is worked even more, then it reaches the point of fracture since plastic deformation cannot be increased beyond this point.

Cold work

The strain hardening at room temperature during mechanical processing is called cold working. The cold work is measured as,

$$CW = \Delta A / A_o \qquad (14.20)$$

where ΔA is the change in cross-sectional area A_o during deformation. The stress-strain behaviour changes under the influence of cold work, Fig.14.10. Though the strength and hardness of the material increases, but its ductility decreases due to cold work. Ductility is measured as the percentage elongation of the material in the direction of stress. Annealing the material can reverse the change in microstructure and mechanical properties produced by cold working. The properties of material change during deformation and care must be taken so that no crack develops during strain hardening. This process requires a greater input of power as the deformation of material increases.

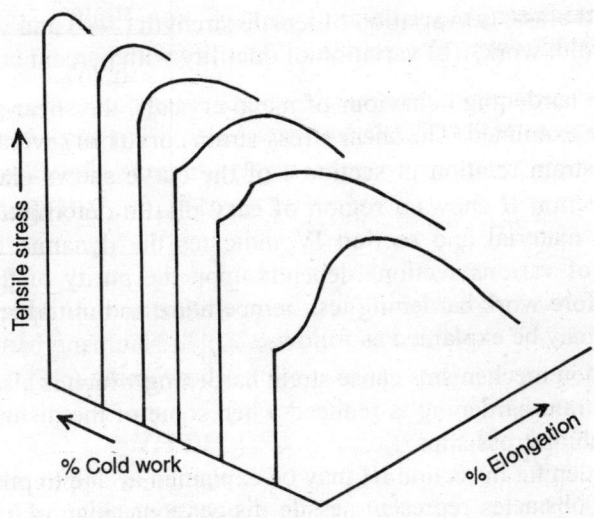

Fig.14.10 The stress-strain behaviour of a ferrite under the influence of cold work

Example 14.7

Calculate the tensile strength and ductility of a cylindrical copper rod, Fig 14.11, if it is cold-worked such that the diameter is reduced from 3.15 mm to 2.56 mm.

Solution:

The percentage cold work resulting from the deformation

$$= \frac{\pi(3.15^2 - 2.56^2)/4}{\pi(3.15^2)/4} \times 100\,\% = 34\%$$

The tensile strength of the given sample as read from the Fig 14.11 is 327 MPa and ductility is approximately 8%EL.

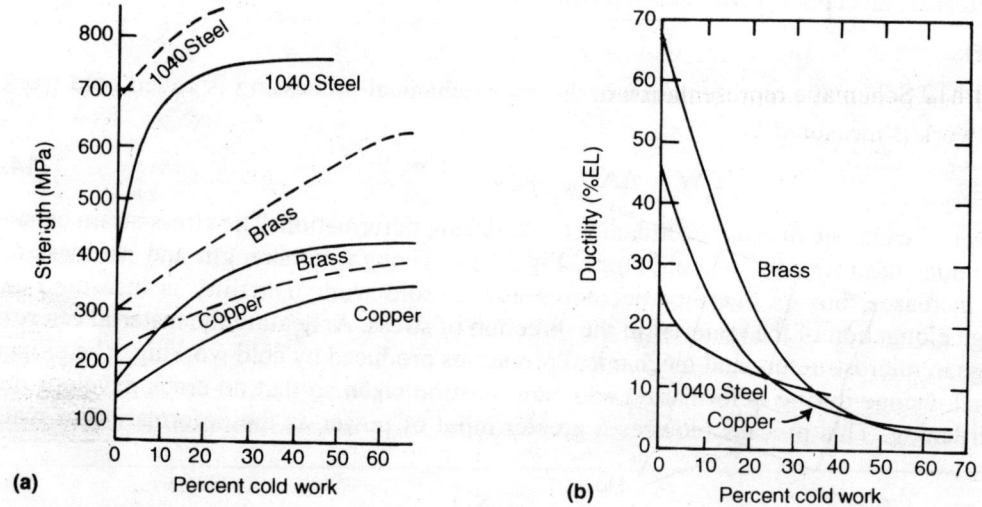

(a)   Percent cold work

(b)   Percent cold work

Fig 14.11 For copper and brass, (a) variation of tensile strength (-----) and yield strength (—) with percent cold work, (b) variation of ductility with percent cold work

To understand the strain hardening behaviour of metal crystals, the shear stress-strain behaviour of the material should be examined. The shear stress-strain curve has several distinct regions, Fig. 14.12, the linear stress-strain relation in section I of the curve shows elastic behaviour until τ reaches the value τ_c. Section II shows a region of easy plastic deformation. Section III shows linear hardening of the material and section IV indicates the dynamic recovery or parabolic hardening. The extent of various sections depends upon the purity of the crystal, dislocation density in the crystal before work hardening test, temperature and initial crystal orientation. The work hardening process may be explained as follows:

(i) Dislocation interaction mechanisms cause strain hardening which is also called work hardening. The effect of strain hardening is reduced when some of the dislocations move to other planes through climb or cross-slips.

(ii) The rapid strain hardening in section III may be explained as due to piling up of dislocations at obstacles. These obstacles represent sessile dislocations, Fig. 14.13, which obstruct the motion of other dislocations on their respective slip planes.

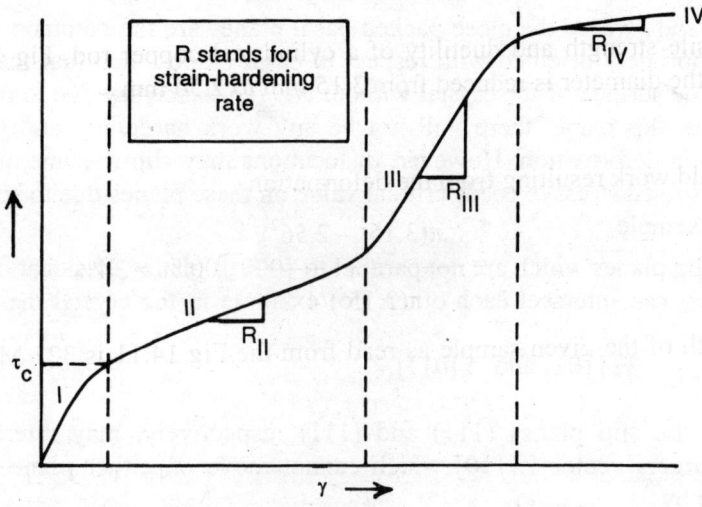

Fig. 14.12 Schematic representation of shear stress-strain curve for single crystal of metal.

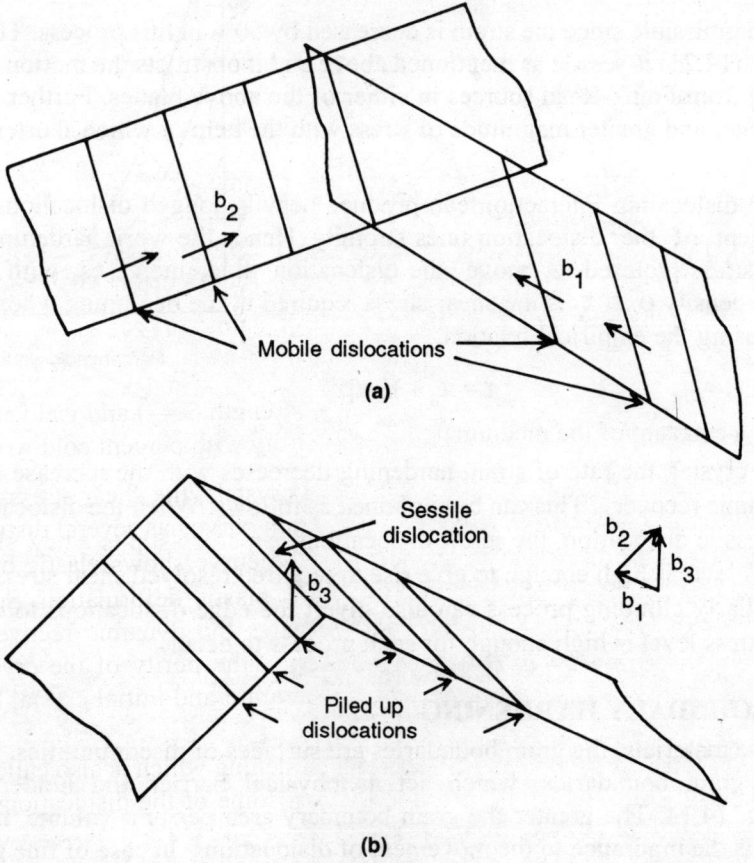

Fig. 14.13 Formation of (a) mobile dislocation and (b) sessile dislocation

Consider an hcp crystal, where the close-packed basal planes are the common slip planes. If the tensile axis is appropriately oriented with respect to basal planes such that the dislocations can easily move out of the surface of the crystal without any hindrance say due to presence of Frank-Read sources on the slip plane, there will not be any work hardening and the stress will be independent of plastic deformation. However, dislocations may slip in some other less common slip planes, if the resolved stress exceeds critical value on these planes due to specific orientation of tensile axis, for example

$\{10\bar{1}1\}$ or $\{11\bar{2}0\}$ slip planes which are not parallel to $\{0001\}$ planes. Dislocations, while moving along the slip planes, can intersect each other. For example in fcc crystal the dislocations with Burgers vectors

$$\frac{1}{2}\,[10\bar{1}]\text{ and }\frac{1}{2}\,[011],$$

which move along the slip planes (111) and $(11\bar{1})$ respectively, may interact to produce a dislocation with Burgers vector $\frac{1}{2}[110]$ which cannot move on either plane. This dislocation interaction, as given by

$$\frac{1}{2}\,[01\bar{1}] + \frac{1}{2}\,[101] = \frac{1}{2}\,[110], \tag{14.21}$$

is energetically permissible since the strain is decreased by 50% in this process. The product dislocation line in Eq. (14.21) is sessile as mentioned above and it obstructs the motion of other dislocation lines arising from Frank-Read sources in either of the above planes. Further plastic deformation requires greater and greater magnitude of stress with the help of which shorter sources may be activated.

(iii) Dislocation-dislocation interaction can produce heavily jogged dislocations, which impede the movement of the dislocation lines strongly. Hence the work hardening rate increases. The shear stress required to move the dislocation line increases with the increase of dislocation density ρ. If τ_o is the shear stress required at the beginning when $\rho = 0$, τ can be expressed using the empirical relation,

$$\tau = \tau_o + K\,\sqrt{\rho} \tag{14.22}$$

where K is a constant of the medium.

(iv) In a single crystal, the rate of strain hardening decreases with the increase of strain. This is called dynamic recovery. This can be explained as follows: When the dislocation lines pile up against a sessile dislocation, the screw dislocations can cross-slip to a new slip plane as soon as the stress level is high enough to give rise to required resolved shear stress on the new slip plane. Similarly climbing process can also divert the edge dislocations to other slip planes when the stress level is high enough for such process to occur.

14.9 GRAIN BOUNDARY HARDENING

In polycrystalline materials, the grain boundaries are surfaces of discontinuities. The dislocations pile up at the grain boundaries, which act as physical barrier and hinder the motion of dislocations, Fig. 14.14. The greater the grain boundary area per unit volume, more is the yield stress and more is the hindrance to the movement of dislocations. In case of fine grained material, strengthening can be achieved faster than coarse grained materials since the grain boundary area per unit volume is greater for the former case. The yield stress is related to the grain dimension (δ) as given by Petch equation,

$$\sigma_y = \sigma_i + K_y\,\delta^{-1/2} \qquad (14.23)$$

where σ_y is the yield stress and σ_i is the inherent strength (or intrinsic resistance to the plastic flow) of the material, K_y is the constant at fixed temperature related to the operation of dislocation sources. A linear relationship between the yield stress and $\delta^{-1/2}$ is experimentally observed for a variety of metals.

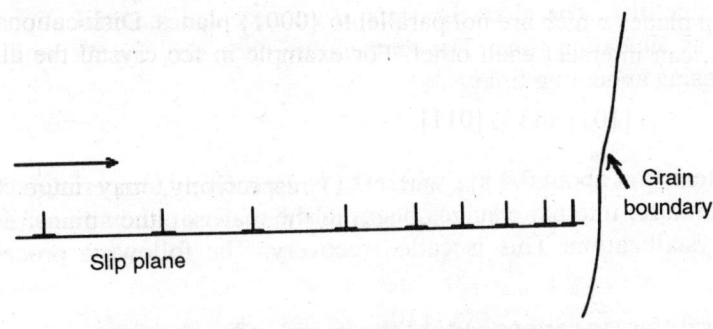

Fig. 14.14 Dislocations pile up at grain boundary

The slope of the linear graph for σ_y versus $\delta^{-1/2}$ gives the value of K_y. The intercept of the straight line on y-axis determines the value of σ_i.

Table 14.4 Values of σ_i and K_y for certain elements

	Al	Cu	Fe	Mo	Nb	Zn
σ_i, MNm^{-2}	16	26	48	110	126	33
K_y, MNm$^{-3/2}$	0.07	0.11	0.71	1.77	0.034	0.22

Example 14.8

The yield strength for an iron sample increases from 135 MNm^{-2} to 396 MNm^{-2} when the average diameter of a grain decreases from 8×10^{-2} mm to 5×10^{-3} mm. At what grain diameter the yield stress is 200 MNm^{-2}?

Solution:

Using Eq (14.23),
$$135 = \sigma_i + K_y\,(8 \times 10^{-5})^{-1/2}$$
$$396 = \sigma_i + K_y\,(5 \times 10^{-6})^{-1/2}$$

Hence
$$K_y = 0.778 \text{ MNm}^{-3/2}$$
$$\sigma_i = 48 \text{ MNm}^{-2}$$

Now
$$\sigma_y = \sigma_i + K_y\,\delta^{-1/2}$$
$$200 = 48 + 0.778\,\delta^{-1/2} \text{ MNm}^{-2}$$

The grain diameter $\quad \delta = 2.6 \times 10^{-2}$ mm

14.10 ANNEALING, RECOVERY AND RECRYSTALLIZATION

Materials in service, often have to withstand high temperature during service. The strength of most materials decreases at high temperature and hence work hardening is not a suitable mechanism for strengthening the material in such cases. Sometimes heat treatment is done to a material to increase its ductility, which also decreases its strength. This is due to the fact that dislocations and point defects, associated with high stress field are nonequilibrium defects and these defects are removed at high temperature.

Annealing

Annealing is a heat treatment process in which a material is heated at an elevated temperature for a period of time and then it is cooled slowly to room temperature. As a result, dislocation density decreases. Annealing yields a microstructure consisting of equiaxed grains. This process softens metal and increases ductility. The characteristics of an annealed material depend on annealing temperature as well as annealing time. The temperature necessary for recrystallization process decreases with increasing annealing time.

Recovery

A material starts softening at about 0.4 T_m, where T_m is the melting temperature. When annealing is done at this temperature, internal changes begin in the material and strain-free properties are detected before recrystallization. This is called recovery. The following processes take place during recovery:

(i) The point defects diffuse to various parts of the crystal and some of these disappear in dislocations. They are also absorbed at the surface or grain boundaries as a result of recombining with the vacancies. There are partial annihilations of vacancies with interstitial atoms. Point defects help in the climbing process of dislocations. Sometimes point defect complexes are formed.

(ii) The dislocations of opposite sign annihilate each other while those of same sign can repel to attain lower energy configuration. The tilt and twist boundaries can be created in this process. The dislocations are also redistributed by climb, which in combination with slip helps in disentangling of the dislocation arrays and sub-boundaries.

(iii) When the dislocations are redistributed due to slip, cross-slip and climb, small-angle boundaries are formed.

(iv) Grain boundaries move in between recrystallized grains forming larger grains.

The yield stress and the internal energy of the system decrease during recovery. This is also called polygonization. When recovery takes place, there are small changes in grain structure, dislocation density and a little decrease in hardness but there is a large reduction in residual stress, Fig. 14.15.

Recrystallization

During recrystallization atoms move freely to grow strain-free grains of lower energy at the cost of grains of higher energy. Recrystallization takes place in a material with a pronounced softening effect at a specific temperature in the range $0.4T_m$-$0.5T_m$. This temperature is called recrystallization temperature. The nucleation process involves local motion of grain boundaries in case of recrystallization. This is a thermally activated diffusion process where atoms jump from deformed grains with higher energy to new strain free grains of lower energy. As a result, elongated grains of higher dislocation density are consumed and equiaxed grains having lower dislocation density are formed. Thus, the thermal energy is a must for recrystallization. Since the dislocation density decreases, the strain energy of the material also decreases during recrystallization but there is no change in crystal structure. Therefore recrystallization is not strictly a phase transformation.

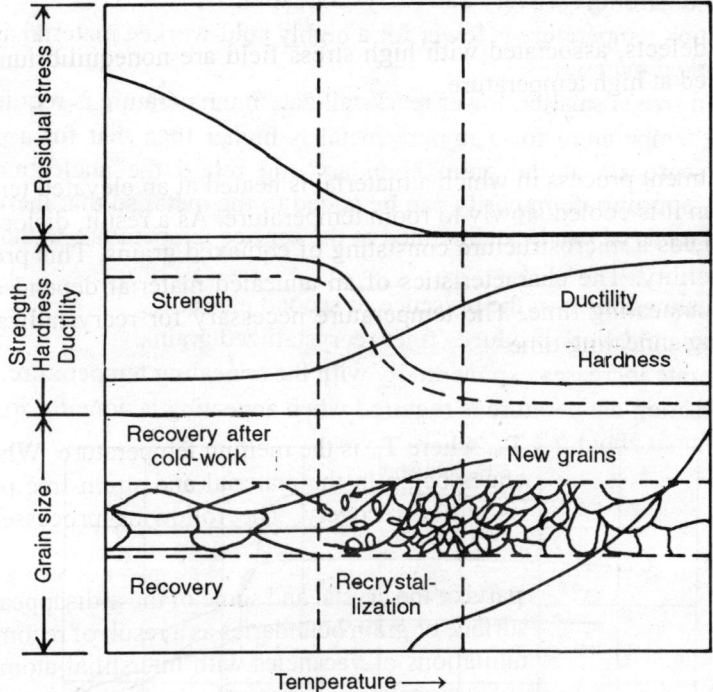

Fig. 14.15 Recovery and recrystallization of plastically deformed crystal during heat treatment.

The reduction in strain energy provides energy for recrystallization. Thus, a heavily deformed material with large dislocation density recrystallizes more rapidly. So, the recrystallization temperature is less for a highly cold-worked metal as compared to that of a lightly cold-worked metal, since in the former case more dislocation energy is available for recrystallization, Fig. 14.15. Rate of recrystallization depends on recrystallization temperature, the amount of cold work done and the time interval during which the material has been kept at elevated temperature. The electrical conductivity decreases during cold work, Fig 14.16, whereas it increases during recrystallization.

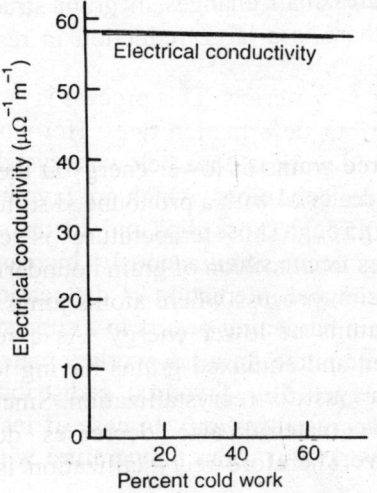

Fig. 14.16 The electrical conductivity starts decreasing due to increase in cold work.

Certain observations regarding recovery and recrystallization:

- The recrystallization temperature is lower for a highly cold-worked material as compared to a lightly cold worked material.
- When initial grain size is smaller, lower recrystallization temperature is required.
- Recrystallization temperature for a impure metal is higher than that for a pure one. The impurity atoms segregate at the grain boundary and retard the nucleation and growth processes. Thus, appropriate impurity can be added to the metal so that increased strength of a cold worked material can be maintained at the service temperature without letting it to recrystallize.
- Recrystallization slows down in the presence of second phase.
- Increase amount of cold work produces finer recrystallized grains.
- Recrystallization rate increases exponentially with the annealing temperature.
- Lower recrystallization temperature is required when annealing is done for longer time.

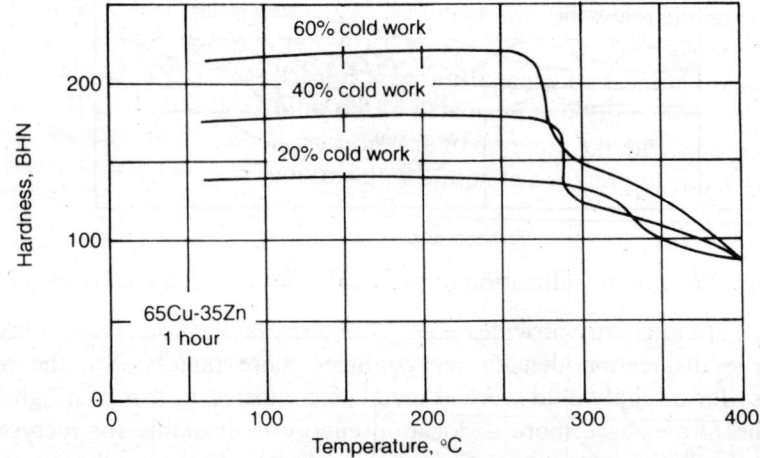

Fig. 14.17 Hardness versus temperature for cold-worked brass (65w/oCu-35w/oZn)

Fig 14.17 shows hardness versus temperature curves during recrystallization and recovery for 65 w/o Cu-35 w/o Zn for various percentage of cold working. These curves show hardness is nearly the same above recrystallization temperature for all the cases.

Hot working

Hot working is also a mechanical process of strengthening of materials. This process is performed in a wide range of temperatures ($0.5\ T_m - 0.3\ T_m$) at various deformation rates. Hot working is performed above recrystallization temperature whereas cold work is done below recrystallization temperature. It differs from cold working as it is done under conditions, which are favourable for recrystallization. During hot working, dislocations move through cross slips and diffusion climb. Two competing processes, strain hardening and softening, occur simultaneously. Increase in the density of dislocations due to application of external force and interaction of dislocations with one another, result in strain hardening. An additional strain hardening occurs in a supersaturated solid solution due to interactions of dislocations with precipitate. In a hot worked material, the density of dislocation decreases accompanied by energetically favoured redistribution of dislocations and recrystallization (also called dynamic recrystallization). In case of lead, room temperature is approximately $0.5\ T_m$, so it can be hot worked at room temperature while mild steel is hot worked at 900°C. The cold worked metal becomes stronger, harder and less ductile. Cold working requires more power for deformation. Cracking can occur more easily during cold working. Metal softens and anneals itself during the mechanical working at a temperature greater than the recrystallization temperature.

14.11 SOLUTION HARDENING

The strength of the material can be increased by introducing solute atoms in the solvent lattice, thus forming a solid solution. In this process, hardness as well as yield stress of the material increases. This is called solution hardening. Substitutional or interstitial solutes generate strains in the material, which in turn impede the movement of the dislocation lines. The dislocations get locked onto these solute atoms very easily since this process decreases the strain energy of the system. Smaller size atoms often get bonded to the edge of the extra half-plane where they become stable. The larger size atoms move to the region of tensile stress field below the dislocation line and decrease the strain energy. So, a greater shear force is required to move such dislocations.

Yield stress and hardness of a metal increase linearly with the concentration of solute. These properties also depend on the size mismatch of the solute and the solvent atoms. The hardness of the material increases with the increase of size mismatch. Fig. 14.18 shows how the yield strength of copper changes due to presence of solute with varying concentration.

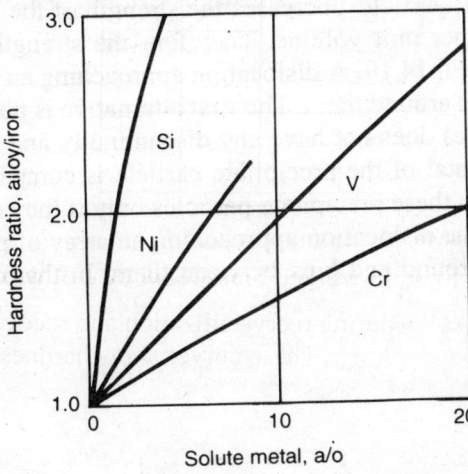

Fig. 14.18 Variation of yield strength ratio with concentration of solute.

In addition to the size difference of solute and solvent atoms and concentration of solute atoms, another factor that affects the strength and solution hardening is the deformability of the solute atoms. The solutes with higher shear modulus cause more hardening of solution than those with lower shear modulus when the sizes of solutes are comparable (e.g. Ni and Zn in copper). It requires less shear force to move a dislocation when Zn is present as solute in copper than when Ni is present, since the shear modulus for Zn is 3.4×10^4 MPa, that of Ni is 7.6×10^4 MPa and that of Cu is 4.6×10^4 MPa.

In solid solution, the stress fields of the solute atoms and dislocations interact elastically. The elastic interactions of the stress fields of solute atoms and dislocations are most important ones, which take place when the shear and hydrostatic stress fields of dislocations interact with the stress field associated with the solute atom. Stress field associated with a screw dislocation is distortional type whereas the edge dislocations have both distortional and hydrostatic stress field surrounding it. As an example, if a chromium atom substitutes for an atom of fcc nickel or bcc iron, a symmetrical (hydrostatic) stress develops due to difference in size dr between solute and solvent atoms and is proportional to ε, where

$$\varepsilon \propto \frac{dr}{dc} \qquad (14.24)$$

where c is solute concentration. Thus the distortion, which is created uniformly in all directions of the host lattice, contributes to the strengthening of material. Hydrostatic stress field of a substitutional solute atom interacts with the hydrostatic stress field associated with edge dislocation. The hardening depends upon μ^{-1} ($d\mu/dc$). When the stress fields associated with the solute atoms or impurities are nonsymmetrical for example when carbon or oxygen atom occupies the nonsymmetrical octahedral interstitial site of bcc iron, they interact with the nonsymmetrical or distortional stress components of both edge and screw dislocations. Thus, the strengthening of solid solution depends on whether the stress field associated with the solute atom is symmetrical or nonsymmetrical.

14.12 DISPERSION HARDENING

Some materials can be hardened when small particles of different phases are distributed throughout the material. The distribution of small particles increases the strength of the material depending on the size and density of particles per unit volume. Therefore the strength of the material depends on the inter-particle spacing L, Fig14.19. A dislocation approaching an array of precipitate particles may cut through them or bend around them. The first alternative is possible if the slip plane containing these precipitate particles does not have any discontinuity and also the energy required to move a dislocation in the crystal of the precipitate particle is comparable to that in the matrix. The dislocation can cut through these precipitate particles only if they are very small. When above conditions are not satisfied, the dislocation approaching an array of particles cannot always cut the particles, it has to bend around and bow between them. In that case the maximum stress required is

$$\tau = \mu b/R = 2\ \mu b/L \qquad (14.25)$$

if the radius of curvature of bow is L = R/2.

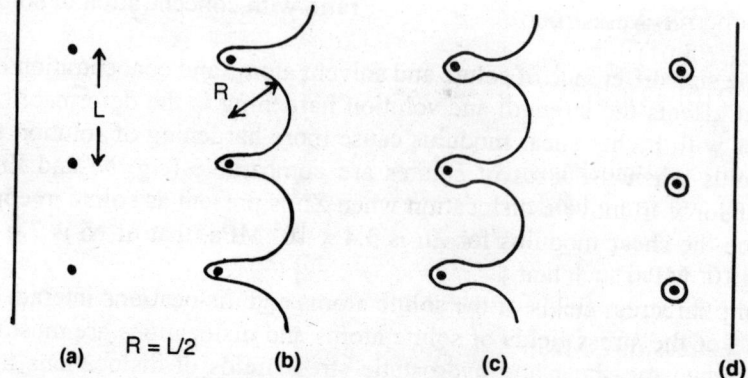

Fig. 14.19 Orowan mechanism for movement of dislocation lines bypassing precipitate particles

Dislocations move through the particles leaving loops around them, Fig 14.19. This process is called Orowan mechanism.

Dispersion of silica particles in copper is an example of dispersion hardening. The dispersion of particles in solid solution can be produced by internal oxidation e.g. when oxygen is diffused in a solid solution of Cu-Si, the dispersed phase of silica is formed. The particles in the precipitation-hardened systems may be metastable and have a coherent interface with the matrix.

EXERCISES

14.1   A tensile stress of 6.90 MNm^{-2} is applied in the [112] direction of tungsten. Calculate the shear stress in the [100] direction of (001) plane.

Ans: 2.30 MNm^{-2}

14.2   A tensile stress is applied in the [110] direction of an aluminium crystal until slip starts. Which of the twelve {111}<101> sets develop a shear stress? If the critical shear stress for slip is 0.828 MNm^{-2}, find out the applied tensile stress.

Ans: (111) [$\bar{1}$01], (11$\bar{1}$) [101], (111) [01$\bar{1}$], ($\bar{1}$11) [01$\bar{1}$]; 2.028 MNm^{-2}

14.3   When tensile test is carried out for a copper crystal, then the normal to its slip plane makes an angle 60° with the tensile axis and the slip direction makes an angle 35°. Find the stress necessary to cause the single crystal to yield, if τ_c for copper is 0.5MNm^{-2}.

Ans: 1.22 MNm^{-2}

14.4   The lowest tensile stress of 1.72 MPa is necessary to produce slip on slip system (110)[$\bar{1}$10] in NaCl crystal, calculate the critical shear stress and the direction of force.

Ans: 0.86 MPa, [100] or [010]

14.5   Distinguish between resolved shear stress and critical resolved shear stress. For what values of θ and φ resolved shear stress is maximum?

14.6   Slip system for hcp crystal is {0001} <11$\bar{2}$0>. Indicate three different slip directions <1120> within the plane (0001).

Ans: [11$\bar{2}$0], [1$\bar{2}$10], [$\bar{2}$110].

14.7   Explain three different mechanisms of strengthening of material—grain size reduction, solution hardening and strain hardening.

14.8   A hcp crystal is oriented such that the normal to basal plane makes an angle 60° with tensile axis and the slip direction makes an angle θ with the tensile axis. If the plastic deformation is observed at a stress of 2 MNm^{-2}, find the value of θ for which critical resolved stress is 0.79 MNm^{-2}.

Ans: 37.8°

14.9 If the width of dislocation in aluminium is 2.5 Å, calculate the shear stress to initiate the plastic deformation. Given $\mu = 2.6 \times 10^4$ MNm^{-2}.
Ans: 107 MNm^{-2}

14.10 A copper cylinder is cold-worked so that its ductility becomes 12%. If its radius becomes 15 mm after cold working, calculate its initial radius before cold working.
Ans: 15.99 mm.

14.11 A brass (35% Zn) sheet of thickness 2 mm is annealed before it is cold-worked such that its thickness becomes 1.6 mm (and nearly no change in width), Fig 14.11. Find
(i) the ductility and the tensile strength of the brass after cold working and
(ii) estimate the recrystallization temperature for this sample, Fig14.17.
Ans: (i) 23%, 420 MNm^{-2}; (ii) 325°C

14.12 The yield strength of a 70 Cu-30 Zn brass sample is 151.80 MNm^{-2} when grain size is 0.01 mm and is 62 MNm^{-2} when grain size is 0.1 mm. Estimate the yield strength of the alloy if the average grain size is 0.001 mm.
Ans: 435.77 MNm^{-2}

14.13 The shear stress, τ of a copper crystal is 0.69 MNm^{-2} at a dislocation density of 10^4 mm^{-2} and is 0.131 MNm^{-2} at a dislocation density of 10^2 mm^{-2}. Find the dislocation density of a copper crystal if τ is 6.9 MNm^{-2}.
Ans: 1.2×10^6 mm^{-2}.

14.14 Copper is cold-worked to attain minimum tensile strength of 300 MNm^{-2} and minimum ductility of 10%, Fig 14.11. How much cold work should be done to this material?
Ans: 30%.

15

Fracture

15.1 INTRODUCTION

When a material is under stress, sometimes cracks are formed. These cracks propagate through the material as the stress is increased and the material fails by breaking into two or more pieces. This phenomenon is called fracture. Fracture can occur in materials under various service conditions. It can occur due to fatigue when the material is subjected to alternating or cyclic loading. Fracture can also arise following creep if the material undergoes deformation at high temperature. Fracture can be broadly classified as:

(i) Brittle fracture   (ii) Ductile fracture   (iii) Fatigue fracture   (iv) Creep

15.2 BRITTLE FRACTURE

When we drop a glass it breaks into pieces. This is an example of brittle fracture. If the broken pieces of the glass are fitted together, the original shape and size can be restored. This will not happen in case of an aluminium mug, which undergoes deformation when it is dropped because aluminium is ductile and malleable. The characteristic of a brittle fracture is that the material fractures before any significant plastic deformation can occur. The surface of the fractured material looks shiny and granular because of the reflection of light from the individual grain surfaces. Thus the fracture takes place along particular crystallographic planes within grains or along grain boundaries. The former type of fracture is transcrystalline fracture and the latter is called intercrystalline fracture.

The crystallographic planes, along which fracture can occur easily compared to other planes, are called cleavage planes. The phenomenon of fracture can be visualized as the breaking of atomic bonds. The force required for this purpose is therefore a function of bond energy. The theoretical fracture stress for an ideal crystal is approximately equal to a major fraction of Young's modulus, whereas for a brittle material breaking stress is a much smaller fraction of Young's modulus, say Y/1000. The difference in fracture stress of real and ideal crystals can be explained due to the existence of tiny cracks in real crystals. The cracks are more commonly found at the surface of the crystal than in the bulk. They are usually unstable. Cracks can cause fracture more easily in glass and ceramics than in metals since there is no mechanism to stop the crack propagation in the former. As the stress increases, the plastic deformation also increases in metals, which strengthens the material. This effect hinders the advancement of the crack.

Griffith has proposed a model for fracture in brittle solids, according to which the fracture arises due to the presence of fine flaws in the surface of the material. The surface changes can arise due to the rubbing with fingers or reactions with the water vapour. These flaws are often small elliptical cracks. If the length of the internal crack is 2C in a thin plate under stress, then the stress concentrates at the tip such that maximum stress is given by

$$\sigma_{max} = 2\sigma \sqrt{(C/\rho)} \qquad (15.1)$$

where σ is the applied stress, C is the semi-major axis, Fig. 15.1, ρ is the radius of curvature at the tip. The ratio σ_{max}/σ (=$2\sqrt{(C/\rho)}$) is the stress concentration factor k_σ, which is a measure of the amplification of external stress at the tip of the crack.

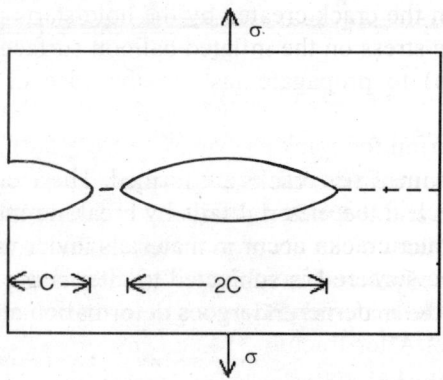

Fig. 15.1 Cracks in a thin plate under plane stress.

If the radius of curvature is of the order of the atomic radius, then σ_{max} is much higher than σ i.e. the stress concentration ratio σ_{max}/σ may be in the range 10^2 to 10^3. Thus, the localized stress at places may be very high, whereas the material as a whole is under a relatively small stress. As the crack propagates new surfaces are created which require surface energy. Consider a plate of unit thickness where a crack of length 2C is formed perpendicular to the tensile force. When the applied stress is increased to a high value, the surface cracks propagate. The crack propagation results in relaxation of the material since in this process elastic strain energy stored in the material is released i.e. the elastic energy in the crack zone decreases which is expressed as

$$\Delta U_e = - [\sigma^2/(2Y)][2\pi C^2] = - \pi\sigma^2 C^2/Y \qquad (15.2)$$

where elastic energy per unit volume is ($\sigma^2/2Y$). An additional energy $4\gamma C$ is consumed in forming two new surfaces of the crack, γ being the energy per unit area. Thus the change in energy, as the crack is formed, can be given by

$$\Delta U = 4\gamma C - \pi\sigma^2 C^2/Y \qquad (15.3)$$

A crack grows if the growth results in decrease in potential energy i.e. the condition for propagation of cracks without any increase in energy corresponds to $d(\Delta U)/dC < 0$. If $d(\Delta U)/dC$ is greater than zero the spontaneous growth of crack is not possible. Thus the critical stress (σ_{cr}) for propagation of a crack can be determined from the condition $d(\Delta U)/dC = 0$. We find the value of σ_{cr} as

$$\sigma_{cr} = (2\gamma Y/\pi C)^{\frac{1}{2}} \qquad (15.4)$$

This is also called fracture stress. Fracture stress is also breaking load per unit cross-section. When the critical stress is applied to a material, the crack starts growing spontaneously. The stress required for crack propagation decreases as the length of the crack increases. An applied stress σ causes instability in cracks that are larger than a critical size C* such that

$$C^* = 2Y\gamma/(\pi\sigma_{cr}^2) \qquad (15.5)$$

Example 15.1

What happens when you stick a pin on the surface of an inflated balloon?

Solution:

There will be a big bang when the crack created by the hole starts propagating very fast through the rubber. In this situation the stress on the inflated balloon surface is high enough for a crack of small length (hole dimension) to propagate fast. In the case of a deflated balloon no such phenomenon will occur.

Griffith's critical stress criterion for crack propagation, Eq.(15.4), also explains the discrepancy between the observed and the ideal strengths of brittle materials. This equation is also valid for surface crack of depth C which is capable of producing similar effect as the internal crack of length 2C and hence the surface cracks are more effective in producing fracture. If the brittle material is held with grips, the surface cracks do not propagate in the bulk region easily. When the length of the longest crack is perpendicular to the tensile stress axis, it is then the most effective one to cause fracture. After fracture, any of the broken pieces has higher strength than the original sample since the most effective crack is removed after the fracture.

Griffith's theory is based on the assumption that the crack propagation and hence fracture occurs without any prior plastic deformation. In a realistic situation, plastic deformation actually helps in nucleating a crack. Such plastic deformation is also present during crack propagation. Plastic deformations blunt the crack by increasing the radius of curvature ρ and hence more energy is required for the fracture to occur. Orowan modified Griffith's relation as follows:

$$\sigma_{cr} = (1/g)\,(2\gamma Y/\pi C)^{\frac{1}{2}} \qquad (15.6)$$

where $g = (a/\rho)^{\frac{1}{2}}$ is a geometric constant and it has different expression for different geometries. 'a' is the interatomic distance. Thus as ρ increases, the degree of blunting i.e. ρ/a also increases.

The fracture of a metal starts at a location, where the stress is highest. The stress intensity at the crack tip depends on applied stress and crack length. From experimental observation, the stress intensity factor K_I can be expressed as

$$K_I = g\sigma(\pi C)^{\frac{1}{2}} \qquad (15.7)$$

The critical value of K_I that causes failure is called the fracture-toughness K_{IC}, which is a function of fracture stress. Comparing Eqs.(15.6) and (15.7), we obtain

$$K_{IC} = (2\gamma Y)^{\frac{1}{2}} = g\sigma_{cr}(\pi C)^{\frac{1}{2}} \qquad (15.8)$$

Fracture-toughness value in SI unit is MPa√m.

Example 15.2

Silicon nitride has strength of 300 MNm^{-2} and it can support the largest size internal crack 45 μm. Calculate the fracture toughness of the material if its g factor is 0.9.

Solution:

$$K_{IC} = 0.9 \times 300 \times\sqrt{(\pi \times 22.5 \times 10^{-6})} = 2.27 \text{ MPa }\sqrt{m}$$

Example 15.3

A 30cm long glass fiber of diameter 0.05mm is broken when exposed to atmosphere for 6 hours. The breaking load is estimated to be 0.16 N. Given $Y = 7 \times 10^{10}$ N.m^{-2} and $\gamma = 0.6$ J.m^{-2}. Calculate the following:

    a) Fracture stress
    b) Crack depth
    c) Stress at the tip of the crack assuming tip radius to be 1.5 Å just prior to fracture.

Solution:

a) Fracture stress $\sigma_{cr} = \dfrac{0.16 \text{ N}}{\pi(2.5 \times 10^{-5})^2} = 8.15 \times 10^7$ N.m^{-2}

b) Crack depth $C = \dfrac{2 \times 0.6 \text{ J.m}^{-2} \times 7 \times 10^{10} \text{ N.m}^{-2}}{\pi(8.15 \times 10^7 \text{ N.m}^{-2})^2} = 4.025 \times 10^{-6}$ m ≈ 4 μm

c)
$$\sigma_{max} = \sigma_{cr} \times 2\sqrt{(4.0 \times 10^{-6}/(1.5 \times 10^{-10}))}$$
$$= 8.15 \times 10^7 \text{ N.m}^{-2} \times 2\sqrt{(4.0 \times 10^{-6}/(1.5 \times 10^{-10}))}$$
$$= 2.66 \times 10^{10} \text{ N.m}^{-2}$$

Example 15.4

An aluminium alloy 7075-T651 is being used for engineering design, which can support 200 MPa in tension. The alloy is being used in the form of a plate. Calculate the largest crack size that this material can support before fracture starts. Given that $K_{IC} = 24.2$ MPa $\sqrt{m}$.

Solution:

Considering $g = 1$,
$$C = (1/\pi)[K_{IC}/\sigma_{cr}]^2 = (1/\pi)[24.2 \text{ MPa}\sqrt{m}/ 200 \text{ MPa}]^2 = 4.66 \text{ mm}$$
The largest crack size = 9.32 mm

15.3 CRACK NUCLEATION IN CRYSTALS

Experimental observations show that in many crystalline solids, plastic deformation precedes brittle fracture. Although there is no specific evidence for the movement of dislocations prior to crack propagation, but there are dislocation interactions, which are responsible for the formation of crack nuclei. According to a simple model of crack nucleation, a series of edge dislocations piles up at a grain boundary, Fig. 15.2, or at any other obstacle, which ultimately coalesce at the head of the pile up. The stress concentration at the head of the pile up leads to the formation of crack nucleus, Fig. 15.3. When such a crack nucleus is formed on a plane normal to slip plane, the tensile stress may reach the ideal fracture stress.

Stroh and Mott found out that a crack could also initiate in a plane not perpendicular to the slip plane and showed that the tensile stress σ at which fracture would occur, becomes maximum at an angle $\theta = 70.5°$ to the slip plane.

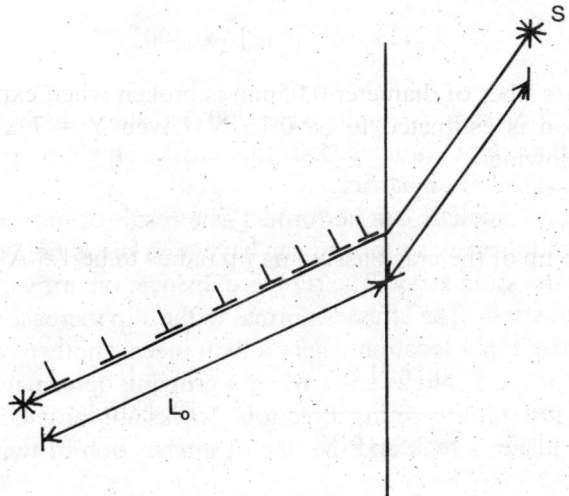

Fig. 15.2 Piling up of edge dislocations

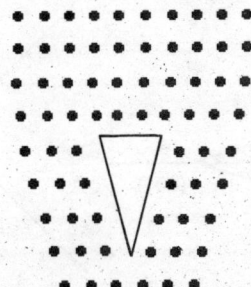

Fig. 15.3 Formation of crack nucleus

Thus, the applied shear stress τ, in the plane where dislocations pile, is related to the maximum stress as,

$$\sigma_{tmax} = 2 \, (L_o/3C)^2 \, \tau \tag{15.9}$$

where L_o is the length of the slip line along which dislocations pile up. The number of dislocations (n), which pile up in the slip plane is given by Eshelby, Frank, and Nabarro as

$$n = (\pi L_o \tau (1- \nu)) \, / \mu b \tag{15.10}$$

where μ is shear modulus, b is Burgers vector and ν is Poisson ratio. The value of n may be as high as 10^2 to 10^3 when the local stresses at the tip of the pile up are of the order of 0.7μ. Such large number of dislocations can form a jog of dimension of the order of few thousands nanometer on the surface of a crystal.

Cottrell proposed another model for crack formation in which dislocations move in two intersecting slip bands, combine along the intersection line. The resultant dislocations feed in to enlarge the cavity. In bcc iron, dislocation reaction is given by

$$\frac{1}{2}[1\bar{1}1] + \frac{1}{2}[111] \rightarrow [001] \qquad\qquad (15.11)$$

Hence in this case, the crack is produced in (001) plane, which is the cleavage plane. The repeated reaction of above type initiates Griffith's crack. This type of crack formation does not require any obstacle or dislocation barrier.

In hcp metals the crack nucleus can be formed as a result of movement of dislocations in the basal plane. The dislocation array so formed can have high stress field such that shear stresses can cause rupture in the dislocation array. The ruptured dislocation arrays, Fig. 15.4, are shifted from each other as a result of stress. The stresses normal to the slip increase and initiate a crack.

Cracks are also initiated in a location where a twin meets another twin or a grain boundary. A favourable condition for crack initiation is when a growing deformation twin intersects another existing twin having a different twinning direction. Cracks are initiated when the resolved normal stress on the cleavage plane is high and the line of intersection of twins is close to the cleavage plane.

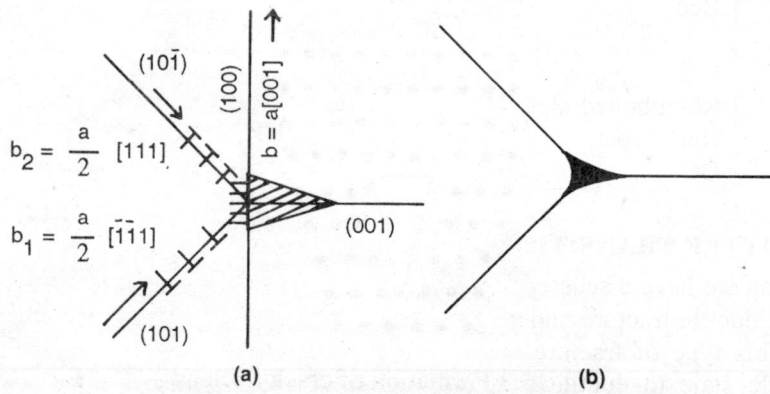

Fig. 15.4 Initiation of Griffith's crack

The formation of crack nuclei can be due to grain boundary sliding which produces angular or[7] wedge shaped microcracks, Fig. 15.5. When vacancies coalesce at the grain boundary, rounded microcracks are produced. In conclusion we can say that the twin and grain boundaries are favourable locations for the initiation of cracks.

15.4 BRITTLE FRACTURE IN SINGLE CRYSTAL

Brittle or cleavage fractures are quite common in ionic and covalent crystals at low temperatures. This type of fracture is uncommon in fcc metals e.g. copper, aluminium etc. Such fractures are found often in bcc metals where interstitial impurities are present e.g. iron, molybdenum, chromium, vanadium etc. A number of hcp metals, like Zn, Mg, Zr, Be etc., undergo brittle fracture along basal plane at a low temperature. There are other metals also which show cleavage or brittle fracture, Table 15.1 usually along a plane where surface energy is lowest.

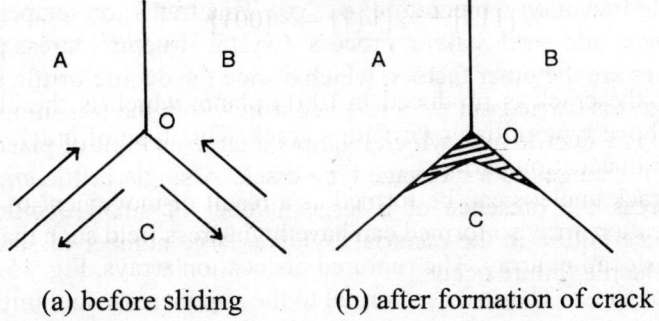

(a) before sliding   (b) after formation of crack

Fig. 15.5 Crack initiation by grain-boundary sliding in a triple point

Table 15.1 Cleavage planes in crystals

Crystal Structure	Metals	Cleavage plane
Bcc	Fe, Mo, Cr, W	{001}
	Ta, V	{110}
hcp	Zn, Be, Mg	{0001}
Rhombohedral	Bi	{111}
NaCl type	NaCl, MgO	{001}
DC	Diamond	{111}

15.5 DUCTILE-BRITTLE TRANSITION

In the previous section we have discussed about certain metals undergoing brittle fracture. These materials can exhibit ductile fracture under changed conditions. Usually brittle fracture occurs at low temperatures. This type of fracture cannot occur at a high deformation temperature. The transition from brittle state to ductile state of a metal occurs usually in a narrow range of temperature. The ductile-brittle transition temperature is a temperature within this range, which depends on other factors like grain size, purity, melting point and type of heat treatment. If we plot the yield stress σ_y versus temperature, Fig. 15.6, we find that σ_y increases rapidly whereas the fracture stress σ_{cr} increases very slowly with lowering of temperature.

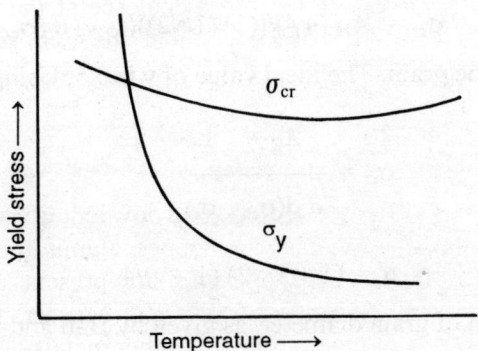

Fig. 15.6 Stress versus temperature

At ductile-brittle transition temperature, $\sigma_y = \sigma_{cr}$. This transition temperature for steel depends on the heat treatment and steel making process. Crystal structure, stress pattern and presence of dissolved impurities are the other factors, which decide the ductile-brittle transition temperature. When the impact test is carried out for a notched iron within the transition temperature range, the first crack to form is a ductile one, which requires a large amount of plastic deformation. As this crack propagates, it changes to a cleavage type crack. A single brittle crack can cause failure of the material whereas the presence of a large number of small ductile crack regions is not sufficient to produce failure in the material. When a large number of small ductile fractures are linked, Fig. 15.7, ductile failure occurs.

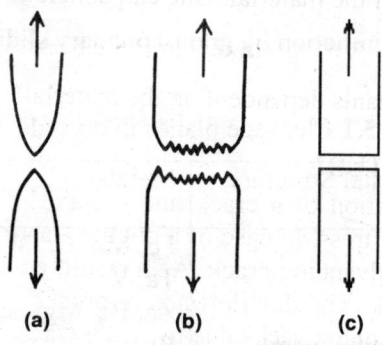

(a)   (b)   (c)

Fig. 15.7 (a) Fracture of a highly ductile material on which the necks become pointed,
(b) Fracture of a moderately ductile material,(c) Brittle fracture without any
plastic deformation

If in a specimen σ_y is greater than σ_{cr}, it undergoes brittle fracture. Stroh has given the expression for fracture stress

$$\sigma_{cr} = 4\,\gamma' / [n\,b\,(1+(1/\sqrt{2}))] \tag{15.12}$$

where brittle crack is formed due to coalescence of n dislocations, and γ' is the effective surface energy. The effective stress operating on the dislocation is $\sigma_{cr}-\sigma_i$, after friction stress is overcome, where σ_i is the friction stress of unlocked dislocation. If the crack nucleus is formed at the end of a slip band, the back stress or the effective stress is expressed as:

$$(\sigma_{cr} - \sigma_i) = 4\mu n\, b/[(1-v)d] \tag{15.13}$$

Thus,
$$\sigma_{cr} = 4\mu\,(4\gamma')/[(1+(1/\sqrt{2}))(1-v)\,d\,(\sigma_{cr}-\sigma_i)] \tag{15.14}$$

where d is the diameter of the grain. The ideal value of v is 0.5. Using Eqs.(15.13) and (15.14) we obtain,

$$\sigma_{cr} = \frac{4\mu\gamma'}{\alpha\,d(\sigma_{cr} - \sigma_i)} \tag{15.15}$$

where
$$\alpha = (1 + 1/\sqrt{2})\,(1 - v)/4 \tag{15.16}$$

The yield stress is a function of grain diameter, as given by Hall and Petch,

$$\sigma_y = \sigma_i + kd^{-\frac{1}{2}} \tag{15.17}$$

In case $\sigma_y = \sigma_{cr}$, we get

$$(\sigma_{cr} - \sigma_i) = kd^{-\frac{1}{2}} \tag{15.18}$$

from Eq (15.17). Hence the criterion for transformation from ductile to cleavage fracture is given by

$$\sigma_{cr} = 4\mu\gamma' / \alpha \, kd^{\frac{1}{2}} \tag{15.19}$$

This equation is approximate, since here we assume that the formation of crack is due to the coalescence of dislocations under shear stress. We know that hydrostatic tensile stress also increases the brittleness of mild steel. Moreover, σ_i has strong dependence on temperature as well as on impurity content of the material. One can conclude from experimental observations that,

$$\ln \sigma_i = \ln B - cT \tag{15.20}$$

where B, and c are constants dependent on the material.

15.6 DUCTILE FRACTURE

Fracture involves formation of a crack nucleus and subsequently, the propagation of crack in response to the applied stress. In case of a ductile fracture, there is extensive plastic deformation in the region near the advancing crack. As a result, a stress field is created which resists further propagation of the crack. The ductile fracture proceeds slowly as the length of crack increases. Such a crack is a stable one. Cracks due to brittle fracture are unstable since they propagate once started. There is very little resistance in the propagation of such cracks.

Highly ductile materials yield due to application of tensile stress causing almost 100% reduction of cross-sectional area, such that it necks down to a point fracture. This type of fracture occurs in metals like gold, lead etc. at room temperature and in certain other metals and polymers at elevated temperatures. The ductile fracture in polycrystalline substances occurs normally in stages:

(i) Initially as necking begins, small cavities or voids are formed. These are formed at the neck in the case of tensile stress.

(ii) In the second stage, as deformation continues, the voids or cavities become large and coalesce to form an elliptical crack which has its major axis perpendicular to the tensile axis. The crack normally grows in the direction parallel to the major axis.

(iii) Finally, the rapid propagation of crack spreads to the outer surface of the neck by shearing in a direction approximately 45° to the tensile axis to give a cup and cone type fracture.

As per the relationship given by Hall and Petch, the yield stress increases with the decrease of grain size of metals and alloys. The finer grains in materials increases the rupture stress. The rupture stress is greater than yield limit for ductile materials. The rupture stress can be attained only on overcoming the hardening effect. When the grain size is above a certain limiting value, the rupture stress is equal to or lower than the yield limit, which indicates the brittle nature. So the grain coarsening in metals and alloys results in loss of ductility. The plastically worked metals usually exhibit anisotropy, which is the characteristic of ductility. Sometimes alloying can raise the ultimate strength and increase the ductility. Relative reduction of cross-section in tensile test increases at an elevated temperature because a larger number of slip systems become operative and diffusion processes are enhanced. The process of softening also decreases the chance of fracture and hence increases ductility.

15.7 REMEDIES AGAINST FRACTURE

The nature of surface of the engineering material is one of the most important factors, which can decide the mechanical strength of a material. The presence of cracks, small ridges and recesses in a badly finished surface of a metal or alloy may act as stress concentrators. The following methods can be adopted to improve the surface so that brittle fracture can be avoided:

(i) Rock salt crystals are usually brittle. They can be made plastic by immersing them in warm water, which dissolves the imperfect surface layer (Ioffe's effect).

(ii) Polishing of the surface of metals and alloys can improve the surface and hence improves the ductility of the material.

(iii) The surface of a glass is etched with HF so that the cracks and other nonuniformities are removed. This improves the strength of the glass. But the etched glass surface should be protected from the other mechanical abrasions as well as from the corrosion and other environmental effects.

(iv) The rupture stress for a material with rough surface can be raised by introducing compressive stresses in the surface layer. This method prevents the propagation of surface cracks and increases the service life of metal parts. There are methods like shot-blasting, roll forming and also some heat treatment procedures like tempering which can improve the surface of the test pieces.

(v) Chemical reaction can produce compressive stresses in the surface layers and prevent the crack propagation. In the ion exchange method, Na^+ ions in the surface of a sodium silicate glass can be replaced by larger ions like K^+ and hence compressive stresses in the surface are introduced.

(vi) Tensile stress often introduces ductile or brittle fracture. High hydrostatic pressure reduces the tensile stress in a deformed body and can suppress the formation of cracks and their propagation. It can also remove the micro-defects, existing in the metal during plastic deformation.

Under high uniform hydrostatic pressure, contact bridges are formed between the opposite sides of a fracture. The particles from one side of the defect jump to the other side as a result of high stresses and thus this process restores the continuity of the deformed metal. This mechanism is similar to cold welding. Local heating at points of contact can be produced due to local plastic deformation or surface energy liberation. Surface energy is liberated as a result of protrusions of a defect moving towards each other by a distance equal to the lattice parameters. This heat energy enhances the chance of activating diffusion processes.

15.8 FATIGUE FAILURE

Fatigue failure can occur in a structure due to rapidly varying stress or repeated application of a load, which is much smaller than the yield or tensile strength for static load. The name fatigue is given to the state of structure because the material seems to tire out after repeated application of dynamic stress. About 90% of all mechanical failures are caused by fatigue. This type of failure can occur in metals, polymers and ceramics (except for glasses) catastrophically i.e. without any major distortion preceding the collapse. The cracks that are formed in the structure by the fatigue process are not usually detected during routine check up.

The fatigue test of a material is carried out by subjecting it to stress cycles until it fails. The number of stress cycles is measured corresponding to the stress of amplitude S. The fatigue behaviour of a material can be determined after carrying out test on a series of specimens. The maximum amplitude S_{max} of the stress cycle is equal to or less than the tensile strength. The stress fluctuation has a particular range and usually the mean stress is more than half the stress range. The mean stress value may be even zero. The number of cycles of stress N, which causes failure

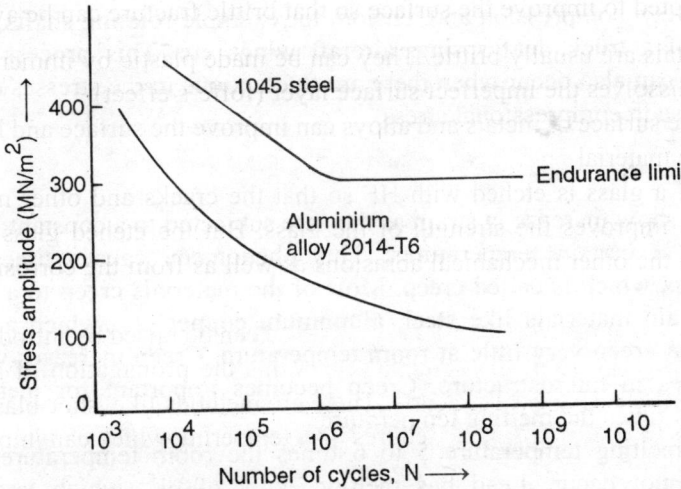

Fig. 15.8 Stress versus logarithm of number of cycles N to fatigue failure for
(a) aluminium alloy, (b) mild steel.

is counted. The test is repeated at progressively decreasing stress amplitudes S. The curves of S versus the $\log_{10} N$ are plotted for steel and aluminium alloys to show the progressive fatigue in these materials, Fig. 15.8.

These curves show the number of cycles of stress, which a structure can tolerate before failure, increases with decreasing amplitude of stress. Certain important engineering materials like steel, titanium etc. have S-N curve with horizontal tail at certain limiting stress amplitude. Such materials will tolerate an indefinite number of stress cycles without any changes in the stress amplitude This is also called fatigue limit S_m or endurance limit, which depends on surface preparation, residual stresses and inclusions. Any stress amplitude greater than the fatigue limit S_m, can cause failure if the number of stress cycles is sufficiently large. There are materials which have no fatigue limit i.e. for such cases larger the stress amplitude, smaller the number of cycles of stress. The fatigue limit of a material in service can be improved by several methods:

(i) Fine grain sizes in alloys or metals improve the fatigue resistances.
(ii) Polishing the surface of the components can reduce the possibilities of crack initiation and propagation.
(iii) The other methodologies, which are adopted for protection against fracture, can also be applied for protecting components from fatigue failure.

While designing an engineering structure, it is desirable to reduce stress concentrations and make use of beneficial compressive residual stress such that fatigue performance improves. The stress concentration factor depends upon the presence of notches, tool marks, decarburization and other surface conditions. The strain hardening, surface coating, introducing compressive residual stresses in the surface layer, can effectively increase the fatigue strength. The fatigue strength of nonferrous metals and annealed steel increases with decrease in grain size. For example, the coarse pearlite has a lower fatigue life than that of the spheroidal microstructure, even when the steel samples of same strength are chosen in the above cases. In aircraft industry where a fatigue failure can lead to a disaster, the large scale models or structures are put to fatigue testing under simulated service conditions.

The materials in service are subjected to very large number of cyclic stresses. Usually these stresses alternate between compression and tension, for example rotating shafts, wheels of a car, loaded components of a truck, leaf spring, aircraft wings etc. This process deteriorates the materials. The fatigue can also occur when there are fluctuations in the stresses of same sign i.e., say increase or decrease in compressional stress.

15.9 CREEP

Engineering materials show increase in strain as they are subjected to a constant load or stress for an interval of time, at certain temperatures. This phenomena causes time dependent and permanent deformation, which is called creep. Most of the materials creep to a certain extent at all temperatures. Certain materials like steel, aluminium copper etc. which are often used in engineering application creep very little at room temperature. Creep increases with temperature and can cause changes in microstructure. Creep becomes important for metals or alloys at temperatures above 0.4 times the melting temperature.

Most metals have melting temperature 5 to 6 times the room temperature hence at room temperature creep cannot occur. Lead has melting point 600°K, which is twice the room temperature 300°K. Hence creep is observed in lead at room temperature. Creep of lead is seen in roof since it occurs at ambient temperature under its own weight.

Analysis of creep phenomena

When stress in excess of yield stress is applied suddenly to a material, an initial strain say ε_0 is developed in the material. If the stress is maintained constant, the strain ε goes on increasing with time until it attains a steady state value. The attainment of steady state value occurs at temperature much lower than the melting temperature, Fig. 15.9. Creep is time dependent strain and it is plotted against elapsed time to obtain creep curve. Creep tests are conducted for a material to measure time dependent strain at controlled temperatures and under constant load. Commonly conducted creep tests are tensile test within elastic limit. The creep curve becomes parallel to time axis after certain interval of time at a low temperature.

At an elevated temperature the creep curve does not remain parallel to time axis at a later time. The strain goes on increasing maintaining same gradient and finally turns upward prior to failure. There can be high-temperature creep when the material in service is maintained at high temperature. The creep at low temperature as observed in lead, is called low- temperature creep.

According to Andrade, the constant – stress creep curve is the resultant of two creep processes – a transient creep, which has a decreasing creep rate with time and a viscous creep, which is a constant rate creep process. Andrade represented the creep curve by the following empirical equation:

$$\varepsilon = \varepsilon_0 (1 + \beta t^{1/3}) \exp \alpha t \qquad (15.21)$$

where ε : strain at time t

   ε_0 : strain at initial time

   β : constant for transient creep

If $\alpha = 0$ $\varepsilon = \varepsilon_0 (1 + \beta t^{1/3})$

hence transient creep rate $\quad \dot{\varepsilon} = (d\varepsilon/dt) = (1/3)\,\varepsilon_o\beta t^{-2/3}$ $\qquad$ (15.22)

When $\beta = 0$ $\qquad\qquad\qquad\qquad \varepsilon/\varepsilon_o = \exp\alpha t$

or $\qquad\qquad\qquad\qquad\qquad\qquad \varepsilon = \alpha\,\varepsilon$ $\qquad\qquad\qquad\qquad$ (15.23)

α represents the creep rate per unit strain. It is the viscous component of creep. The creep curve exhibits three well defined stages, Fig.15.9:

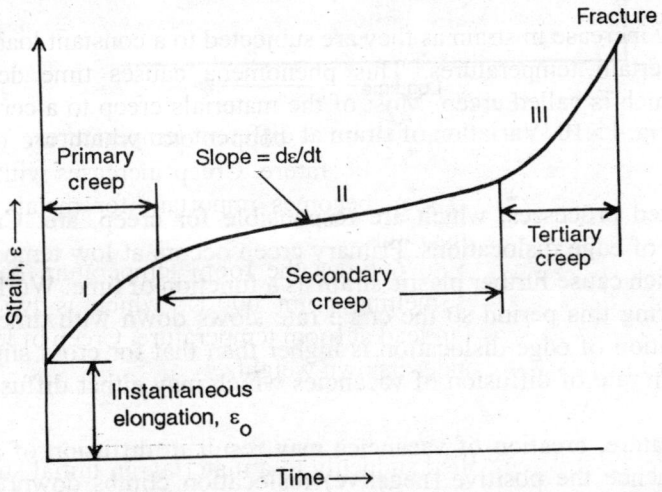

Fig. 15.9 Variation of strain with time.

(i) In stage I, the creep rate decreases with time until it becomes constant and this is called primary creep. During primary creep, the work hardening effect is more than the effect of recovery. This part of the curve has same shape at low as well as high temperature. Primary creep is a period of predominantly transient creep.

(ii) In stage II, the creep rate is approximately constant. The effect of work hardening is balanced by that of recovery. This is referred to as steady state or secondary creep. The secondary creep arises due to the combined effect of movements of grain boundary and dislocations within the grains. The average value of the creep rate during this stage is called minimum creep rate. This type of creep occurs in gas turbines, reactors, and boilers etc. where the stress is applied at high temperature.

(iii) The creep rate increases rapidly with the increase of elapsed time in stage III and failure starts. The necking of the specimen is observed during this stage. This is called tertiary creep.

Tertiary creep occurs during creep tests under constant load at high stress and high temperature. Tertiary creep is not only due to necking of specimen but also due to structural changes occurring in the metal. During tertiary creep, the growth of voids at the grain boundaries leads to extensive cracks and intergranular creep fracture. Fig. 15.10 shows the effect of applied stress on creep curve at constant temperature. The various stages of creep phenomena are explained in detail.

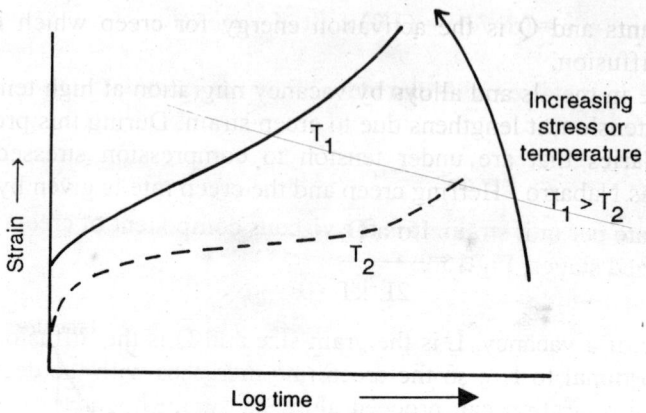

Fig. 15.10 Variation of strain at different temperatures.

(i) Primary creep

The thermally activated processes, which are responsible for creep, are: Cross slip of screw dislocation and climb of edge dislocations. Primary creep occurs at low temperatures due to the cross slip or climb which cause further plastic strain as a function of time. Work hardening occurs at a very fast rate during this period so the creep rate slows down with time. The temperature required for climb motion of edge dislocation is higher than that for cross slip since the former process requires a high rate of diffusion of vacancies which may either diffuse to or away from edge dislocation line.

At a higher temperature, creation of vacancies may result in diffusion of atoms to the edge dislocation line and hence the positive (negative) dislocation climbs down(up). Alternatively, atoms may diffuse away from dislocation line resulting in climb up(down) of positive(negative) edge dislocation and annihilation of vacancies. Dislocations while gliding on slip plane can pile up against an obstacle. At an elevated temperature or due to applied stress, dislocations can move to other parallel slip planes by climb or by cross-slip. The equations, which describe the primary creep, are:

$$\varepsilon = a \log t \quad \text{for very low temperature,} \tag{15.24}$$

and
$$\varepsilon = b\, t^{p} \quad \text{at an elevated temperature and higher value of stress} \tag{15.25}$$
where a, b and p are constants; p is a fraction depending upon the material.

(ii)Secondary Creep

Secondary creep is also called steady state creep. The rate of recovery is very fast in this temperature range such that it balances the rate of work hardening. During recovery, there is annihilation of edge dislocations, which results in decrease in dislocation density as well as strain energy. There is also rearrangement of dislocation lines giving rise to a configuration of lower energy known as polygonization. Polygonization and grain boundary sliding are important processes that occur during deformation by steady state creep The equation for secondary creep is

$$\varepsilon = \kappa\, t \tag{15.26}$$

where κ is a stress and temperatures dependent constant. The expression for κ is given by

$$\kappa = d\varepsilon /dt = A\, \sigma^{n} \exp(-Q/RT) \tag{15.27}$$

where A, n are constants and Q is the activation energy for creep which is nearly equal to activation energy for diffusion.

Creep can take place in metals and alloys by vacancy migration at high temperatures. When a rod specimen is under tension, it lengthens due to creep strain. During this process vacancy flow from the grain boundaries that are under tension to compression stressed boundaries. This mechanism, is known as Nabarro - Herring creep and the creep rate is given by

$$\dot{\varepsilon} = \frac{\sigma a^3 D}{2L^2 kT} \tag{15.28}$$

where a^3 is the volume of a vacancy, L is the grain size and D is the diffusion coefficient. Since the creep rate is proportional to L^{-2}, so the creep rate increases with the decrease of grain size. Sometimes the vacancy migration can proceed along the grain boundaries, this is called coble creep and it leads, to the dependence of creep rate on L^{-3}. Materials of coarse grain size should be chosen for high temperature application to avoid failure due to diffusion creep.

iii) Tertiary Creep

Formations of microcracks and voids are associated with the tertiary creep. During tertiary creep, the voids are created either by collection of vacancies at the grain boundaries or by grain boundary sliding. Angular or wedge shaped voids are created by sliding of grain boundaries. The grain boundaries are noncrystalline material and hence they have less viscosity as compared to that of the grain material.

The grain boundary sliding increases at higher temperature since the viscosity of the grain boundary decreases. At a temperature above $0.5\ T_m$, the viscosity becomes very low such that the grain boundaries behave like liquid, which facilitate the neighbouring grains to slide against each other. At lower temperature, the grain boundaries provide effective obstacles to dislocation motion; thus they increase yield strength.

Mechanisms of creep

The discussion of creep process reveals that it is temperature as well as time dependent process. The creep mechanisms in crystal have the origin in atomic processes. The mechanisms of creep can be explained in terms of the following processes:

(a) Dislocation climb
(b) Vacancy diffusion
(c) Grain boundary sliding

Dislocation climb

If the temperature is high enough to provide activation energy for vacancy diffusion, they may diffuse to or away from the edge dislocations resulting in climb motion. In case the gliding motion of a dislocation line is hindered due to presence of impurity or other obstacles, then the dislocation line can climb to another parallel slip plane and continue gliding in response to stress.

Vacancy diffusion

Diffusion of vacancies also controls the creep rate without causing climbing motion of dislocation lines under tensile stress, the vacancies move from the surfaces of the specimen perpendicular to the stress axis to the surface parallel to the stress axis. As a result of such motion, the specimen gets elongated, parallel to the stress axis. This gives rise to diffusional creep.

Grain boundary sliding

As the temperature rises, grain boundaries lose strength faster than the grains themselves since the viscosity of noncrystalline grain boundaries decreases more than that of bulk material in the grain. At a temperature $0.4T_m$, the grain boundaries behave like a viscous liquid separating neighbouring grains and allowing the grains to slide against each other. At low temperatures, they act as obstacles to dislocation motion. Thus, at high temperature, creep rate becomes high due to grain boundary sliding, whereas at low temperatures, yield strength increases since grain boundaries hinder the movement of dislocation.

The creep rate has to be kept at the minimum level for any engineering design, for example jet-engine alloys require creep rate to be 0.0001% per hour. The minimum creep rate is the most important design parameter. Creep rate becomes significant at a temperature $> 0.4T_m$.

The materials in service are quite often used at high temperatures. They should be creep resistant at these temperatures, for example turbine blades. Choice of these materials should be such that the working temperature is less than $0.4T_m$ i.e. they should have high melting point so that creep rate is insignificant in service condition. Refractory oxides are suitable for high temperature applications but these materials are unsuitable for the applications, where compressive stresses are required since they are brittle.

Iron base, nickel base and cobalt base alloys are creep resistant and found to have wide spread applications in machine design. Thoria dispersed nickel maintains its strength up to $0.9\ T_m$. In this dispersion hardened materials, oxide particles are spread in a metallic matrix with very little solubility in the later. Strength of an alloy may reduce due to coarsening of solute particles. Normally coarsening occurs only by the process in which small solute particles dissolve in the solvent matrix and reprecipitate as coarse particles. Coarsening effect of oxide particles is very much less due to their negligible solubility in metal matrix. There are other mechanisms by which coarsening effect is reduced, for example, the precipitate particles of $Ni_3(Ti, Al)$ form a grain boundary of very low energy with the matrix, which prevents these particles from coarsening in nickel base superalloys. Hence such alloys are also useful for high temperature applications.

Effect of creep in materials

The creep response of a material for machine parts and structure components depends on the properties of the material as well as external factors. When the materials in high temperature applications, are under stress, creep may cause serious damage in the structure, for example, material used in reactor, turbine blades, and in petroleum industry. These materials must have high melting point since the creep becomes important at temperature greater than $0.4T_m$. Some refractory oxides like MgO, Al_2O_3 have high melting point but these materials are brittle so they are not suitable for applications where only compressive stresses are encountered. Some newly developed ceramics like Si_3N_4 etc. are suitable for such applications. The low creep rate can be ensured by strengthening the material using various technique like solution hardening, dispersion hardening etc.

Creep-rupture test

Creep-rupture test is essentially creep test, which is carried out to the failure of the specimen to determine creep-rupture time. The anticipated lifetime of a component is specified as the time interval during which the strain developed in the component does not exceed a specific limit depending upon the application. It is called creep rapture time (t_r) if fracture is the limit of straining specified for a particular application. While selecting a material for a specific application, one requires to correlate σ, τ and t_r. Creep-rupture data are plotted as log(stress) versus log(rupture time) which are essentially straight lines.The correlation among σ, t_r and τ has been formulated using Larsen-Miller Parameter (LM). Stress value for creep is expressed as function of LM

$$\sigma = f(LM) \tag{15.29}$$

where
$$LM = T\,[C_r + \log t_r] \tag{15.30}$$

LM has a fixed value for a given stress where C_r is a constant. Master curve for a material is drawn by plotting σ versus LM for various temperatures, T, using creep-rupture data. It is possible to determine the creep rupture time for any combination of stress and temperature for the same material from this master curve.

Example 15.5

Consider the stress versus LM data, Table 15.2 for an alloy of nickel. Larson-Miller parameter LM is given as:

$$LM = T[20 + \log t_r]$$

Calculate the creep rupture time of the alloy at a stress of 62 MN/m^2 and a temperature of 900°C.

Table 15.2 Showing LM vs stress data

LM	Stress in MNm^{-2}
23 x 10^3	270
25 x 10^3	100
27 x 10^3	38

Solution:

From Table 15.2 one can find that LM $= 26 \times 10^3$ corresponding to stress value 62 MN/m^2. Also given that T= 900+273 °K

$$LM = 1173\,(20 + \log t_r)$$

So
$$26 = 1.173\,(20 + \log t_r)$$

$$\log_{10} t_r = \frac{26}{1.173} - 20 = 2.17$$

$$t_r = 146 \text{ h}$$

EXERCISES

15.1 Calculate the stress concentration factor at the tip of an internal crack having radius of curvature of 2.5×10^{-4} mm and a crack of length 2.5×10^{-2} mm. If the tensile stress of 100MPa is applied, find the maximum stress on the tip of internal crack.

Ans: 1.41×10^3 MPa

15.2 The fracture toughness of Titanium alloy Ti-6Al-4V is 55MPa√m. The structural plate of the above alloy can support 400 MPa in tension. What is the largest flaw size in the material? Assume g = 1.

   Ans: 1.2 cm

15.3 Calculate the fracture stress of a glass plate, which has a crack length of 10^{-6} m in its surface. The tensile stress is perpendicular to the surface. The Young's modulus of the glass plate is 0.5×10^5 MN/m^3 and its surface energy is 0.4 J/m^2.

   Ans: 113 MPa

15.4 Distinguish between a ductile and brittle fracture.

15.5 (a) A sheet of glass has an internal crack of length equal to 4 μm. If the crack length increases to 6μm, calculate the change in fracture stress. Y_{glass} is equal to 70 GN/m^2. The surface energy of the cracks per unit area is equal to 0.5 J/m^2 and the stress is perpendicular to the surface.

   (b) If a surface crack of 0 6 μm deep is introduced in the above sample by scratching the surface, which of the cracks will propagate first on increasing the applied stress? If the glass is etched with HF, what way does it change the fracture stress?

   Ans: (a) Decrease of fracture stress by 13.7 MPa; (b) surface cracks.

15.6 (a) In bcc iron, the applied shear stress in the slip plane causes piling up of dislocations along the slip line of length 1mm. If shear stress is 70 GN/m^2 and half-length of the internal crack is 1μm, calculate the value of maximum tensile stress?

   (b) If the shear modules of the steel is 83 GN/m^2, calculate the number of dislocations that pile up assuming Poisson ratio to be 0.5. Explain the mechanism of ductile to brittle transition.

   Ans: (a) 1.56 x 10^{16} Nm^{-2}; (b) 5.5 x 10^6

15.7 If yield strength of 70 Cu-30 Zn is 65.55 MPa for the sample with the diameter of grain equal to 0.1 mm and it is 1.5×10^2 MPa, when the diameter of grain is 0.01 mm, calculate the value of constant κ and σ_i . Now predict the yield strength when the average grain diameter is 5×10^{-3} mm.

   Ans: 0.414 GNm^{-2}, −0.414 MNm$^{-3/2}$

16

Deterioration of Materials

16.1 INTRODUCTION

Most materials are exposed to diverse environmental conditions while in service. This results in deterioration of surfaces, mechanical properties, physical properties and appearance of the material. Internal changes in a material are caused due to absorption of thermal energy, electromagnetic and nuclear radiation. When internal or structural changes in the material are undesirable, from the point of view of applications and properties, we call them to be thermal or radiation damage depending upon the source of damage.

Material engineering deals with materials that are suitable for specific designs and at the same time are unaffected in the anticipated environment. The surface of material gets deteriorated mainly due to interactions with gaseous substances or with liquids. Metals form oxide or carbonate in the surface layer due to interactions with air e.g. rusting of iron or formation of greenish $CuCO_3$ layer on copper. Deteriorative processes for metals and alloys are called corrosion. Deterioration of ceramic materials is much less as compared to that of metals and alloys. Deterioration in ceramics occurs at elevated temperatures. Polymers swell and become hard when exposed to the atmosphere. Polymers may dissolve when exposed to a suitable liquid solvent. Electromagnetic radiation and heat energy can break the molecular bond. Deterioration of polymers is also called degradation.

16.2 THERMAL DAMAGE

Strength of a material is affected at elevated temperatures. Overaging and overtempering at elevated temperatures are the two types of thermal damages, which affect the strength. Polymers degrade by a process termed scission — the severance or rupture of molecular bonds. The chain segments are separated at the point of scission. Hence the molecular weight of the polymer is reduced. The mechanical strength and resistance to chemical attack of a polymeric material depend on its molecular weight.

Thermoplastic polymers show creep above glass transition temperature. These polymers soften and lose dimensional stability above 100-150°C. They lose their strength at elevated temperatures due to weakening of secondary bonding between the chain polymers. Thermoset polymers have covalent bonding between atoms. When the thermoset polymers are subjected to high temperatures, they also degrade by interacting with atmosphere. These materials have a maximum-use temperature above, which the materials degrade.

16.3 RADIATION DAMAGE

Materials in general are severely affected by radiations (x-rays; α-, β- and γ- rays; ultraviolet radiations). Radiation causes structural changes, which in turn severely affect mechanical properties and corrosion resistance. Orbital electrons in an atom are excited due to absorption of radiation which can cause ionization and breaking of bonds. Depending on the chemical structure of the polymer and dose of radiation, there may be scission or cross-linking at the ionization site. Ultraviolet radiation interacts with polymers and breaks some of the covalent bonds along the molecular chains. Radiation causes physicochemical deterioration of polymers. Radioactive irradiation can trigger radioactive processes resulting in unwanted changes in structure and properties. The strongest effect is produced by neutron irradiation. The effects of α-particles, protons, and β-particles are less strong whereas that of γ-rays are not strong. Since the neutrons possess lower energy than the protons or α-particles, they are absorbed more easily as compared to protons or α-particles. Another possible reason is that the neutrons carry no electric charge and hence are not easily deflected by charged particles such as protons, electrons and nuclei. A neutron collides with a number of particles while passing through the material, before eventually being 'captured' by the atomic nucleus. This leads to emission of α-, β- and γ-rays by the nucleus and hence the nucleus may be transformed into a radioactive isotope of the same or of some other element.

Sometimes, a small volume of a crystal may get heated when exposed to radiation. This is because the neutrons besides displacing atoms into interstitial positions also transfer a part of their energy to them, which increases its oscillations. This event is accompanied by a local rise in temperature. The phenomenon of this rise in temperature is called radiation annealing. When there is heavy neutron flow, the formation of aggregates of vacancies may change to dislocation loops. This is known as radiation swelling.

Effect of radiation on structure and properties

When materials are exposed to electromagnetic radiation of various frequency ranges, they develop structural deformities, which include point defects and linear defects. Primarily, the effect of irradiation is to excite electrons to higher unoccupied energy states. When the electrons fall back to lower energy states, photons of appropriate energies are emitted. For gaseous substances sharp spectral lines are observed as a result of these electronic transitions. In liquids and solids, the ions are closer together, so there are bands of allowed states. Consequently, the absorption bands are observed instead of spectral lines. Certain impurity atoms may be present in materials, which contribute donor or acceptor levels giving rise to electronic transitions. They are called impurity centers. For example F centres are generated when alkali halides are heated in alkali vapour. Similar colour centres are also produced by x-ray or neutron bombardment.

Irradiation also leads to the displacement of atoms of exposed materials into interstices and previously unoccupied position thereby creating vacancies. In oxides, impurity centers are commonly observed with oxygen deficient materials. TiO_2 and ZrO_2 change colour when they become nonstoichiometric. MgO crystal becomes blue in colour when it is heated in magnesium vapour. When oxides and glasses are exposed to ionizing radiations optical absorption bands are produced in the ultraviolet and visible part of the spectrum.

Effect of irradiation on corrosion resistance

When structural materials operating in a corrosive medium like water are subjected to radiation, chemical corrosion may also occur in addition to electrochemical corrosion. The radiation produces visible damage in the form of destruction of protective layers on the surface of the metal. This damage decreases the electrochemical potential of the metal and hence increases the rate of corrosion destruction. For example, the corrosion rate of Aluminium in water doubles when exposed to thermal neutrons at 190°C. This effect can be attributed to an increase in the concentration of hydroxyl ions leading to destruction of surface films. Austenitic Cr-Ni steels also show this type of increase in corrosion.

16.4 CORROSION

Corrosion is defined as destructive chemical or electrochemical reaction between a metal and its environment, which results in the formation of oxide or some other compounds. Corrosion begins at the surface and then spreads in the interior of the material. It depends on the temperature, mechanical stress, erosion and concentration of the reactants. Most metals get corroded due to chemical attack by other materials, water or environment. The familiar examples include rusting of radiator, automobile body panels and exhaust components. Corrosion being an electrochemical reaction leads to transfer of electrons from one chemical species to another. Water and atmosphere corrode most metals. Metals are found in nature in the form of oxides, carbonates, sulphides or silicates since the energy of each of these compounds is less than that of pure metal. For example, Fe_2O_3 or FeO has less energy than that of metallic iron. Hence the pure metal returns to oxide form spontaneously in the presence of atmosphere. Nonmetallic materials like ceramics and polymers usually deteriorate by direct chemical attack. Polymers are degraded by chemical attack of organic solvents. Polymers swell after absorption of water.

Factors responsible for corrosion

Environmental factors, which are responsible for corrosion of materials, are the following:

  (i) Velocity of the fluid in contact with material surface
  (ii) Temperature
  (iii) Composition
  (iv) Anode potential

  As the velocity of the fluid increases, the rate of corrosion also increases due to erosive effects. Normally the temperature of the material increases with the increase in the rate of chemical reaction. This is also true for corrosive reactions. The composition plays a very important role in deciding corrosive properties of the materials. The corrosion increases rapidly if the corrosive contents are increased but sometimes these reactions lead to passivation of the material. The ductile metals are cold worked or plastically deformed to increase its strength. The cold worked metal is more susceptible to corrosion than the annealed metal.

Forms of corrosion

Forms of corrosion that are observed in engineering materials depend upon the type of usage and environment. Various types of corrosions are given below:

Uniform attack

Uniform attack corrosion occurs due to electrochemical or chemical reaction, which causes corrosion of uniform intensity over the entire exposed surface and often deposits a scale on the surface. This happens due to random occurrence of oxidation and reduction reactions over the entire surface. Rusting of an iron sheet, scaling on the inner surface of a boiler are examples of this type of corrosion.

Galvanic corrosion

Galvanic corrosion develops at a place where two dissimilar metals or alloys are in electrical contact in presence of electrolyte. In this case, one of these metals becomes anode while the other becomes cathode in direct analogy with a galvanic cell. The metal, which is anodic in a specific medium because of equilibrium electrode potential of the metal being more negative as compared to the other metal, starts showing corrosive effects. For example, if steel and copper tubing are joined in a water tank, the corrosion may occur in the vicinity of the junction. The galvanic corrosion can also take place in concentration cells, which may arise because of the variations in ionic concentration or dissolved gas concentration in the electrolytic medium. The galvanic corrosion can be reduced, if the following precautions are taken:

(i) The surface area of the anodic material that is exposed to electrolyte should be large compared to that of the cathode e.g. copper rivets can be used with steel sheet so that small area of copper accepts very few electrons and the rate of anode corrosion slows down.

(ii) If the coupling of two dissimilar materials is present, the two materials should be chosen such that their electrode potentials should not differ much.

(iii) Two dissimilar metals should be electrically insulated from each other. For example, if steel pipe is connected to a brass tap, galvanic cell is formed between steel and brass and the steel corrodes. However, this can be avoided by introducing an intermediate plastic filling that can insulate brass from steel and stop corrosion of steel.

(iv) A third anodic metal can be connected to the other two metals to provide "cathodic protection". Galvanic corrosion is discussed in more detail in section 16.6.

Crevice corrosion

Crevice corrosion is an electrochemical type of corrosion, which occurs in crevices, recesses or under-covered regions where there is stagnant solution. Crevice corrosion is also observed under gaskets, rivets, and in porous deposits. Crevice occurs due to formation of concentration cells. There is localized depletion of dissolved oxygen gas in the crevices. The metal at this point becomes anodic and hence it is oxidized. The adjacent external metallic region consumes electrons from the oxidized region. Sometimes solution within crevice becomes rich in H^+ and Cl^- ions in aqueous environments. These ions destroy the passivity of many alloys. Crevice corrosion may be prevented by using welded joints instead of riveted joints, or by using nonabsorbing gaskets. Crevice corrosions can also be prevented by removing deposits, which get accumulated. Designing of vessels should be done in the following way:

- Allow complete drainage and not to permit stagnation of solutions.
- Fluid systems should be closed so that stagnant pools do not form and gases from environment do not dissolve in closed systems. This reduces the chance of corrosion.

Pitting corrosion

The mechanism of pitting corrosion is similar to that of crevice corrosion i.e. pitting arises due to formation of concentration cells. Pitting is a form of localized corrosion, which attacks a very small area forming pits or holes. They usually start from the top of a horizontal surface and move vertically downward (may be due to gravity). Pitting can cause perforation of metals in an engineering structure. As a result there can be serious damage to the structure. Small pits are sometimes not detected since they are usually covered by corrosion. Depending upon the number and depth of pits the seriousness of the damage can be decided. Pitting, if not detected, can sometimes cause sudden failure of the structure. A pit may be initiated at surface defects e.g. scratches or variations in compositions etc. The oxidation occurs within the pit due to the formation of concentration cells. As the pit grows downward, the solution at the pit tip becomes denser. It involves the dissolution of metal in the pit. The stainless steels are quite susceptible to this form of corrosion. The steel containing 18w/o Cr, 8w/o Ni and 2w/o Mo has better resistance against pitting. The polishing of the surface may reduce pitting. Depending upon the application, corrosion resistant properties of alloys should be tested in the specific environment before making the final selection.

Intergranular corrosion

Intergranular corrosion occurs preferably along the grain boundaries because the atoms or ions in the grain boundary region are more reactive than those in the bulk region, the former having higher energy than the later. Alloys with precipitated phases are more susceptible to intergranular corrosion. The net result of intergranular corrosion is the macroscopic disintegration of the sample. This type of corrosion occurs in some stainless steels (e.g. steel containing 18w/o Cr, 8w/o Ni and 0.06w/o C) when heated to a temperature between 500°C and 800°C for sufficiently long time. The alloy at this temperature remains in sensitized condition, which helps the formation of precipitate particles of complex chromium carbide ($Cr_{23}C_6$) along the grain boundaries, Fig. 16.1. Both Cr and C atoms diffuse to grain boundary to form the complex carbide and a chromium depleted zone is created adjacent to the grain boundary. Hence the grain boundaries become susceptible to corrosive attack by various media (e.g. Cl^- ions are specially bad). Intergranular corrosions cause severe problem in welding of stainless steel. The failure of weld occurs due to the precipitation of chromium carbide as described above. This type of failure is called weld-decay. Thus, a weld-decay is formed away from the central line of the welded

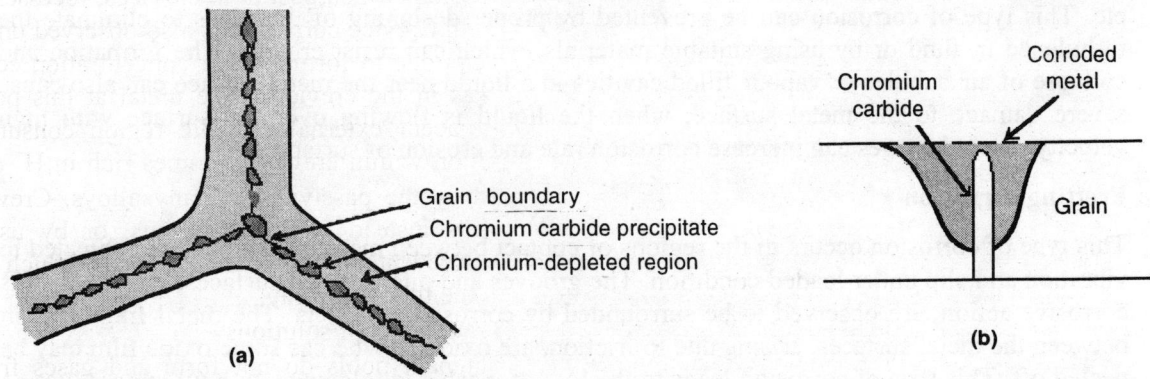

Fig. 16.1 Intergranular corrosion

region. To avoid the formation of weld-decay zone, the welded joint in sensitized condition should be reheated at 500°C to 800°C followed by water quenching so that the chromium carbide again dissolves in steel and formation of intergranular corrosion can be stopped. Adding some specific element with stainless steel such that $Cr_{23}C_6$ cannot be formed can prevent the intergranular corrosion.

Metals like columbium and titanium, which have more affinity for carbon than chromium, are added so that these elements combine with carbon. Hence, complex chromium carbide precipitates cannot form. Lower carbon content in the steel can also prevent intergranular corrosion.

Selective leaching

Selective leaching is a corrosion process in which one of the elements of a solid solution is removed preferentially e.g. zinc is selectively removed from the brass containing more than 15 w/o Zn. This is called dezincification. When brass is in contact with aqueous solution, zinc and copper ions dissolve in the solutions at high temperature. Later on copper ions again get deposited on brass whereas zinc remains in solution. So the brass becomes porous. The mechanical properties of the alloy are deteriorated significantly since it becomes porous. The colour of the material changes from yellow to copper red. Selective leaching also occurs with other alloys where Al, Fe, Co, Cr and other elements are present since these elements are preferentially removed from the alloys. Due to selective leaching of iron in buried grey cast iron gas pipes, the pipes become corroded and this may lead to explosion.

Erosion corrosion

Erosion corrosion arises due to combined effect of chemical attack and mechanical abrasion when there is relative motion between corrosive fluid and a metal surface. When the relative speed of the fluid is very high compared to the metal, the corrosion of the metal surface can be very severe. The formation of grooves, pits, valleys, and rounded holes on the metallic surface in the direction of the flow of corrosive fluid confirms the damage due to erosion corrosion.

Erosion corrosion occurs at bends of piping, elbows, abrupt changes in pipe diameter etc. where flow becomes turbulent due to change in the direction of flow. There are metals, which passivate by forming protective coating. The abrasive action of the corrosive fluid can erode away the protective surface film. This process corrodes metals like Cu and Pb since these metals are relatively soft ones. Normally this type of corrosion affects turbine blades, propellers, pumps etc. This type of corrosion can be prevented by proper designing of the pipes to eliminate the turbulence in fluid or by using suitable materials, which can resist erosion. The formation and collapse of air bubbles or vapour filled cavities in a liquid near the metal surface can also cause severe damage to the metal surface, when the liquid is flowing over the surface with high velocity. Such damages can increase corrosion rate and erosion of surface.

Fretting corrosion

This type of corrosion occurs in the regions of contact between materials, which are subjected to vibration and slip under loaded condition. The grooves and pits in metal surfaces, created by this corrosive action, are observed to be surrounded by corrosion products. The metal fragments in between the metal surfaces, arising due to friction, are oxidized whereas some oxide film may be peeled off. This type of corrosion leads to the loss of machined tolerance with ultimate initiation of fatigue failure at fretting pits. Fretting corrosion sometimes occurs in automobile axles or ball bearings in ships or in boats.

Stress corrosion

The combined effect of applied tensile stress and corrosive environment causes stress corrosion. Some of the materials may be inert in corrosive environment but if some cracks or surface defects are produced in the material due to applied stress, the cracks may propagate and eventually cause failure of the material. Sometimes high residual stresses can cause stress corrosion cracking.

Stresses arising out of unequal cooling rate, poor mechanical design for stresses, cold working, welding and phase transformation can cause stress corrosion cracking. Certain combination of alloys and corrosive media cause stress corrosion e.g. aluminium alloys develop stress-corrosion in NaCl solutions or air or water vapour media, whereas brass starts cracking in ammoniacal solutions but not in chloride environment. Stainless steel cracks in chloride environments but not in ammonia vapour or solution.

Best way to prevent stress corrosion is to reduce the stress and to eliminate the corrosive environment. The material can be annealed appropriately so that it can be relieved of stress. Depending upon the environment the material can be chosen appropriately such that stress corrosion cracking is avoided. For example titanium can replace steel for heat exchangers, which are in contact with air and water. Inhibitors can also be added to reduce corrosive effects. Cathodic protection can also be applied using suitable consumable anode.

16.5 GALVANIC CORROSION AND GALVANIC SERIES

A simple electrical cell with copper and zinc electrodes, dipped into an electrolyte, conducts electricity since there is a 1.1 volts potential difference between the two metals. After using the cell for a period of time, the copper electrode remains unchanged but zinc electrode is badly corroded. Zinc is anode in this electrochemical cell. This electrode supplies electrons to the external circuit while copper, the cathode receives the electrons through the external circuit. The potential of each cell is estimated with respect to hydrogen. The zinc electrode in the HCl solution has the following reactions.

$$2HCl \rightarrow 2H^+ + 2Cl^- \tag{16.1}$$

$$Zn + 2H^+ \rightarrow Zn^{2+} + H_2 \tag{16.2}$$

In the above reaction, Zn atom loses two electrons and can be called as the oxidation half-cell reaction whereas hydrogen ions (H^+) gains electrons to form H_2 molecules and is the reduction half cell reaction. The oxidation half-cell reaction is called anodic reaction. Similarly, reduction half-cell reaction is called cathodic reaction. Electrochemical corrosion reactions involve both oxidation and reduction where anode gets corroded due to formation of cations and electrons produced remain in the metal.

For a metal to get corroded, a particular environment is required. Zn corrodes in dilute HCl but gold does not. Thus, as a result of oxidation, the metal ions may either go into corroding solution as ions or they may form an insoluble compound as happens in case of rusting of iron. The rusting of iron occurs in two steps. Fe is oxidized to Fe^{2+} and then forms Fe $(OH)_2$ or Fe $(OH)_3$ as follows:

$$Fe + \frac{1}{2}O_2 + H_2O \rightarrow Fe^{2+} + 2OH^- \rightarrow Fe\,(OH)_2. \tag{16.3}$$

$$2Fe(OH)_2 + \frac{1}{2}O_2 + H_2O \rightarrow 2Fe(OH)_3 \tag{16.4}$$

Different metals oxidize to varying extents under similar oxidizing environment. Consider an electrochemical cell consisting of electrode of pure iron dipped into a suitable electrolyte containing Fe^{2+} ions as one half-cell and pure copper electrode in an electrolyte of same strength containing Cu^{2+} ions as another half-cell. If now the iron and copper electrodes are connected electrically, following reaction takes place.

$$Cu^{2+} + Fe \rightarrow Cu + Fe^{2+} \tag{16.5}$$

i.e. copper will deposit on the copper electrode while Fe^{2+} ions dissolve in the electrolyte and thus iron electrode corrodes. Therefore, the electroplating occurs at the copper electrode but corrosion occurs at the iron electrode. The electrons generated as a result of oxidation of iron electrode flow to the copper cell and reduce Cu^{2+} ions. Moreover there is some net ion motion from cell to cell through the separating membrane, Fig. 16.2. This combination forms a galvanic couple. Electrons flow from iron electrode to copper electrode clearly indicates the existence of potential difference between the electrodes. The potential difference is 0.78 V for copper-iron galvanic cell at temperature 25°C. In case of iron-zinc galvanic cell, the zinc electrode gets corroded whereas iron electrodeposits. The potential difference in this cell is 0.323 V. This galvanic cell differs from the previous one because the iron electrode is anode in the Cu-Fe galvanic cell whereas the same electrode is cathode in Fe-Zn galvanic cell. It is convenient to define the potential of each electrode with respect to a reference cell. The reference cell is usually chosen as the standard hydrogen electrode cell. This reference cell consists of inert platinum rod immersed in 1M solutions of H^+ ions saturated with hydrogen gases at 1 atmosphere and a temperature of 25°C. The electromotive series or galvanic series of metal is generated by coupling standard half-cells of various electrodes to the standard hydrogen electrode.

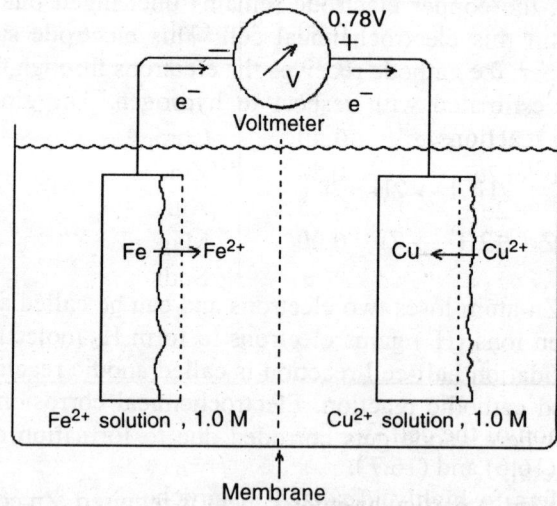

Fig. 16.2 Iron-copper electrochemical cell

The hydrogen electrode potential is taken as zero. The metals, which have higher positive electrode potential, are chemically inert i.e. these metals do not corrode easily.

The metals with higher negative electrode potential are more susceptible to oxidation and corrosion, Table 16.1. The table clearly shows that the noble metals have higher positive electrode potential taking hydrogen has zero electrode potential. The metals appearing after hydrogen have negative electrode potential indicating that these metals oxidize i.e. corrode easily. Metals, having electrode potential of higher negative value, are more reactive. Consider a cell consisting of electrodes M_A and M_C where there is oxidation reaction at A and reduction reaction at C. Thus the reactions are:

$$M_A \rightarrow M_A^{n+} + ne^- \quad ; \quad V_A^\circ = V_1^\circ \qquad (16.6)$$

$$M_C^{n+} + ne^- \rightarrow M_C \quad ; \quad V_C^\circ = -V_2^\circ \qquad (16.7)$$

where $V°$'s are the standard electrode potentials as given in Table 16.1. Adding Eqs. (16.6) and (16.7) we get :

$$M_A + M_C^{n+} \rightarrow M_A^{n+} + M_C \;; \quad V°_{cell} \qquad (16.8)$$

where

$$V°_{cell} = V_C° + V_A°$$

Table 16.1 Galvanic series

Metal	Electrode reaction	Standard electrode potential in volts (More cathodic)	Metal	Electrode reaction	Standard electrode potential in volts (More anodic)
Gold	$Au \rightarrow Au^{3+} + 3e^-$	1.5	Lead	$Pb \rightarrow Pb^{2+} + 2e^-$	-0.126
H_2O	$2H_2O \rightarrow O_2 + 4H^+ + 4e^-$	1.23	Tin	$Sn \rightarrow Sn^{2+} + 2e^-$	-0.136
Platinum	$Pt \rightarrow Pt^{2+} + 2e^-$	1.2	Nickel	$Ni \rightarrow Ni^{2+} + 2e^-$	-0.25
Silver	$Ag \rightarrow Ag^+ + e^-$	0.80	Cobalt	$Co \rightarrow Co^{2+} + 2e^-$	-0.28
Mercury	$Hg \rightarrow Hg^{2+} + 2e^-$	0.79	Cadmium	$Cd \rightarrow Cd^{2+} + 2e^-$	-0.40
Ferrous ion	$Fe^{2+} \rightarrow Fe^{3+} + e^-$	0.77	Iron	$Fe \rightarrow Fe^{2+} + 2e^-$	-0.44
$(OH)^-$	$4(OH)^- \rightarrow O_2 + 2H_2O + 4e^-$	0.40	Chromium	$Cr \rightarrow Cr^{3+} + 3e^-$	-0.74
Copper	$Cu \rightarrow Cu^{2+} + 2e^-$	0.34	Zinc	$Zn \rightarrow Zn^{2+} + 2e^-$	-0.76
Stannous ion	$Sn^{2+} \rightarrow Sn^{4+} + 2e^-$	0.15	Aluminium	$Al \rightarrow Al^{3+} + 3e^-$	-1.66
Hydrogen	$H_2 \rightarrow 2H^+ + 2e^-$	0.00	Magnesium	$Mg \rightarrow Mg^{2+} + 2e^-$	-2.36
			Sodium	$Na \rightarrow Na^+ + e^-$	-2.71
			Potassium	$K \rightarrow K^+ + e^-$	-2.92
			Lithium	$Li \rightarrow Li^+ + e^-$	-2.96

For the spontaneous reaction of the cell, $V_C° > V_A°$. If $V_C° < V_A°$, the spontaneous cell direction is just the reverse of Eqs. (16.6) and (16.7).

The galvanic series refers to highly idealized electrochemical cells (i.e. pure metals in 1M solutions of their ions at 25.5°C). The overall cell potential varies with the absolute temperature and the molar ion concentrations C_A^{n+} and C_C^{n+} for M_A^{n+} and M_C^{n+}, respectively, and is given by the Nernst equation:

$$V_{cell} = V_C + V_A = (V_C° + V_A°) + (RT/nF) \ln(C_A^{n+}/C_C^{n+}) \qquad (16.9)$$

where V_A is the potential for anodic half cell and V_C is that for cathodic half cell. These potentials are given by

$$V_C = V_C° - (RT/nF) \ln C_C^{n+} = V_C° - (0.0592/n) \log C_C^{n+}$$

and
$$V_A = V_A° + (RT/nF) \ln C_A^{n+} = V_A° + (0.0592/n) \log C_A^{n+} \qquad (16.10)$$

n is the number of electrons participating in either of the half-cell reactions and F is Faraday constant (96500 C/mole), at 25.5°C. Hence, we can also express

$$V_{cell} = V°_{cell} + (0.0592/n) \log (C_A^{n+}/C_C^{n+}) \qquad (16.11)$$

If $C_A{}^{n+} = C_C{}^{n+} = 1$, then $\qquad V_{cell} = V^o{}_{cell}$ $\hspace{4cm}$ (16.12)

Most metals and alloys are oxidized or corroded in a wide variety of environments. So they are more stable in ionic state than as metals. This is due to the fact that there is decrease in free energy in going from the metallic state to the oxidized state. Only noble metals like platinum and gold are exceptional cases where oxidation is not favourable in most of the environments.

Example 16.1

Write down the electrochemical reactions at cathode and anode in the following cases:
  (a) Copper and zinc electrodes in a dilute cupric sulphate ($CuSO_4$).
  (b) A copper electrode immersed in an oxygenated water solution.
  (c) Magnesium and cadmium electrodes connected by an external wire immersed in an oxygenated 1 w/o NaCl.

Solution:

Using Table 16.1 we get:
  (a) At anode $\qquad$ Zn $\rightarrow$ Zn^{2+} + 2e$^-$; $\quad V^o = -0.76$ V
  $\quad$ At cathode $\quad$ Cu^{2+} + 2e$^-$ $\rightarrow$ Cu ; $\quad V^o = -0.34$ V
  $\quad$ Overall reaction: Zn + Cu^{2+} $\rightarrow$ Zn^{2+} + Cu ; $\quad V^o{}_{cell} = -1.10$ V
  $\quad$ Zinc has electrode potential more negative than copper. Hence zinc is oxidized.

  (b) When oxidation takes place at the copper electrode, the reaction is given by
  $$Cu \rightarrow Cu^{2+} + 2e^- ; \quad V^o = 0.34 \text{ V}$$
  i.e. 0.34 V potential is required for this process. The reduction occurs at local cathodes and is given by
  $$O_2 + 2H_2O + 4e^- \rightarrow 4OH^- ; V^o = -0.40 \text{ V}$$
  The oxidation of copper is not possible in this case since copper has much higher positive oxidation potential as compared to –0.40 V.

  (c) Anode reaction: Mg $\rightarrow$ Mg^{2+} + 2e$^-$; $V^o = -2.36$ V
  Cathode reaction: Since there are no Cd^{2+} ions in the electrolyte to be reduced to cadmium atoms and the electrolyte is oxygenated 1 w/o NaCl, oxygen and water molecules will react to form OH$^-$ ions. Thus, the cathode reaction becomes
  $$O_2 + 2H_2O + 4e^- \rightarrow 4OH^- ; V^o = -0.40 \text{ V}$$
  Magnesium has more negative oxidation potential and is thus the anode. Mg is oxidized.

Example 16.2

A galvanic cell at 25°C consists of an electrode of Zinc in a 0.10M $ZnSO_4$ solution and another of Cu in a 0.05 M $CuSO_4$ solution. The two electrodes are separated by a porous wall and connected by an external wire. What is the emf of the cell when a switch between the two electrodes is just closed?

Solution:

Assume that the dilution of 1M solutions does not affect the order of electrode potentials of Zn and Cu. So Zn will act as anode and Cu as cathode. To calculate the emf of the cell, the standard equilibrium potentials are modified using Nernst equation under the changed environment i.e.,

At anode : $V_A = -0.76V + (0.0592/2) \log 0.10 \; V = -0.76V - 0.0296V = -0.79V$
At cathode: $V_C = -0.34V - (0.0592/2) \log 0.05 \; V = -0.34V + 0.0385V = -0.30 \; V$
Emf of the cell $V_A + V_C = -0.79V - 0.30V = -1.09 \; V$

16.6 GALVANIC CELLS IN METALS, ALLOYS AND COMPOSITES

Galvanic cells are formed in metals and alloys due to variations in composition, structure and stress concentrations. These cells can seriously affect the corrosion resistance of a metal. These microscope galvanic cells are usually created near the following regions:

i) Grain boundary
ii) Phase boundary of multiphase alloys

Grain boundary electrochemical cells

Grain boundaries are more chemically active than the bulk region of a grain. This is due to the fact that within the grain boundary region, which has a width of just several atomic distances, there is some atomic mismatch. Thus, the grain boundary region has higher energy as compared to that of the bulk region. That is why precipitated solutes and impurities migrate to the grain

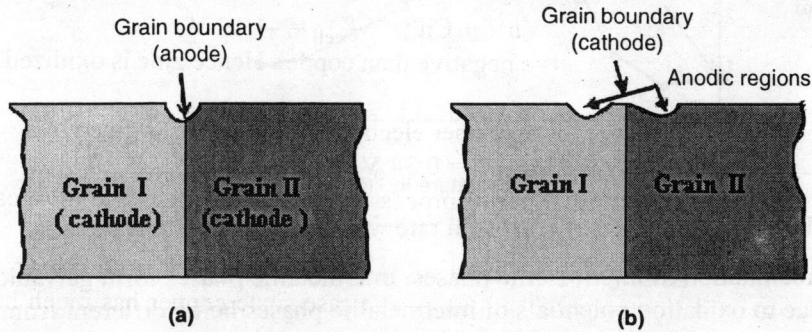

(a)   **(b)**

Fig. 16.3 Grain boundary electrochemical cell

boundaries and corrode the later. Thus, the grain boundary region shows anodic behaviour, Fig. 16.3(a) in case of certain alloys, whereas in some other cases, grain boundaries become more cathodic or noble as compared to the bulk regions adjacent to the grain boundaries, Fig. 16.3(b). In this case, the region adjacent to grain boundaries corrodes more as compared to the grain boundary.

Phase boundary electrochemical cells

Sometimes electrochemical cells are generated near the phase boundary of a multiple-phase alloy in which one phase acts as anode and the other acts as cathode. So the corrosion rates are higher for these multiphase alloys. This type of galvanic corrosion occurs in pearlitic gray cast iron. The microstructure of pearlitic gray cast iron has graphite flakes in the matrix of pearlite. The graphite is much more cathodic than the pearlite matrix. So quite active galvanic cells are formed with graphite flakes as cathode and pearlite matrix as anode. The galvanic corrosion of pearlitic gray cast iron can be very high which may ultimately become a network of interconnected graphite flakes.

We know that martensite is an extremely hard and brittle phase of steel. In this phase, the corrosion rate is low because the martensite is a single phase supersaturated solid. During tempering, ferrite and cementite are formed from martensite. Since the cementite particles are cathodic with respect to ferrite, galvanic cells are formed and under corrosive conditions ferrite corrodes away. The finer are the ferrite and cementite particles, the lesser the corrosion resistance to the formation of large number of galvanic cells with cementite and ferrite. Corrosion rate increases with increasing of tempering temperature. Above 500°C, the finer cementite particles coalesce into large particles; as a result of which there is decrease in the phase boundary. Consequently the number of galvanic cells decreases at the phase boundary. Hence there is decrease in the corrosion rate, Fig. 16.4. The presence of metallic impurities in metals or alloys

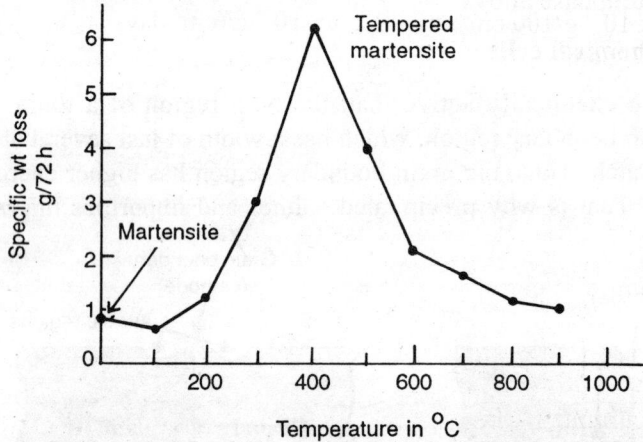

Fig. 16.4 Variation of corrosion rate with tempering temperature

can lead to the precipitation of intermetallic phases. Intermetallic phases form galvanic cells with the host matrix due to oxidation potentials of intermetallic phases being different from that of the host matrix. Thus, pure metals have higher corrosion resistance as compared to their alloys.

Corrosion rate

The kinetics of corroding materials is very complex since the corroding systems are not in equilibrium and hence the thermodynamic potentials do not tell us about the rates of corrosion reactions. According to Faraday's equation, the amount of corrosion (W) of anode material or the amount of material electroplated on a cathode in an aqueous solution during the time period t can be expressed as

$$W = (I \, tM/nF) \tag{16.13}$$

where
   W : weight of metal (in g) corroded or electroplated in aqueous solution in time t seconds
    I : corrosion current in amperes
   M : atomic mass of metal in g/ mol
    n : number of electrons or atoms produced or consumed in the process.
    F : Faraday's constant = 96,500 C/ mol
Corrosion penetration rate CP is expressed as

$$CP = kW/ \rho At \tag{16.14}$$

where CP is expressed in mm/year, $k = 3.16 \times 10^8$, W is in g, ρ, density of the metal is in g/cc, A, area of cross-section of the material is in cm^2 and t is in second. Thus, CP is also given by

$$CP = kIM/\rho AnF = kJM/\rho nF \qquad (16.15)$$

J being the current density.

Example 16.3

The steel tank containing aerated water is corroding at a rate of 62. 5 mg per decimeter square of surface area per day. Calculate CP of tank.

Solution:

Corrosion rate = 62.5×10^{-3} g/100 cm^2 day = 62.5×10^{-5} g/(cm^2 day)
The density of Fe = 7.87 g/cm^3.
Thus, the depth of corrosion per day = $(62. 5 \times 10^{-5}/ 7.87)$ cm/day
or   CP = $(62.5 \times 10^{-5} \times 365 \times 10/7.87)$ = 0.29 mm/year

Example 16.4

A steel tank 1.25m high and 35 cm in diameter contains aerated water up to 50 cm level and shows a loss in weight due to corrosion of 320 g after 15 days. Calculate: (a) Corrosion current, (b) Current density. Assume uniform corrosion of the inner surface of the tank and no corrosion from outer surfaces due to environment.

Solution:

The expression for corrosion current: I= WnF/tM
Putting W = 320 g, n = 2 for Fe $\rightarrow$ Fe^{2+} + 2e$^-$, F= 96,500 C /mole, M=55.85 g/mole for Fe.
And t =15 x 24 x 3600 = 1.296×10^6s we get:
$$I = (320 \times 96,500 \times 2)/(1.296 \times 10^6 \times 55.85) = 0.853 \text{ A}$$

Area of the corroding surface of the tank
= area of sides + area of bottom = $\pi Dh + \pi r^2$ = $\pi((35)(50) + (35)^2/4)$ cm^2 ≈ 6460 cm^2
$$J = 0.853/6460 = 1.32 \times 10^{-4} \text{ A/cm}^2$$

16.7 POLARIZATION IN ELECTROCHEMICAL REACTIONS

The formation and short-circuiting of microscopic galvanic cells in metals and alloys lead to corrosion. This also results in the oxidation and reduction reactions on the surface of metals. Consequently there are changes in anodic and cathodic potentials. Consider the electrochemical reactions for zinc electrode in hydrochloric acid. The reactions are:

Anodic reaction: Zn $\rightarrow$ Zn^{2+} + 2e$^-$ (16.16)

Cathodic reaction: $2H^+$ + 2e$^-$ $\rightarrow$ H_2 (16.17)

where zinc acts as external metallic connection between the electrodes on the surface. When zinc begins to react with HCl, zinc surface remains at a constant potential V_a since zinc is good electrical conductor. As the zinc electrode corrodes, the potential of the cathodic areas becomes more negative and that of anodic region becomes more positive. At a particular value of current density J_c, the rate of zinc dissolution is equal to the rate of hydrogen evolution. Hence, J_c is equal to the rate of zinc corrosion.

Normally, the electrochemical reactions in cells are reversible i.e. these reactions can be reversed by application of suitable potential across the electrodes. In a galvanic cell, applying a voltage that is equal and opposite of the cell voltage can stop an electrochemical reaction. When a galvanic cell delivers electric current, the cell voltage is less than the difference between the equilibrium electrode potential since galvanic cell has internal resistance. When an electrolysis cell is forced to operate, the applied voltage should be greater than the difference between the equilibrium electrode potentials. This results in nonequilibrium effects along with the current flow. The current causes the displacement of the electrode potentials from their equilibrium values such that the cathode potential shifts towards the active side and the anode potential shifts towards the noble end, Fig. 16.5. In the above example of anodic half-cell with zinc electrode

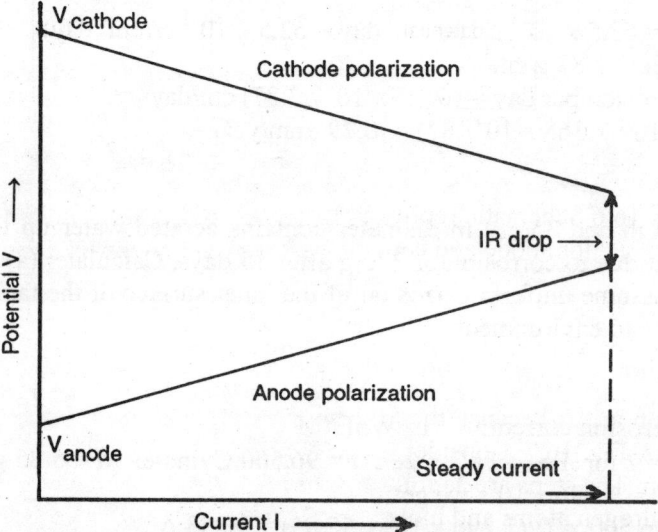

Fig. 16.5 Schematic representation of variation of anode and cathode potential with current I

in HCl, the internal resistance is nearly zero since the anode and cathode regions superimpose. The current increases to a maximum value I_{cor}, when the cathode potential equals anode potential. This potential is called corrosion potential V_{cor}, Fig. 16.6. The limited rates of the anodic and cathodic reactions result in the potential difference of smaller value as compared to the standard potential difference between two electrodes. The establishment of a net current flow as a result of a constant potential difference between the anode and cathode is called polarization. The polarization is of the following types:

(a) Activation polarization, which arises from the restricted reaction rates at an electrode.
(b) Concentration polarization, which arises from changes in the electrolyte concentration around electrode.

These polarizations occur due to change in potentials of each electrode, which is called overpotential. The overpotential is given by the difference of actual electrode potential and the equilibrium electrode potential. If we consider V to be the applied voltage, then the current density J in the circuit is related to the value J_c, as given by

$$J = J_c \exp(V^{act}/B) \tag{16.18}$$

where V_{act} is also called activation polarization overpotential and B is Tafel constant.

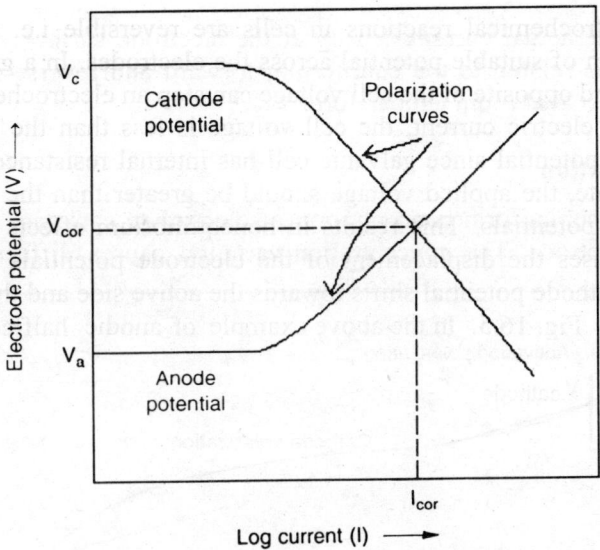

Fig. 16.6 Schematic representation of potential-current relationship
for anode and cathode reactions during corrosion

Activation polarization

Activation polarization arises because the activation energy barrier controls the electrochemical
reactions at electrodes, Fig. 16.7. Consider the hydrogen reduction reaction at zinc cathode under
activation polarization. The steps are as follows: Hydrogen ions migrate to the zinc surface which
are converted into hydrogen atoms and molecules as given by,

$$H^+ + e^- \rightarrow H \tag{16.19}$$

$$H + H \rightarrow H_2. \tag{16.20}$$

and H_2 molecules combine to produce bubble of hydrogen gas. The slowest of these steps in the
reaction sequence will control the half-cell reaction. In case of most of the metallic electrodes
other than Pt, Pd, Ir and Rh, the activation energy barriers for hydrogen gas formation are
appreciably high. On the other hand, the activation polarization overpotential required for
deposition of metals at cathode is small i.e.

$$Zn + 2H^+ \rightarrow Zn^{2+} + H_2$$

$$Zn^{2+} + 2e^- \rightarrow Zn \tag{16.21}$$

It is observed that many metals are electroplated in acidic electrolytic solution where the
expected evolution of hydrogen gas does not take place. Activation polarization overpotential for
many practical cases can be generally written as

$$V_c^{act} = -\eta_c \ln[j/j_o] \quad \text{at cathode}$$

and

$$V_a^{act} = \eta_a \ln[j/j_o] \quad \text{at anode} \tag{16.22}$$

where $[j/j_o] > 3$, j is the actual current density at the electrode surface and j_o is the exchange current density which is related to the equilibrium forward and reverse reaction rates. η is an empirical constant ~ 0.05 V at room temperature.

Concentration polarization

In a galvanic cell, the nonuniform concentration of electrolyte can arise locally in a region in contact with a metal electrode. The concentration variation causes diffusion of ions to obtain

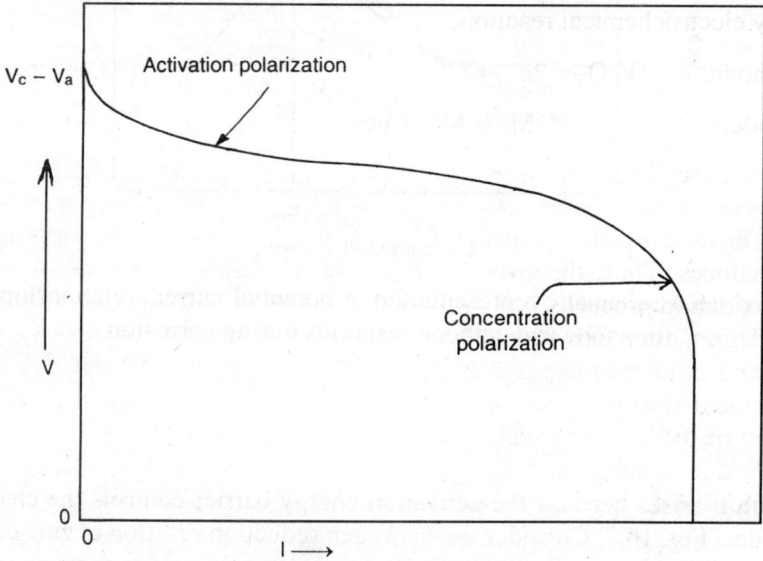

Fig. 16.7 Variation of cell voltage with current for a galvanic cell

uniform electrolyte concentration. In case of zinc-copper galvanic cell, the concentration of Cu^{2+} ions decreases around the cathode, whereas the concentration of Zn^{2+} ions builds up near anode when appreciable cell current flows. The part of the electrode in contact with more concentrated electrolyte acts as a cathode while the part in contact with more dilute electrolyte acts as an anode and such a cell is called concentration cell. This type of cell produces back voltage that opposes cell voltage. As a result of concentration polarization at the cathode, the equilibrium electrode potential is decreased by V_c^{conc}, which is called overpotential. The overpotential is expressed as

$$V_c^{conc} = (RT/nF) \log(1 - (j/j_l^c)) \qquad (16.23)$$

where j is the actual current density and j_l^c is a limiting value of current density determined by the Cu^{2+} ion diffusion rate. Similarly the overpotential arising at the anode due to concentration polarization is given by

$$V_a^{conc} = -(RT/nF) \log (1 - (j/j_l^a)) \qquad (16.24)$$

Usually j_l^a for anode reaction is greater than j_l^c for cathode reaction. As temperature increases, diffusion proceeds more rapidly and this leads to a significant increase in j_l. So concentration polarization decreases at higher temperature. Stirring of the electrolyte can minimize or remove the concentration polarization.

16.8 OXIDATION

At room temperature, most of the metals form oxide layers at the outer surface. Initially the oxygen molecules decompose and form layer of oxygen atoms at the surface of the metal. These oxygen atoms easily ionize and then form oxides with metal ions. Any natural process leads to equilibrium state i.e., the final state of spontaneous reaction possesses lower energy than the initial state. Therefore, energy is released when iron forms iron oxide. Oxidation rates of certain metals and polymers are slow and they can be controlled to reduce scaling of the surface. The slow rate of oxidation can be explained due to scale formation. The oxide formation can be visualized as a dry electrochemical reaction.

At cathode, $\qquad \frac{1}{2} O_2 + 2e^- \rightarrow O^{2-}$ $\qquad$ (16.25)

At anode, $\qquad M \rightarrow M^{n+} + ne^-$ $\qquad$ (16.26)

As the oxide layer (scaling) grows, the metal and oxygen are no longer in direct contact. The metal atoms or oxygen atoms should diffuse through surface oxide layer (scaling) to interact and form oxides. The diffusion distance increases with the growth of scaling and the electric field across the layer reduces. Thus, the growth of oxide layer slows down because ion diffusion is decreased as the oxidation proceeds.

The rate of oxidation, after formation of thin oxide film, is usually measured by weighing a test specimen before and after it is exposed to air. The amount of oxide formation increases with temperature since the diffusion rate of ions is faster at higher temperature. It also depends on the metal and structure of oxide. The weight gain due to oxide formation increases with time. This variation of weight due to oxide formation may be linear, parabolic, logarithmic or intermittent with time, Fig. 16.8.

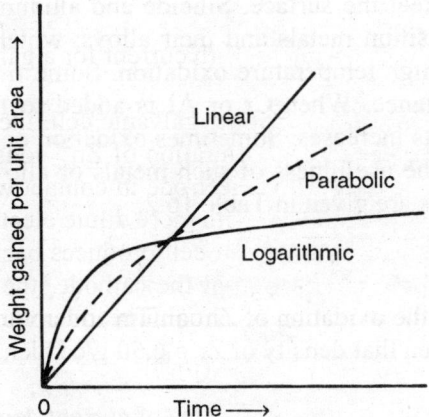

Fig. 16.8 Variation of weight gain due to oxide formation with time.

The linear law for oxidation rate is expressed as,

$$W/A = C_L t$$ $\qquad$ (16.27)

where

$\qquad$ W/A : weight gain per unit area

$\qquad\quad$ t : time

and $\qquad$ C_L : Linear rate constant

Potassium and tantalum are examples of metals, which follow nearly linear law. When ion diffusion is the main mechanism in the oxidation process pure metals follow parabolic relation,

$$(W/A)^2 = C_P t + K \tag{16.28}$$

where C_P = parabolic rate constant and K is another constant. Usually Fe, Cu and Co etc. show parabolic oxidation behaviour when they form thick coherent oxides. At ambient temperature Fe, Cu and Al form thin films on metal surface and they follow the logarithmic oxidation rate law. Metals in service may be subjected to high temperatures and thus there is a tendency to form oxide layer at the surface. The tendency for an oxide film to be protective can be measured by a parameter called Pilling Bedworth (PB) ratio which is given by

$$PB = \frac{\text{Oxide volume per metal atom}}{\text{Metal volume per atom}} \tag{16.29}$$

When PB is less than unity the oxide layer becomes porous and unprotective. Under these conditions, tensile stresses are set up at the surface layer, which cause cracking of the oxide layer. When PB $\approx$ 1, an intact and protective oxide layer is formed at the surface. For values of PB greater than unity, the oxide layer is dense and flakes off the metal surface and as a result oxidation layer breaks away. This happens because compressive stresses act on the oxide layer. At higher temperature, oxidation proceeds faster, thus the adherent oxide layer with parabolic growth rate may not be able to protect the surface. Silicide and aluminide are commonly used oxidation resistant coatings for transition metals and their alloys, which are used as refractory materials and are poor resistant to high temperature oxidation. Some times alloying of suitable metals can improve oxidation resistance. When Cr or Al is added to Fe, Ni or Co, oxidation resistance of these transition elements increases. Sometimes oxidation occurs at grain boundaries of metals or alloys, which reduces the usefulness of such metals or alloys at high temperatures. Some of the oxidation resistant alloys are given in Table 16.2.

Example 16.5

Calculate Pilling-Bedworth ratio for the oxidation of Zirconium and comment on whether ZrO_2 forms protective coating or not. Given that density of Zr = 6.50 g/cc, density of ZrO_2 = 5.60 g/cc.

Solution:

$$Zr + O_2 = ZrO_2$$

Molecular weight of Zr is 91.22 g/mol and that of ZrO_2 is 123.22 g/mole
Consider that the weight of Zr is 100g. Then the mass of ZrO_2 produced by the oxidation of 100 gm of Zr is equal to (100/91.22) x 123.22 g = 135.08 g of ZrO_2
Volume of above amount of ZrO_2 = M/ρ = (135.08/5.60) cm^3 = 24.12 cm^3
Volume of Zr = (100/6.50) cm^3 = 15.38 cm^3
PB ratio = (24.12 cm^3/15.38 cm^3) = 1.57
1.57 is only slightly greater than one, hence ZrO_2 form protective coating on Zr metal.

Table 16.2 Oxidation Resistant Alloys

Name of the alloy	Composition	Uses
Stainless steel	10–12w/o Cr and rest Fe	Oil refinery components, turbine blades, furnace parts, and valve for IC engines
18-8 stainless steel	18w/o Cr, 8w/o Ni, and 74w/oFe	Furnace parts
Kanthal	24w/oCr, 5.5w/oAl, and 70.5w/oFe	Furnace windings for use up to 1300°C
Nichrome	80w/oNi, 20w/oCr	Good mechanical property
Inconel 600	75.5w/oNi, 15.5w/oCr,and 9w/oFe	Good mechanical property
Chromel	10w/oCr, 90w/oNi	Thermocouple for use up to 1100°C
Alumel	2w/oAl, 2w/oMn, 1w/oSi and 95w/oNi	Thermocouple for use up to 1100°C

Example 16.6

A sample of 99.90 weight percent of Ni having area 2 cm² and thickness 0.5 mm, is oxidized in oxygen at 1 atm pressure at 500°C. After 6 hours, the sample shows a 60 µg/cm² weight gain. If the oxidation process follows parabolic behaviour, what will be the weight gain after 2 days of oxidation? Assume K = 0.

Solution:

Initially one should calculate rate constant C_p

$$C_p = \frac{(W/A)^2}{t} = \frac{(60 \ \mu g/cm^2)^2}{6h} = 600 \ \mu g^2/(cm^4 \ h)$$

For time t = 2 days = 2 x 24 = 48 h, the weight gain in microgram per cm² is given by

$$\sqrt{(C_p \ t)} = \sqrt{(600 \times 48)} \ \mu g \ /cm^2 = 169.7 \ \mu g \ /cm^2$$

Example 16.7

Consider a steel pipe coupled to a magnesium anode of 3.2 kg. If the anode eorrodes completely in 120 days, what is the average current produced by the anode during this period?

Solution:

Using Eq. (16.13), we get

Since $Mg \rightarrow Mg^{2+} + 2e^-$

$I = WnF/tM$
$W = 3.2 \ kg = 3200 \ g$
$n = 2, F = 96,500 \ C/ \ mole$
$t = 120 \times 24 \times 3600 = 1.037 \times 10^7 \ s$
$M = 24.31 \ g/mole$

$$I = \frac{3200 \times 2 \times 96500}{1.037 \times 10^7 \times 24.31} = 2.45 \ A$$

Rust

When iron is exposed to alkaline or neutral solution, the layer of iron oxide and hydroxide is formed. This is a corrosion product called rust. The cathode reaction that accompanies the crevice corrosion of iron in an alkaline or neutral solution is

$$\tfrac{1}{2} O_2 + H_2O + 2e^- \rightarrow 2OH^- \qquad (16.30)$$

at cathode. The oxygen dissolved in water gets depleted due to above reaction within the crevice so cathode reaction slows down here. Outside the crevice, enough oxygen is available so that the cathode reaction proceeds quickly. Consequently, crevice region becomes anodic and corrosion occurs due to the anode reaction

$$Fe \rightarrow Fe^{2+} + 2e^- \qquad (16.31)$$

at anode. The overall reaction is a summation of these half-cell reactions and is given by

$$Fe + \tfrac{1}{2} O_2 + H_2O \rightarrow Fe^{2+} + 2OH^- \rightarrow Fe(OH)_2 \qquad (16.32)$$

The ferrous hydroxide is oxidized further to ferric hydroxide due to reaction with dissolved oxygen. The ferric hydroxide is the principal component of rust. More generally, the rust consists of hydrated Fe_2O_3.

16.9 PASSIVITY

Active metals and alloys lose their chemical reactivity under particular environmental conditions. Certain active metals such as chromium, iron, nickel, titanium and their alloys form oxide layer on the surface when exposed to suitable oxidizing agents at high concentrations. As a result of which, these metals and alloys become passive or more noble. These metals do not corrode easily after the formation of oxide film on the surface. Corrosion of anode can be prevented by passivating it electrically. Increasing the potential of the anode beyond a critical value can do passivation. As the anode potential increases, the current density also increases. However, after attaining a critical value, the current density falls to a low value when the potential increases further. This happens due to the formation of a passive film on the surface of the anode e.g. anodized aluminium. As a result, strong anodic polarization develops. Some metals and alloys like titanium and stainless steel form oxide layer when exposed to atmosphere and form more protective layer when dipped in oxygenated aqueous solution for a long time. An iron electrode becomes passive when immersed in aqueous solution of potassium dichromate. The reactions are:

$$K_2Cr_2O_7 \rightarrow 2K^+ + Cr_2O_7^{2-} \qquad (16.33)$$

$$Fe \rightarrow Fe^{3+} + 3e- \qquad (16.34)$$

$$3\,Cr_2O_7^{2-} + 2Fe^{3+} \rightarrow Fe_2O_3 + 3Cr_2O_6 \qquad (16.35)$$

16.10 PREVENTION AGAINST CORROSION

Prevention measures against corrosion are summarized below:

(i) Selection of material is very important for any engineering structure. If the material is noble metal, it will definitely prevent corrosion but noble metals have very limited applications. The corrosion resistant materials can be selected by consulting corrosion handbooks such that other required properties of the material are also present.

(ii) While designing a structure, the physical contact between dissimilar metals with different electrode potential should be avoided so that a galvanic couple should not form. Sometimes it is not possible to avoid such contacts, so in such situations the surface area of the anode in contact with the electrolyte should not be small as compared to that of the cathode. The small surface area of the anode implies high current density at anode and hence high rate of corrosion occurs at that place.

(iii) Materials having microstructure with two or more phases, are sometimes found to have better mechanical properties as compared to pure metal e.g. duralumin has better strength than pure aluminium but it corrodes faster than aluminium since it has two-phase structure. The two phases form galvanic cells at the microstructure level. The corrosion of duralumin is prevented by making Alclad, which consists of duralumin sheet sandwiched between two pure aluminium sheets.

(iv) Selection of corrosion resistant material for specific engineering applications involves specific combination of material and environment. For example, nickel and copper alloys are often used in reducing atmosphere of air-free acid and aqueous solution. Similarly, for oxidizing conditions, chromium alloys, titanium and its alloys are used.

(v) Rate of corrosion can be reduced by decreasing electrode reaction rates. This can be achieved by adding inhibitors to the corrosive environment. Some of the inhibitors can inhibit cathode reaction by increasing activation energy, others form film on cathode or anode surface thereby blocking the activities of electrode. Inhibitors produce concentration or resistance polarization at electrode. There are other types of inhibitors e.g. certain polar organic compounds, which get adsorbed on the metal surfaces. The concentration of inhibitors should be small so that they cannot contaminate the environment. Chromate salts act as inhibitors in automobile radiators. There are certain chromate, phosphate, molybdate and nitrite that form protective films on anodes or cathodes in heat-exchangers, power plants etc. Sometimes removal of reactants can also reduce the electrode reactions e.g. deaerated water can improve the life of water heating system.

(vi) Out of nonmetallic substances, polymeric materials are usually soft and less resistant to corrosive effects of strong inorganic acids so they are unsuitable for corrosion resistant applications. Ceramics have resistance against high temperature as well as corrosion since these are mostly complex metallic oxides. These materials are brittle with low tensile strength and hence have limited applications due to poor mechanical properties. They are mostly used as gaskets, coatings etc.

Coatings

Suitable coatings can protect the surface of metals and alloys from corrosion. Coatings can be metallic, nonmetallic inorganic or organic. Metallic coatings are often used for preventing the surface of a material to come in contact with the environment e.g. paints, varnish, lacquer etc. These coatings can be selected such that they are noble with respect to the underlying material. They prevent the material from coming in direct contact with the environment e.g. tin coated steel containers are normally used for food packaging. The tin coating on steel can be produced either by dipping it in molten tin or by electroplating it, where the object to be coated i.e. steel form cathode and is placed in a suitable electrolytic solution. The electrolytic solution is a salt of metal (say Sn), which is to be deposited. Many food materials contain organic acids, which may form complexes with tin, and as a result, tin may become anodic with respect to iron. Any flaws in the coating may expose a part of the anode (i.e. material) where corrosion starts since coating and material form galvanic couple. Cathodic protection is another method of limiting the rate of corrosion, which is accomplished by suitable galvanic coupling and by passing appropriate current. For example, consider an underground steel pipe, which can be protected from corrosion by electrically connecting it to magnesium or zinc anode. Such anodes corrode preferentially to the steel and are called sacrificial anodes. Zinc coated steel (galvanized steel) is protected from corrosion in a similar manner. If there is any puncture in the coating the exposed metallic surface corrodes slowly. The applied current in case of cathodic protection should be equal to the difference between the cathode current and corrosion current.

Another method of protecting the metal from corrosion is that by raising its potential such that it becomes passive. This is termed as anodic protection. The potentiostat is a device, which can maintain metal at a constant potential (i.e., within the passive range of metal) with respect to a reference electrode. In this case the applied current is given by the difference between minimum corrosion current and cathode current. Passivation of material by suitable alloying can prevent them from corrosion. When 12 w/o or more of chromium is added during formation of steel, the alloy requires very small current density to become passive.

EXERCISES

16.1 Explain radiation damage and thermal damage.

16.2 Define corrosion and explain the origin of corrosion. How can crevice corrosion be prevented? Which corrosion is responsible for problems in welding? Explain.

16.3 What is oxidation reaction called in which a metal or nonmetal is oxidized in valence charge in electrochemical corrosion reaction? Are electrons produced or consumed in this reaction?

16.4 Write down the name of five metals, which are cathodic to iron. Give reasons why are they cathodic to iron.

16.5 Consider a standard galvanic cell, which has electrodes of zinc in 1 M $ZnSO_4$ and tin in 1M $SnCl_2$. A porous wall separates these half-cells and the whole cell is at 25°C. When the electrodes are externally connected, answer the following:
   i)   In which direction will the current flow in the external circuit?
   ii) In which direction the anions move in the solution?
   iii) Write the reactions at anode and at cathode.
   iv) Which electrode corrodes and what is the emf of the cell?
   Ans: i) From cathode to anode; ii) Anions move towards anode; iii) At anode:
       $Zn \rightarrow Zn^{2+} + 2e^-$; At cathode: $Sn^{2+} + 2e^- \rightarrow Sn$; iv) Anode corrodes; emf = -0.624 V

16.6 A Copper electrode is immersed in an electrolyte 0.5 M $CuSO_4$ and is connected to another copper electrode in 0.02 M $CuSO_4$. A porous wall separates the two electrolytes.
   i)   Find the cell potential.
   ii) Which of these electrodes will act as anode?
   Ans: i) -0.00826V; ii) Copper electrode in 0.02 M $CuSO_4$ solution acts as anode

16.7 In a process of electroplating a copper cathode, the current of 15 A flows and copper anode dissolves in electrolytes. If the reaction goes for 30 min calculate the amount of copper that corroded away from anode?
   Ans: 8.88 g

16.8 How much copper is to be dissolved in 1 kg of water to make 0.02 M solution of Cu^{2+}? Calculate the electrode potential of the copper-half-cell in this electrolyte.
Ans: 1.27 g; 0.289 V

16.9 How much time will be required to electroplate a 1mm thick layer of silver on to a 50 mm^2 area of cathode surface, if the current is 5A?
Ans: 93.8 s

16.10 A copper can of 100 mm diameter is filled to a height of 100 mm with a corrosive liquid. After 15 days the container has decreased in weight by 50 g. Calculate the corrosion current and current density.
Ans: 0.117 A; 0.0298 A/cm^2

16.11 Consider a copper-zinc electrochemical cell. If the current density at copper cathode is 600 A-m^{2}, calculate the weight loss in one day if the cathode area is 200 cm^2 and anode area is 2 cm^2. If anode area is 200 cm^2 and cathode area is 2 cm^2 what will be the loss in a day?
Ans: 3.41 g; 341 g

16.12 A galvanized steel (zinc-coated steel) is found to corrode a thickness of 1 mm of zinc coating uniformly in 5 years. Find the average current density.
Ans: 1.34×10^{-5} A/cm^2

16.13 Consider a sample of pure nickel 0.5 mm thick is oxidized at 1 atm pressure and at 600°C. After 5h, the sample showed weight gain of 100 µg/cm^2. After how much time its weight gain is 200 µg/cm^2.
Ans: 20 h

16.14 A tin surface is corroding in seawater at a current density of 4.9×10^{-7} A/cm^2. What is the corrosion penetration rate?
Ans: 0.013 mm/year

16.15 An aluminium can develop pits in 6 months by pitting corrosion. If the container wall is 1mm thick and the diameter of the average pit is 0.14 mm, what is the average current associated with the formation of a single pit? How much time it will take for a pit to corrode through the wall assuming the pit to be of cylindrical in shape?
Ans: $I = 1.337 \times 10^{-9}$ A; t = 10.56 years

16.16 Calculate the ratio of the oxide volume to the metal volume (Pilling-Bredworth ratio) for the oxidation of vanadium to vanadium oxide (V_2O_5). The density of vanadium is 5.96 g/cc and that of vanadium oxide is 3.36 g/cc.
Ans: 3.166

Electrical Properties of Materials

17.1 INTRODUCTION

Electrical properties of a material are derived from its responses to the applied electric field. Electrical conduction is one of the most important electrical properties. It is the mechanism of flow of electrons or ions in the presence of an applied electric field. The flow of electrons or ions constitutes electric current I that is defined as charge flowing through the conductor per unit time. The charge flowing through unit cross-section per unit time is defined as current density J. The ease, with which electrical conduction can take place, depends upon the structure of the material. The materials in which electrical conduction takes place easily are called conductors. In metallic solids, metal ions are visualized as embedded in the sea of electrons, contributed from the valence shell of metal atoms. These are called valence electrons or conduction electrons, which move freely in metals and are shared by many atoms so that metallic bonds are formed.

At room temperature, positive metal ions vibrate about their lattice positions. Movement of valence electrons is impeded due to collisions with metal ions. At higher temperature, when ions and electrons acquire thermal energy these collisions become more frequent. The motion of the electrons is mostly random when they acquire thermal energy and hence there is no net flow of current. In the presence of an applied electric field (E) electrons get accelerated towards the positive electrode, thereby acquiring drift velocity v_d in the direction opposite to E. The acceleration a of an electron can be expressed as

$$a = eE/m \tag{17.1}$$

As the electrons move under the influence of electric field they collide with each other or with metal ions. If we assume that the electrons acquire the maximum velocity v_{max} in between two collisions, then

$$v_{max} = aT = (eE/m) T \tag{17.2}$$

where T is the average time between collisions and is called as collision time.

428

17.2 ELECTRICAL CONDUCTION

Electronic conduction obeys Ohm's law, which states that the current I, flowing through a conductor is proportional to the voltage V applied across it i.e.

$$I = V/R \qquad (17.3)$$

where R is resistance of the conductor. R increases linearly with the length L and decreases linearly with the area of cross-section, A of the conductor. Thus,

$$R = \rho L/A \ \Omega \qquad (17.4)$$

where constant of proportionality ρ is called the resistivity and Ω is ohm, the unit of resistance. The resistivity for a material is constant i.e. independent of the shape or size of the conductor, but it does depend on the temperature. Resistivity can be expressed as

$$\rho = \frac{RA}{L} = \frac{VA}{IL} \ \Omega\text{-m.} \qquad (17.5)$$

A schematic diagram for experimental set up to measure electrical resistivity is shown in Fig. 17.1. Electrical conductivity is defined as the reciprocal of electrical resistivity i.e.

$$\sigma = 1/\rho \ (\Omega\text{-m})^{-1}, \qquad (17.6)$$

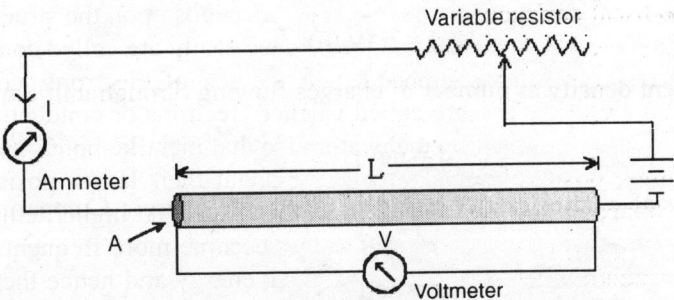

Fig. 17.1 Schematic diagram for experimental set up to measure resistivity.

where unit of conductivity $(\Omega\text{-m})^{-1}$ is also called Siemens(S). σ can be expressed as

$$\sigma = (I\,L/VA) \qquad (17.7)$$

I can also be expressed as rate of flow of net charge q through cross-sectional area A i.e. I = dq/dt. The current density J is the current flowing through unit cross-section i.e.

$$J = I/A \qquad (17.8)$$

Thus, from Eq. (17.7)

$$\sigma = JL/V = J/E$$

or

$$J = \sigma E \qquad (17.9)$$

where the electric field E=V/L. The current density and the electric field are vector quantities. Eq. (17.9) is microscopic form of Ohm's law.

When a uniform electric field **E** is applied across a conductor, the electrons acquire drift velocity v_d in the direction of **E**. They periodically collide with ion cores in the lattice and lose kinetic energy. These electrons are accelerated again due to **E** and regain kinetic energy. The drift velocity increases linearly in between two consecutive collisions and becomes zero after collision i.e. it varies with time in saw-tooth manner.

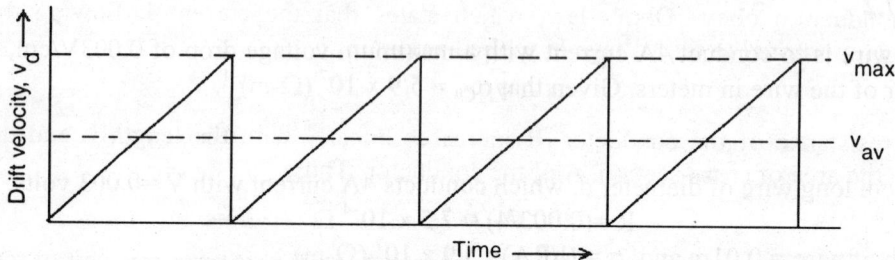

Fig. 17.2 Electron drift velocity versus time.

The time interval between two collisions is relaxation time, which is given by T/2. Variation of v_d of electron with time can be represented as shown in Fig. 17.2. The drift velocity increases linearly with electric field. The average velocity $v_{av} = v_{max}/2$. Assuming v_{av} to be v_d,

$$v_d = (eT/2m) E$$

$$v_d = \mu E \qquad (17.10)$$

where μ (=eT/2m) is called mobility. In other words, mobility is defined as the drift velocity of the electron per unit electric field, so that,

$$J = \sigma (v_d/\mu) \qquad (17.11)$$

We can express current density as number of charges flowing through unit cross-section per unit time i.e.

$$J = nev_d \qquad (17.12)$$

where n is density of charge carrier and e is electronic charge. Comparing Eqs.(17.11) and (17.12), we get

$$\sigma = ne\mu \qquad (17.13)$$

The electrical conductivity of pure metals are usually of the order of 10^7 $(\Omega\text{-m})^{-1}$ whereas that of insulators e.g. teflon (PTFE), is of the order of 10^{-16} $(\Omega\text{-m})^{-1}$. Thus, the conductivity of metals is about 10^{23} times that of insulators. Silicon, germanium etc., have conductivity in between that of metals and insulators and are thus classified as semiconductors. The conductivity of some selected materials are listed in Table 17.1.

Example 17.1

A wire of diameter 0.2 cm carries a current 30A. If the maximum power dissipation along the wire is 0.025W/cm, what is the minimum allowable electrical conductivity of the wire?

Solution:

Consider the wire of 1 m length. The power dissipation is given by

$$P = I^2R = J^2RA^2 = J^2(RA/l) A = (J^2/ \sigma)A$$

Given that
$$P = 0.025\text{W/cm} = 2.5 \text{ W/m}$$
$$A = 3.14\text{x} (0.001)^2 = 3.14\text{x}10^{-6}\text{m}^2$$
$$J = 30/(3.14\text{x} 10^{-6}) \text{ A/m}^2 = 9.55\text{x}10^6\text{A/ m}^2$$

$$\sigma = (J^2 A/P) = (9.12 \times 10^{13} \times 3.14 \times 10^{-6}/2.5) = 1.15 \times 10^8 \ (\Omega\text{-m})^{-1}$$

Example 17.2

If a copper wire is to conduct 4A current with a maximum voltage drop of 0.003V/cm, calculate the diameter of the wire in meters. Given that $\sigma_{Cu} = 5.9 \times 10^7 \ (\Omega\text{-m})^{-1}$.

Solution:

Consider 1 cm long wire of diameter d, which conducts 4A current with V =0.003 volts. Then,

$$R = (0.003/4) = 7.5 \times 10^{-4} \ \Omega$$

Since length of wire = 0.01m and $\sigma = (l/RA) = 5.9 \times 10^7 \ (\Omega\text{-m})^{-1}$

So $\quad\quad\quad 5.9 \times 10^7 = 0.01/ (7.5 \times 10^{-4} \times A)$

The area of cross-section: $A = (10^{-5}/5.9 \times 7.5) \ m^2 = 2.26 \times 10^{-7} \ m^2$

Hence diameter d of the wire = $(4A/\pi)^{1/2} = 0.536$ mm

Table 17.1 Electrical conductivity of materials.

Metals & alloys	$\sigma(\ (\Omega\text{-m})^{-1}) \times 10^7$	Nonmetals	$\sigma((\Omega\text{-m})^{-1})$
Silver	6.8	Graphite	10^5
Copper	5.9	SiC	10
Gold	4.25	Germanium	2.2
Aluminium	3.8	Silicon	4.3×10^{-4}
Al -1.2w/oMn alloy	3.0	Phenol formaldehyde	$10^{-7} - 10^{-11}$
Sodium	2.0	Glass	$<10^{-10}$
Molybdenum	2.0	Mica	$10^{-11} - 10^{-15}$
Tungsten	1.77	Polymethyle methacrylate	$<10^{-12}$
Brass(70w/oCu–30w/oZn)	1.6	Polyethylene	$<10^{-14}$
Platinum	0.92	Polystyrene	$<10^{-14}$
Nickel	1.46	Diamond	$<10^{-14}$
Plain carbon steel	0.6	Silica Glass	$<10^{-16}$
Tantalum	0.64	Teflon	$<10^{-16}$
Titanium	0.24		
Stainless steel	0.2		
Nichrome (80w/oNi-20w/oCr)	0.09		
Manganin	0.21		
Kanthal	0.007		

17.3 ENERGY BAND STRUCTURE IN SOLIDS

Conductivity of a solid depends on the number of conduction electrons in the outermost shell of atoms. According to the free electron theory, outermost electrons of an atom in metals are loosely bound to the nucleus. These electrons move freely through the lattice of metal ions forming what is called as free electron gas or Fermi gas. Potential field due to metal ions is assumed to be uniform i.e. the potential energy of free electrons is the same throughout the crystal. Considering this constant potential energy to be zero of the energy scale for free electrons, one can express the energy of electron to be purely kinetic and is given by ½ mv².

To derive the concept of energy band in case of solid, we introduce de Broglie hypothesis. According to this hypothesis, electrons have both particle-like and wave-like characteristics.

When a beam of electron falls on a crystal at an appropriate inclination, the diffraction of electron beam is observed which is the characteristic of a wave. The de Broglie wavelength λ of an electron is given by

$$\lambda = h/mv \qquad (17.14)$$

where h is Planck's constant, mv is the momentum of the electron. The wave number k for the de Broglie wave of the electron is given by

$$k = 2\pi/\lambda = 2\pi mv/h = mv/\hbar \qquad (17.15)$$

As per free electron theory, the energy of the electron in terms of wave number k is

$$E = \tfrac{1}{2} mv^2 = \hbar^2 k^2 /2m \qquad (17.16)$$

If E is plotted against k, the curve is parabolic. Thus, the allowed energy values vary continuously with k, from zero to infinity. However, according to Bohr's theory, electrons of an isolated atom can occupy only discrete energy states, called quantum states. The de Broglie wavelength λ_n of an electron in the nth energy state should be such that

$$2\pi r_n = n\lambda_n \qquad (17.17)$$

where r_n is the radius of nth circular orbit.

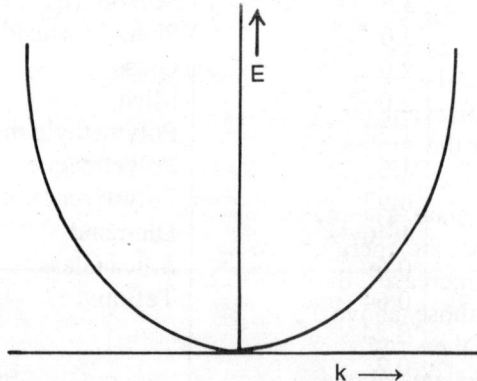

Fig. 17.3 The Energy versus wave number k for free electron

According to Pauli exclusion principle, no two electrons can have same set of quantum numbers i.e. no two electrons can occupy the same quantum state. Valence electrons fill up the energy states such that each energy state can be occupied by a pair of electrons of opposite spin in the absence of external field.

Movement of valence electrons in an atom gets modified due to their interaction with the electrons of other atoms in the solid. This results in the modification of the electronic energy states. Since Pauli exclusion principle holds for electrons in a solid, the number of electronic energy states corresponding to an atomic energy level becomes enormous but their energy values differ minutely from one another. These electronic energy states form an energy band in a crystalline solid, called valence band which consists of all the energy states of valence electrons contributed by interacting atoms in the solid. Thus, each band contains as many discrete energy states as there are atoms in the solid. Hence in a solid containing n atoms, each band can accommodate 2n electrons.

The electron occupying specific energy state has fixed energy E and has a momentum p. The wave number k can have only discrete positive or negative values that are

$$p = \sqrt{(2mE)} = \hbar k$$

arising from the periodic boundary conditions of the crystal. The discrete values of k are such that its components are given by

$$
\begin{aligned}
k_x &= 0, \pm \pi/L, \pm 2\pi/L,... \ n_x\pi/L \\
k_y &= 0, \pm \pi/L, \pm 2\pi/L,... \ n_y\pi/L \\
k_z &= 0, \pm \pi/L, \pm 2\pi/L,... \ n_z\pi/L
\end{aligned}
\tag{17.18}
$$

considering the motion of electrons in three-dimensional crystal lattice. L is the repeat distance and n is the number of lattice points along each direction. Hence, the energy expression in Eq.(17.16) can be modified as

$$E = \frac{\hbar^2}{2m} (k_x^2 + k_y^2 + k_z^2) = \frac{h^2}{8mL^2} (n_x^2 + n_y^2 + n_z^2) \tag{17.19}$$

Each state is designated by a combination of n_x, n_y and n_z. In each state there can be two electrons with opposite spins. There can be a number of states having different combinations of n_x, n_y and n_z which have same energy. Such states are called degenerate states. If the energy states calculated from Eq.(17.19), are arranged in the increasing order of magnitude, they are found to be so closely placed, such that the energy E can be taken as nearly continuously varying with **k**. Since the electrons obey Pauli exclusion principle, they fill up energy states in the increasing order of energy starting from the lowest energy state such that the total energy of the system becomes minimum. A pair of electrons of opposite spin, which have all other quantum numbers same, occupy the same energy state. The highest energy state filled up by electron at 0°K is called the Fermi energy level (E_f). All the energy states below Fermi level are filled up and are empty above E_f. As the temperature increases, there is probability that electrons from the energy states below E_f can be excited to those above E_f. When the electrons are excited due to external electromagnetic field or thermal energy, they make transition to higher energy states from valence states. These allowed energy states above Fermi level form conduction band. The probability f(E,T) for electron to be in the energy state E at temperature T is given by Fermi-Dirac statistics,

$$f(E,T) = \frac{1}{1 + \exp\ [(E - E_f)\ /kT]} \tag{17.20}$$

where f(E,T) is also called Fermi-Dirac distribution function. If E_f does not vary with temperature, the variation of f(E,T) versus E is shown in Fig. 17.4. At 0°K, f(E,0) remains constant with value unity upto $E = E_f$ and just after that it becomes zero as per definition of E_f. At higher temperature, the electrons below E_f are thermally excited to energy levels above E_f, that is why f(E,T) becomes less than unity for $E < E_f$ and above E_f, it has values greater than zero. As the temperature increases further, occupancy at energy states beyond E_f increases. At the Fermi level, the probability of occupation of an electron is half at any temperature greater than 0°K as can be seen from Eq.(17.20).

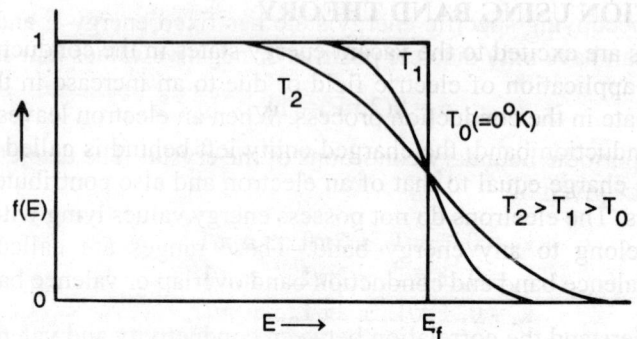

Fig. 17.4 Variation of Fermi-Dirac distribution function for free electrons at different temperatures.

Other occupancy values are given below:

$$E = E_f \text{ at } T > 0°K \qquad \exp((E - E_f)/kT) = 1; \qquad f(E_f, T) = \frac{1}{2} \qquad (17.21)$$

$$E > E_f \text{ at } T > 0°K \qquad \infty > \exp((E - E_f)/kT) > 1; \qquad 0 \le f(E, T) < \frac{1}{2} \qquad (17.22)$$

$$E < E_f \text{ at } T > 0°K \qquad 1 > \exp((E - E_f)/kT) > 0; \qquad \frac{1}{2} < f(E, T) \le 1 \qquad (17.23)$$

$$E > E_f \text{ at } T \approx 0°K \qquad \exp((E - E_f)/kT) \to \infty; \qquad f(E, T) \to 0 \qquad (17.24)$$

$$E < E_f \text{ at } T \approx 0°K \qquad \exp((E - E_f)/kT) \to 0; \qquad f(E, T) \to 1 \qquad (17.25)$$

Example 17.3

Find the probability that the electronic states at energy $E_f - 0.05$ eV are occupied at temperature 500°C.

Solution:

Using Eq (17.20), $E = E_f - 0.05$eV and k= 8.63×10^{-5} eV/°K, the probability that the energy state be occupied is given as

$$f(E, T) = [1 + \exp\{(-0.05)/(8.63 \times 10^{-5} \times 773)\}]^{-1} = 0.679$$

Example 17.4

Tabulate the fraction of energy states occupied at energy E= E_f+ 0.04eV as function of temperature for T = 0°K, 100°K, 200°K,....1000°K. Find the temperature at which f (E,T) =½.

Solution:

$$f(E, T) = [1 + \exp(0.04/kT)]^{-1}$$

T in °K	0	100	200	300	400	500	600	700	800	900	1000
f(E,T)	0	0.0096	0.0897	0.176	0.239	0.284	0.316	0.340	0.359	0.374	0.386

To find the temperature T at which f(E,T) = $[1 + \exp(0.04/kT)]^{-1}$ = ½, we get T = $0.04/[k \ln(1)]$ which is undefined, and we can find out that f(E,T) = ½, only if E= E_f.

17.4 CONDUCTION USING BAND THEORY

Valence electrons are excited to the vacant energy states in the conduction band above the Fermi level, due to the application of electric field or due to an increase in the thermal energy. These electrons participate in the conduction process. When an electron leaves valence band after being excited to the conduction band, the charged entity left behind is called hole. The hole carries an effective positive charge equal to that of an electron and also contributes to conductivity, mostly in semiconductors. The electrons do not possess energy values lying within certain energy ranges, which do not belong to any energy band. These ranges are called band gaps. In case of conductors, the valence band and conduction band overlap or valence band is only partially filled.

In order to understand the correlation between conductivity and valency of element in terms of band structure, consider a linear crystal having lattice points at distance 'L' apart. According to Eq.(17.18) the allowed values of k vectors are given by

$$k = 0; \pm \pi/L; \pm 2\pi/L; ... n\pi/L;$$

where n is the number of lattice points and is equal to the number of allowed k vectors. Corresponding to each lattice point there is one independent value of k (i.e. energy level) for each energy band. Taking account of the two independent orientations of the electron spin, there are 2n independent electronic states in each energy band. If the basis consists of a single atom having one valence electron, then there will be n electrons contributed by the linear lattice. Thus, the energy band can only be half-filled in this case. In case, each atom contributes two valence electrons, the band can be exactly filled.

If the basis consists of two atoms, each having one valence electron, the valence band can be exactly filled up. When valence electrons completely fill one or more bands and the valence band is separated by a large energy gap from the next higher conduction band, the crystal behaves as an insulator. In an insulator, the valence electrons are excited to unfilled higher energy states only if the excitation energy is greater than the energy gap. Electric field of very high strength is required for this purpose so that these electrons can contribute to conduction process. So there is no continuous way to change the total momentum of electrons in an insulator in presence of electric field since every accessible energy state in a band is filled up.

If the number of valence electrons corresponding to a basis atom is even, the crystal can be a conductor or an insulator depending upon whether the valence and conduction bands overlap or not. In case the bands overlap, there can be more than one partially filled bands as observed in case of a conductor, Fig.17.5 (b). The alkali metals have one valence electron per atom i.e. per lattice point, so they are conductors whereas alkaline earth metal, which has two valence electrons per atom, could be a insulator. Due to overlapping of bands, the alkaline earth metals behave as conductors but not very good ones. Thus in sodium (Fig. 17.6), which has only one valence electron per atom, there is first valence band only half-filled. In case of aluminium ($[Ar]3s^2 3p^1$) the first valence band is completely filled up but its second band is half filled. The detailed investigation reveals that the first and second energy bands of alkali and alkaline earth metals overlap, whereas there is energy gap between second and third energy bands. In case of pure materials, there is no electronic energy state within this gap.

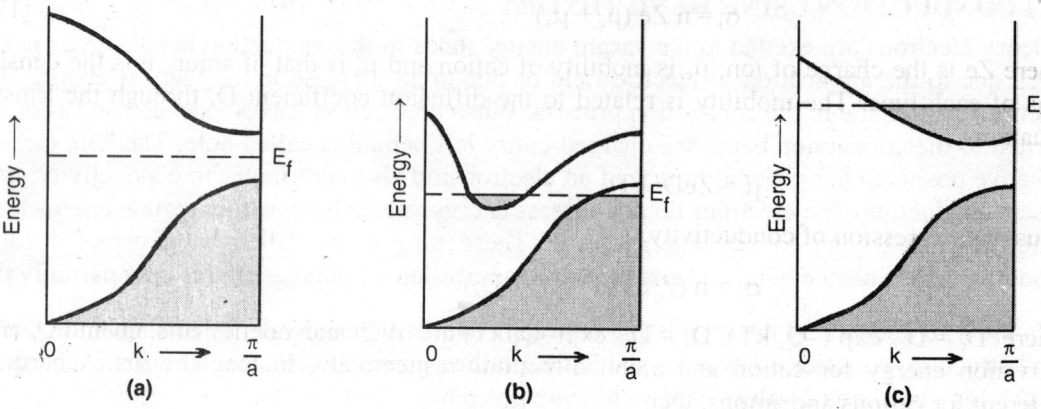

Fig. 17.5   Band structures for (a) an insulator (b) a conductor having band overlap and (c) a conductor of partially filled band.

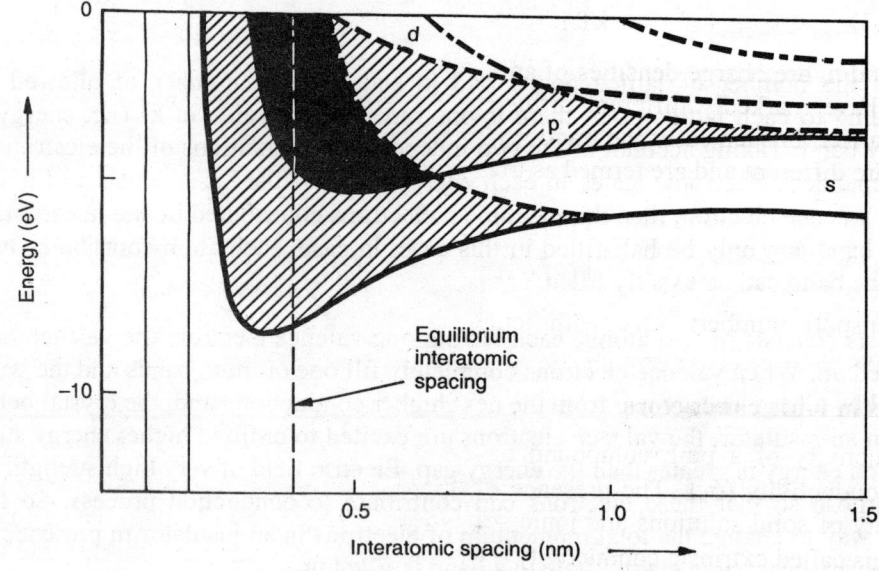

Fig. 17.6 Energy levels of Sodium.

17.5 IONIC CONDUCTIVITY

Ionic solids are normally insulators. Because of the presence of large energy gap in these materials, the electrons cannot be excited into the conduction band unless the temperature is increased a lot or the required magnitude of electric field is applied. The basic mechanism of conduction in ionic solid is diffusion of ions in the presence of the electric field. The diffusion becomes faster in the presence of Schottky or Frenkel defects. When Frenkel defects dominate in an ionic crystal, the cation interstitials carry the diffusion flux. When Schottky defects are present, the conduction takes place because the vacancies of cations and anions drift in the opposite directions under the influence of the electric field. Thus, the ionic conduction takes place due to the diffusion of ions.

The electrical conductivity of ionic solid having equal number of cations and anions can be expressed as [using Eq.(17.13)],

$$\sigma_i = n \, Ze \, (\mu_c + \mu_a) \tag{17.26}$$

where Ze is the charge of ion; μ_c is mobility of cation and μ_a is that of anion; n is the density of ions of each type. The mobility is related to the diffusion coefficient D, through the Einstein's equations,

$$\mu = ZeD/kT \tag{17.27}$$

Thus, the expression of conductivity,

$$\sigma_i = n \, (Z^2 e^2 / kT) \, (D_c + D_a) \tag{17.28}$$

where $D_c = D_{co} \exp(-Q_c/kT)$, $D_a = D_{ao} \exp(-Q_a/kT)$ are diffusion coefficients, Q_c and Q_a are the activation energy for cation and anion diffusions respectively. In case the ionic charges are different for cations and anions, then

$$\sigma_i = \frac{n_c \, Z_c^2 e^2}{kT} D_c + \frac{n_a \, Z_a^2 e^2}{kT} D_a \tag{17.29}$$

where n_a and n_c are charge densities of anion and cation respectively. The positive and negative ions may have markedly different diffusion coefficients, such that one of them becomes the majority carrier. In such cases the fractions of conductivity arising from cation and anion vacancy diffusion are different and are termed as transport numbers t (+) and t (−) respectively. Thus,

$$t\,(+) = \sigma_c / \sigma_i \tag{17.30}$$

$$t\,(-) = \sigma_a / \sigma_i \tag{17.31}$$

The transport numbers vary with temperature since the number of defects varies with temperature.

Impurities in ionic conductors

The conductivity of a pure compound or element depends on temperature and is termed as intrinsic conductivity (σ_{in}). The presence of impurities in some form or other may increase the conductivity of solid solutions and ionic solids. The increase in conductivity due to presence of impurities is called extrinsic conductivity (σ_{ex}). The total ionic conductivity is expressed as

$$\sigma_i = \sigma_{in} + \sigma_{ex} \tag{17.32}$$

The host solid and impurities have different set of σ values; each is a function of temperature. If the ions of different valencies form a substitutional solid solution, then vacancies are created to maintain charge balance. This is also called nonstoichiometric compound. For example, a Cd^{2+} ion, in NaCl crystal creates one cation vacancy. Even a small concentration, such as 0.1% of the above impurity can increase the conductivity of NaCl significantly.

Example 17.5

Electrical conductivity in cubic $Fe_{0.85}O$ is mainly due to cation movements through cation vacancies at 828°C. This unit cell has four anion sites that are all filled up and four cation sites partially vacant. If the diffusion coefficient for Fe^{2+} is $10^{-12} m^2/s$, and lattice constant a = 4.27 Å, calculate the electrical conductivity at 828°C.

Solution:

$\sigma_i = nZ^2e^2D/kT$

$n = (4 \times 0.85)/(4.27 \times 10^{-10}m)^3 = 4.367 \times 10^{28}/m^3$

$\sigma_i = 4.367 \times 10^{28}/m^3 \times (2 \times 1.6 \times 10^{-19} C)^2 \times 10^{-12}m^2/s/(1.38 \times 10^{-23} \times 1101) = 0.294 \ \Omega^{-1}.m^{-1}$

17.6 ELECTRICAL RESISTIVITY IN METALS

Metals have low electrical resistivity. Electrical resistivity in a metal depends on the number of valence electrons per atom and their binding energy. Resistivity in a metal also depends on temperature and presence of defects. Defects act as scattering centres for conduction electrons and hence increase the resistivity. Sometimes defects are present in the form of interstitial or substitutional impurities. Defects can be vacancies, dislocation lines or grain boundaries, which are developed during cold working, heat treatment or at the time of crystallization. The total resistivity of a metal is the sum of the contributions from thermal vibrations (ρ_{th}), impurities (ρ_{im}) and plastic deformations (ρ_d). Thus we can express

$$\rho_{total} = \rho_{th} + \rho_{im} + \rho_d \qquad (17.33)$$

Effect of temperature on resistivity

The resistivity ρ_{th} for pure metals increases linearly with temperature as given by

$$\rho_{th} = \rho_o (1 + \alpha_t (t-t_o)), \qquad (17.34)$$

where ρ_o is the resistivity at $t_o°C$, α_t is the temperature coefficient of resistivity, in $°C^{-1}$, $t-t_o$ is the rise in temperature of the material in °C.

The thermal component of resistivity arises from the vibrations of ions about their equilibrium positions in crystal lattice. Vibrations of ions increase with the temperature and as a result large number of phonons (thermally excited quantized elastic waves) are generated. These phonons collide with electrons more frequently with the rise of temperature, thereby reducing mean free path and relaxation time between collisions. The schematic representation of variation of electrical resistivity of a pure metal is shown in Fig. 17.7. The components ρ_{im} and ρ_d are almost independent of temperature and contribute significantly only at low temperatures.

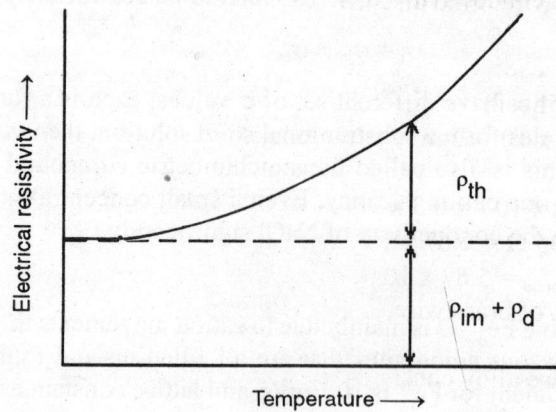

Fig. 17.7 Variation of electrical resistivity of a metal with absolute temperature.

Example 17.6

An alloy of metal is found to have a resistivity of 10^{-6} Ω-m at 0°C. When it is heated to a temperature 700°C, its resistivity increases by a factor of 5.5, calculate the temperature coefficient of resistivity of the alloy.

Solution:

$$(\rho_{700} - \rho_o)/ \rho_o = \alpha_t \times 700$$
$$\alpha_t = (4.5/700)/°C = 0.006/°C$$

Effect of impurities

Resistivity in a metal increases due to the presence of impurities. If f_i is the fraction of impurity atoms present in a metal, then

$$\rho_{im} = Af_i (1 - f_i) \tag{17.35}$$

where A is the resistivity coefficient of the metal which depends on the size of the host as well as that of the impurity atoms.

When the valency of the impurity and the metal atoms differ, the resistivity of the metal increases. The resistivity also increases if the size of the impurity atom differs by more than 15% of that of the metal atom. In the dilute solution ($f_i \ll 1$),

$$\rho_{im} = Af_i \tag{17.36}$$

Resistivity of a disordered alloy is more than that of the ordered one. For example, in gold and copper alloy the resistivity decreases below 380°C since it becomes ordered with fcc structure having gold atoms at corners and copper atoms at face centers ($AuCu_3$).

Example 17.7

A brass (70 a/o Cu-30 a/o Zn) wire is to replace a copper wire, which has a resistance of 0.05 Ω per 4m to make it economical. Calculate the length of brass wire with same cross-section, a and resistance as that of copper wire. Find also the cross-section. Given $\rho_{Cu} = 1.67 \times 10^{-8}$ Ω-m.

Solution:

$$\rho_{im} = Af_i (1 - f_i)$$
where $\quad A = 0.2 \times 10^{-6}$ Ω-m $\quad$ for brass
and $\quad f_{Zn} = 0.3,$
Hence $\quad \rho_{70Cu-30Zn} = 0.2 \times 10^{-6} \times 0.3 \times 0.7 + \rho_{Cu} = 5.87 \times 10^{-8}$ Ω-m
$$R_{Cu} = 0.05\ \Omega = 1.67 \times 10^{-8}\ \Omega\text{-m} \times 4\text{m}/a$$
$$= R_{brass} = 5.87 \times 10^{-8}\ \Omega\text{-m} \times L /a$$
Required length L of brass wire, $L = 1.14$ m and $a = 1.336 \times 10^{-6}$ m^2

The resistivity of a two phase alloy consisting of α and β phases can be approximated as

$$\rho_{im} = \rho_\alpha V_\alpha + \rho_\beta V_\beta \tag{17.37}$$

where ρ_α, ρ_β are resistivities and V_α, V_β are the volume fractions of α and β phases, respectively (assuming that the phases are series connected).

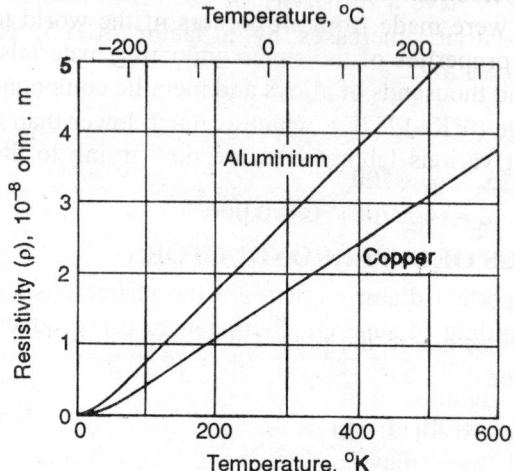

Fig. 17.8 The electrical resistivity versus temperature for copper and aluminium.

17.7 USES OF CONDUCTORS

The conductors with high conductivity have following uses:

 (i) Copper and aluminium are used in transmission and distribution cables where low I^2R loss, fabricability and mechanical strengths are prime considerations. Aluminium cable with large cross-section is the most likely choice for transmission. Aluminium conductor steel rein-forced (ACSR) cables are considered to be even better choice for transmission line since the above material has improved elastic modulus enabling the increased distance between successive poles. Copper is used for distribution lines and bus bars. Oxygen free high conductivity (OFHC) copper is often used for many electrical applications. Impurities like Fe, P etc., in copper decrease its electrical conductivity.

 (ii) Normally copper is used for electrical contacts in switches, brushes and relays where high electrical and thermal conductivity is needed along with the property of high melting point and good oxidation resistance. High thermal conductivity of the contacts helps in dissipating the heat. High melting point prevents the contact material from melting at an elevated temperature otherwise contact points may fuse. Silver is more suitable for such applications but not usually used because of high cost. Sometimes CdO particles are dispersed in silver to improve the mechanical strength.

(iii) Alloys like manganin (87 w/o Cu and 13 w/o Mn), constantan (60 w/o Cu and 40 w/o Ni) are used as resistors since these alloys have uniform resistivity, stable resistance (i.e. aging and effect of residual stresses are avoided), low value of temperature coefficient of resistance (α_t) and low thermoelectric potential with respect to copper.

(iv) Other alloys like kanthal (69 w/o Fe, 23 w/o Cr, 6 w/o Al and 2 w/o Co), nichrome (80 w/o Ni, and 20 w/o Cr) are used for heating elements, since they have the properties like high melting point, high electric resistance, good corrosion resistance, low thermal expansion, low elastic modulus etc. Graphite and SiC are also used for heating elements. Tungsten is used in incandescent lamps. Platinum is used in resistance thermometer since it has a high temperature coefficient of resistance.

17.8 SUPERCONDUCTIVITY

Since the discovery of superconductivity in mercury at 4.2°K by Dutch physicist, Kamerlingh Ones, in 1911, the efforts were made from all corners of the world to investigate the origin of superconductivity and the properties of the superconducting materials. By now it is known that nearly twenty six metals and thousands of alloys and metallic compounds show superconductivity in a wide temperature range (0°K–125°K), which is much lower than the room temperature. The scientists and engineers at various laboratories are now trying to discover superconductors at room temperature.

17.9 CHARACTERISTICS OF SUPERCONDUCTORS

Infinite conductivity and perfect diamagnetism are the characteristics of a superconductor. An electromagnet with the winding of superconducting wire can produce very high magnetic field without any I^2R loss. It is expected that motors and generators with superconducting wire winding would be smaller, lighter and more efficient than those built with copper winding. Very high magnetic field is required for firing projectiles, floating molten metal in a steel mill where superconducting material can play an important role. Powerful electromagnets using superconducting wire are commercially used in magnetic resonance spectrometers and medical imaging systems. Superconducting magnets have very important role in frictionless high speed transportation systems since strong magnetic fields are required for levitating trains.

As we increase the temperature beyond a specific value, a superconductor exhibits electrical resistance like a normal conductor, when direct current flows. This is called transition or critical temperature, T_c. Below critical temperature, a steady direct current flows forever through a superconductor, as there is no loss of power. If there is a variation in the direct current or an alternating current flows through a superconductor it dissipates energy i.e. a superconductor shows ac losses. The highest critical temperature known up to 1986 was 23°K for superconductor Nb_3Ge. Later in the same year, Alex Müller and George Bednorz discovered a new class of superconductors (perovskites), which showed superconductivity at 35°K– 40°K. That was the beginning of a new era. Just after that, a new material $YBa_2Cu_3O_{7-\delta}$ (or called as 1-2-3 material) was found having critical temperature T_c = 95°K. More recently, copper oxide ceramics, containing bismuth or thallium are found to have T_c in the range of 110°K – 125°K. So the efforts are on and on to reach the ultimate goal i.e. room-temperature superconductor.

Meissner effect

Superconductors show perfect diamagnetism. Meissner and Ochsenfeld discovered this property of superconductors in 1933. This is called Meissner effect, Fig. 17.9. They observed that when a superconductor is placed in a magnetic field the magnetic lines of force do not pass through it. When a magnetic field is applied, current flows in the outer skin of the material leading to an induced magnetic field that exactly opposes the applied field. The material is strongly diamagnetic as a result. In this experiment, a magnet floats above the surface of the superconductor. Magnetic flux lines, already present in a sample of the superconducting material at temperature $T > T_c$, are expelled out, as it is cooled below T_c. The experimental observation shows magnetic induction **B** decreases exponentially from the surface to the interior of the superconductor, Fig. 17.10. This phenomenon cannot be explained by perfect conductivity, which would tend to trap flux in when it is cooled down to near absolute zero. If the magnitude of the magnetic field is increased beyond a critical value, the magnetic flux lines start penetrating the superconductor and the material changes to a normal conductor. This magnetic field is called critical magnetic field (H_c). It has been observed experimentally that the lower the temperature of the superconductor below T_c, the higher the value of H_c, Fig. 17.11.

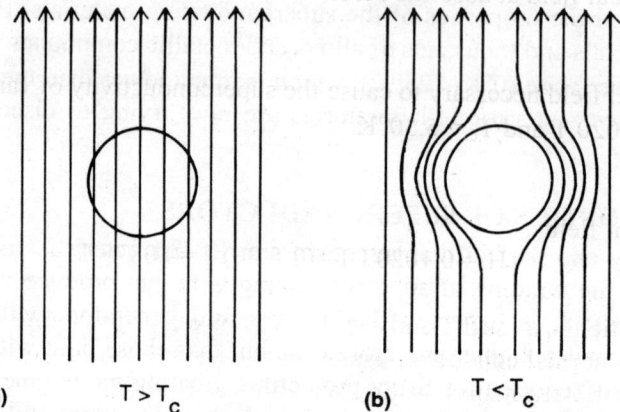

(a) $T > T_c$ **(b)** $T < T_c$

Fig. 17.9 Meissner effect (a) penetration of field in superconducting material at T > Tc
(b) expulsion of magnetic field at T < Tc

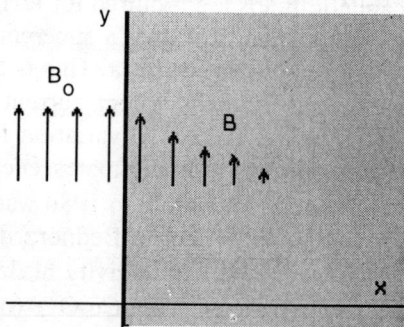

Fig. 17.10 Schematic representation of penetration of magnetic induction inside the semi-
infinite superconductor

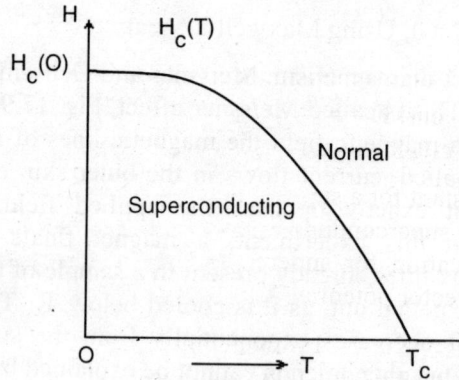

Fig. 17.11 Schematic representation of variation of critical magnetic field with temperature
T < Tc.

We can express,

$$H_c = H_o \left[1 - (T/T_c)^2 \right] \tag{17.38}$$

where H_o is the critical field at absolute zero.

Example 17.8

Calculate the critical field necessary to cause the superconductivity of vanadium to disappear at 2°K. Given $H_o = 0.1020$ T and $T_c = 5.30$°K.

Solution:

The value of critical field

$$H_c = 0.1020 \left[1 - (2/5.30)^2 \right] = 0.0875 \text{ T}$$

Like magnetic field, a sufficiently high value of current density can also change a superconductor to a normal conductor. This is called the critical current density (J_c). The value of H_o is different for different materials. So the curves of H_c vs T as well as J_c vs T for $T < T_c$ vary for different materials. We know that the magnetic induction **B** in a superconducting material is zero, hence

$$\mathbf{B} = \mathbf{B_a} + \mu_o \mathbf{M} = \mu_o (\mathbf{H} + \mathbf{M}) = 0 \tag{17.39}$$

where $\mathbf{B_a}$ is the applied magnetic field, M is the magnetization vector and H is auxiliary field in the medium. From (17.39),

$$\mathbf{H} = -\mathbf{M} \tag{17.40}$$

We know that

$$\mathbf{M} = \chi \mathbf{H} \tag{17.41}$$

so that the magnetic susceptibility (χ) of pure superconductor is -1, which shows that a pure superconductor behaves like perfect diamagnetic material below T_c.

London equations

Electrical conduction in the normal metal obeys Ohm's law

$$\mathbf{J} = \sigma \mathbf{E} \tag{17.42}$$

In case of superconductors $\mathbf{E} = 0$. Using Maxwell's equation

$$\text{curl } \mathbf{E} = -\frac{d\mathbf{B}}{dt}, \tag{17.43}$$

So $d\mathbf{B}/dt = 0$, i.e. $\mathbf{B}$ = constant for a superconductor. This is not true since B decreases from the surface to the interior of the superconductor according to Meissner effect, Fig. 17.10. Therefore, Ohm's law requires modification for superconductors. For superconducting state, the current density J is proportional to vector potential **A**,

$$\mathbf{J} = -\frac{1}{\mu_o \lambda_L^2} \mathbf{A} \tag{17.44}$$

This is one of the London equations. Taking curl of both sides and using Maxwell's equation curl $\mathbf{A} = \mathbf{B}$,

$$\text{curl } \mathbf{J} = -\frac{1}{\mu_0 \lambda_L^2} \text{ curl } \mathbf{A} = -\frac{1}{\mu_0 \lambda_L^2} \mathbf{B} \tag{17.45}$$

where λ_L is a constant with dimension of length and μ_0 is permeability in vacuum. To show London equation leads to the Meissner effect, we make use of Maxwell's equation (Ampere's Law)

$$\text{curl } \mathbf{B} = \mu_0 \mathbf{J} \tag{17.46}$$

and

$$\text{curl curl } \mathbf{B} = \mu_0 \text{ curl } \mathbf{J} \tag{17.47}$$

So

$$\nabla(\nabla.\mathbf{B}) - \nabla^2\mathbf{B} = \mu_0 \text{ curl } \mathbf{J}$$

Since $\nabla.\mathbf{B} = 0$,

$$-\nabla^2\mathbf{B} = \mu_0 \text{ curl } \mathbf{J} \tag{17.48}$$

$$\nabla^2\mathbf{B} = \mathbf{B}/\lambda_L^2 \tag{17.49}$$

This equation leads to a nonuniform magnetic induction B in the superconductor. In case of a semi-infinite superconductor, Fig. 17.10, the magnetic induction at a point x inside the material

$$\mathbf{B}(x) = \mathbf{B_0} \exp(-x/\lambda_L) \tag{17.50}$$

where $\mathbf{B_0}$ is the magnetic induction at the boundary. The magnetic induction is taken as parallel to the boundary. λ_L is the measure of the depth of penetration of magnetic induction in the superconductor, which is also called London penetration depth. In case of superconductors, the effective carriers are pairs of electrons with charge $q = -2e$, the concentration n_s of the carrier is one half of the concentration of the conduction electrons. The mass of the carrier is twice the electronic mass. It can be shown that

$$\lambda_L^2 = m/(\mu_0 n_s q^2) \tag{17.51}$$

where n_s is zero at T_c. λ_L is a function of temperature and is given by

$$\lambda_L(T) \approx \lambda_L(0)[1- (T/T_c)^4]^{-1/2} \tag{17.52}$$

An applied magnetic induction can penetrate a thin film uniformly if the thickness of the film is much less than λ_L; so in a thin film Meissner effect is incomplete.

The response of a superconductor to the electric field can be formulated using two basic equations of electric conduction

$$d(mv_s)/dt = q\mathbf{E} \tag{17.53}$$

and

$$\mathbf{J_s} = n_s q \mathbf{v_s} \tag{17.54}$$

Canonical momentum of the carrier in a superconductor is $\mathbf{p} = m\mathbf{v} + q\mathbf{A}$. In the absence of an applied field the ground state has zero net momentum. Hence, the local average of velocity in the presence of field is

$$<\mathbf{v_s}> = \frac{-q\mathbf{A}}{m} \tag{17.55}$$

Using Eqs. (17.54) and (17.55),

$$\mathbf{J_s} = \frac{-n_s q^2 \mathbf{A}}{m} \tag{17.56}$$

In the absence of electrostatic potential,

$$\mathbf{E} = -\frac{\partial \mathbf{A}}{\partial t}$$

Hence

$$\mathbf{E} = \frac{m}{n_s q^2} \frac{\partial \mathbf{J_s}}{\partial t} \tag{17.57}$$

This is another London equation.

Stabilization energy of the superconducting state

The transition between the normal and superconducting states is thermodynamically reversible. Meissner effect also shows this reversibility property. The energy of superconducting state is lower than that of the normal state below T_c. The energy difference of the two states at absolute zero is called stabilization energy. The stabilization energy can be estimated from the minimum value of the applied magnetic induction, which will destroy superconductivity. The work done on a superconductor to bring it from infinity where magnetic induction is zero to a position where it is $\mathbf{B_a}$ can be expressed by

$$W = -\int_0^{\mathbf{B_a}} \mathbf{M}.d\mathbf{B_a} \tag{17.58}$$

per unit volume. Using $\mu_o \mathbf{M} = \mathbf{B_a}$ and Eq.(17.58) we get change in potential energy density as

$$dU_s = TdS - \mathbf{M}.d\mathbf{B_a}$$

$$= TdS + \mathbf{B_a}.d\mathbf{B_a}/\mu_o \tag{17.59}$$

The change in the energy density of the superconductor at absolute zero when the magnetic induction increases from 0 to $\mathbf{B_a}$ is

$$U_s(\mathbf{B_a}) - U_s(0) = B_a^2/2\mu_o \tag{17.60}$$

If the metal is nonmagnetic i.e. M=0, the energy of the normal metal does not depend on field. Hence at critical value of magnetic induction $\mathbf{B_{ac}}$, $U_N(\mathbf{B_{ac}}) = U_N(0)$. At the critical magnetic induction $\mathbf{B_{ac}}$ the energies of superconducting and normal states are equal,

$$U_N(\mathbf{B_{ac}}) = U_s(\mathbf{B_{ac}}) = U_s(0) + B_{ac}^2/2\mu_o$$

Thus, the specimen is stable at either state at $\mathbf{B} = \mathbf{B_{ac}}$. Since the energy density of the normal conductor remains unchanged due to application of magnetic field, we can also write

$$U_N(0) - U_s(0) = B_{ac}^2/2\mu_o \tag{17.61}$$

which is the stabilization energy density of the superconducting material at absolute zero.

Example 17.9

For aluminium, B_{ac} is 0.0105T at absolute zero, find the stabilization energy density.

Solution:

Stabilization energy density = $(0.0105)^2/(2 \times 4\pi \times 10^{-7}) = 43.87$ J/m^3

Effect of magnetic field on type-I and type-II superconductors

Magnetization versus applied magnetic field for a pure superconductor shows Meissner effect i.e. perfect diamagnetism, Fig. 17.12. A superconductor with this type of characteristic is called type I superconductor. As the magnetic field increases above H_c, the specimen becomes normal conductor.

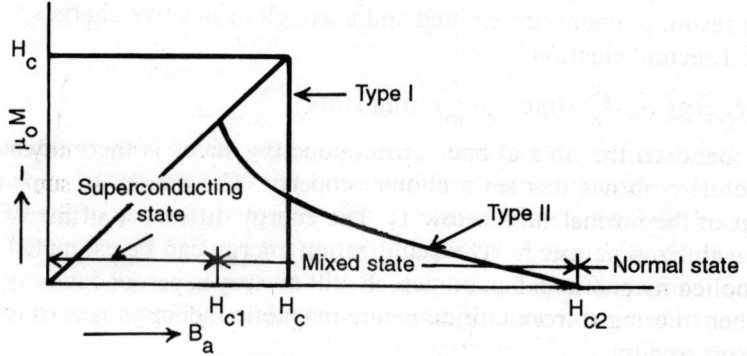

Fig. 17.12 Magnetization curves for (a) type I and (b) type II superconductors.

In case of type II superconductors, the magnetic flux starts penetrating the specimen at a field H_{c1}, which is lower than the thermodynamic value H_c. This type of superconductors have superconducting electrical properties up to the field H_{c2} but they do not show perfect diamagnetism for $H > H_{c1}$. In the range of magnetic fields H, $H_{c1} < H < H_{c2}$, the specimen is in vortex or mixed state wherein the superconductors are threaded by flux lines. Type II superconductors are also called hard superconductors in which there is a large amount of magnetic hysteresis or flux pinning and are used as core for magnet, e.g. an alloy of Pb- 8.23w/o In at 4.2°K.

High-current high magnetic-field material

The magnetic flux penetrates the ideal type II superconductors, when the applied magnetic field lies between H_{c1} and H_{c2} but these superconductors have very small J_c values. In certain superconductors flux is allowed to penetrate through the cores. These are also called vortices. Currents around the normal cores generate magnetic fields parallel to the applied field. These are fluxoids, which have magnetic moments due to which they repel each other and arrange themselves in a lattice called fluxon lattice.

The small value of J_c is actually due to the fact that the fluxoids are weakly tied to the crystal lattice and thus can move inside the crystal lattice causing resistance to the current flow. These fluxoids can be pinned by dislocations, grain boundaries and fine precipitates so that the current density is increased in these materials. Heavily cold worked and heat-treated materials have high J_c values. The examples of high current and high field superconducting magnetic materials are

45w/o Ni – 55w/o Ti, at $T_c = 9°K$ and Nb_3Sn at $T_c = 18°K$ and $H_{c2} = 11$ T

High current and high magnetic field superconductors are also used in magnetic resonance imaging (MRI) systems for medical investigations and for magnetic levitation of vehicles. Superconducting magnetic rings are used in particle accelerators.

17.10 BCS THEORY AND MICROSCOPIC PROPERTIES

The phenomena of superconductivity was first explained in 1957 by three American physicists John Bardeen, Leon Cooper and John Schrieffer, through their theory known as BCS Theory. Cooper proposed that the electrons paired into teams with the help of atomic lattice vibrations. The quanta of lattice vibrations are termed as phonons. According to this theory, as an electron passes by a positively charged ion in the lattice of the superconductor, the lattice becomes distorted. As a result, phonons are emitted and a trough of positive charges is formed around the electron. Then a second electron is drawn to the trough before the lattice goes back to its normal position. This process helps the two electrons to form a pair, called Cooper pair. The interaction between electrons and phonons is responsible for this linking up of two electrons, which remain separated by some distance. So a Cooper pair has a spatial extension. In terms of electron-phonon interaction one electron has emitted a phonon and the other electron has absorbed the phonon. The Cooper pairs are constantly breaking and reforming while the ions in the lattice are oscillating. The electron pairing is favourable because it has the effect of decreasing the energy of the system. The teams of Cooper pairs pass all the obstacles, which can cause resistance in the conductor. When the superconductor remains at very low temperatures, Cooper pairs remain intact. As a result of attractive interaction between electrons, the energy of the system is lowered and the entire electronic system have a ground state, which is below the lowest energy state of normal electrons by a discrete amount. This is known as superconducting energy gap. According to this theory a minimum energy $E_g = 2\Delta(T)$ is required to break a Cooper pair, creating two quasiparticle excitations. As the temperature increases the vibrations in the lattice also increase. As a result, the Cooper pairs may break and superconductivity decreases. The energy gap decreases monotonically with the increase of temperature and becomes zero as $T \to T_c$. In the vicinity of T_c,

$$\Delta(T) \approx 1.74 \, \Delta(0) \, [1 - (T/T_c)]^{1/2} \tag{17.62}$$

where
$$\Delta(0) = 1.764 \, k \, T_c \tag{17.63}$$

Thus, there is continuous decrease of superconducting gap parameter $\Delta(T)$ to zero value with the increase of temperature, which clearly indicates the characteristic of a second order phase transition. The superconducting materials become normal conductors above the transition temperature T_c.

Josephson Effect

Another important microscopic property of superconductors is the phenomenon of electron tunneling in superconductors. Brian Josephson discovered the tunneling of Cooper pair between two superconducting materials separated by thin layer (10Å – 20Å) of insulating material (e.g. oxide) without any resistance. This is known as dc Josephson effect. This is in contrary to what happens in ordinary materials, where a potential difference must be there for a current to flow. The above device is called Josephson junction. As long as the current through the junction is below a critical value the superconducting electrons tunnel through the barrier and there is no voltage drop across the junction. The critical current depends on the characteristics of the junction materials and its geometry. Josephson junction is a very fast switching device. Switching of voltages as performed by this device is ten times faster than the ordinary semiconducting circuits.

17.11 HIGH T_c SUPERCONDUCTORS

Superconductors with $T_c > 25°K$ are normally called as high T_c superconductors. Ba-La-Cu-O was the prototype high T_c superconductor, the forerunner of more remarkable advances that were yet to come.

Bednorz and Müller described possible high T_c superconductivity in Ba-La-Cu-O. The precise formula that gave the landmark of success was $Ba_xLa_{5-x}Cu_5O_{5(3-y)}$, where x varies from 0.75 to 1.0, with y >0. They observed that at about 80°K–100°K, the resistance began to decrease in an exponential fashion as the temperature was lowered further. When the temperature reaches 25°–30°K range, the resistance drops dramatically by as much as three orders of magnitude. A small diamagnetic effect is also seen.

The effect of high pressure in the processing of superconductors has the ultimate effect of increasing T_c. The highest T_c was observed at 30°K. The phenomenon was thought to be related to 2-D superconducting fluctuations of double perovskite layers of one of the phases. When a complex superconducting compound is discovered, the researchers almost· immediately be in substituting other elements. A substitution of strontium for barium in La-Ba-Cu-O was notably successful. Superconductivity was reported in $La_{1.8}Sr_{0.2}CuO_4$ at 36°K. This sample showed a smaller transition width of 1.4°K. These new samples showed Meissner effect of the order of 60–70% of the level expected for perfect diamagnet. These are clearly type II superconductors. The value of the slope of H_{c2} as a function of temperature near T_c suggests that the value of H_{c2} at 0°K may be of the order of $7.96 \times 10^6 - 1.194 \times 10^7$ Am^{-1}.

Later on, a compound $Y_{1.2}Ba_{0.8}CuO_{4-\delta}$ was discovered that showed zero resistance transition at a temperature just above 80°K; it showed 24% Meissner effect. The original sample was in mixed phase. With the discovery of ceramic oxide superconductors $YBa_2Cu_3O_{7-\delta}$, a gate of new hopes was opened 1987. The material was studied thoroughly. $YBa_2Cu_3O_{7-\delta}$ has a defective perovskite structure with three perovskite unit cells stacked on top of each other, Fig. 17.13, whereas δ varies from 0.35 to 0.1, the material shows the superconducting property. At $\delta = 0.1$, the critical temperature T_c is 90°K and at $\delta = 0.35$ the material is an insulator. Oxygen vacancies play an important role to decide the superconducting behaviour of the material.

When $YBa_2Cu_3O_{7-\delta}$ compound is slowly cooled from above 750°C in the presence of oxygen, it undergoes a transformation from tetragonal to orthorhombic crystal structure. If $\delta = 0.1$, the unit cell has lattice parameters a = 3.82 Å, b = 3.89 Å and c =11.69 Å. The study of correlation of oxygen vacancies with T_c shows that to achieve high T_c values, oxygen atoms on the (001) plane are arranged in such a way as to have oxygen vacancies in the 'a' direction. The superconducting behaviour in these compound arise because of the CuO_2 planes, which have oxygen vacancies providing the electron coupling between the CuO_2 planes. A recently found superconductor $HgBa_2Ca_2Cu_3O_{8+x}$ has T_c value 133°K.

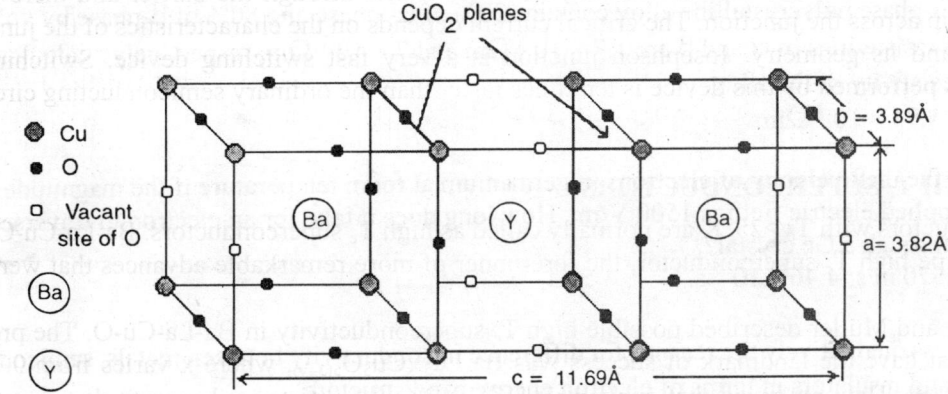

Fig. 17.13 $YBa_2Cu_3O_{7-s}$, a defective perovskite structure.

Future applications of high temperature superconductors include electrical transmission lines using superconducting wires, magnetic shielding devices, superconducting quantum interference devices (SQUIDS), signal processing devices, and infrared sensors. Wherever there is need of powerful magnet, superconducting magnets are of use there e.g. generators, levitated vehicle transportation, magnetic separators. The superconducting electronics has a great future with the advent of logic delays of 13 picoseconds and switching time of 9 picoseconds. The use of new superconducting films may result in more densely packed computer chips, which can transmit information several order faster than the existing computer chips. Applications of new high-temperature superconductors in multi-prong directions are expected to cross the door of scientific research laboratories very soon and enter the commercial world to fulfill the human need.

EXERCISES

17.1 (a) Compare the number of electrons at $E = E_f + 0.5eV$, for the temperatures 10°C and 40°C.
(b) Repeat for the number of electrons at $E = E_f - 0.5$ eV
Ans: (a) 0.14; (b) 1

17.2 A 0.72 mm diameter copper wire carries 0.96 A current. If the charge is carried by 1% of the 4s electrons, what is the drift velocity?
Ans: 1.74×10^{-2} m/s

17.3 A 0.16 inches diameter copper wire carries a current of 1.6 A. Calculate the electron flux.
Ans: 7.709×10^{23} electrons/m².s

17.4 The electronic conductivity and mobility of Cu are 6×10^7 $(\Omega m)^{-1}$ and 0.0030 m²/V.s, respectively at 27°C.
a) Compute the number of free electrons/m³ in Cu at room temperature.
b) What is the number of free electrons/Cu atom?
Ans: (a) 1.25×10^{29}/m³; (b) 1.5 electrons/Cu atom

17.5 A two-phase polycrystalline alloy contains 20% of α- phase and 80% of β-phase by volume. If the resistivities of α and β are 1.7×10^{-8} Ωm and 7×10^{-8} Ωm, respectively, calculate the net resistivity of the alloy.

Ans: 5.94×10^{-8} Ωm

17.6 Find the drift velocity of electrons in germanium at room temperature if the magnitude of the applied electric field is 1500 V/m . How long does it take for an electron to traverse 1 inch through a Ge crystal?

Ans: 570 m/s; 4.46×10^{-5} s.

17.7 Discuss qualitatively the reasons for difference in conductivity between metals, semiconductors and insulators in terms of electron energy band structure.

17.8 The composition of tin bronze is 81w/o Cu and 19w/o Sn. It consists of two phases at room temperature namely α containing very small amount of tin in solid solution and ε consisting of approximately 37 w/o of Sn. Calculate the conductivity of this alloy at room temperature. Given that:

Phase	Resistivity (Ωm)	Density (g/cm³)
α	1.88×10^{-8}	8.92
ε	5.32×10^{-7}	8.43

Ans: $3.407 \times 10^{6}(\Omega m)^{-1}$

17.9 The resistivity ratio of a conductor is defined as $\rho_{300}/\rho_{4.2}$ where ρ_{300} is the resistivity of He at 300°K and $\rho_{4.2}$ is its resistivity at the boiling temperature 4.2°K. Explain why the resistivity ratio may be used for analyzing the purity of a metallic material.

17.10 A current of 25A is to pass through a wire with a diameter of 0.25 cm, given the maximum power dissipation along the wire to be 6 W /m. Calculate the minimum allowable conductivity of the wire in $(\Omega m)^{-1}$.

Ans: 2.12×10^{7} $(\Omega m)^{-1}$

17.11 For an application involving 15A, the maximum voltage drop is to be 0.5 V/ m. What is the minimum diameter of Cu wire that is required for this application. Given that σ for commercially pure Cu is 5.85×10^{7} $(\Omega m)^{-1}$.

Ans: $D_{min} = 8.08 \times 10^{-4}$ m

17.12 Explain the variation of conductivity with temperature in metals as well as in semiconductors.

18

Semiconductors

18.1 INTRODUCTION

The electrical conductivity of a semiconductor is generally in the range of 10^{-7} to 10^4(ohm-m)$^{-1}$ at room temperature, which is intermediate between that of the conductors 10^7(ohm-m)$^{-1}$ and of the insulators (10^{-20} to 10^{-12} (ohm-m)$^{-1}$). The conductivity depends on the concentration of the current carriers. In case of metals, carriers are electrons whose concentration usually varies from 10^{28} to 10^{30}/m^3, whereas in case of semiconductors, carrier concentration is in the range of 10^{19} to 10^{23}/m^3. Thermal excitations, impurities or nonstoichiometry can produce changes in the electrical characteristics of conductors, semiconductors and insulators.

Lack of free carriers in the insulator is responsible for its high resistivity. The electrons are tightly bound to their bonding atoms by ionic or covalent bonding in insulators and are not free to conduct electricity unless they gain enough energy. The energy band structure of an insulator consists of a filled valence band and an upper empty conduction band. These bands are separated by a large energy gap E_g. The electrons in filled valence band cannot be excited to higher energy state i.e. cannot conduct when electric field is applied since there are no empty energy states in the band where they can jump. When electrons are given enough energy such that they are excited to upper conduction band, they contribute to the electrical conduction. The energy gap for diamond, an insulator, is 5 to 6 eV whereas that for silicon, a semiconductor, is 1.1eV and for metals, usually the valence band and conduction band overlap. In diamond, the sp^3 bonding keeps the electrons tightly bonded to atoms.

18.2 INTRINSIC SEMICONDUCTORS

Intrinsic semiconductors are pure semiconductors belonging to group IVA e.g. Si, Ge etc. There are semiconductors that are compounds like GaAs, CdS, InP etc. Pure semiconductors have diamond cubic structure with sp^3 covalent bonding. The bonding orbital consists of electron pairs. In this type of semiconductors, each atom contributes four valence electrons—one s and three p-electrons. Two-dimensional representation of the DC lattice of Si or Ge can be represented as shown in Fig. 18.1

The black circles in this representation are positive ion cores of Si atoms, each of which is bonded to four atoms through covalent bonding. The bonded electrons cannot contribute to the conductivity unless sufficient amount of energy is supplied to excite them from their bonding positions. When required amount of energy is supplied to the electron, it is excited from the valence band to the conduction band leaving a hole in the valence band. The holes are positively charged carriers and contribute to the conductivity.

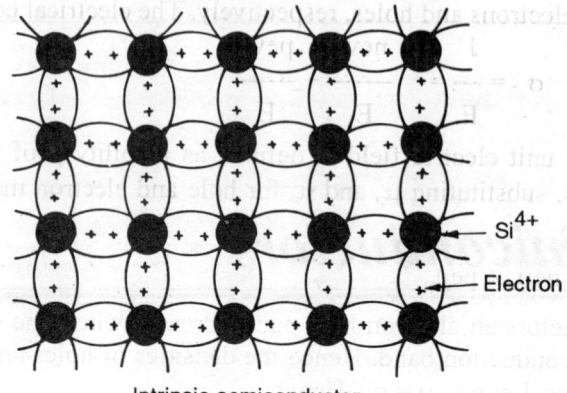

Intrinsic semiconductor

Fig. 18.1 Schematic representation of two-dimensional DC lattice of Si

Electrical conductivity

We have already discussed the mechanism of conduction in metals. When an electric field is applied, the free electrons in a metal are accelerated and contribute to electrical conduction. In case of intrinsic semiconductors, the applied electric field cannot always cause electrical conduction, since the electrons in the valence band have to gain enough energy, Fig. 18.2, which is equivalent to the band gap, to get excited to the conduction band i.e.

$$eV > \text{band gap} \tag{18.1}$$

where e is the electronic charge, and V is the potential difference through which electrons are excited. The valence electrons of covalently bonded semiconductor crystal fill up the energy levels in the valence band, following the Pauli exclusion principle. There is band gap above the valence band, where there are no allowed energy states for electrons. This band gap is usually absent or is of negligible value in case of conductors whereas in case of semiconductors it is of the order of electron volt.

The electric current in semiconductors arises due to   both the carriers electrons and holes and thus current density J is expressed as,

$$J = nev_e + pev_p, \tag{18.2}$$

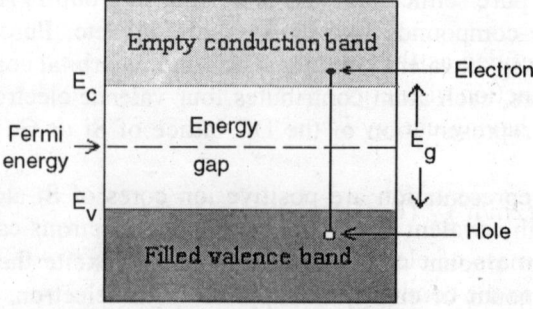

Fig. 18.2 Schematic representation of energy band diagram of intrinsic semiconductor with Fermi energy at $0°K$

where n, p are the number of conduction electrons and holes per unit volume, respectively; v_e, v_p are drift velocities of electrons and holes, respectively. The electrical conductivity is given as

$$\sigma = \frac{J}{E} = \frac{nev_e}{E} + \frac{pev_p}{E} \tag{18.3}$$

The drift velocity per unit electric field is defined as mobility μ of the carrier. The unit of mobility is $m^2/V.s$. Thus, substituting μ_p and μ_e for hole and electron mobilities, the expression for conductivity becomes

$$\sigma = ne\mu_e + pe\mu_p \tag{18.4}$$

In intrinsic semiconductors an electron-hole pair is created when one electron is excited from the valence band to the conduction band. Hence the densities of holes and electrons are equal in an intrinsic semiconductor. Let $n = p = n_i$. Hence

$$\sigma = n_i e (\mu_p + \mu_e) \tag{18.5}$$

The mobility of an electron is much larger than that of a hole. For example, the mobility of an electron is $0.14 \ m^2/V.s$ whereas that of a hole is $0.05 \ m^2/V.s$ in silicon. The number of electrons excited across the gap can be calculated from the Fermi-Dirac probability distribution, Eq.(17.20):

$$f(E,T) = \frac{1}{1 + \exp[(E - E_f)/kT]}$$

where Fermi energy E_f for intrinsic semiconductor lies midway in the band gap at $0°K$, Fig. 18.2. The probability of finding the electron in the valence band is 100% if the energy of the electron E $< E_f$ at $0°K$. Hence at $0°K$

$$f(E,0) = 1 \qquad \qquad \text{if } E \leq E_f$$
$$= 0 \qquad \qquad \text{if } E > E_f$$

As the temperature increases the probability of an electron to make transition from valence band to conduction band increases. The probability of finding the electron in the conduction band $(E > E_f)$ increases and it is observed to be 50% at E_f for $T > 0°K$ although the Fermi energy level lies in the forbidden band gap, Fig. 17.4. To find the relationship between conductivity and temperature let us first define density of states.

Density of states

The density of states is defined as the number of energy states per unit energy interval and per unit volume. It is also a function of energy. From quantum mechanical considerations, the density of states per unit volume having energy lying between E and E + dE in the conduction band is given by

$$N(E) = (2m/\hbar^2)^{3/2} (E - E_C)^{1/2} /2\pi^2 \tag{18.6}$$

where m is the effective mass of electron. As the temperature increases, some of the electrons from valence band excite to the conduction band. The energy of these electrons varies from E_g to ∞, when the zero of the energy scale is chosen to be at the bottom of the energy gap i.e. at E_V. Hence E_C is equal to E_g. These excited electrons are called conduction electrons. The number of conduction electrons in the conduction band is equal to the number of holes created in the valence band. With this choice of energy scale, the number of conduction electrons at temperature T is given by

$$n = \int_{E_g}^{\infty} N(E)\, f(E)\, dE$$

$$n = \int_{E_g}^{\infty} [(2m/\hbar^2)^{3/2}\, (E-E_g)^{1/2}/2\pi^2]/[1+\exp\{(E-E_g+E_g-E_f)/kT\}]\, dE \qquad (18.7)$$

When $(E - E_f) \gg kT$,

$$n \approx \int_{E_g}^{\infty} [(2m/\hbar^2)^{3/2}\, (E-E_g)^{1/2}/2\pi^2]\, \exp-\{(E-E_g)/kT\}\, \exp-\{(E_g-E_f)/kT\}\, dE \qquad (18.8)$$

As E_g is smaller than ∞, we take the lower limit of integration to be zero. Substituting $y=\{(E-E_g)/kT\}$, and using the integral

$$\int_{0}^{\infty} y^{1/2}\, \exp(-y)\, dy = \sqrt{\pi}/2 ,$$

we get, $\quad n = 2[mkT/2\pi\hbar^2]^{3/2}\, \exp-\{(E_g-E_f)/kT\} = 2[mkT/2\pi\hbar^2]^{3/2}\, \exp-\{(E_C-E_f)/kT\}$

$$\approx N_C \exp(-E_g/2kT), \qquad (18.9a)$$

where $E_f = E_g/2$ near $0°K$ and $N_C = 2[mkT/2\pi\hbar^2]^{3/2}$ is known as the effective density of states in the conduction band. Putting the values of various constants we get $N_C \approx 4.82 \times 10^{21}\, T^{3/2}$ electrons/m³. If we assume $N(E)$ to be constant and is equal to N, then

$$n = NkT \exp[-E_g/(2kT)] \qquad (18.9b)$$

Now $\qquad \ln n = \ln N_C - E_g/(2kT) \approx -E_g/(2kT), \qquad (18.10)$

since $\ln N_C$ is small compared to the term containing $1/kT$. The size of the energy gap can be calculated by plotting $\ln n$ versus $1/T$ graph. The slope of the graph is $E_g/2k$.

Considering the conductivity σ arising out of conduction of electrons, one can get the expression of σ combining Eqs. (18.4) and (18.9a)

$$\sigma = ne\mu_e = N_C\, e\mu_e\, e^{-E_g/2\,kT} \qquad (18.11)$$

$$= NkTe\mu_e\, e^{-E_g/2\,kT}$$

So $\qquad\qquad \ln \sigma = \ln \sigma_0 - E_g/2kT$

where $\sigma_0 = eN\mu_e kT$ can be determined experimentally from the intercept along y-axis of $\ln \sigma$ versus $1/T$ graph, neglecting its slight dependence on temperature. We consider σ_0 to be constant.

Table 18.1 Electrical conductivity of group IVA elements

Element	Electrical conductivity $(\Omega.m)^{-1}$	Energy gap (E_g) (eV)	Electron mobility (μ_e) $(m^2.V^{-1}.s^{-1})$	Hole mobility (μ_p) $(m^2.V^{-1}.s^{-1})$
C(Diamond)	$< 10^{-16}$	5.4	0.18	0.14
Si	5.0×10^{-4}	1.11	0.19	0.05
Ge	2.0	0.67	0.38	0.18
Sn	9.0×10^{6}	0.8	0.25	0.24

Similarly, the hole concentration in the valence band can be calculated as

$$p = \int_{-\infty}^{E_v} N(E)(1 - f(E)) \, dE \qquad (18.12)$$

$$= \int_{-\infty}^{E_v} [(2m/\hbar^2)^{3/2} (E_v - E)^{1/2}/2\pi^2] \left[\frac{\exp\{(E - E_f)/kT\}}{[1 + \exp\{-(E_f - E)/kT\}]} \right] dE$$

$$p \approx \int_{-\infty}^{0} [(2m/\hbar^2)^{3/2} (E_v - E)^{1/2}/2\pi^2] \exp\{-(E_v - E + E_f - E_v)/kT\} dE \qquad (18.13)$$

Taking $E_v = 0$ in the energy scale and $(E_f - E) >> kT$. Thus,

$$p \approx N_V \exp(- E_g/2kT), \quad N_V = 2[m_h kT/ 2\pi\hbar^2]^{3/2},$$

where m_h is the effective mass of hole. The value of N_V is $4.82 \times 10^{21} \, T^{3/2}$ holes/m^3. Conductivity due to holes is given by

$$\sigma = pe\mu_p = N_V e\mu_p e^{-Eg/2 kT} = 4.82 \times 10^{21} \, T^{3/2} \, e\mu_p e^{-Eg/2 kT}$$

The product np is given by

$$np = [4.82 \times 10^{21}]^2 \, T^3 \exp(-E_g/kT) = n_i^2$$

If we assume N(E) to be constant and is equal to N, $np = [NkT]^2 \exp(- E_g/kT) = n_i^2$

$$n_i = NkT \exp(- E_g/2kT)$$

This is valid for intrinsic semiconductor since $n = p = n_i$, where n_i is concentration of intrinsic carriers.

Example 18.1

The energy gap of pure silicon is 1.1 eV. Compare the number of conduction electrons at 27°C and 180°C (k = 8.63×10^{-5} eV/ °K). Calculate the conductivity at 180°C. Given the electrical conductivity to be 4.3×10^{-4} $(\Omega m)^{-1}$ at room -temperature. Assume N(E) to be constant.

Solution:

Using Eq. (18.9b) we get

$$\frac{n_e \text{ (at } 27^0 \text{ C)}}{n_e \text{ (at } 180^0 \text{ C)}} = \frac{Nk \text{ (300) } exp(-0.55/(k \times 300))}{Nk \text{ (453) } exp(-0.55/(k \times 453))}$$

$$= \frac{300}{453} \ exp \left[-\frac{0.55}{k} \left[\frac{1}{300} - \frac{1}{453} \right] \right] = 5.069 \times 10^{-4}$$

$$\frac{\sigma_{300^\circ K}}{\sigma_{453^\circ K}} = \frac{\sigma_o \ e^{-Eg/2k(300)}}{\sigma_o \ e^{-Eg/2k(453)}}$$

$$= exp \left[-\frac{0.55}{k} \left[\frac{1}{300} - \frac{1}{453} \right] \right] = 7.65 \times 10^{-4}$$

$$\sigma_{453^\circ K} = \frac{4.3 \times 10^{-4}}{7.65 \times 10^{-4}} \ (\Omega m)^{-1} = 0.562 \ (\Omega m)^{-1}$$

Example 18.2

Consider a 1 cm x 2 mm x 1 mm bar of pure silicon at 27^0C. The mobility of electrons and holes in silicon are 0.135 m^2/V.s and 0.048 m^2/V.s, respectively. Calculate the number of charge carriers in the bar, fraction of electrons excited to the conduction band. Given that $\sigma = 4.3 \times 10^{-4}$ $(\Omega m)^{-1}$ and a for silicon unit cell is 5.43×10^{-10}m.

Solution:

We know from Eq. (18.5)

$$n_i = \frac{\sigma}{e \ (\mu_p + \mu_e)} = \frac{4.3 \times 10^{-4}}{1.6 \times 10^{-19} \times 0.183} = 1.47 \times 10^{16}/m^3$$

In this bar since $n_i = n = p$, we get

$$n = (1.47 \times 10^{16})/m^3 \times (2 \times 10^{-8} \ m^3) = 2.94 \times 10^8$$

Total number of electrons in the valence band of silicon

$$= \frac{(8 \ \text{atoms/unit cell}) \ x \ (4 \ \text{electrons/atom})}{(5.43 \times 10^{-10} m)^3} = 2.00 \times 10^{29} /m^3$$

$$\text{Fraction excited to conduction band} = \frac{1.47 \times 10^{16}}{2.00 \times 10^{29}} = 7.35 \times 10^{-14}$$

18.3 EXTRINSIC SEMICONDUCTORS

Suitable impurities are doped in a pure semiconductor, to increase its conductivity such that it is useful for device applications. These atoms have valencies different from that of the host atoms and are called dopants. The solid solutions so formed are called extrinsic semiconductors, which have energy levels in the energy gap due to presence of dopants, Table 18.2 where zero of energy scale is chosen at $E=E_v$. So, the position of the Fermi level changes in the extrinsic semiconductor as compared to the intrinsic semiconductor. Fig. 18.3 shows the impurity energy level in silicon considering top of the valence band as zero of the energy scale.

Table 18.2 Impurity energy levels

Impurity Element	Energy level in eV	Type
Boron	0.045	Acceptor
Aluminium	0.057	--do--
Gallium	0.065	--do--
Antimony	1.07	Donor
Phosphorus	1.056	--do—

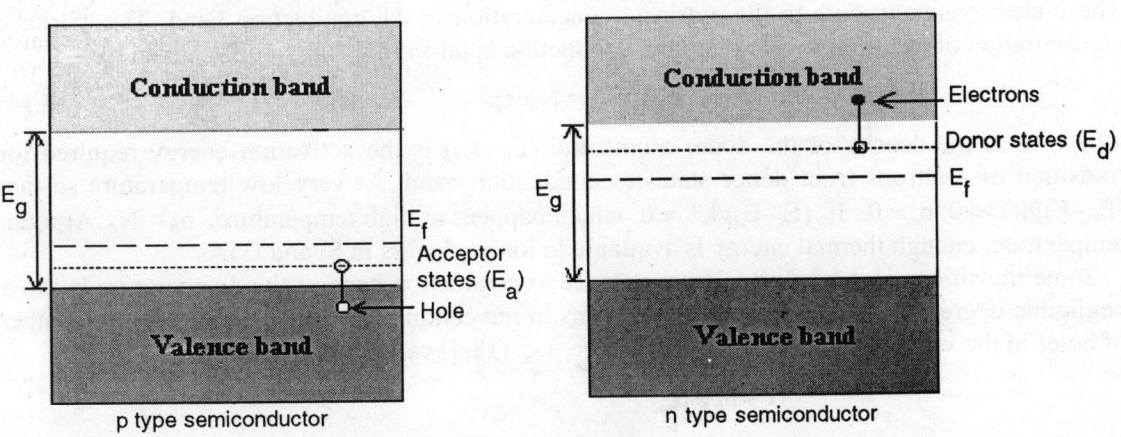

Fig. 18.3 Energy bands of extrinsic silicon semiconductor at T > 0°K

n-type semiconductors

If a small amount of phosphorous or some other group V element is doped in pure silicon, the phosphorous atoms occupy the positions of silicon atoms in the host matrix. Phosphorous atoms form covalent bonding with near by silicon atoms, Fig. 18.4, but one extra electron with each phosphorous atom, which is only loosely bound to the phosphorous atom, contributes to the conductivity. Such dopants are donors. The number of excess electrons is equal to the number of impurity atoms added. This type of material is called n-type semiconductor. Since these excess electrons have more energy than those electrons forming covalent bonding, they occupy energy levels called donor energy states close to conduction band. The Fermi energy level in n-type semiconductor lies midway between the topmost donor state and lowest energy state of the conduction band at 0°K. As the temperature increases, the Fermi energy increases first very little and then decreases, moving towards the centre of the energy gap. At a specific temperature when the donor atoms are completely ionized, the Fermi energy level approaches the intrinsic Fermi energy value.

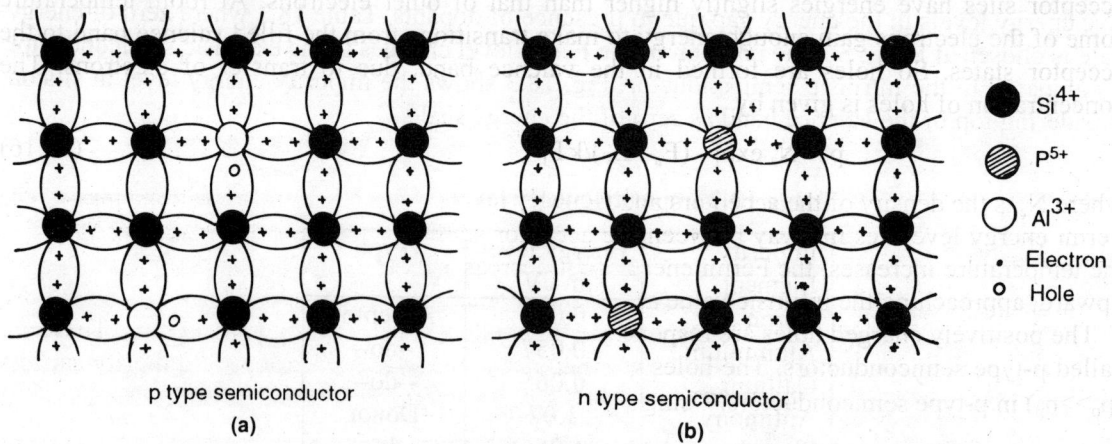

p type semiconductor
(a)

n type semiconductor
(b)

Si^{4+}

P^{5+}

Al^{3+}

Electron

Hole

Fig 18.4 Schematic representation of extrinsic semiconductor (a) p-type (b) n-type

The electrons in the donor states are excited to conduction band even at the room temperature. These electrons contribute to the extrinsic concentration in the conduction band. The extrinsic concentration of electrons excited into the conduction band from donor states is given by

$$n_n = N_d \exp[-(E_c - E_d)/kT] = N_d \exp[-(E_g - E_d)/kT] \ , \tag{18.14}$$

where N_d is the density of the donor atoms and $(E_c - E_d)$ is the activation energy required for transition of electrons from donor states to conduction band. At very low temperature so that $(E_c - E_d)/kT \gg 0$, $n_n \approx 0$. If $(E_c - E_d)/kT \approx 0$, which happens at high temperature, $n_n = N_d$. At room temperature, enough thermal energy is available to ionize donors in Si and GaAs.

Some transitions also take place from valence to conduction band at this temperature but to a negligible degree. Thus, the number of electrons in the conduction band far exceeds the number of holes in the valence band (ie $n_n \gg p_n$). Hence, Eq. (18.4) can be approximated as,

$$\sigma = n_n e \mu_e,$$

$$\ln \sigma = \ln N_d e \mu_e - (E_g - E_d)/kT \tag{18.15}$$

where n_n and p_n are electron and holes densities, respectively, in n-type semiconductor. The negatively charged carriers (i.e. electrons) are majority carriers in this type of semiconductors and holes are minority carriers.

p-type semiconductors

When elements like aluminium, boron or gallium of group III A are doped in silicon or germanium the substitutional solid solution thus produced is again extrinsic semiconductor. Each of these impurity atoms forms covalent bonds with four silicon atoms such that one of the covalent bonds of each impurity atom is deficient in an electron. The above can be viewed as a hole, which is weakly bound to the impurity atom. An electron from an adjacent bond may be attracted towards the hole and occupy the hole position under the influence of external energy and the hole, in turn, gets shifted to the electron position in the bond, which is vacated. Such dopants are acceptors. In essence, the electron and hole exchange positions i.e. they move in the opposite directions. A moving hole is considered to be an excited carrier, Fig 18.4(a). Group III A elemental atoms provide acceptor sites for electrons which come from silicon atoms. The acceptor sites have energies slightly higher than that of other electrons. At room temperature some of the electrons gain enough energy to make transitions from the filled valence band to the acceptor states. So holes are formed in the valence band, due to transfer of electrons. The concentration of holes is given by

$$p = N_a \exp[-(E_a - E_V)/kT] \qquad (18.16)$$

where N_a is the density of the acceptors and $(E_a - E_V)$ is the corresponding activation energy. The Fermi energy level lies midway between the acceptor states and the valence band at $0°K$. When the temperature increases, the Fermi energy first decreases by a small amount and then increases upward, approaching the intrinsic value as the acceptor atoms are fully ionized.

The positively charged holes are responsible for conduction in this type of materials. These are called p-type semiconductors. The holes are majority carriers and electrons are minority carriers ($p_p \gg n_p$) in p-type semiconductors. Using Eq. (18.4), we get the expression of σ as

$$\sigma = p_p e \, \mu_p, \qquad (18.17)$$

where p_p and n_p are holes and electron densities in p-type semiconductors. In case of n-type semiconductors electrons get exhausted from the donor state, whereas in p-type semiconductor acceptor state gets saturated with electrons.

To summarize, the conduction and valence bands have two sources of charge carriers – one from thermal generation and other from ionized donor or acceptor. In case of intrinsic semiconductors the temperature dependence of the electron concentration in the conduction band is expressed as

$$n = 4.82 \times 10^{21} \, T^{3/2} \exp[-(E_g - E_f)/kT] \qquad (18.18)$$

and hole concentration in the valence band is expressed as

$$p = 4.82 \times 10^{21} \, T^{3/2} \exp(-E_f/kT) \qquad (18.19)$$

For an intrinsic semiconductor, $n = p = n$. Equating $n = p$ we get $E_f = E_g/2$. As the temperature increases in the extrinsic semiconductor, generation and recombination of conduction electrons and holes takes place.

18.4 FERMI DISTRIBUTION IN EXTRINSIC SEMICONDUCTORS

In the extrinsic semiconductor, the concentration of carriers arises from intrinsic as well as extrinsic sources. The Fermi distribution is not symmetrical across the energy gap in case of extrinsic semiconductors. This is because of the presence of impurity levels near the valence or conduction bands. Hence the Fermi energy is not in the middle of the gap, Fig. 18.5.

460 Material Science for Engineers

The Fermi distribution of electrons in various energy states occurs as a result of equilibrium between the rate of generation and recombination of conduction electrons and holes. The equilibrium constant K, for above equilibrium reaction is given by

$$K = np. \qquad (18.20)$$

where K depends on the energy gap, the temperature and the amount of incident radiation. The concentrations n, p, refer to that of extrinsic semiconductors. The above expression is also called law of mass action where $K = n_i^2$ i.e.

$$np = n_i^2 \qquad (18.21)$$

n_i being the concentration of carriers in the intrinsic semiconductor.

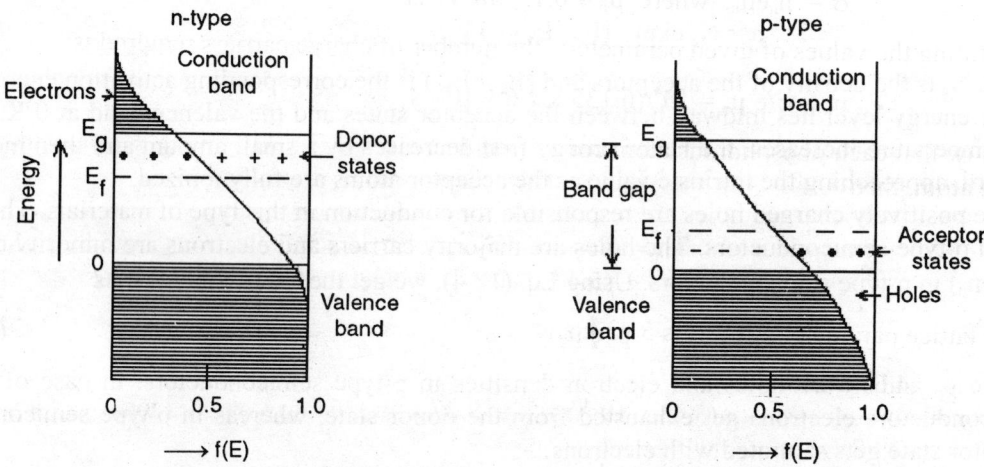

Fig 18.5 Fermi distribution in extrinsic semiconductor

Charge density in impurity semiconductors

We know that any compound or element is electrically neutral which implies that the negative charge density is equal to positive change density. In case of a p-type semiconductor, there are electrons and acceptor ions which carry negative charges and holes carry positive charges. In case of for n-type semiconductors donor ions and holes are positive charge carriers and electrons are negative charge carriers. In general, at equilibrium for an extrinsic semiconductor,

$$N_a + n = N_d + p, \qquad (18.22)$$

where N_a and N_d are densities of acceptor and donor ions respectively. For n-type semiconductor the relation reduces to

$$n_n = N_d \qquad (18.23)$$

Using the relation of law of mass action, Eq. (18.21), we get

$$p_n = n_i^2/N_d \qquad (18.24)$$

Similar relation for p-type semiconductor is

$$p_p = N_a \qquad (18.25)$$

and $\qquad\qquad n_p = n_i^2/N_a \qquad\qquad$ (18.26)

Example 18.3

Design a n-type semiconductor based on silicon which provides constant conductivity of $10^4 \; \Omega^{-1}.m^{-1}$.

Solution:

The elements of group V like P, As etc. are doped in silicon to design n-type semiconductor.

$$\sigma = n_n e \mu_n, \quad \text{where} \quad \mu_n = 0.135 \; m^2 V^{-1}.s^{-1}$$

Substituting the values of given parameters, the number of charge carriers required is:

$$n_n = \sigma/e \; \mu_n = 10^4/(0.135 \times 1.6 \times 10^{-19}) = 4.63 \times 10^{23} \; \text{electrons/m}^3$$

We know that each phosphorous atom contributes one electron and if there are x doped atoms per silicon atom, then

$$n_n = \frac{8x}{(5.43 \times 10^{-10}m)^3}$$

where lattice parameter a for Si is 5.43Å and there are 8 atoms per unit cell.

$$x = \frac{4.63 \times 10^{23}}{8} \times (5.43 \times 10^{-10}m)^3 = 9.27 \times 10^{-6} \; \text{dopants per Si atom}$$

There is one dopant per 1.08×10^5 Si atoms.

Example 18.4

A silicon based p-type semiconductor material has majority carrier concentration equal to 5.0×10^{21} holes/m^3 and total impurity (consisting of boron and phosphorus atoms) concentration is $2.5 \times 10^{22}/m^3$, calculate (a) The equilibrium electron concentration, (b) The concentration of boron and phosphorus atoms. $n_i($ Si$) = 1.5 \times 10^{16}$ electrons/m^3.

Solution:

a) The equilibrium electron concentration

$$n_p = \frac{n_i^2}{p_p} = \frac{(1.5 \times 10^{16}/m^3)^2}{(5 \times 10^{21}/m^3)} = 4.5 \times 10^{10} \; \text{electrons/m}^3$$

b) The total concentration of boron (N_a) and phosphorus (N_d) ions $= N_a + N_d = 2.5 \times 10^{22}/m^3$, where boron atoms act as acceptors and phosphorus as donors..

The concentration of holes $= N_a - N_d = 5.0 \times 10^{21}/m^3 \qquad$ as $p_p \gg n_p$

Concentration of boron atoms, $N_a = 1.5 \times 10^{22}/m^3$

Concentration of phosphorus atoms, $N_d = 10^{22}/m^3$

18.5 EFFECT OF TEMPERATURE ON CONDUCTIVITY OF EXTRINSIC SEMICONDUCTORS

The density of charge carriers is directly proportional to the conductivity. As the temperature increases, more and more number of impurity atoms are ionized, thereby increasing the number of charge carriers. Thus, the electrical conductivity of extrinsic semiconductor increases with the temperature.

In case of intrinsic semiconductors, the energy required to excite an electron is equal to E_g, whereas it is only $E_g - E_d$ in case of n-type semiconductors, where topmost level of valence band is taken as zero and donor level is E_d and lowest level of conduction band is E_g. Thus, relatively smaller amount of energy is required to excite electrons in n-type semiconductor than that for intrinsic one. One can plot ln σ versus (1/T) °K^{-1} to obtain Arrhenius relationship, which is a straight line and from which E_g, E_d etc. can be calculated. The slope of this line is - $(E_g - E_d)/$ k. When all the impurity atoms of the extrinsic semiconductor are ionized, the rise in temperature cannot affect the conductivity much because the carrier density does not change. In case of n-type semiconductors the range of temperature for which the electrical conductivity remains constant is referred to as the exhaustion range. In p-type semiconductors, the acceptor impurity level (E_a) is above the valence band. The energy required to excite an electron from valence band to the acceptor level is given by the energy gap $E_a - E_v = E_a - 0$. The slope of ln σ vs 1/T °K^{-1} curve for p-type semiconductor is $-E_a/k$. For a certain temperature range after all the impurity atoms are ionized, the electrical conductivity does not change since there is no change in the carrier density. This is referred to as the saturation range for p-type semiconductors. This temperature range provides constant electrical conductivities for operation. Hence exhaustion and saturation ranges are important for a device.

As the temperature is increased beyond these ranges, the electrons in the valence band get enough activation energy to jump to the conduction band so as to contribute to electrical conductivity. At this temperature range the intrinsic conduction becomes predominant and the ln σ vs (1/T)°K^{-1} curve, has got much steeper slope given by $-E_g/2k$, Fig. 18.6.

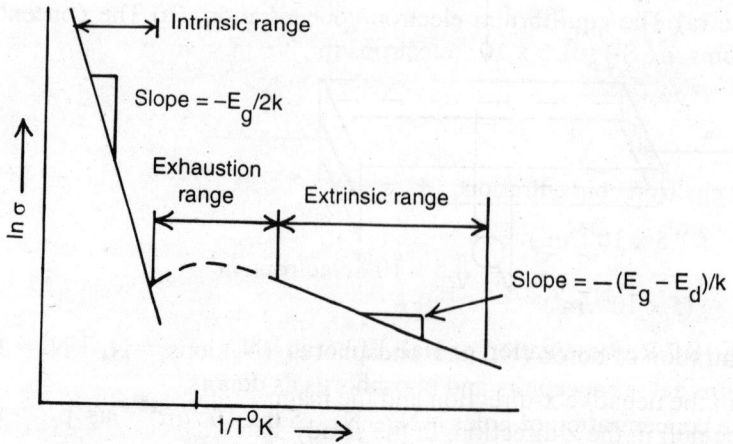

Fig. 18.6 Schematic representation of ln σ vs 1/T for a n-type semiconductor

18.6 EXCESS CARRIER LIFETIMES

When light or thermal energy is incident on a semiconductor crystal, electron-hole pairs are generated continuously. The number of electron-hole pairs generated per unit volume per unit time due to thermal breakage of covalent bonds is thermal generation rate. The other process, which also occurs simultaneously, is the recombination of electrons and holes. In case of thermal equilibrium, the thermal generation rate is equal to the recombination rate. When these rates are not equal the number of electron-hole pairs will decrease or increase with time.

When the source of radiation is removed suddenly, the number of conduction electrons and holes i.e. np becomes higher than the new equilibrium constant K. Now some of the electrons may jump from the acceptor level to valence band and recombine with holes in case of p-type semiconductor. Also electrons from conduction band may be de-excited to the donor level to recombine with holes for n-type semiconductor. The reverse reactions take place till np equals K. It takes time for excess carriers to decrease. The rate of decrease of electrons is given by,

$$-\frac{dn}{dt} = \frac{n}{\tau_c}$$

or $$n = n_0 \exp[-t/\tau_c]$$ (18.27)

The factor n_0 is the excess carrier density at time $t = 0$ and τ_c is the lifetime of the carriers.

18.7 HALL EFFECT

Study of Hall effect reveals the nature of charge carriers in the material. While deriving the expression of Hall effect, the sign of charge carrier is explicitly taken into account. When a charge flows in a semiconducting rectangular bar in a direction perpendicular to the magnetic field, it is deflected due to the Lorentz force given by

$$\mathbf{F} = e \, \mathbf{v} \times \mathbf{B}$$ (18.28)

where $\mathbf{v}$ is the velocity of the moving charge, Fig.18.7. When a current I_x flows in the x- direction

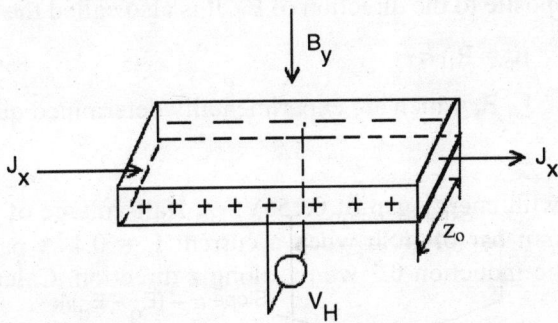

Fig 18.7 Representation of Hall Effect in a n-type semiconductor

i.e. electrons flow in the negative x-direction and the magnetic field points in the y-direction, the electrons experience pull in the z-direction. In the steady state, there is no net force acting on the electrons, so an electrostatic field E_z is generated which exactly balances the Lorentz force. Hence the net force along z-direction is zero,

$$F_z = -ev_xB_y - eE_z = 0$$

so
$$E_z = -v_xB_y,$$ (18.29)

E_z is called the Hall field, E_H. The current density J_x is given by

$$J_x = n_eev_x$$ (18.30)

Thus, using (18.29) and (18.30)

$$E_H = -\frac{J_x}{n_ee}B_y$$ (18.31)

or
$$\frac{E_H}{J_x B_y} = -\frac{1}{n_ee}$$ (18.32)

where the factor on the right-hand side of Eq.(18.32) is called the Hall coefficient R_H. So

$$R_H = -1/n_ee$$ (18.33)

In case of holes in p-type semiconductor

$$R_H = 1/p_ee$$ (18.34)

The Hall field can be calculated by measuring the Hall voltage V_H which develops across the sample due to deflection of charge carriers under the influence of Lorentz force,

$$V_H = E_Hz_0$$ (18.35)

where z_0 is the thickness of the sample along z-direction. The sign of V_H changes when the charge carriers are holes or any other type of positive charges. Thus, the nature of the majority charge carrier in a semiconductor can be identified from the sign of the Hall voltage. If E_x is the applied field to cause the flow of current, the mobility of the electron can be expressed as,

$$\mu_e = v_x/E_x = -E_z/\{E_xB_y\} = -E_z\sigma/\{J_xB_y\}$$ (18.36)
$$= -R_H\sigma$$

considering the flow of electrons opposite to the direction of E_x. It is also called the Hall mobility.

Similarly for holes $\qquad \mu_p = R_H\sigma$ (18.37)

where R_H can be estimated from E_H, J_x, B_y which are experimentally determined quantities.

Example 18.5

InSb is an intrinsic semiconductor with energy gap of 0.15eV. A Hall voltage of – 4.07 mV is developed across 1cm x 1mm x 1mm bar of InSb when a current $I_x = 0.1$ A passes along x-direction in the presence of magnetic induction 0.1 wb/m^2 along z-direction. Calculate the Hall coefficient and density of carrier.

Solution:

$$E_H = V_H/10^{-3} \text{ V.m}^{-1} = -4.07 \text{ Vm}^{-1}$$

Also
$$\frac{E_H}{J_x B_z} = -R_H$$

So $\qquad$ $R_H = 4.07/\{(0.1/10^{-6}) \times 0.1\}$ m^{3}C^{-1} = 4.07 $\times$ 10^{-4} m^{3}C^{-1}

Since $\qquad$ $R_H = 1/ne$

we get $\qquad$ $n = 1/(4.07 \times 10^{-4} \times 1.6 \times 10^{-19})/m^3$ = 1.54 $\times$ 10^{22} m^{-3}

18.8 SEMICONDUCTING DEVICES

The electronic structure of semiconductor is responsible for its unique electrical properties. Diodes and transistors are the two important devices of semiconductors.

p-n junction diode

Ideally it is made up of p and n-type semiconductors, which are joined together to form a junction, Fig. 18.8. Both types of semiconductors are electrically neutral before the formation of p-n junction. In n-type semiconductors, the electrons are majority carriers and Fermi level shifts near donor states from the middle of the energy gap. In p-type, the holes are majority carriers and

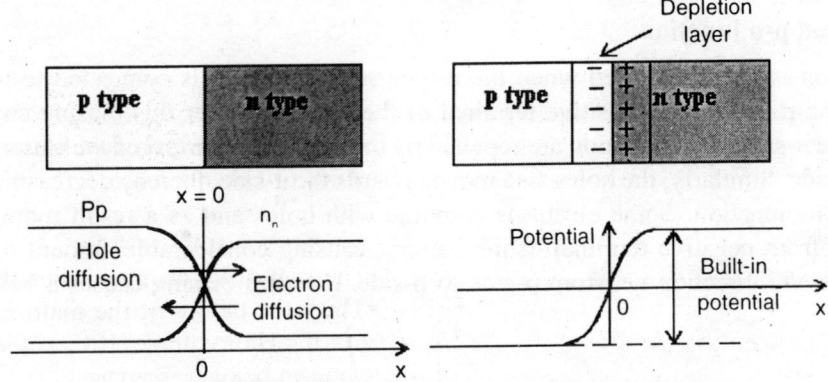

Fig. 18.8 Representation of the contact potential

the Fermi level shifts toward the acceptor states from the middle of the energy gap. At thermal equilibrium, the Fermi level has to be constant throughout a p-n junction. The p-n junction is usually formed by doping different types of impurities from two sides of the same crystal wafer such that one side becomes n-type other side becomes p-type. At thermal equilibrium, the electronic energy level at the bottom of the conduction band in the n-side is lower than that in p-side by an amount eV_o. V_o is the built-in-voltage of the diode. Formation of such potential difference can be explained as follows:

As soon as the junction is formed, the majority carriers i.e. electrons from n-side diffuse to the p-side whereas the holes from p-side diffuse to the n-side. Since donor and acceptor ions are heavier and larger, they cannot diffuse easily through p-n junctions. As a result, there is depletion of electrons in a small region near the junction on the n-side and similar depletion of holes on the p-side. Thus, a thin layer of n-region next to the junction becomes positively charged due to excess of ionized donors whereas that of the p-side becomes negatively charged due to presence of ionized acceptors. Under open circuit condition, a potential difference is created across p-n junction due to depletion of majority carriers, which resists the flow of majority carriers. So there is no net current flow in the open circuit condition. This potential difference is the measure of built-in voltage of diode and is also called contact potential. The layer near the junction, where there is depletion of majority carriers, is called the depletion layer.

Reverse-biased p-n junction

When the external voltage is applied across the p-n junction, it is said to be biased. The junction is said to be reverse-biased, if the potential applied across the junction is of same sign as that of the potential barrier. This happens when n-type semiconductor side is connected to the positive terminal and p-type material is connected to the negative terminal of the battery. Since the majority carriers move away from the junction under this biasing condition, the width of the depletion layer increases. The current density of the majority carriers decreases. However, the minority carriers (holes in n-type and electrons in p-type semiconductors) flow across the junction so that they combine with oppositely charged carriers and cause a small reverse current flow. The magnitude of the minority current is of the order of microamperes, Fig. 18.9. This is also called leakage current. Small reverse biases of a few kT/e reduce the minority carrier concentrations to zero at the edges of the depletion region.

Forward-biased p-n junction

The p-n junction is forward biased when the n-type semiconductor is connected to the negative terminal and the p-type to the positive terminal of the battery. Under this biasing condition, the electrons on the n-side of the junction are repelled by the negative terminal of the battery and move towards the p-side. Similarly, the holes also move towards the n-side, thereby decreasing the potential barrier at the junction. Some electrons combine with holes and as a result more number of electrons flow from negative terminal of the battery, causing considerable amount of current to flow in the forward direction i.e. from p-side to n-side. Forward biasing causes a building up of

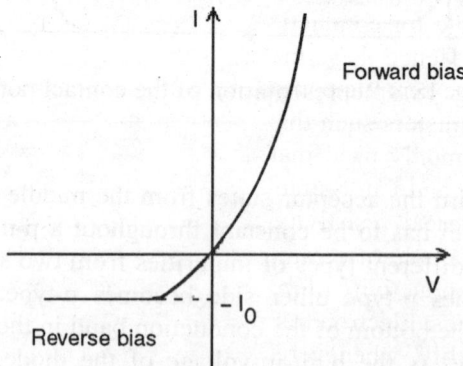

Fig 18.9 I-V characteristics of p-n junction diode

building up of minority carriers near the edges of the depletion region. The electron flow continues from n- to p-side as long as the p-n junction is forward biased.

The variation of diode current I with applied V, Fig. 18.9 can be expressed as,

$$I = I_o(\exp(eV/kT) - 1), \tag{18.38}$$

where
$$I_o = e[D_n n_p/L_n + D_p p_n/L_p]A$$

is the reverse saturation current, A is the cross-section of the sample, k is the Boltzmann constant, D_n, D_p are diffusion coefficients and L_n, L_p are diffusion lengths for n-and p-type carriers, respectively. In case of forward biasing, the current increases exponentially with voltage whereas in the reverse bias case the current becomes of opposite sign as compared to forward biasing and increases slowly to reach the saturation current I_o. This is because exponential term becomes dominant when $V > 0$ for forward biasing and the constant term in the diode equation becomes dominant in case of reverse biasing.

18.9 APPLICATIONS OF SEMICONDUCTORS

The magnetic flux density in a semiconductor is directly proportional to the Hall voltage V_H. The semiconducting device, which uses V_H as a measure of magnetic flux density is called magnetometer. Photoconducting material like ZnO has widespread applications in electrophotography. A number of other semiconductor devices have been developed based on the bulk properties of the material. Some of these are given below:

Thermistors

Thermistor is an electrical device, which consists of thermally sensitive resistor. It is used for accurate temperature measurement and control. There are some ceramic semiconductors, which are useful for such applications. Carrier density in a semiconductor increases with temperature. Since the mobility is not a sensitive function of temperature and charge per carrier is constant, the resistivity, ρ, given by

$$\rho = 1/ne\mu \tag{18.39}$$

decreases with the increase of temperature. To select a material for thermistors, one should choose a material with a high melting point and its resistivity should be sensitive to the changes of temperature. The materials having covalent bonding are suitable for this purpose. As more and more number of covalent bonds are broken with the increase of temperature, a large number of electrons become available for conduction i.e. the resistivity of the material decreases with the increase of temperature. Hence these materials have negative temperature coefficient (NTC) of resistivity. There are some ceramic compounds, which have high negative value of $d\rho/dT$. These materials can act as thermistors such that the temperature differences as small as 10^{-6} °C can be detected. The most commonly used materials for such applications are sintered oxides of the elements Mn, Ni, Fe, Co and Cu. Suitable solid solutions of these oxides are used to obtain the necessary range of electrical resistivity with temperature changes. Magnetite Fe_3O_4 (inverse spinel structure) is an example of ceramic semiconductor, which has very low resistivity 10^{-5} Ω m. The good electrical conductivity of magnetite is attributed to the random locations of Fe^{2+} and Fe^{3+} ions in octahedral sites. Electron transfer can take place easily from Fe^{2+} to Fe^{3+} ions, without destroying charge neutrality, whenever thermal energy is available. The resistivity of Fe_3O_4 is increased by adding it to suitable solid solutions of metal oxides like $MgCr_2O_4$. Thus, it is possible to obtain the required thermistors with controlled temperature coefficient of resistivity. Silicon is suitable for such application since it is covalently bonded and it melts at 1410°C.

Pressure transducer

When pressure is applied on certain semiconducting materials, the band gap of the material decreases since the atoms are forced to come closer. Decrease in energy gap results in increase in conductivity. The pressure acting on the material can be found out by measuring the change in the conductivity of the material.

Rectifier

The p-n junction diodes are used as rectifier i.e. they convert alternating voltage into direct voltage. When alternating voltage is applied to a p-n junction diode, the diode will conduct only during forward biasing. Thus, the output signal is only obtained during the half of input voltage cycle when p-type material is connected to positive and n-type to the negative of the input voltage. These are called half-wave rectifier, Fig. 18.10. The output signal can be smoothed out using full wave rectifiers and other electronic devices. Rectifiers are used in a wide range of current as well voltage capacities

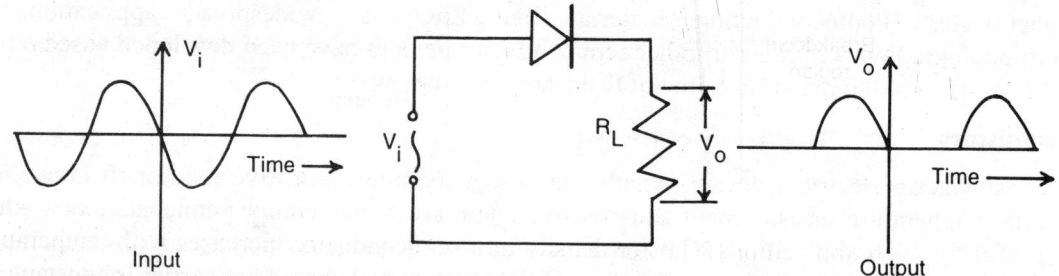

Fig 18.10 Rectifier diode.

Zener diodes

Zener diodes are basically rectifiers in which small reverse current saturates at specific reverse bias voltage. If the reverse bias voltage is increased beyond this point, zener breakdown occurs at a voltage called breakdown voltage. If the voltage is higher than the breakdown voltage the electric field in the diode becomes so high that it attracts electrons directly out of the covalent bonding of the semiconducting crystal. This phenomenon produces a large number of electron hole pairs, which cause a very large reverse current to flow. This is called avalanche effect, during which electrons gain sufficient energy between collisions to dislodge more electrons from the covalent bonds, which in turn gain enough energy to conduct electricity, Fig 18.11. Zener diodes can be fabricated with breakdown voltages from a few to several hundred volts. Zener diodes are used for voltage stabilizing while current is fluctuating and also to protect circuitry from very high voltages.

Transistors

Transistors are very important semiconducting devices, which are used as:
    i) current amplifiers,
    ii) switch devices,

These devices are useful for processing and storing information in computers. Transistors are broadly classified as bipolar junction transistors (BJT) and metal-oxide semiconductor field effect transistors (MOSFET).

Junction transistors

The junction transistors are composed of two pn junctions arranged back to back in either npn-type or pnp-type configurations. These junctions are formed in a single crystal of semiconductor

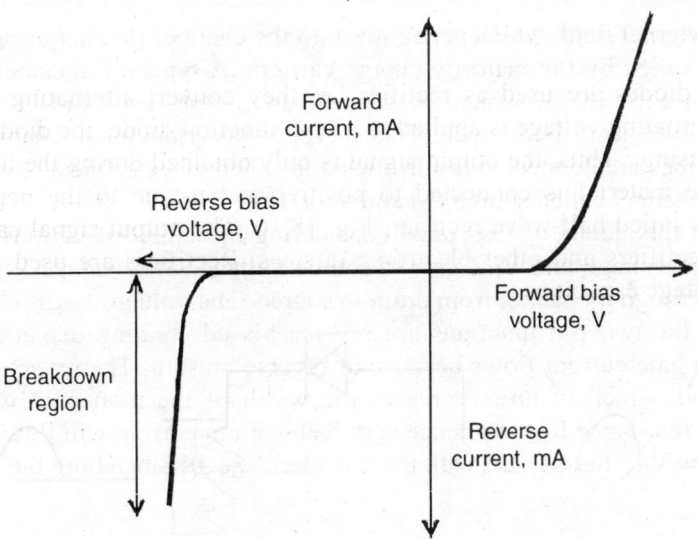

Fig 18.11 Zener diode characteristic curve

e.g. silicon. In a typical npn transistor, a thin p-type base region is sandwiched between n-type emitter and collector regions. The emitter-base np junction (junction 1) is forward biased whereas base-collector pn junction (junction 2) is reverse biased, Fig. 18.12. As junction 1 is forward biased, large number of electrons from n-type material move to p-type material. These injected electrons become minority carriers in the p-type base. Only a few of these electrons recombine with the majority carriers i.e. the holes in the p-region. This is because the base is thin and lightly

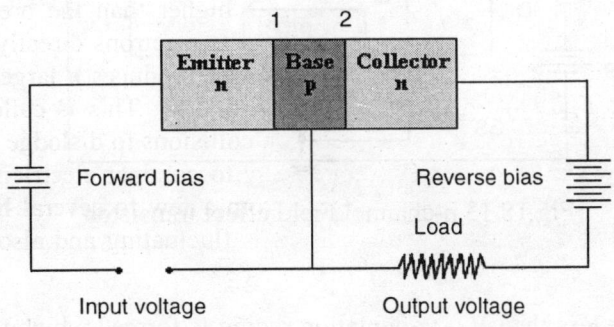

Fig 18.12 npn-type junction transistor.

doped. Hence most of these electrons move across the base without recombination and cross the junction 2 because of reverse biasing of junction 2. The minority carriers from p-type base are accelerated towards n-type collector, causing a large increase in the collector current. The same is reflected by a large increase in the collector voltage across the load resistor. Thus, an input signal gets amplified while passing through the junction. Similar reasoning applies to the operation of pnp-type transistor excepting that in this case basically flow of holes takes place.

Field effect transistors

Field effect transistor is often used in modern microelectronic systems. In this type of transistor the flow of charge carriers between the two terminals, drain and source, can be controlled by the application of an external field, which penetrates into the channel for charge carriers. The current flow is totally controlled by the majority charge carriers. A typical n-channel junction FET (n-JFET) is shown in Fig. 18.13, which consists of an n-type semiconductor bar with p-type semiconductor diffused from both sides. This n-type semiconductor is channel and the p-type materials on both sides are connected to form what is called as gate. Metal contacts are made on the both ends of n-channel called source (S) and drain (D).

The operation of this device is explained considering the effect of various bias voltages e.g. V_{DS}, drain to source voltage and V_{GS}, gate-to-source voltage. The voltage V_{DS} causes the drain current I_D to flow through n-channel from drain to source. The voltage V_{GS} is chosen to be zero or negative such that the two p-n junctions are reverse biased creating depletion regions around these junctions. No gate current flows because of reverse biasing. The reverse biasing increases the depletion region, which in turn decreases the width of the n-channel. This results in the increase of channel resistance R_{DS} and hence very feeble drain current will flow. The drain current is thus, controlled by V_{GS}. Let us start with the potential $V_{GS}=0$. Since both the p-n junctions are

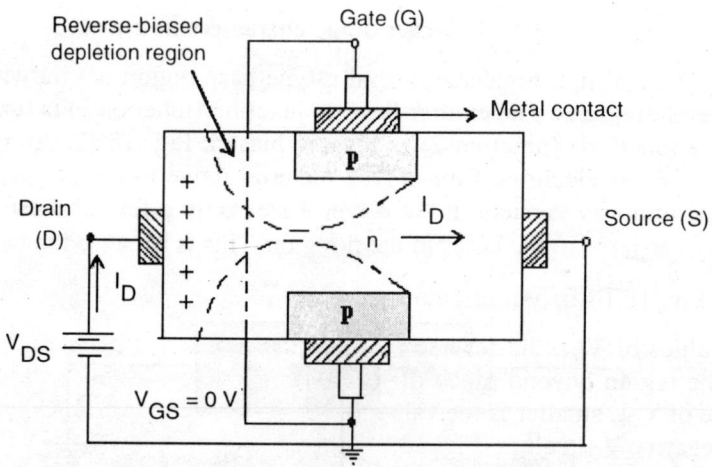

Fig 18.13 n-channel Field effect transistor

reverse biased by voltage less than V_{DS}, a depletion region is formed which restricts the channel width. Starting from a low value, as V_{DS} increases I_D also increases linearly i.e. R_{DS} is more or less constant, Fig 18.14(a). As V_{DS} increases channel width decreases more since the depletion region increases and beyond a specific value of V_{DS}, the channel width restriction by the depletion regions is greatly increased so that I_D saturates. The I_D versus V_{DS} curve bends sharply at this value of V_{DS} and then becomes horizontal. This happens due to merging of the depletion regions which causes pinch off and the corresponding V_{DS} is called pinch off voltage V_P (usually 5V). If V_{DS} is increased beyond this pinch off value V_P, the tendency to increase I_D is counteracted by that to restrict the channel width. Thus, I_D remains constant even when V_{DS} increases. The constant value of I_D at $V_{GS}=0$ V is referred to as I_{DSS}, the Drain-to-Source current with the gate source shorted. When V_{DS} is increased to a value such that break down occurs at p-n junctions, the channel current I_D increases very sharply.

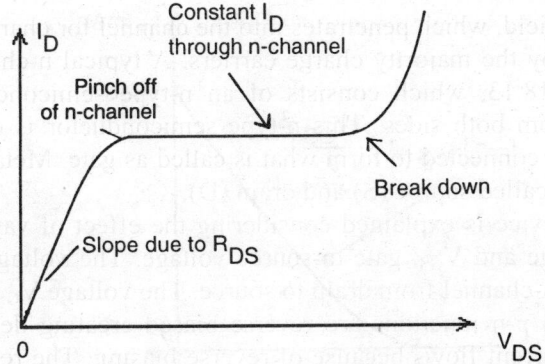

Fig. 18.14(a) Variation of I_D with V_{DS}

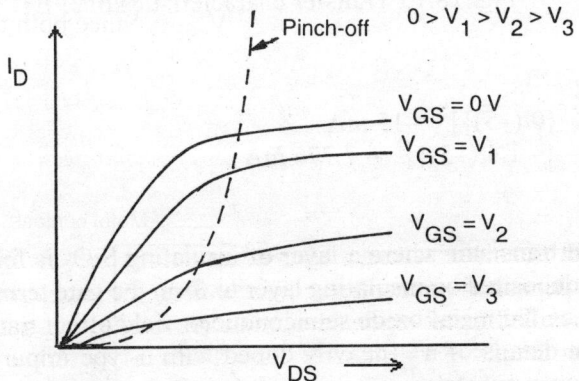

Fig. 18.14(b) Variation of I_D with V_{DS} for different values of V_{GS}

For negative values of V_{GS}, the reverse biasing increases causing pinch off to occur at lower values of V_{DS}. The region beyond pinch off is called the active region where I_D saturates. More negative the value of V_{GS}, smaller is the value of I_D. The active region changes progressively with more and more negative V_{GS} values, Fig. 18.14(b).

The gate current remains zero since no current will pass through the reverse-biased gate-source junction. The transfer characteristic of JFET is a plot of drain current I_D as a function of V_{GS} for a constant value of V_{DS}, Fig 18.15. Two important points in this curve are values of I_{DSS} and V_P. The relation between I_D and V_{GS} can be given by

$$I_D = I_{DSS} (1 - V_{GS}/V_P)^2 \tag{18.40}$$

A small change in the gate-source voltage near $V_{GS} = 0$ causes appreciable change in drain current. So JFET is a voltage sensitive device. In the linear region i.e. where V_{DS} is less than pinch off value, this device can be used as a voltage controlled variable resistance. When JFET is used as an amplifier, it is usually operated in the active region (saturation region).

Example 18.6

Find the drain current of an n-channel JFET having pinch off voltage $V_P = -5$ V and $I_{DSS} = 15$ mA at the following V_{GS} values: (a) 0 V (b) –1.4 V.

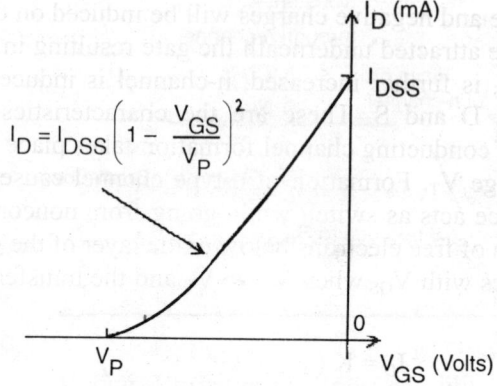

Fig. 18.15 Transfer characteristics of JFET

Solution:

Using Eq (18.40), we get
(a) $I_D = 15 [1- \{0/(-5)\}]^2 = 15$ mA
(b) $I_D = 15 [1- \{- 1.4/(-5)\}]^2 = 7.776$ mA

MOSFET

MOSFET is a field effect transistor where a layer of insulating SiO_2 is formed on to the channel and then a metal layer is deposited on insulating layer to form the gate terminal. This is the reason why such type of FET is called metal oxide semiconductor field effect transistor (MOSFET). Fig 18.16 shows construction details of n+ (heavily doped with n-type impurity) regions, which are formed, at the two ends in a substrate of p-type silicon. Proper contacts are formed on n^+-type regions to create D and S terminals of the device. The SiO_2 layer is formed on the middle part of the p-type region between the n^+-type silicon of the source and drain. The metallic gate G is formed over the layer of SiO_2. There is no direct conducting channel between D and S with $V_{GS} = 0$. No current can flow from the source to the drain irrespective of the value of V_{DS}. The n+-type source, p-type substrate and n^+ type drain behave as two diodes connected back to back and hence one of the diodes will always be reverse biased, preventing the conduction.

The set up of the oxide layer sandwiched between the metallic gate and the p substrate acts as capacitor. Whenever a positive voltage applied at the gate (i.e. V_{GS} is positive), the positive

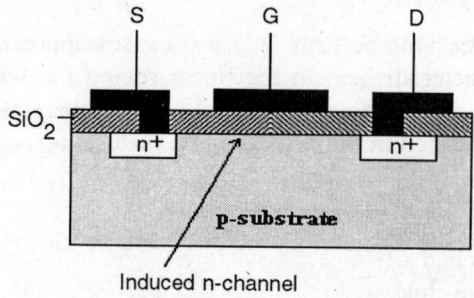

Fig.18.16 Enhancement type n-channel MOSFET

charges will be on the gate and negative charges will be induced on the p substrate near the oxide layer i.e. free electrons are attracted underneath the gate resulting in the formation of a depletion layer. As the voltage V_{GS} is further increased, n-channel is induced below the gate creating a conducting path between D and S. These are the characteristics of n-channel enhancement MOSFET. The process of conducting channel formation takes place above a certain value of V_{GS} called the threshold voltage V_T. Formation of n-type channel causes the electrons to flow from source to drain. The device acts as switch while going from nonconducting to conducting state. The layer of concentration of free electrons below oxide layer of the gate is called inversion layer. The drain current increases with V_{GS} when $V_{GS} > V_T$ and the transfer characteristic, Fig 18.17a is given by

$$I_D = K(V_{GS} - V)^2 \qquad (18.41)$$

where K is given as 0.3 mA/V^2, a property of the device construction. There is no I_{DSS} associated with an enhancement MOSFET since $I_D = 0$ if $V_{GS} = 0$ V. This type of device is very useful in large scale integration circuits because of its simpler construction and smaller size.

The drain current increases nearly linearly as V_{DS} is raised, since channel resistivity remains constant until V_{DS} exceeds a certain value when the D end of the induced n-channel becomes reverse biased and constrict the channel width. As a result the drain current cannot increase any more. I_D saturates for a value of $V_{GS} > V_T$, Fig 18.17b. Since the gate is insulated from the channel, the gate current is negligible.

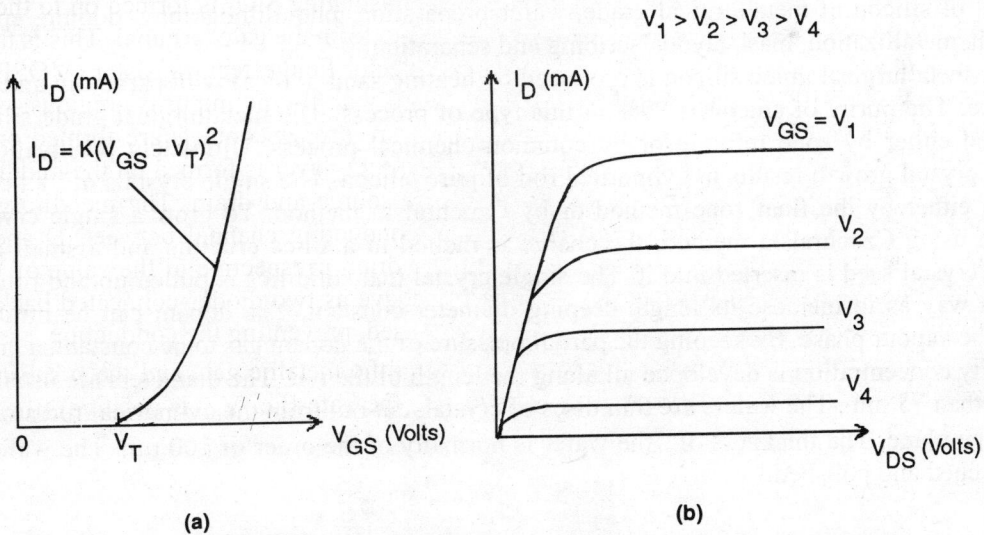

(a)   (b)

Fig. 18.17 Characteristics of n-channel enhancement MOSFET

Another type of MOSFET device is called depletion mode transistor, which is turned on initially, but becomes off during the operation. In this device the area under the gate is now doped to create a permanent n-type channel so that initially there is current. When $V_{GS} < 0$, holes are attracted towards the SiO_2 interface and invert the n-type channel to p-type. At a critical value of V_{GS} the electron channel is pinched off and the current flow stops; the device switches from conducting to nonconducting state. Just as there are NMOS devices as described above, there are also PMOS devices where p-channels are formed.

Circuits containing NMOS and PMOS devices are called complementary MOS or CMOS circuits. This type of circuits can be made by separating all NMOS devices with islands of p-type material. In CMOS circuits, low power consumption can be achieved by suitable arrangement of the MOS devices. The CMOS circuits are used in modern electronic watches, calculators, microprocessors and computer memories.

Example 18.7

What value of I_D is expected if in an n-channel enhancement MOSFET (a) $V_{GS} = 5$ V, (b) $V_{GS} = 7.5$ V, where K= 0.3 mA/V^2 and threshold voltage is 2.5 V.

Solution:

(a) $I_D = 0.3(5 - 2.5)^2 = 1.875$ mA   (b) $I_D = 0.3(7.5 - 2.5)^2 = 7.5$ mA

18.10 FABRICATION OF INTEGRATED CIRCUITS

Large scale integrated (LSI) digital memory circuits are developed starting from MOSFET fabrication technique, since the individual MOSFET occupies less silicon chip area than the bipolar transistor. The cost of fabrication of MOSFET LSI is less than that for bipolar transistor. The various steps of fabrications of integrated circuits are – production of cylindrical shape single crystal of silicon of metallurgical grade, wafer preparation, photolithography, doping, epitaxial growth, metallization, mask layout, scribing and separation.

The metallurgical grade silicon is produced by heating sand (silica) with carbon in an electric furnace. The purity of silicon is 99% in this type of process. The metallurgical grade silicon is purified either by zone refining or by common chemical process. Ultimately purification and single crystal growth results in cylindrical rod of pure silicon. The single crystals of silicon are grown either by the float zone method or by Czochralski method. To grow a single crystal of silicon using Czochralski method, the charge is melted in a silica crucible and a small cooled single crystal seed is inserted into it. The single crystal that solidifies is pulled up and rotated in such a way as to increase its length-keeping diameter constant. The dopant can be introduced from the vapour phase. By keeping the partial pressure of the dopant gas to be constant, a uniform impurity concentration is developed all along the length of the rod. The diameters are usually not larger than 75 mm. The wafers are thin discs of crystal, cut out from the cylindrical rod along the cleavage plane. The thickness of the wafer is normally of the order of 100 µm. The wafers are then etched and polished.

Oxidation

The surface of the wafer forms an oxide layer when heated in oxygen or steam. The oxide layer or layer of silica produces a chemically stable, protective and insulating layer on the surface of wafer. It passivates the surface of the wafer electrically and chemically. Thermal oxidation is carried out between 900 and 1300°C in dry oxygen or steam. Doping can be done after formation of oxide layer. The selective diffusion of dopant occurs over areas of the wafer surface not covered by the oxide layer. In case of doping, done before the formation of oxide layer, the dopants tend to distribute itself in different concentrations in silicon and silica during the oxidation process.

Photolithography

Once the wafer is ready, the desired geometrical patterns can be transferred on a mask to the wafer surface using the technique of photolithography i.e. printing with light. The microelectronic integrated circuit is first laid out on a large scale, using computer. The mask is then prepared. The design pattern is generated and transferred to a photosensitive glass called the photomask using electron beam machine. A set of photomasks is prepared from the layout. Each photomask contains the pattern for a single layer of the multi-layer finished integrated circuit. The silica-coated wafer is coated with a layer of a light-sensitive material called photoresist. The photomask is put on the wafer, which is then exposed to ultraviolet (UV) radiation. Exposure to UV radiation makes photomask insoluble in a developer solution. After development, a pattern of photoresist is left where no light has fallen due to photomask. The wafer is then immersed in hydrofluoric acid, which removes exposed silicon dioxide and not the photoresist. In the final step of the process, the left over layer of photoresist is removed by another chemical treatment and the pattern layer of silica is left behind.

Active circuit elements of integrated circuits such as bipolar and MOS transistors, rectifiers etc., are formed by selectively introducing dopants into the silicon substrate to create localized n- and p-type regions using diffusion or ion implantation techniques. Predeposition of the dopant on the silicon surface is done via a vapour phase, e.g. PH_3, B_2. H_6 or B_2O_3 in an atmosphere of hydrogen passing over the heated substrate. This is followed by drive-in mechanism, when the impurity atoms already deposited are redistributed.

Epitaxial growth

Integrated circuits are confined to the thin film of thickness few microns on the surface of the wafer. That is why a layer of silicon is grown on a substrate, such that its crystal structure is a continuation of the crystal structure of the substrate. This is called epitaxial growth. The surface of the substrate has to be cleaned so that atom-to-atom contact is formed during epitaxial growth. Epitaxial growth is done using chemical vapour deposition technique. Advantage of having epitaxial layer is Si film is purer and more defect free than the substrate wafer and can be doped independently of substrate. Silane, dichlorosilane, trichlorosilane vapour are used as source gas along with carrier gas as hydrogen for epitaxial growth. Dopant gases are also used to give desired extrinsic characteristics. Sometimes the deposition of epitaxial compound semiconductor films like GaAs is required. This needs to control the composition of the compound critically. The film material and substrate must have nearly same values of lattice constants to reduce the effect of lattice mismatch. Lattice mismatch is defined as

$$f = (a_s - a_f)/ a_f,$$
(18.42)

where s and f refer to substrate and film respectively.

There are commercially important methods of epitaxial growth e.g. Liquid Phase Epitaxy (LPE), Molecular Beam Epitaxy (MBE) methods other than CVD. LPE method is basically single crystal growth technique used for fabrication of compound semiconductor (e.g. GaAs). Take a melt of Ga rich alloy (containing 10 atomic percent of As). As this melt is cooled slowly below 930°C it enters the two-phase region (L + GaAs) of the alloy. If a single crystal of GaAs wafer substrate is inserted into the melt, thin single crystal epitaxial layer is grown. In MBE method, simultaneous thermal evaporation of the involved atoms takes place in a highly controlled way such that these atomic beams from different heated sources are directed to the heated substrate wafer. Temperature and resulting vapour pressures are selected in such a way that a micron thick film is typically grown within an hour.

Metallization

Metallization is the process of forming electrical connections between different elements or parts of the circuit. Aluminium and copper are used for this purpose. In recent days, a new way to inlay copper wires in semiconductor wafers has been developed to create integrated circuits. This is an ion-assisted trench filling method. The standard materials used for interconnections are aluminium and its alloys and in interlayer connectors, tungsten. As the devices are becoming smaller in size, the requirement of new materials becomes obvious. Copper is more suitable for this purpose as compared to aluminium since it is more conductive, less vulnerable to electromigration and less likely to fracture under stress than aluminium. Unfortunately, copper is poisonous to silicon as it readily diffuses into silicon and causes deep level defects. To stop the diffusion of copper into substrate and poisoning it, a diffusion barrier is used, which lines the trench walls between the copper and the substrate. The barrier materials are titanium nitride, tantalum, tantalum alloys and tantalum nitride. IBM and Motorola produced copper wired chips by electroplating the copper over diffusion barrier. In the new ion assisted technique, a substrate wafer etched with trenches is placed under a plasma source. A pulsed-bias voltage is applied to the substrate and can be tuned to accelerate ions towards both the sides and bottom of the trench. The layer of copper builds up uniformly filling the trench from the bottom up. The process is continued till the desired thickness of the material has been obtained. Using different cathode materials, the desired connections are obtained.

The steps described above from oxidation to metallization can be repeated a number of times for translating actual circuit design. The fabrication of circuits is done by transferring images of various layers of a mask set on the front side of the wafer by photolithography and then followed by other steps as required in between two oxidation and masking steps. The circuit is built in each of the squares of the wafer, which are separated into chips later. The miniaturization of the components, their interconnections and external connections are made so well that the number of components per chip has become of the order of a million. After completion of various layers of a circuit design, the backside of the wafer may be just polished or gold might be deposited on the backside. The scribes are usually made in square shapes at the boundaries of the repeating basic pattern with the help of diamond tip or by using laser beam. Breaking along the scribes does the separation.

18.11 SOLID STATE LASER

When a p-n junction is forward biased, a large number of electrons flow from n-type to p-type material where these electrons recombine with the holes on p side. This process causes emission of a coherent, monochromatic beam of light. The wavelength of this radiation is 8900Å in case of GaAs junction. If the junction meets the sidewalls of the crystal, the light beam is intense. In solid state laser the mirrors are placed on the opposite sides of the crystal to cause multiple reflections of the beam thereby increasing the intensity. Ga-As/Ga-Al-As optoelectronic interface units provide high data rate optical links. A number of p-n junction lasers are discovered producing beams having wavelength ranging from far infrared to ultraviolet region. Laser beams have variety of applications e.g. communications, fabrication of electronic devices, medical applications, metal cutting etc.

18.12 PHOTOCELL

Photocells operations are based on the photoconduction arising from photon activation. A photon can activate an electron in the valence band causing transition of electron from valence band to conduction band. Both the electron and the hole become charge carrier and thus an insulator may become semiconductor when a photon of energy greater than or equal to the band gap incident on it. These are called photoconductors. In case of a p-n junction, if electron hole pairs are produced in the junction region, the electrons move in the n-region and holes in the p-region, thereby changing bias favourable for conduction. Illuminated junction serves as photocell or solar cell. This type of device does not need any biasing for conduction. These devices are useful for generation of power, exposure meter etc. Solar cells are being used largely in our country for producing power especially in remote areas or in some villages.

EXERCISES

18.1 Estimate the density of silicon.

18.2 A copper wire with cross-sectional area of 2×10^{-6} m^2 carrying a current of 1.5A offers a resistivity of 1.73×10^8 Ωm at 20°C. Calculate the average drift velocity in the conductor.
Ans: 5.5×10^{-5} m/s

18.3 Find the ratio of concentration of silicon atoms and electron hole pairs at room temperature. Determine the intrinsic resistivity. Given that the atomic weight of Si = 28.09 amu, density of Si=2.33 Mg/m^3, $\sigma = 5 \times 10^{-4}$ $(\Omega m)^{-1}$ μ_n =0.135m^2/V-s, μ_p = 0.04m^2/V-s.)
Ans: 2.798×10^{12}, 2000 Ωm

18.4 Compare the resistivity of intrinsic silicon with that of silicon doped with 10 Indium atoms for every 10 million silicon atoms. Given density = 2.33 Mg/m^3 for doped silicon, intrinsic carrier concentration = 1.7×10^{16}/m^3, $\mu_n = 0.135$m^2/V-s, $\mu_p = 0.048$ m^2/V-s.
Ans: $\rho_{si} = 2009$ Ωm, $\rho_{doped} = 2.6 \times 10^{-3}$ Ωm; $\rho_{si}/\rho_{doped} = 7.7 \times 10^5$

18.5 Silicon has eight atoms per unit cell. The unit cell is cubic with lattice parameter a= 5.43 Å. If there are 1.5×10^{13} conduction electrons/cm^3. Find the fraction of electrons that are excited to the conduction band.
Ans: 7.5×10^{-11}

18.6 Find the percentage increase in conduction electron when pure germanium ($E_g = 0.72$ eV) is heated from 20°C to 40°C.
Ans: 148%

18.7 Calculate the current produced in a germanium plate of area 1.5 cm^2 and thickness 0.2 mm, when a potential difference of 2 V is applied across the faces. Given: concentration of free charge carriers is 2×10^{19}/m^3 and the mobilities of electrons and holes are 0.36 m^{2}V^{-1}s^{-1} and 0.17 m^{2}V^{-1} s^{-1}, respectively.
Ans: 2.54 A

18.8 Compare the position of the Fermi level in intrinsic as well as both types of extrinsic semiconductors

18.9 Calculate the temperature at which intrinsic silicon has the same carrier density as that of doped p-type silicon of 10^{18} acceptors/ m^3 at room temperature. Analyze the result with reference to the temperature dependence of conductivity, where $E_g = 1.1$ eV for Si.
Ans: 367°K

18.10 Calculate the reverse saturation current density across an abrupt Si p-n junction with $N_D = 10^{22}$ atoms /m^3, $N_A = 10^{21}$ atoms/m^3, $n_i = 1.5 \times 10^{16}$ carriers/m^3, $L_p = 3.5 \times 10^{-4}$ m, $L_n = 7.1 \times 10^{-4}$ m, $D_n = 0.0035$ m^2/s and $D_p = 0.0012$ m^2/s. Assume all impurities are ionized.
Ans: 1.898×10^{-7} A/m^2

18.11 A semiconductor XY is cut into a small bar of 2cm x 2mm x 2 mm with a lengthwise resistance of 1.5 Ω. A hall field of +2.2V/m develops when a current of 0.16A is carried lengthwise and the magnetic flux density is 1000G. (G = 10^{-4} V.s/m^2).
(a) Is the semiconductor p-type or n-type?
(b) What is the carrier density?
(c) What is the mobility of the carriers?
Ans: (a) p type; (b) 1.136×10^{22}/m^3; (c) 1.833 m^2/V.s

18.12 Consider a sample of pure silicon, which has conductivity 5 x 10^{-4} ohm^{-1}m^{-1}. Its conductivity was improved to 200 ohm^{-1}m^{-1} by adding aluminium as an impurity (p-type). Calculate the mass of aluminium to be added per m^3 of the crystal in order to get this conductivity.
$\mu_p = 0.0485$m^{2}V^{-1}s^{-1}; . $\mu_n = 0.135$ m^2 V^{-1}s^{-1}.
Ans: 1.15 g /m^3

18.13 The phosphor of a television tube glows because conduction electrons release light photons as they return across the energy gap. What minimum energy is required for a red phosphor (the corresponding wave length is 6700 Å)?
Ans: 1.85 eV

18.14 A p-n junction is made as follows:
(a) 150 g of silicon is melted.
(b) 0.005 g of an alloy of Ga and Si is added which contains 0.7 w/o gallium.
(c) The total alloy is solidified as a single crystal and cut into small wafers.
(d) Antimony is vapour coated onto the surface of the wafer and the wafer is heated, so that Sb fuses into silicon.
Find w/o of Ga throughout wafer and w/o of Sb in the surface layer to provide it with as many carriers/ m^3 from Sb as there are in the wafer from gallium.
Ans: w/o of Ga = 2.33×10^{-5} , w/o of Sb = 4.06×10^{-5}

18.15 A silicon semiconductor at 27°C is doped with 1.4×10^{16} boron atoms/cm^3 and 1.2×10^{16} phosphorous atoms/cm$^{3.}$ Find the majority and minority carriers and calculate their respective concentrations. If the mobilities μ_p and μ_n are 295 cm^2/(V-s) and 880 cm^2/(V-s), respectively, where n_i (Si)= 1.5×10^{16}carriers/ m^3, find the electrical resistivity of doped semiconductor.
Ans: 10.6×10^{-2}Ω-m

18.16 The following electrical characteristics have been observed for both intrinsic and n-type extrinsic InP, at 27°C.

	σ(ohm-m)$^{-1}$	n(m^{-3})	p(m^{-3})
Intrinsic	2.5×10^{-6}	2.5×10^{13}	2.5×10^{13}
Extrinsic	3.6×10^{-5}	6×10^{14}	1.8×10^{12}

Calculate electron and hole mobilities.
Ans: $\mu_p = 0.251$ m^2/V.s; $\mu_n = 0.374$ m^2/V.s

18.17 A CdS photodetector receives radiation of wavelength 6300 Å over an area of 2.5×10^{-6} m^2 with an intensity of 100 W/m^2 and energy gap of 2.4 eV. Calculate the number of electron-hole pairs generated per second if each quantum generates a pair.
Ans : 7.9×10^{14}

19

Magnetic Materials

19.1 INTRODUCTION

Magnetic materials acquire permanent magnetic moment in the presence of magnetic field e.g. iron, nickel, cobalt and some of their alloys and compounds. Permanent magnets have innumerable applications in industries e.g. magnetron oscillators, loud speakers, telephone receivers, synchronous and brush less motors and automotive starting motors.

19.2 MAGNETIZATION IN MATERIALS

When we keep a bar magnet, its influence can be observed by keeping a piece of iron near it. The iron piece gets attracted towards the magnet. The force experienced by the iron in the neighbourhood of a bar magnet is due to the presence of the magnetic field in the region. Even though a magnetic monopole does not exist, it is customary to define, in analogy with the electrical field, that the force at a point experienced by a monopole of unit strength is called the magnetic field strength H at that point. The magnetic field H, generated by a solenoid of n turns and length l carrying current I, is

$$H = \frac{nI}{l} \tag{19.1}$$

The unit of H is ampere-turns per meter or amperes per meter and is equal to $4\pi \times 10^{-3}$ Oersteds. The direction of the magnetic field is along the axis of the solenoid. The magnetic lines of force or field lines surrounding a bar magnet can be visualized by the arrangement of iron filings on a drawing sheet which is placed above the bar magnet. Fig.19.1 illustrates the distribution of magnetic lines of force around a bar magnet and a solenoid. These lines of force are also called flux lines. The number of flux lines passing through any particular area can be arbitrarily fixed. However, once the number of lines per unit flux is fixed, the number of flux lines passing through any given area is a measure of the flux passing through that area.

The flux density or magnetic induction **B** is defined as the flux per unit area perpendicular to the direction of lines of force. The magnetic induction is proportional to the magnetic field strength. If the magnetic field H produces magnetic flux density **B** within the material, then

$$B = \mu H \tag{19.2}$$

where μ is the permeability of the medium; the unit of **B** is Wb/m^2 or Tesla. The magnetic induction in vacuum, **B$_0$** can be expressed as,

$$\mathbf{B_0} = \mu_0 \, \mathbf{H} \tag{19.3}$$

where μ_0 is the permeability of vacuum, a universal constant which has the value $4\pi \times 10^{-7}$ Wb A^{-1}m^{-1} or H/m. The ratio

$$\mu_r = \mu/\mu_0 \tag{19.4}$$

for any medium is called the relative permeability of the medium. The relative permeability is a measure of the degree to which a material can be magnetized. When an external magnetic field **H** is applied to a magnetic material, magnetic moment is induced in it. The magnetic moment per unit volume of the material is called magnetization **M,** which is a vector parallel to **H** or **B**. **M** is proportional to the external field **H** and is given by

$$\mathbf{M} = \chi_m \mathbf{H}, \tag{19.5}$$

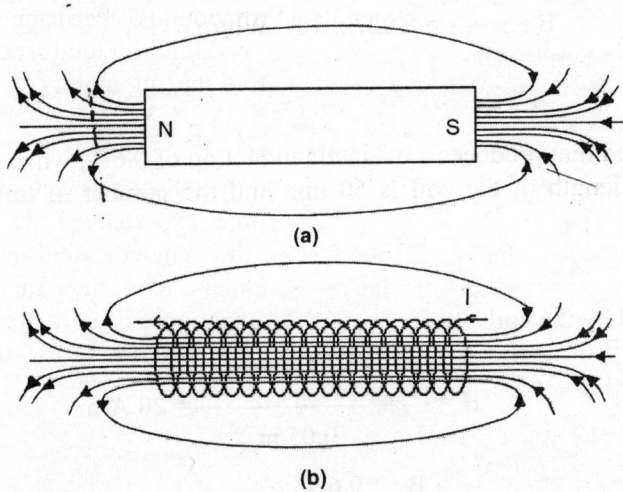

(a)

(b)

Fig.19.1 The magnetic lines of force around a) bar magnet b) solenoid

where χ_m is called the magnetic susceptibility. The magnetic induction **B** arises not only due to the external field **H** but also due to the induced magnetization **M** and hence is given by

$$\mathbf{B} = \mu_0 \, (\mathbf{H} + \mathbf{M}) \tag{19.6}$$

Combining the Eqs.(19.2) and (19.6), we get

$$\mu\mathbf{H} = \mu_0 \, (\, \mathbf{H} + \mathbf{M} \,)$$

i.e.

$$\mu_r\mathbf{H} = \mathbf{H} + \mathbf{M}$$

$$(\mu_r - 1) \, \mathbf{H} = \mathbf{M}$$

Thus, the relation between relative permeability and susceptibility is given by

$$\chi_m = \mu_r - 1 \tag{19.7}$$

Example 19.1

Compute the flux density B due to a coil 0.5 m long, having 400 turns and carrying current of 5A. Also compute the flux density inside a bar of titanium that is kept within the coil. The susceptibility of titanium is 1.81×10^{-4}.

Solution:

(a)

$$B_o = \frac{\mu_o \, nI}{L} = \frac{4\pi \times 10^{-7} \times 400 \times 5}{0.5} = 5.026 \times 10^{-3} \text{ Wb/m}^2$$

(b)

$$\chi_m = \frac{\mu}{\mu_o} - 1 = 1.81 \times 10^{-4} \quad \text{for titanium}$$

So,

$$\mu = 1.2569 \times 10^{-6} \text{ H/m} \quad \text{for titanium}$$

$$B = \frac{\mu \, nI}{l} = 5.027 \times 10^{-3} \text{ Wb/m}^2$$

Example 19.2

A solenoid coil is required that produces a magnetic induction of 0.6 T, when a current of 10 mA flows through it. If the length of the coil is 50 mm and the number of turns is 100, select a suitable core material.

Solution:

The value of H produced by the coil

$$H = \frac{nI}{l} = \frac{100 \times 0.01 \text{ A}}{0.05 \text{ m}} = 20 \text{ A/m}$$

$$\mu = \frac{B}{H} = \frac{0.6 \text{ T}}{20 \text{ A/m}} = 0.03 \text{ T m A}^{-1}$$

The suitable core material should be such that its relative permeability is

$$\mu_r = \frac{\mu}{\mu_o} = \frac{0.03}{4\pi \times 10^{-7}} = 23873$$

When we examine Table 19.3, we find that relative permeability of 45 Permalloy is 25,000. So this will be suitable material for the core.

The macroscopic magnetic behaviour has its origin in magnetic moment of atoms. The spin magnetic moments of electrons, nucleus and also magnetic moment arising from the orbital motion of electrons contribute to the atomic magnetic moment. The latter two sources have only minor contributions to the magnetization of materials. An atom possesses a magnetic moment if the net magnetic moment of electrons is nonzero. The type of magnetism can be classified as,

* Diamagnetism
* Paramagnetism
* Ferromagnetism
* Antiferromagnetism
* Ferrimagnetism

19.3 DIAMAGNETISM

The origin of diamagnetism has its source in magnetic moment due to the orbital motion of electrons. When a magnetic field is applied, the orbital motion of electrons in an atom gets modified in such a way that a small magnetic dipole is created within the atom, which opposes the external field. Diamagnetism is a universal property of elements and it is usually very feeble. The diamagnetic susceptibility is negative since the induced magnetization opposes the applied magnetic field, Fig 19.2. Perfect diamagnetic susceptibility is –1. Diamagnetism can be detected only when other types of magnetism are absent.

19.4 PARAMAGNETISM

Paramagnetic substances have atoms with unpaired electrons in the outer incomplete shell, by the virtue of which the atoms possess magnetic moment. The atomic magnetic moments are randomly oriented at room temperature, so that the resultant magnetic moment of the material is zero in the absence of magnetic field. In the presence of a magnetic field, atomic magnetic moments in the material try to align along the magnetic field. This results in overall magnetization of the material provided its temperature is not so high that the orientations of atomic magnetic moments are randomized due to thermal energy. The magnetization disappears on removal of the magnetic field. Magnetic moment of a paramagnetic material is a function of temperature, and the corresponding magnetization is expressed as,

$$M = N\mu_o\mu_\beta^2\,(H/3kT) \tag{19.8}$$

where μ_β is the spin magnetic moment of an electron. It is also called Bohr magneton. The magnitude of Bohr magneton is $9.24 \times 10^{-24}\,Am^2$. N is the density of atoms with unbalanced spins. The susceptibility of a paramagnetic substance is a function of temperature and is given by

$$\chi_m = \frac{M}{H} = \mu_r - 1 = \frac{N\mu_o\mu_\beta^2}{3kT} = \frac{C}{T} \tag{19.9}$$

This is called Curie law, where C is a constant. The paramagnetic susceptibility is less than 10^{-3} at room temperature for a typical paramagnetic material with 5×10^{28} atoms per cubic meter. Paramagnetic materials e.g. $CrCl_3$; Fe_2O_3 etc., have very low values of susceptibility. These are mainly used for the study of paramagnetic resonance.

19.5 FERROMAGNETISM

Ferromagnetism is a kind of magnetism where large induced magnetization can be retained even when magnetic field is removed. The most important ferromagnetic elements are transition elements Fe, Co and Ni. Also Gd, a rare earth element, is ferromagnetic below 16°C.

The ferromagnetic properties of transition elements arise due to the spins of the unpaired electrons in the incomplete 3d shell of atoms that are aligned in their crystal lattices. The other inner shells of atom have paired electrons and hence have no resultant magnetic dipole moments.

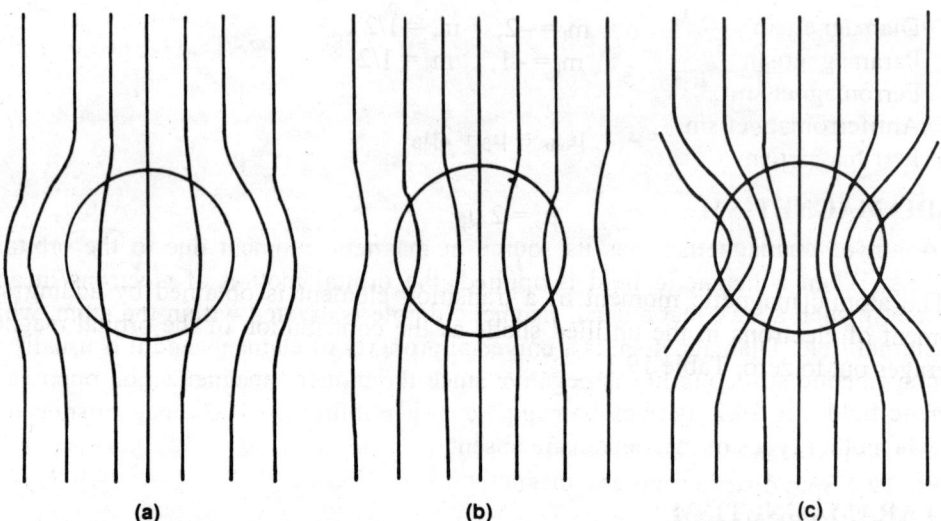

(a)   (b)   (c)

Fig.19.2 (a) Diamagnetic sample repels external magnetic field
(b) Paramagnetic sample attracts external magnetic lines of force feebly
(c) Ferromagnetic sample attracts them strongly.

As the temperature of a ferromagnetic material is increased, its spin alignment is disturbed and at a specific temperature, spin alignment is randomized in such a way that ferromagnetic materials become paramagnetic. The temperature at which this happens is called Curie temperature T_c.

The magnetic moment of an electron is calculated considering its orbital and spin angular momenta. Electrons are arranged in atomic levels according to Hund's rule. This arrangement results in maximum possible total spin angular momentum consistent with Pauli's principle. The magnetic moment μ of a current carrying loop is given by

$$\mu_{orb} = I\pi r^2 \qquad (19.10)$$

where r is the radius of the loop and I is the current. Similarly, the magnetic moment of an electronic orbit is expressed as

$$\mu_{orb} = - \frac{ep}{2\pi rm}\pi r^2 = - \frac{erp}{2m} \qquad (19.11)$$

where p is linear momentum and rp is angular momentum for a circular electronic orbit. The angular momentum of an atomic electron has quantized values ($m_l\hbar$) in the presence of magnetic field, where magnetic quantum number, m_l = 1, 1–1, 1–2,....– (1–2),– (1–1), – 1; 1 being angular momentum quantum number. Hence,

$$\mu_{orb} = - m_l\, eh/(2m) \qquad (19.12)$$

and spin magnetic moment is given by

$$\mu_s = - m_s\, [\, 2eh/(2m)] \qquad (19.13)$$

where eh/(4πm) is Bohr magneton, μ_β. The spin quantum number m_s can have values + 1/2 or –1/2. In case of titanium there are two unpaired electrons having quantum numbers,

$$m_l = -2, \quad m_s = 1/2$$

and

$$m_l = -1, \quad m_s = 1/2$$

So

$$\mu_{orb} = \mu_\beta + 2\mu_\beta \qquad (19.14)$$

and

$$\mu_s = 2\,\mu_\beta \qquad (19.15)$$

The atomic magnetic moment of a transition element is obtained by adding spin magnetic moment of electrons in the unfilled shells as the contribution of the orbital magnetic moments averages out to zero, Table 19.1.

Table 19.1 Spin distributions and magnetic moments of neutral atoms of 3d transition elements.

Elements	Number of 3d electrons	Electronic Configuration 3d orbitals	Magnetic Moments In terms of μ_β		$\dfrac{d_{atomic}}{d_{3d}}$
			Calculated	Observed	
Sc	1	↑	1	0	-
Ti	2	↑ ↑	2	-	1.12
V	3	↑ ↑ ↑	3	-	-
Cr	5	↑ ↑ ↑ ↑ ↑	5	-	1.18
Mn	5	↑ ↑ ↑ ↑ ↑	5	-	1.47
Fe	6	↑↓ ↑ ↑ ↑ ↑	4	2.22	1.63
Co	7	↑↓ ↑↓ ↑ ↑ ↑	3	1.72	1.82
Ni	8	↑↓ ↑↓ ↑↓ ↑ ↑	2	0.61	1.98
Cu	10	↑↓ ↑↓ ↑↓ ↑↓ ↑↓	0	0	-
Zn	10	↑↓ ↑↓ ↑↓ ↑↓ ↑↓	0	0	-

The number of Bohr magnetons per atom as calculated above is found to be different from the experimental value. This can be explained from the peculiar metallic bonding of these transition metal ions. In ferromagnetic metals, the 3d electrons are able to wander through the lattice, which results in the reduction of effective magnetic moment of each electron. In another approach, the effective magnetic moment of an atom gets modified in the solid state due to overlap of the valence energy bands. The electronic energy levels of the valence electrons interact to form energy bands. In transition metals, there is overlapping between 3d and 4s bands, as a result, some of the spins of unpaired 3d electrons may align antiparallel to that of 4s electrons to form pairs. So, the spin magnetic moment of atom decreases in solid state. For example, a free iron atom $(3d^6 4s^2)$ has a moment of 4 units but it is only 2.2 in crystal. Other magnetic substances like cobalt and nickel have moments of 1.7 and 0.6 units in crystalline state. The rare earth metal gadolinium has net magnetic moment of seven units remains unchanged in crystal since there is no overlap of energy band due to 4f orbitals with other energy bands. The discrepancies in measured and calculated values are given in Table 19.1.

Example 19.3

Find the magnetization **M** for cobalt whose density is 8.9×10^3 kg/m^3.

Solution:

Atomic magnetic moment of cobalt is $1.72\mu_\beta$ and atomic weight is 58.93; the magnetic moment per kg is $(6.023 \times 10^{26} \times 1.72\mu_\beta)/ 58.93$ Am^2/kg. Hence

$$M = \frac{6.023 \times 10^{26} \times 1.72 \times 9.27 \times 10^{-24} \times 8.9 \times 10^3}{58.93} = 14.5 \times 10^5 \text{ A/m}$$

Ferromagnetic domains

In a ferromagnetic material, the magnetic moments of adjacent atoms align in a specific direction due to exchange interaction in microscopic regions. Consequently, the resultant spin magnetic moment in a given microscopic region points in the same direction. This phenomenon is also called spontaneous magnetization and the microscopic regions are called domains. Each domain has a resultant magnetic moment but the total magnetization of the sample is zero in the absence of external magnetic field, since the magnetic dipoles of domains are not aligned.

The exchange interaction might try to make the entire ferromagnetic sample one large domain, but this would create magnetic poles on the surface of the ferromagnetic material. As a result of this, the associated spatial magnetostatic energy would be very high. The magnetostatic energy density is defined as the density of magnetic flux lines (J/m^3) in free space. So such alignment would not be favoured and hence a compromise between these two energies i.e. exchange energy and magnetostatic energy, is reached such that the energy of the system is minimum. The magnetic domains with various directions of magnetization form close loops, so that there are a few magnetic poles on the surface or in the bulk region. The boundary between two domains is called Bloch wall. There are certain crystallographic directions along which magnetization can occur very easily, which are often referred to as the directions of easiest magnetization. An additional energy is required for magnetization in any direction other than these easy directions of magnetization. This additional energy is called anisotropy energy. The directions of easy magnetizations are <100> type in an iron single crystal, i.e. these directions coincide with crystal axes. Easy directions of magnetization are [111] in nickel and [0001] in cobalt.

Consider two adjacent vertical domains as shown in Fig. 19.3(a). The direction of magnetization must turn out to be the direction of easy magnetization while making a transition from one domain to the other. This transition of magnetization from one domain to another is gradual and takes place in the region of Bloch wall or domain wall, Fig.19.3(b). As a result the anisotropy energy increases the energy of such a domain boundary.

In transition metals, the incomplete 3d shell is responsible for magnetic characteristics. At the beginning of the transition series, i.e. in Sc, there is only one 3d electron. The size of 3d-orbital is large when there is less number of 3d-electrons. This results in good overlapping of 3d-orbitals of neighbouring atoms in the crystalline state. The overlapping of these orbitals gives rise to 3d band, which contains all paired electrons, and hence net magnetic moment of this element is zero. In case of heavier elements of transition series, the 3d-orbitals become smaller due to greater attraction of 3d electrons from the increasing charge on nucleus. As a result, there is decrease in overlap of the 3d-orbitals.

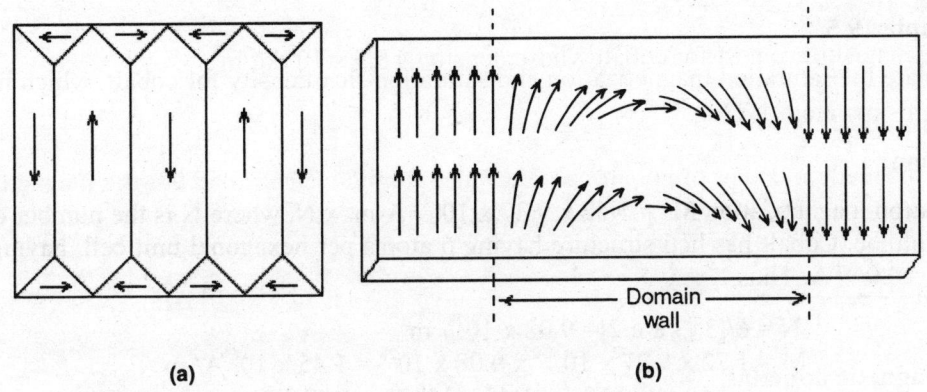

Fig. 19.3 (a) Domain structure in a single crystal of iron.
(b) Bloch wall in ferromagnetic material

Some elemental crystals of first rare earth series such as gadolinium, terbium and dysprosium have exchange energy due to which the spins of 4f electrons align in parallel. The unpaired electrons in the neighbouring atoms align their spins in parallel so that there is decrease in spin-dependent electrostatic energy. The alignment of spins in parallel fashion also results in increase of energy due to higher energy bands start filling up even though the lower energy bands are not completely filled up since there are fewer electrons with antiparallel spins. Hence, there will be raising of Fermi energy level and increase in the average kinetic energy. If the net change in energy is negative, the magnetized state is energetically more favoured than the unmagnetized state. This alignment occurs in a material, if the ratio of the atomic spacing d_{atomic} to the diameter d_{3d} of the 3d orbital is in the range 1.12 to 1.98, Table 19.1. This condition is satisfied for iron, cobalt and nickel but not for chromium and manganese. That is why chromium and manganese are not ferromagnetic materials.

Example 19.4

Calculate theoretical values for the saturation magnetization **M** and magnetic flux density **B** for bcc pure iron.

Solution:

$$\mu_s = \tfrac{1}{2} \times 2 \, (e\hbar/2m) = \mu_\beta = 9.27 \times 10^{-24} \text{ A m}^2$$

There are four unpaired electrons per atom. The magnetic moment of atom is $4 \times 9.27 \times 10^{-24}$ A m^2. The atomic density of Fe is given by

$$n = 2/[2.87 \times 10^{-10} \text{ m}]^3$$

Thus magnetization is given by

$$M = \frac{2 \times 4 \times 9.27 \times 10^{-24}}{(2.87)^3 \times 10^{-30}} = 3.14 \times 10^6 \text{ A/m}$$

$$\mathbf{B} = \mu_o M = 4\pi \times 10^{-7} \times 3.14 \times 10^6 \text{ T} = 3.94 \text{ T},$$

assuming $\mu_o \mathbf{H}$ is negligible.

Example 19.5

Calculate the saturation magnetization and saturation flux density for cobalt, which has magnetic moments per atom $1.72\mu_\beta$.

Solution:

Saturation magnetization $M_s = 1.72 \times 9.27 \times 10^{-24}$ A-m^2 x N, where N is the number of atoms per unit volume. Cobalt has hcp structure having 6 atoms per hexagonal unit cell, having a= 2.50 Å and c = 4.069 Å. Thus,

$$N = 6/[3\sqrt{3}\ a^2 c/2] = 9.08 \times 10^{28}/\ m^3$$

So
$$M_s = 1.72 \times 9.27 \times 10^{-24} \times 9.08 \times 10^{28} = 1.45 \times 10^6\ A/\ m$$
$$B = \mu_o M = 4\pi \times 10^{-7} \times 1.45 \times 10^6\ T = 1.82\ T, \text{ assuming } \mu_o H \text{ is negligible.}$$

In a macroscopic piece of ferromagnetic material, there are a large number of domains. The magnitude of M can be worked by taking the vector sum of the magnetization of all the domains. The contribution of each domain is proportional to its volume fraction.

Process of magnetization

Ferromagnetic materials get magnetized when an external magnetic field, **H** is applied. As the magnetic field is increased, the magnetization as well as magnetic induction, **B** increases. In order to study the process of magnetization, a sample of unmagnetized ferromagnetic material like iron is chosen and is placed in a magnetic field, **H**. The field is gradually increased and the magnetic induction, **B** in the material is measured to study the nature of **B-H** curve in ferromagnetic material. **B** increases (dotted line in Fig 19.4) with the increase in value of **H** and eventually attains a constant value B_s, the saturation magnetic induction. As the external field **H** is decreased, the flux density **B** also decreases (as shown by solid line) and it attains the value B_r, the residual magnetic induction when **H** = 0. B_r is also called remnant induction.

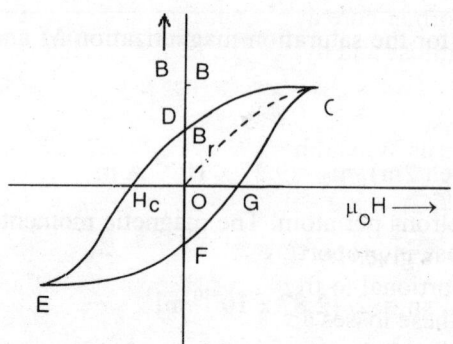

Fig.19.4 B-H curve of a ferromagnet

As the magnetic field is reversed and increased in the opposite direction, **B** decreases more and attains the zero value. The reverse field necessary for this purpose is called coercive field, H_c. As the reverse magnetic field is increased further, the magnetic induction **B** also increases in the opposite direction and eventually reaches the value $-B_s$. The remaining half of the B-H loop can be obtained by decreasing the reverse magnetic field to zero and reversing it again. The **B-H** loop produced is also called hysteresis loop as indicated by the complete path CDEFGC.

Hysteresis and eddy current losses

The energy consumed per cycle is equal to the **B-H** area within the loop. During the magnetization and demagnetization of ferromagnetic materials, the domain walls are pushed back and forth. This results in dissipation of energy and it is called hysteresis loss. Point defects and dislocations act as barrier to hinder the movement of domain wall during the magnetization cycle and hence increase the hysteresis losses.

There is change in length of a ferromagnetic material along the direction of magnetization. This is called magnetostriction. Mechanical stress due to magnetostriction increases hysteresis loss. In case of a transformer using frequency 50 Hz, the hysteresis loop is executed 50 times per second in the iron core. Since for each cycle there is hysteresis loss, there is more loss with the increase of AC input frequency. Besides hysteresis loss, there is another type of energy dissipation in magnetic material. AC input to the transformer sets up a fluctuating magnetic field, which in turn produces transient voltage gradient. The voltage gradient results in induced current in the conducting magnetic core. This is called eddy current, which is also a source of energy loss. The eddy current produces joule heating and power losses. The eddy current loss is high if high frequency AC source is used. Laminated or sheet structure in magnetic core can reduce the energy loss due to eddy current. The laminated layers are insulated from each other, which prevents the flow of eddy current from one sheet to another, and hence the energy loss is decreased.

Soft magnetic materials produced from ceramics have high resistivity. Use of these soft magnetic insulation materials in the core can reduce the loss due to eddy currents. This loss is proportional to square of induced emf, whereas hysteresis loss and induced emf are proportional to frequency. Total core loss can be small if the sum of hysteresis loss and eddy current loss is small. Ferromagnetic materials are used in transformer cores, electromagnets, magnetic sensors etc.

Example 19.6

Total core loss in a transformer core at normal flux density is 1000 W at 25 Hz and 2800 W at 50 Hz. Calculate eddy current loss and hysteresis loss at 50 Hz and at 60 Hz.

Solution:

Consider eddy current loss as W_e and hysteresis loss as W_h at 25 Hz.

$$W_e = K \, \mathcal{E}^2$$

since, the eddy current loss is proportional to square of induced emf $\mathcal{E}$, whereas hysteresis loss and induced emf are proportional to frequency. Hence the eddy current loss is $4W_e$ and hysteresis loss is $2W_h$ at 50 Hz and these losses are $5.76W_e$ and $2.4W_h$ respectively at 60 Hz. Given that

$$W_e + W_h = 1000 \text{ W} \quad \text{and} \quad 4W_e + 2W_h = 2800 \text{ W}$$

Solving we get $W_e = 400$ W, $W_h = 600$ W. Hence,

at 50 Hz $W_e = 1600$ W $W_h = 1200$ W   and

at 60 Hz $W_e = 2304$ W $W_h = 1440$ W

19.6 SOFT AND HARD MAGNETS

When soft iron is magnetized, we get a magnet with low retentivity, which is also called soft magnet. Consequently, a soft magnetic material is magnetized and demagnetized easily and has narrow hysteresis loop with low H_c value. On the other hand, hard magnets have wide hysteresis loop, Fig 19.5. Soft magnetic materials are used in cores of transformers, generators and motors, whereas hard magnetic materials are used for making permanent magnets.

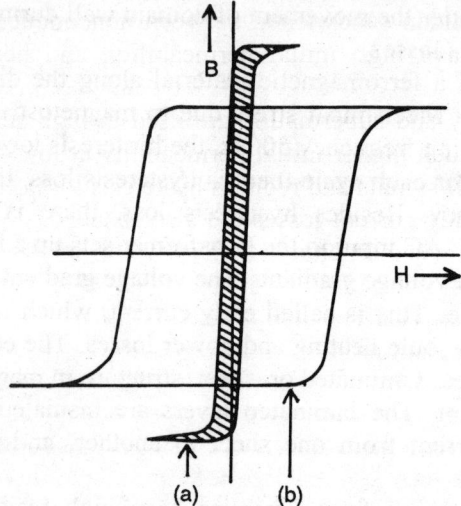

Fig.19.5 Schematic representations of **B-H** loop for (a) soft magnet and (b) hard magnet

Soft magnetic materials

Examples of soft magnetic materials are iron and its alloys e.g. iron-silicon alloy with 3w/o to 4w/o silicon. High saturation value of magnetic induction is preferred in nearly all applications. Soft magnetic materials have the following special properties:

   i)   A high value of saturation magnetization allows the material to do work.
   ii) Small value of remnant induction ensures little magnetization left, when the external field is removed.
   iii) A larger value of relative permeability μ_r produces high value of **B** with a smaller magnetic field intensity **H**.
   iv) Coercive field, H_c should be less so that demagnetization can be done with smaller values of reverse magnetic field.
   v)   Small hysteresis loop makes hysteresis loss to be very small.
   vi) High electrical resistivity
   vii) Rapid response to high frequency magnetic fields: Magnetic materials for high frequency applications should be such that their domains are realigned in each cycle, at an exceptionally rapid rate.

Uses of soft magnetic materials

Iron-silicon alloy reduces core losses. This is primarily due to the fact that the electrical resistivity of this alloy is more than that of soft iron, thereby decreasing the eddy current losses. Additionally the hysteresis core losses also decrease since the relative permeability of the alloy is more than that of soft iron. The alloy has less magnetostriction effect compared to pure iron. This reduces hysteresis loss and transformer noise. The disadvantages of the alloy are that its ductility and Curie temperature are less than that of iron. Iron-silicon alloys are used for making transformer cores in the form of laminated structures. They are suitable for operation at the power frequencies in the range of 50-60Hz. The oriented iron-silicon alloys are used for this purpose since they are electrically conductive and have small hysteresis loops. The laminated structure consists of thin sheets of Iron-silicon sandwiched between nonconducting dielectric materials. However these alloys do not have high initial permeability, and hence are not suitable in communication equipments.

Fe-Ni alloys such as permalloy and supermalloy are used in high sensitivity communication equipments. Fe-Ni alloys have much higher initial permeability at low field, which makes them suitable to detect or transmit small signals and reduces hysteresis loss. Since the resistivity of the alloy is substantially high, the eddy current losses are also reduced. When the communication frequency exceeds more than the megahertz range then eddy current losses become very high for metals and alloys. If a thin film of permalloy is vapour deposited on a glass substrate, it can be used in the above range of frequency. These alloys are specifically used in audio transformers, relays and for rotor and starter laminations. The initial permeability of Fe-Ni alloy (58w/o Ni) can be increased three to four times, first by heating it at a high temperature and then annealing the alloy in presence of magnetic field.

Magnetic materials are also used in computer memory. Memory comprises of bits of information in the form of strings of 0 and 1. Bits of information are stored by magnetizing the material in a certain direction; for example, pointing up of north pole is equivalent to storing the bit of information to be 1 and it is 0 if the north pole is down. The magnetic materials for such application should have a small square hysteresis loop i.e. a low remnant induction, a low saturation magnetization and a low coercive field. Alloys containing ferrites, manganese, magnesium or cobalt are suitable for such applications. In such materials a bit of information placed by a field remains stored. Small external field produces an abrupt change in magnetization that is required to change the bit 1 to 0.

Hard magnetic materials

Hard magnetic materials have a high value of coercivity and retentivity. The hysteresis loops of hard magnetic materials are wide and high i.e., there is quite high hysteresis energy loss in these materials. The initial permeability of hard magnetic materials is low. As a consequence of above characteristics, the magnetized hard magnetic material is difficult to demagnetize. The relative hardness of the magnetic material is related to the product of $\mathbf{B_r}$ and $\mathbf{H_c}$, which is roughly twice the demagnetization energy per unit volume. Thus, a higher value of $\mathbf{B_rH_c}$ indicates harder magnetic material. The higher value of coercivity and susceptibility indicates impeding of domain wall motion. The domain wall movement is impeded when small precipitates are present. The hard magnetic materials are used for making permanent magnets. Usually ferromagnetic alloys are used for this purpose. The example of such alloy is tungsten steel with 92.8w/o Fe, 6w/o W, 0.5w/o Cr, and 0.7w/o C. The two elements tungsten and chromium readily combine with carbon in steel to form tungsten and chromium carbide precipitate particles under the proper heat treatment conditions. The magnetic potential energy of a hard magnetic material is expressed by

$$\mathbf{PE} = (\mathbf{BH})_{max} \tag{19.16}$$

where $\mathbf{H}$ is the demagnetizing field and $(\mathbf{BH})_{max}$ is calculated from the demagnetizing curve of the material. This can be estimated as the area of the largest rectangle drawn under the B-H curve in the second or fourth quadrant. The energy product $(\mathbf{BH})_{max}$ is expressed in J/m^3 in SI unit and is also called the power. Permanent magnets like Alnico alloys are used for ammeters and voltmeters. A $Nd_2Fe_{12}B$ alloy has exceptionally high **BH** product and is recommended for magnetic resonance imaging (MRI).

Example 19.7

A rare earth alloy magnet has demagnetization curve expressed as $B = 1.5 - 4.5 \times 10^{-12}H^2$, where B is in Tesla, and H is in A/m. Find the value of H_c and $(BH)_{max}$.

Solution:

a) H_c is the value of H, for B = 0

 i.e. $0 = 1.5 - 4.5 \times 10^{-12} H_c^2$

 or $H_c = -5.8 \times 10^5$ A/m

b) $BH = 1.5 H - 4.5 \times 10^{-12} H^3$.

 Now d(BH)/dH =0 gives the value $(BH)_{max}$. Hence BH is a maximum.

 So $H_{max} = 3.33 \times 10^5$ A/m , $B_{max} = 1$ T.

 $(BH)_{max} = B_{max}H_{max} = 3.33 \times 10^5$ Wb-A/m^3

Some properties of a few important soft magnetic and hard magnetic materials are given in the Tables 19.2, 19.3 and 19.4. Ferromagnetism is observed in a large number of metal alloys, ceramics e.g. ferrites, garnets etc., ionic solids e.g. MnSb. Ferromagnetic materials can be transparent as well as opaque.

Table 19.2 Saturation magnetization and Curie temperature for selected ferromagnetic crystals

Material	Magnetization (A/m)x10^3 at		No. of Bohr, magneton per formula unit	Curie Temperature (°K)
	Room temp	0°K		
Co	1707	1748	2.22	1043
Gd	1400	1446	1.72	1388
Dy	485	510	0.606	627
Fe	-	2060	7.63	292
Ni	-	2920	10.2	88
CrO_2	670	870	3.4	318
$CuOFe_2O_3$	620	680	3.52	630
EuO	710	-	3.5	587
$FeOFe_2O_3$	515	-	2.03	386
$MgOFe_2O_3$	410	-	5.0	573
MnAs	480	-	4.1	858
MnBi	270	-	2.4	858
$MnOFe_2O_3$	135	-	1.3	728
MnSb	110	-	1.1	713
$NiOFe_2O_3$	-	1920	6.8	69
$Y_3Fe_5O_{12}$	130	200	5.0	560

Tale 19.3 Properties of soft magnetic materials

Material	Composition (%)	Initial relative Permeability (μ_r)	Saturation induction(T)	Hysteresis loss(J/m^3)	Resistivity (w/o)
Soft iron (Commercial)	99.9Fe	250	2.15	500	1.0×10^{-7}
Silicon-iron (oriented)	97Fe, 3Si	15000	2.01	40	4.7×10^{-7}
45 Permalloy	55Fe, 45Ni	2500	1.60	120	4.5×10^{-7}
Supermalloy	79Ni, 15Fe, 5Mo, 0.5Mn	10^5	0.80	2	6.0×10^{-7}

Table 19.4 Properties of hard magnetic materials

Material	Composition (%)	B_r (Tesla)	H_c (A-turn/m)	$(BH)max$ (J/m^3)	Curie Temp T_c (°C)	ρ (w/o))
Martensitic steel	98.1Fe, 0.9C, 1Mn	0.95	4000	1600	768	-
Tungsten steel	92.8Fe, 6W, 0.5Cr, 0.7C	0.95	5900	2600	760	3.0×10^{-7}
AlNiCo 1	12Al, 21Ni, 5Co, 2Cu, 60Fe	0.71	35000	11100	780	-
AlNiCo 5	8Al, 14Ni, 25Co, 3Cu, 50Fe	1.31	51000	47700	900	-
AlNiCo 8	7Al, 15Ni, 24Co, 3Cu, 51Fe	0.72	150000	40000	900	-
AlNiCo 12	6Al, 18Ni, 35Co, 8Ti, 33Fe	0.58	75600	2700	-	-
Cunife	60Cu, 20Ni, 20Fe	0.54	44000	12000	410	1.8×10^{-7}
Co_5Sm	35Sm, 65Co	0.90	756000	200000	725	-
Barium ferrite	$BaO.6Fe_2O_3$	0.32	240000	20000	450	10^4
$Nd_2Fe_{12}B$		1.20	875000	360000	310	-
Fe-Cr-Co	30Cr, 10Co, 1Si, 59Fe	1.17	46000	34000	-	-

Example 19.8

Select an appropriate material for permanent magnet with highest possible power for a device to be used in spacecraft, which at the time of entry to earth's atmosphere is exposed to a temperature of 600°C and a magnetic field of 5×10^4 $A.m^{-1}$.

Solution:

The material should have coercive field, H_c greater than 5×10^4 $A.m^{-1}$ and Curie temperature greater than 600°C. If we examine Table 19.4, we find AlNiCo 12, AlNiCo 5 and Co_5Sm are the materials out of which choice can be made. We find Co_5Sm has maximum power i.e. $(BH)_{max}$. So Co_5Sm is the appropriate material although it is very costly.

19.7 ANTIFERROMAGNETISM

The explanation of ferromagnetism involves the exchange interactions between adjacent atoms that result in parallel alignment of spin moments of neighbouring atoms or ions. This interaction is said to be positive since its effect is to add the magnetic moments of neighbouring atoms. In another type of exchange interaction, which is negative, the magnetic moments of neighbouring atoms or ions are aligned in opposite directions (antiparallel). This is called antiferromagnetism, Fig. 19.6.

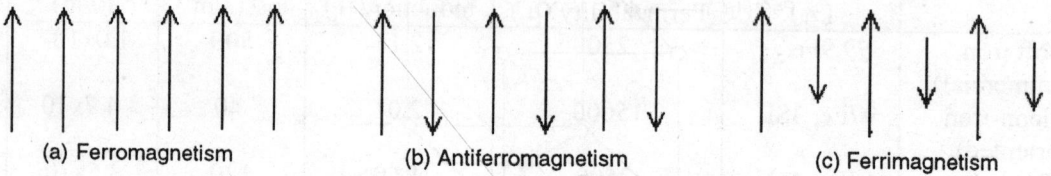

(a) Ferromagnetism   (b) Antiferromagnetism   (c) Ferrimagnetism

Fig. 19.6 Alignment of magnetic dipoles for different types of magnetism

As a result of antiparallel arrangement of magnetic moments of neighbouring atoms, the magnetic moment of the material is zero as a whole. Ceramic like MnO is an example of antiferromagnetic material. In this ionic compound Mn^{2+} and O^{2-} ions are arranged as shown in Fig 19.7. The O^{2-} ions do not have any magnetic moment. whereas Mn^{2+} ions possess a net magnetic moment specifically arising due to spin. The Mn^{2+} ions are arranged in arrays in the crystal structure of MnO such that the moments of adjacent ions are antiparallel, leading to zero magnetic moment of MnO. Antiparallel arrangement of Mn^{2+} in MnO is the source for antiferromagnetism in MnO.

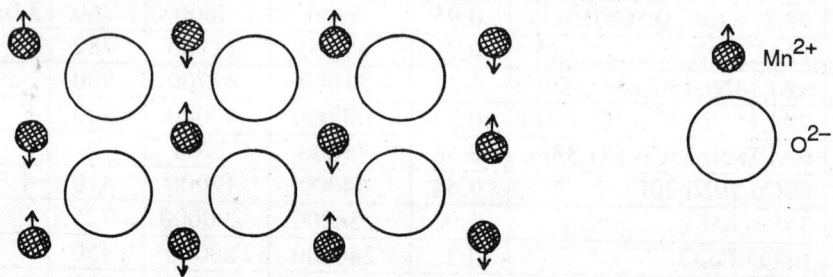

Fig. 19.7 Schematic representation of the antiparallel arrangements of Mn^{2+} in MnO.

19.8 FERRIMAGNETISM

Ferrimagnetism is a permanent magnetization exhibited by some ceramics. Ferrimagnets show large magnetic effects, which disappear above Curie temperature. The ferrimagnets can be magnetically hard or soft varities. They exhibit hysteresis effects like ferromagnets. The origin of ferrimagnetism is different from ferromagnetism. Different spin orientations and magnetization of magnetic ions are responsible for ferrimagnetism. These ceramic magnetic materials possess high electrical resistivity. Ferrites and garnets are important classes of ferrimagnetic materials. The saturation magnetization of a ferrimagnetic substance is much less than that of ferromagnetic substances.

Cubic ferrites

To explain the source of ferrimagnetism, cubic ferrite is considered. The cubic soft ferrite has the formula $MO.Fe_2O_3$, where M is a divalent metal ions e.g. Fe^{2+}, Mn^{2+}, Ni^{2+} or Zn^{2+}. These ionic ferrites have inverted spinel structure, which is a modification of the spinel structure of the mineral spinel $(MgO.Al_2O_3)$. Consider the cubic unit cell of ferrite divided into eight cubes, Fig.19.8 and each one is called as subcell. Each subcell has fcc packing of O^{2-} ions and therefore it has eight tetrahedral and four octahedral interstitial sites. The cations occupying interstitial sites

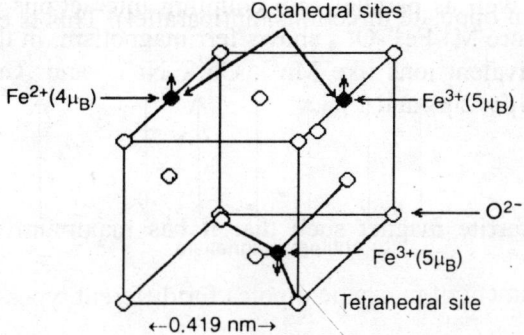

Fig. 19.8 Schematic arrangement of ions in magnetite.

have different spin orientations as specified in Table 19.5. Since O^{2-} ions do not have magnetic moment, the total value of magnetic moment is $4 \times 9.27 \times 10^{-24}$ Am^2 for magnetite ($FeO.Fe_2O_3$) subcell. If one side of the subcell is 4.19Å, the saturation magnetization is

$$M_s = \frac{4 \times 9.27 \times 10^{-24}}{(4.19 \times 10^{-10})^3} A/m = 5.04 \times 10^5 \ A/m$$

There are seven ions altogether in the subcell and $7 \times 8 = 56$ ions in the unit cell of inverted spinel structure. The ferrite unit cell can be visualized as consisting of 8 subcells each of which has fcc

Table 19.5 Ionic arrangement and total magnetic moment per ferrite molecule.

Structure	Ion	Number in subcell	Site	Spin direction	Ion moment in μ_B
Inverted spinel $FeO.Fe_2O_3$	Fe^{3+}	1	O	↑	5
	Fe^{2+}	1	O	↑	4
	Fe^{3+}	1	T	↓	−5
				Overall magnetic moment = $4\mu_B$	
Spinel $ZnO.Fe_2O_3$	Zn^{2+}	1	T	0	0
	Fe^{3+}	1	O	↑	5
	Fe^{3+}	1	O	↓	−5
				Overall magnetic moment = 0	

Note: T stands for tetrahedral interstitial site. O stands for octahedral interstitial site.

packing of O^{2-} ions. Therefore, there are $8 \times 8 = 64$ tetrahedral and $8 \times 4 = 32$ octahedral interstitial sites. In case of inverted spinel structure M^{2+} ions (spin up) occupy 8 out of 32 octahedral sites and 8 Fe^{3+} ions (spin up) occupy other 8 octahedral sites whereas remaining 8 Fe^{3+} ions (spin down) occupy 8 out of 64 tetrahedral sites. In spinel structure, M^{2+} ions occupy 8 out of 64 tetrahedral sites whereas Fe^{3+} ions occupy 16 out of 32 octahedral sites. Out of these 8 Fe^{3+} ions have spin up and other 8 Fe^{3+} ions have spin down. As one can observe from Table 19.5, such that the net magnetic moment is zero for spinel structure.

There are antiparallel as well as parallel spin coupling interactions present in ferrite. Other materials with similar structure $M^{2+}Fe^{3+}_2O^{2-}_4$ shows ferrimagnetism. In these materials instead of Fe^{2+} ions there are other divalent ions like Mn^{2+}, Co^{2+}, Ni^{2+} and Cu^{2+} with respective spin magnetic moments as 5 μ_B, 3μ_B, 2 μ_B and 1 μ_B.

Example 19.9

Select material for cubic ferrite magnet such that it has maximum saturation magnetization 6×10^5 A.m^{-1}.

Solution:

The magnetic moment per meter cube of Fe_3O_4 is given by

$$\frac{32 \mu_B}{(8.37 \times 10^{-10})^3} = 5.06 \times 10^5 \text{ A.m}^{-1}$$

where 8.37×10^{-10} m is lattice parameter for Fe_3O_4. To obtain a higher value of magnetization, we replace some of the Fe^{2+} ions with Mn^{2+} ions, since Mn^{2+} has more Bohr magnetons per atom. Let x be the fraction of Fe^{2+} ions that are replaced by Mn^{2+} ions. So the total magnetic moment per unit volume is given by

$$M_s = \frac{8[(1-x) 4\mu_B + (x) 5 \mu_B]}{(8.37 \times 10^{-10})^3} = \frac{8[5x+4 - 4x] \times 9.27 \times 10^{-24}}{5.864 \times 10^{-28}} = 6 \times 10^5 \text{ A.m}^{-1}$$

x = 0.744 i.e. 74.4 % Fe^{2+} ions are to be replaced with Mn^{2+} ions to design the required material.

Hexagonal ferrites

There are other magnetic materials like hexagonal ferrites, which have a complex hexagonal crystal structure similar to that of the mineral magnetoplumbite. These ferrites have the formula $MO.6Fe_2O_3$, where M is Ba, Sr, or Pb. The common examples of hexagonal ferrites are $BaFe_{12}O_{19}$ and $PbFe_{12}O_{19}$. In this ferrite, ions are arranged as given in Table 19.6. The hexagonal ferrites make permanent magnets.

Table 19.6 Ionic arrangement and total magnetic moment per hexagonal ferrite molecule.

Ion	Number	Site	Spin direction	Ion moment in μ_B	Total moment in μ_B
Fe^{3+}	2	T	↑	5	10
	2	O	↑	5	10
	7	O	↓	−5	−35
	1	F	↓	−5	− 5

Overall magnetic moment $= -20\mu_B$

Note: T stands for tetrahedral interstitial site,
O stands for octahedral interstitial site,
F stands for five-fold coordination.

Garnet

Garnets are alloys of Fe_2O_3 and trivalent oxides having a formula $3M_2O_3.5Fe_2O_3$ or $M_3Fe_5O_{12}$, where M= Y, Gd, Sm etc. One of the most important magnetic garnets is yttrium-iron-garnet (YIG), $Y_3Fe_5O_{12}$. Garnets also show ferrimagnetism. In magnetic garnets, M is a rare earth element like Y. The resulting magnetic moment of this garnet is due to two oppositely magnetized lattices of Fe^{3+} ions. In each formula unit, three Fe^{3+} ions have magnetic moments in a specific direction and the remaining two Fe^{3+} ions have magnetic moments in the opposite direction. This gives a resultant magnetization of $5\mu_B$ per unit. YIG and derivatives based on it e.g. $Gd_3Ga_5O_{12}$ are used in magnetic bubble memory and microwave devices since garnets show magneto-optic as well as magneto-acoustic effects. Thin films of YIG are transparent to visible light. Consider a garnet film in xy plane, and magnetization vector **M** is perpendicular to the film plane and points in the positive z-direction. When plane polarized light passes parallel to the direction of M, the plane of polarization is rotated clockwise; the rotation of plane of polarization is anticlockwise when the direction of plane-polarized light is antiparallel to **M** i.e. **M** points in negative z-direction. This is magneto-optic Faraday effect. With the help of analyzer plate light and dark domain pattern can be observed in YIG thin films. Dark domain has M vector pointing up whereas it is pointing down for light domain.

Magnetic bubble memory devices are made out of garnets. They have capacities in excess of a megabit. These devices have high storage densities (10^9 bits/in^2), no mechanical wear during operation, nonvolatile memory and read-write memory for a wide range of temperatures.

19.9 EFFECT OF TEMPERATURE ON MAGNETIZATION

When the temperature of a magnetic material increases, the thermal agitation of lattice disrupts the spin alignment and hence the intensity of magnetization is decreased, Fig.19.9. For example, the saturation magnetization of nickel is 5.1×10^5 A/m at temperature $0°K$. As the temperature increases the magnetization decreases and becomes zero at $T_c = 631°K$. Similarly, iron has saturation magnetization 3.14×10^6 A/m at $0°K$ and the magnetization becomes zero at $T_c = 1043°K$. At a temperature equal to or greater than T_c, the exchange forces trying to align the atomic magnetic moments are not strong enough to overcome the randomizing effect of thermal agitation and the material behaves like a paramagnet. The Curie temperature for Co is $1388°K$. This temperature is a function of exchange energy. Cobalt has highest exchange energy and highest Curie temperature whereas Gd has low exchange energy and low Curie temperature.

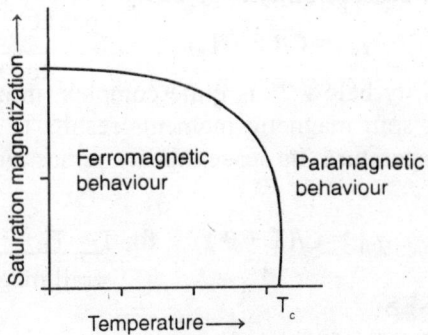

Fig.19.9 Intensity of magnetization versus temperature.

Example 19.10

Explain with reasons why dysprosium has highest saturation magnetization at 0°K and iron has highest saturation magnetization at room temperature (300° K).

Solution:

At 0°K, there is no thermal energy. Therefore, all the spins align. Dy has maximum saturation magnetization since Dy has maximum magnetic moment 10.0 μ_B per atom, Table 19.2. At 300°K, Gd and Dy behave as paramagnetic material since they have Curie temperatures less than 300°K. Out of Fe, Co and Ni, Fe has highest magnetic moment 2.22 μ_B and hence it has highest saturation magnetization at room temperature.

The susceptibility can be expressed as a function of temperature for various magnetic materials, Fig 19.10. The paramagnetic substances obey Curie law and according to this law susceptibility is given by

$$\chi_m = C/T \tag{19.17}$$

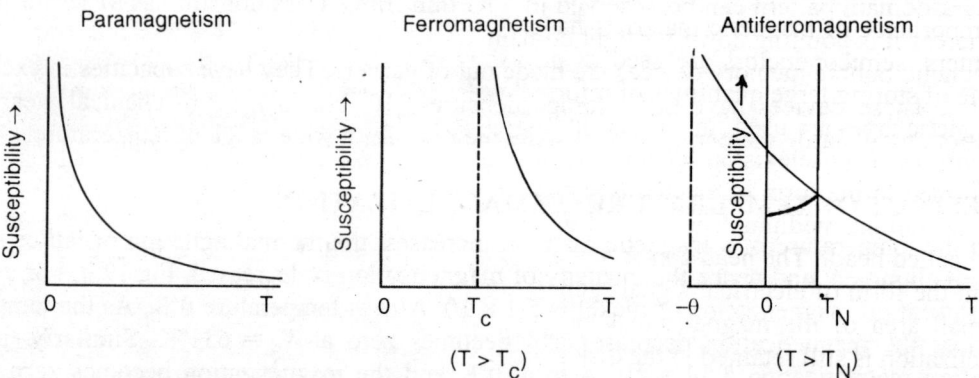

Fig 19.10 Susceptibility as a function of temperature in paramagnets, ferromagnets, and antiferromagnets

where C is Curie constant. This law applies in the range of temperature where magnetic energy is much less than thermal energy per atom. The ferromagnetic substances obey Curie-Weiss law at a temperature $T > T_c$ and their susceptibility is given by

$$\chi_m = C/(T - T_c) \tag{19.18}$$

The variation of susceptibility below T_c is quite complex. In antiferromagnetic substances, the antiparallel arrangement of spin magnetic moments results in zero magnetic moment below a temperature called Neel temperature. Its susceptibility versus temperature curve has a weak cusp at $T=T_N$ and χ_m is given by

$$\chi_m = C/(T + \theta) \qquad \text{for } T > T_N \tag{19.19}$$

19.10 METALLIC GLASSES

Metallic glasses are metal alloys containing basically one or more transition metals e.g. Fe, Co, Ni and 15%– 25% of metalloids e.g. silicon, boron and carbon manufactured in a totally different process so that the dominant characteristics is a noncrystalline structure. Metallic glasses are produced by rapid solidification of molten alloys during splat cooling. Metallic glass is obtained in the form of continuous ribbon of about 2.5×10^{-5}m thick and 1.5×10^{-3} m wide when molten alloy is poured over a continuously rotating chilled roller.

Metallic glass, produced by this method contains large concentration of impurity atoms and hence its electrical resistivity is much higher than that for alloy containing 96% iron and 4% silicon. Increase in resistivity reduces the eddy current losses remarkably. The metallic glass does not contain grain boundaries due to glassy matrix and this reduces hysteresis losses. Metallic glass is very hard and corrosion resistant. Metallic glasses are magnetically soft which is indicated by high saturation magnetic induction, maximum permeability and also easy demagnetization i.e. low coercive field. A typical example of metallic glass is 79a/o Fe, 13a/o B and 8a/o Si. The domain walls in this material are able to move with ease because of absence of grain boundaries and low crystal anisotropy. Metallic glasses are used in low-loss power transformer, magnetic switches in pulse transformers, sensing cores in current transformers, recording head in magnetic sensors (i.e. in audio and computer tape heads) etc.

19.11 MAGNETIC STORAGE

The importance of magnetic materials has become more in this era of information explosion. In computers, semiconductors are used as primary memory whereas magnetic disks and tapes are capable of storing large quantities of information at low costs and are called auxiliary memories.

Magnetic tapes are used for storage and reproduction of sequential audio and video information in recording and television industries. The visual image, sound or computer bytes can be transformed in the form of electrical signal and these are retained within very small segments of magnetic storage medium. The transfer of data to and from the tape or disk is done by means of a device called head. The head consists of a coil wound around a magnetic core with a gap cut in it. Data in the form of electrical signals generate a magnetic field across the gap, which magnetizes the small area of the magnetic disk or tape near the head. When the signal is removed, the magnetization is still retained in the disk and this is the process of storing of electrical signals. The stored information is retrieved with the help of head using similar method. During retrieval of stored data from the magnetic disk or tape a voltage signal is induced (Faraday's law) when there is a change in the magnetic field as the tape or disk passes by the head coil gap. These voltage signals are amplified and converted into the original form i.e. sound, picture or information stored initially in the storage media. Therefore, the magnetic recording systems require combination of the properties of soft as well as hard magnetic materials. The soft magnetic materials e.g. soft MnZn ferrite and permalloy are required for recording and playback heads and hard magnetic materials for storage media e.g. γ-Fe_2O_3, γ-Fe_2O_3 (Co). Usually magnetic media consists of very small needle like particles of γ-ferrite or other doped alloy, which are bonded to a polymeric film in case of a magnetic tape or to a metal or a polymer disk using the techniques of evaporation, sputtering, or plating. These particles are aligned with their long axis parallel to the direction of motion of the tape or disk during manufacturing process. It is possible to magnetize each of these particles as single domain having resultant magnetic moment along its long axis. Two magnetic states are possible for each domain, one with saturation magnetization along its axial direction and other opposite to it. These two states make it possible to store data or information using binary digits '1' and '0'. The magnetic materials used for this purpose should have the following characteristics:

  i) Low remanence
  ii) Low saturation magnetization
  iii) Low coercive field

The hysteresis loops of these materials are relatively large and square. These characteristics ensure permanent storage. The reversal of magnetization can be done for a small range of applied field.

19.12 MAGNETOSTRICTION

When a magnetic material is magnetized, there is elongation or reduction of the material by a factor of the order of 10^{-6} in the direction of magnetization. This magnetically induced reversible elastic strain is magnetostriction. The change in energy arising out of the elastic stresses during the magnetically induced contraction or elongation process is called magnetostrictive energy. In case of iron magnetostriction is positive at low fields and negative at high fields.

During the process of magnetization the atomic magnetic moments rotate to have parallel alignment with the magnetic field. Due to these rotations the bond length between the atoms in a

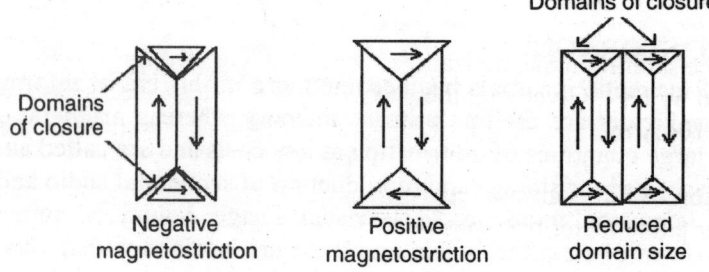

Fig 19.11 Magnetostriction in cubic magnetic materials

ferromagnetic material changes, which cause the dipoles to come nearer or further as compared to their equilibrium distance. As a result, magnetic moments may either attract or repel leading to elongation or reduction of the length of material in the direction of the magnetic field.

Let us consider the magnetostriction on the domain structure of a cubic crystalline material in equilibrium. Such crystals have domains with magnetic moment antiparallel to each other and a triangular-shaped domain as shown in Fig. 19.11. These triangular domains are called domains of closure, since they eliminate the magnetostatic energy in the presence of an external field. As a result of negative magnetostriction, the domains will be shortened and boundaries of domain closure will not fit with the boundaries of domains. In the case of positive magnetostriction also similar mismatch will occur. This will increase magnetostrictive stresses. Domain size reduces as a result of equilibrium. Also the domain wall energies become a minimum.

19.13 CERAMIC MAGNETS

Some ceramic materials are magnetic in nature. The primary reason why ceramic magnets (ceramets) are attractive is that they combine the traditional properties of metallic magnets along with properties exclusive to ceramics such as high electrical resistivity. Since oxygen in ceramic magnets gives fewer ferromagnetic atoms per unit volume (the dilution effect), saturation magnetization and magnetic permeability are lower in them than in metallic magnets. Ceramic magnets are normally ferrimagnetic. Typical examples of ceramic magnets are ferromagnetic oxides, e.g. ferrites.

EXERCISES

19.1   Define magnetic induction **B** and magnetization **M**. Find the relationship between **B** and magnetic field strength **H**.

19.2   How many 3d electrons are there per atom in Cr, Mn, Fe, Co, Ni and Cu?

19.3   The magnetic flux density within a bar of a material X is 0.882 Wb/m^2 when $H = 7 \times 10^5$ A/m. Calculate the magnetic permeability and magnetic susceptibility for this material.
Ans: 1.26×10^{-6} H/m ; 2.68×10^{-3}

19.4   Consider an alloy of Ni and Co having observed magnetization of 0.88×10^6 $A.m^{-1}$. The crystal structure of the alloy is fcc and lattice parameter is 3.544Å. Find the a/o Ni present in the alloy. Calculate the theoretical value of M.
Ans: 60%; 2×10^6 $A.m^{-1}$

19.5   Calculate the saturation magnetization in an inverted spinel $FeO.Fe_2O_3$, if 20 a/o of Fe^{2+} ions are replaced by Ni^{2+}. Given that the lattice parameter of the crystal is 8.37Å.
Ans: 4.5×10^5 A/m

19.6   What fraction of Fe^{2+} ions are replaced by Cu^+ ions in magnetite so that the total magnetic moment per cubic meter becomes 4.68×10^5 $A.m^{-1}$?
Ans: 7.5%

19.7   Manganese ferrite has eight formula units per unit cell. Hence its chemical formula is $(MnFe_2O_4)_8$. The number of Bohr magnetons associated with each Mn^{+2} ion is considered to be 4.6. Calculate the saturation magnetization if the density of the material is $5 g/cm^3$.
Ans: $\cdot 5.57 \times 10^5$ A/m

19.8   Calculate the saturation magnetization and the saturation flux density for cobalt given that the density of C_o is equal to 8.9 g/cm^3.
Ans: 2.53×10^6 A/m; 3.18 T

19.9   A unit cell of Fe_3O_4 has $8Fe^{+2}$, $16Fe^{+3}$ and 32 O^{2-} ions. It is cubic with a = 8.37Å. The ferrous and one-half of their ferric ions have their magnetic moments up; the remainder has their moments down. Compute the saturation magnetization of Fe_3O_4.

19.10 An iron cube of side one meter has an atomic magnetic moment of 2.22 Bohr magnetons and saturation magnetization 1.74×10^6 A/m. Calculate its mass.
Ans: 7833 kg

19.11 A coil of Cu wire has 450 turns and is 0.3 m long. It carries a current of 10 A.
(i)   What is the magnitude of H?
(ii) Compute flux density B, if coil is in vacuum
(iii) Compute B if coil is in a medium of $\mu_r = 3.2$
(iv) Compute magnitude of magnetization M.
Ans: (i) 1.5×10^4 A/m; (ii) 1.88×10^{-2} Wb/m^2; (iii) 6.03×10^{-2} Wb/m^2; (iv) 3.3×10^4 A/m

19.12 How is the Curie temperature related to the exchange energy? Arrange Fe (768°C), Gd (16°C), Co (1127°C) and Dy (−168°C) in order of increasing exchange energy, the temperatures in parentheses being the Curie temperatures.

19.13 Explain qualitatively why copper is diamagnetic even though Cu atom has an outer electron in the configuration $[Ar]3d^{10}4s^1$. Copper has diamagnetic susceptibility of -0.5×10^{-5}. Find the induction B and the magnetization M when the applied field is 121 kA m^{-1}.
Ans: $B = 0.152$ Wb /m^2; $M = -0.605$ A/m

19.14 The variation of B versus H is given by $H(J/m^3) = -2 \times 10^5 B + 3 \times 10^5$ for a magnetic material in second quadrant.
a) Plot BH versus H.
b) Find the power of the magnetic material.
c) Calculate the value of (BH)$_{max}$.
Ans: (b) 1.125×10^5 J /m^3; (c) 1.5×10^5 A/m

19.15 A ferrite core of 15 kg has a rectangular hysteresis loop with $H_c = -0.5$ A.m^{-1} and $B_s = 0.3$ T. Calculate the rise in temperature of the core after one magnetizing cycle assuming no other loss and the process is carried out adiabatically. Given that the density of ferrite is 5000 kg/m^3 and specific heat is 850 J/kg-°C. After how many hours the temperature will rise beyond Curie temperature i.e. 420°C starting from room temperature during the 50Hz operation?
Ans: $\Delta T = 1.4 \times 10^{-7}$ °C; No. of hours $= 1.55 \times 10^4$ hours

20

Dielectric Properties of Materials

20.1 INTRODUCTION

Dielectric materials have very wide applications in electrical and electronic industries as insulators. A large number of ceramics are dielectrics. Dielectrics have polar molecules, which try to align in the direction of the applied electric field. They also possess very interesting optical properties. Dielectrics find wide applications in capacitors. Certain special types of dielectric materials can convert weak fluctuations in pressure into electrical signals. These are piezoelectric materials, which are widely used in transducers.

20.2 DIELECTRIC PROPERTIES

One of the important properties of dielectric materials is that they can be polarized in the presence of electric field. When a dielectric medium is placed in between the capacitor plates, its capacitance increases. Let us consider that an electric field E is applied across the two parallel plates of a capacitor, then one of the plates, which is connected to the positive terminal of the battery, acquires positive charge say $+q$, while the other plate acquires negative charge $-q$. The charge q is directly proportional to the voltage V between the two plates and is expressed as

$$q = CV \qquad (20.1)$$

where C is called the capacitance of the capacitor. The capacitance is the measure for the capacity of the capacitor to hold electric charges. When the area A of the plate is much greater than the distance of separation d between the plates, the capacitance is given by

$$C = \varepsilon_o \, (A/d) \qquad (20.2)$$

where ε_o ($= 8.854 \times 10^{-12}$ F/m) is the permittivity of free space (vacuum). When we place a dielectric material of permittivity ε in between the two plates, the capacitance becomes

$$C' = \varepsilon(A/d) = \varepsilon_o \, \varepsilon_r \, (A/d) \qquad (20.3)$$

where $\varepsilon_r = \varepsilon/\varepsilon_o$ is relative permittivity of the medium. ε_r is also called the dielectric constant of the medium and is denoted by K. Now the charge q', acquired by the capacitance plates in presence of dielectric medium is more than q and is expressed as

$$q' = C'V = Kq \qquad (20.4)$$

The energy stored in a capacitor is given by

$$E = (\tfrac{1}{2})\, C'V^2 = \varepsilon_0 KAV^2/2d \qquad (20.5)$$

Thus, the energy stored in the capacitor is directly proportional to the dielectric constant K. The dielectric constants of some commonly used materials are given in Table 20.1.

Table 20.1 Dielectric constants ε_r of some selected materials
(at room temperature and at $v = 10^6$ Hz)

Materials	ε_r	Materials	ε_r
Air	1	Nylon 66	3.0 – 3.5
Alumina	8.8	Paper	7.0
Barium oxide	3.4	Phenolformaldehyde	4.5 – 5.0
Cordierite ceramics	4.5 -5.4	Polyester resin	3.1
Cellulose acetate	3.5 -5.5	Polyehylene	2.35
Diamond	5.5	Polyisobutylene	2.23
Epoxy resin	3.5-3.6	Polystyrene	2.55
Forsterite (Mg_2SiO_4)	6.22	Polytetrafluoroethyene	2.1
Glass (high-lead)	19.0	Polyvinyle chloride	3.3
Glass (pyrex)	4.0- 6.0	Porcelain	6.0 –7.0
Glass (silica)	3.8	Potassium bromide	4.9
Glass(soda-lime- silica)	6.9	Potassium chloride	4.75
Glass (vycor)	4.0- 6.0	Potassium iodide	5.6
Lithium fluoride	9.0	Rubber (butyle)	2.56
Lucite	2.63	Silica (thin film)	3.8
Magnesium oxide	9.65	Titanates	15– 12000
Mica	4.5-7.0	Titanium dioxide	14– 110
Mullite($3Al_2O_3.2SiO_2$)	6.6	Urea formaldehyde	6.4 – 6.9
Neoprene	6.26	Zircon porcelain	8 – 9

A dielectric medium responds to the electric field differently from free space because in the dielectric medium there is charge displacement in the presence of electric field. This produces an electric field that neutralizes the applied field partially or fully.

Let us consider an electric field **E** applied across a dielectric medium. The positive charges in the dielectric medium are displaced towards the negative electrode and the negative charges in the opposite direction. The charge displacement induces dipole moment in the medium. The dipole moment per unit volume is called the polarization **P** of the dielectric material, Fig. 20.1. The electric flux density D, due to the applied electric field E, within a dielectric medium of relative permittivity ε_r can be expressed as

$$\mathbf{D} = \varepsilon_0\, \varepsilon_r\, \mathbf{E} \qquad (20.6)$$

D can also be written in terms of polarization, **P** as,

$$\mathbf{D} = \varepsilon_0\, \mathbf{E} + \mathbf{P} \qquad (20.7)$$

D is also called displacement vector. **P** is proportional to **E** as given below

$$\mathbf{P} = \varepsilon_0\, (\varepsilon_r - 1)\, \mathbf{E} \qquad (20.8)$$

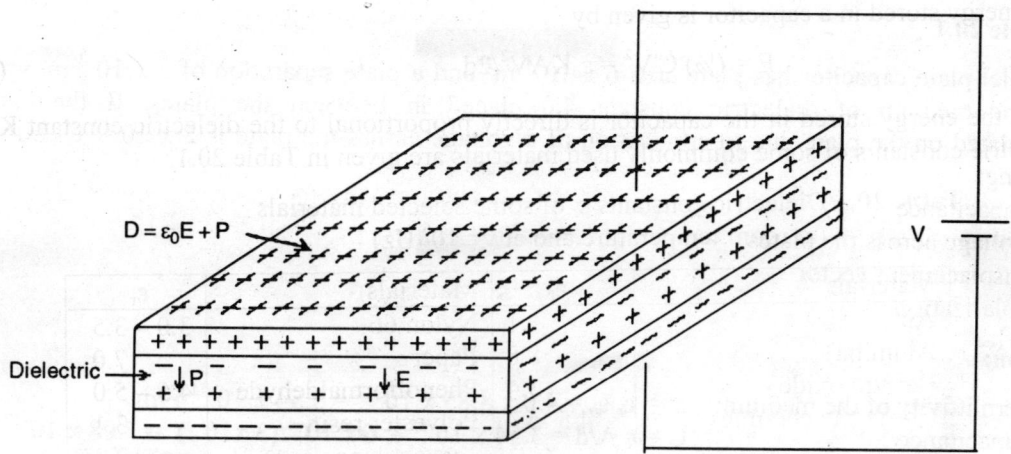

Fig. 20.1 Capacitor with dielectric medium

The electrical susceptibility χ of the medium is expressed as

$$\chi = \frac{P}{\varepsilon_o E} = \varepsilon_r - 1 \qquad (20.9)$$

Consider there are N dipoles per unit volume aligned parallel to one another in the presence of the applied field **E** and p is the average dipole moment. Then,

$$\mathbf{P} = \mathbf{N\, p} \qquad (20.10)$$

The average dipole moment of each dipole is proportional to the local electric field **E'** which is the electric field experienced by a dipole within a dielectric medium and is greater than the applied field **E**. The average dipole moment can be expressed as

$$\mathbf{p} = \alpha\, \mathbf{E'} \qquad (20.11)$$

and $\qquad\qquad\qquad \mathbf{P} = \mathbf{N}\, \alpha\, \mathbf{E'} \qquad (20.12)$

where α is the constant of proportionality and is called the polarizability of the medium. The dipolar polarizability of the medium can also be defined as average dipole moment per unit electric field. The polarizability is constant for a given material. The polarizability can be quadrupolar when we consider average quadrupole moment per unit electric field. Polarizability of certain molecular solid is given in Table 20.2.

Table 20.2 Polarizability of selected dielectric materials

Material	Polarizability $(10^{-40}\ F\ m^2)$	Material	Polarizability $(10^{-40}\ F\ m^2)$
Ar	1.45	Kr	2.18
CCl_4	13.0	LiF	6.8
CH_4	2.7	NaCl	8.9
He	0.18	Ne	0.35
KI	15.3	Xe	3.52

Example 20.1

A parallel plate capacitor has plate area 6×10^{-4} m^2 and a plate separation of 3×10^{-3} m, with a dielectric medium of dielectric constant 4.0 placed in between the plates. If the charge accumulated on the plate due to application of a voltage across the plate is 2×10^{-10} C, find the following:

  i) Capacitance
  ii) Voltage across the plates
  iii) Displacement vector
  iv) Polarization

Solution:

  i) Permittivity of the medium: $\varepsilon = \varepsilon_r \varepsilon_o = 4.0 \times 8.85 \times 10^{-12} = 3.54 \times 10^{-11}$ F/m
     Capacitance : $C = \varepsilon\, A/d = 3.54 \times 10^{-11} \times 6 \times 10^{-4}/(3 \times 10^{-3}) = 7.08 \times 10^{-12}$ F
  ii) Voltage across the plates: $V = Q/C = 2 \times 10^{-10}$ C/ $7.08 \times 10^{-12} = 28.25$ V
  iii) Displacement vector : $D = \varepsilon\, V/d = 3.54 \times 10^{-11} \times 28.25/(3 \times 10^{-3}) = 3.33 \times 10^{-7}$ Cm^{-2}
  iv) Polarization: $P = D - \varepsilon_o V/d = 3.33 \times 10^{-7} - 8.85 \times 10^{-12} \times 28.25/\, 3 \times 10^{-3}$
                        $= 2.50 \times 10^{-7}$ Cm^{-2}

Designing of capacitor

Capacitor can be designed by taking series or parallel combinations of capacitors. Suppose there are n capacitors in series, each of capacitance C. The voltage across each capacitor is V, say. The voltage across the combination of 'n' capacitors is given by

$$V_{eff} = nV = \frac{nQ}{C} \qquad (20.13)$$

The effective capacitance of the combination of capacitors is given by

$$C_{eff} = \frac{Q}{V_{eff}} = \frac{C}{n} = (\varepsilon_o K/n)\,(A/d) \qquad (20.14)$$

In case of parallel combination of 'n' capacitors the charge on the plate of equivalent capacitor is

$$Q_{eff} = nQ = nCV \qquad (20.15)$$

and also

$$Q_{eff} = C_{eff}V \qquad (20.16)$$

Hence, $C_{eff} = nC = n\varepsilon_o K(A/d) \qquad (20.17)$

If there are 'n' parallel plates, the number of capacitors is 'n–1'.

Example 20.2

Design a capacitor with dielectric medium as mica, which is capable of storing 5 μC when a voltage 150 V is applied. Given the thickness of the mica sheet to be 2 μm.

Solution:

                  Dielectric constant K = 7 for mica

$$\text{Capacitance } C = Q/V = (5/150) \times 10^{-6}F$$

i.e. $C = \varepsilon_0 \varepsilon_r (A/d)(n-1) = 3.33 \times 10^{-8} F$

$$A(n-1) = \frac{Cd}{\varepsilon_0 \varepsilon_r} = \frac{3.33 \times 10^{-8} \times 2 \times 10^{-6}}{8.85 \times 10^{-12} \times 7} = 1.076 \times 10^{-3} \text{ m}^2$$

where n is the number of conducting plates. Considering single capacitor one gets n=2. Hence the area of the condenser plate is A = 1.076 x 10^{-3}m^2. One can design a set of capacitors connected parallely to give same capacitance. As the number of capacitors is increased, the area of the plate is decreased as shown in below:

No. of capacitors	Number of conducting plates	Area A in m^2 x 10^{-3}
2	3	0.54
3	4	0.36
4	5	0.27

20.3 TYPES OF POLARIZATION

Polarization can arise due to various interactions between permanent or induced electric dipoles with an applied electric field and varies linearly with the field. Depending upon the interactions, polarizations are classified as follows:

a) Electronic Polarization

When an electric field is applied to a dielectric medium, the electronic charge cloud around the nucleus is displaced towards the positive terminal. As a result, the positive charge center of the atom is shifted with respect to the negative charge center through a distance d and a dipole moment arises in the atom. Such phenomena occur throughout the medium, causing electronic polarization (P_c). Monoatomic gases exhibit electronic polarization, Fig. 20.2.

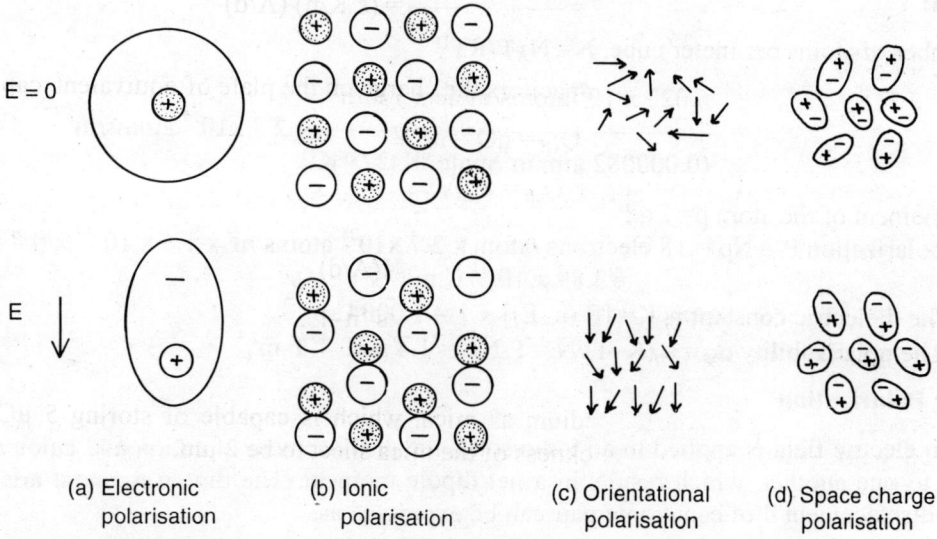

(a) Electronic polarisation   (b) Ionic polarisation   (c) Orientational polarisation   (d) Space charge polarisation

Fig. 20.2 Electronic, ionic, orientational and space charge polarization.

Two opposing forces are acting on the nucleus in the polarized dielectric medium -- one due to applied electric field $\mathbf{E}$, $\mathbf{F}_E$ and the other coulomb force $\mathbf{F}_C$,

$$\mathbf{F}_E = Zq\mathbf{E} = -\mathbf{F}_C$$

$$ZqE = \frac{Zq(Zqd^3/R^3)}{4\pi\varepsilon_o d^2} \qquad (20.18)$$

where R is the radius of the atom and d is the distance of separation between negative and positive charge centers. Hence

$$d = \frac{4\pi\varepsilon_o R^3}{Zq}E \qquad (20.19)$$

The electronic dipole moment of the atom is $p_e = Zqd$ and electronic polarizability

$$\alpha_e = \frac{p_e}{E} = 4\pi\varepsilon_o R^3 \qquad (20.20)$$

Electronic polarization is given by

$$P_e = N\alpha_e E \qquad (20.21)$$

where N is the number of atoms per unit volume.

Example 20.3

Calculate the electronic polarization, which causes the average displacement of the electrons with respect to nucleus of an argon atom by 0.5×10^{-18} m at 1 atmospheric pressure and 0°C. If the applied electric field is 10^4 V/m, find the dielectric constant and polarizability of argon.

Solution:

The number of atoms per meter cube, $N = N_A P/RT$

$$= \frac{6.023 \times 10^{23} \text{atoms/mole} \times 1 \text{ atm}}{(0.000082 \text{ atm.m}^3/\text{mole.°K})273°K} = 2.7 \times 10^{25} \text{ atoms/m}^3$$

Dipole moment of the atom $p = Zqd$
    The polarization $P = Np = 18$ electrons /atom $\times 2.7 \times 10^{25}$ atoms/m^3 $\times 1.6 \times 10^{-19} \times 0.5 \times 10^{-18}$
$$= 3.89 \times 10^{-11} \text{ C.m}^{-2}$$
    The dielectric constant is $K = [P/(\varepsilon_o E)] + 1 = 1.0004$
    The polarizability $\alpha_e = \varepsilon_o(K-1)/N = P/NE = 1.4 \times 10^{-40}$ F.m^2

b) Ionic Polarization

When an electric field is applied to an ionic solid, the displacement of cation and anion occurs in relation to one another, which results in a net dipole moment. The dipole moment arising from relative displacement d_i of each ionic pair can be expressed as,

$$p_i = Z_i qd, \qquad (20.22)$$

where Z_iq is the charge of each ion. This gives rise to ionic polarization P_i. Both ionic polarizability and electronic polarizability are insensitive to the variation of temperature.

c) Orientation Polarization

In certain covalent molecules like CH_3Cl, the centres of negative and positive charges do not coincide. The chlorine atom pulls the bonding electron more strongly than hydrogen atom since chlorine is more electronegative than hydrogen. Such molecules possess dipole moment even in the absence of electric field. These molecules are called polar molecules. When a substance contains polar molecules, the dipole moments of these molecules tend to align along the direction of the applied field. The change of orientation of dipole moments in the presence of applied electric field causes polarization, which is called orientation polarization (P_o). The alignment of these permanent dipole moments gets disturbed as the temperature of the material increases. Thus, the orientation polarization decreases with the increase in thermal vibration of dipoles. Orientation polarization can be expressed as

$$P_o = N (p_p)^2 E/(3kT) \tag{20.23}$$

where p_p is the value of permanent dipole moment, k is Boltzmann constant, E is the applied electric field at temperature $T°K$, and N is the number of dipole moments per unit volume.

d) Space Charge Polarization

The microstructures in materials have been discussed in Chapter 8. Sometimes these microstructures contain phases, which are electrically conductive. Such phases are at times found embedded in a dielectric medium. The space charge polarization occurs in such dielectric medium in the presence of applied electric field. The electrons in the conductive phase move towards the positive end of the electric field. The net polarization effect depends on the size of conducting particles and the volume fraction of the conductive phase. A fine dispersion of conducting or semiconducting phases within glass can produce a high dielectric constant, due to small particle to particle distance and space charge polarization (P_s).

Thus, the total polarization of a material is the sum of the contributions from the various sources described above.

$$P = P_e + P_i + P_o + P_s \tag{20.24}$$

We can therefore define the total polarizability of the dielectric medium as,

$$\alpha = \alpha_e + \alpha_i + \alpha_o + \alpha_s \tag{20.25}$$

In certain media, spontaneous polarization occurs due to the presence of the local field of the neighbouring dipoles, even when there is no external electric field. This type of polarization process is observed in ferroelectric materials.

Example 20.4

The electronic part of the dielectric constant K_e is related to the optical behaviour of the dielectric medium and is expressed as $K_e = n^2$, where n is the refractive index of the medium. The refractive index of Al_2O_3 is 1.76 and its dielectric constant is 9 at 10^6 Hz. It has NaCl structure and its density is 4000 kg/m³. Calculate the ionic polarization and electronic polarization when electric field 10^4 V/m is applied. Also calculate ionic polarizability.

Solution:

$$\text{Mass of } Al_2O_3 = 101.96 \times 1.66 \times 10^{-27} \text{ kg} = 1.69 \times 10^{-25} \text{ kg}$$
$$\text{Ionic density } N = 5 \times 4000/1.69 \times 10^{-25} \text{ ions./m}^3 = 1.18 \times 10^{29} \text{ ions/m}^3$$
$$P_e = [n^2 - 1]\varepsilon_0 E = [1.76^2 - 1]\varepsilon_0 E = 2.1\varepsilon_0 E = 1.86 \times 10^{-7} \text{C-m/m}^3$$
$$P_i = [9.0 - 1]\varepsilon_0 E - 2.1\varepsilon_0 E = 5.9\varepsilon_0 E = 5.22 \times 10^{-7} \text{C-m/m}^3,$$

assuming orientation and space charge polarization to be negligible
$$\alpha_i = P_i / NE = 5.9 \times 8.85 \times 10^{-12} E /(1.18 \times 10^{29} E)$$
$$= 4.43 \times 10^{-40} \text{ F.m}^2$$

20.4 FREQUENCY DEPENDENT POLARIZABILITY

Dielectric materials are often used in ac circuits. They are used in wave guides since it has very little ohmic power dissipation. It is therefore necessary to study the effect of ac field on dielectric polarization. When an alternating electric field is applied to a dielectric medium there is inertia to charge movement. As the direction of the alternating field changes with time, the dipoles try to reorient with the field but require some finite time to do the same. When the reorientation of the dipoles is difficult, dipole friction occurs. This causes energy loss. At certain frequencies, the energy loss becomes very high since the dipole cannot properly reorient to follow the field. Thus, the dielectric constant and the polarization are frequency dependent. At frequencies greater than 10^{16} Hz, no dipole can follow the field, so no polarization is possible. The electronic polarization process rapidly follows alternating fields in the visible part of the spectrum i.e. at frequencies less than 10^{16} Hz. Ionic polarization follows an applied high frequency field and contributes to the dielectric constant at frequencies ranging from ultraviolet to infrared region of the spectrum i.e. at frequencies less than 10^{13} Hz. Dipole reorientation does not occur in ceramics without destroying the crystal structure. It is not possible for an electric field to invert the existing dipoles in certain ceramic materials, which give rise to piezoelectric effect. Orientation and space charge polarization contribute to the polarizability at low frequency field since entire atoms or groups of atoms must be rearranged. Maximum polarization occurs at low frequency fields, where all types of polarizations are possible.

The frequency dependence of polarization is also influenced by the structure of the material. Gases polarize at higher frequencies than liquids; gases and liquids polarize at higher frequencies than solids. Amorphous dielectric polarizes at higher frequencies than crystalline ones.

During charging of a capacitor the rate of change of polarization of the dielectric is given by

$$d(P_t - P_o)/dt = (1/\tau)[(P_s - P_o) - (P_t - P_o)] \qquad (20.26)$$

Hence
$$P_t - P_o = (P_s - P_o)(1 - e^{-t/\tau}) \qquad (20.27)$$

where P_t is the polarization at time t, P_o is the instantaneous polarization on applying the field and P_s is the final value of polarization. τ is the time constant, which is also called the relaxation time. It is a measure of the time lag of the system. The response of dielectric material to an applied field is not well represented by a single relaxation time, since the various types of polarizations have different time lags. One can plot variation of polarizability with frequency of alternating applied field, Fig.20.3. This explains the response of various types of polarization to the varying electric field.

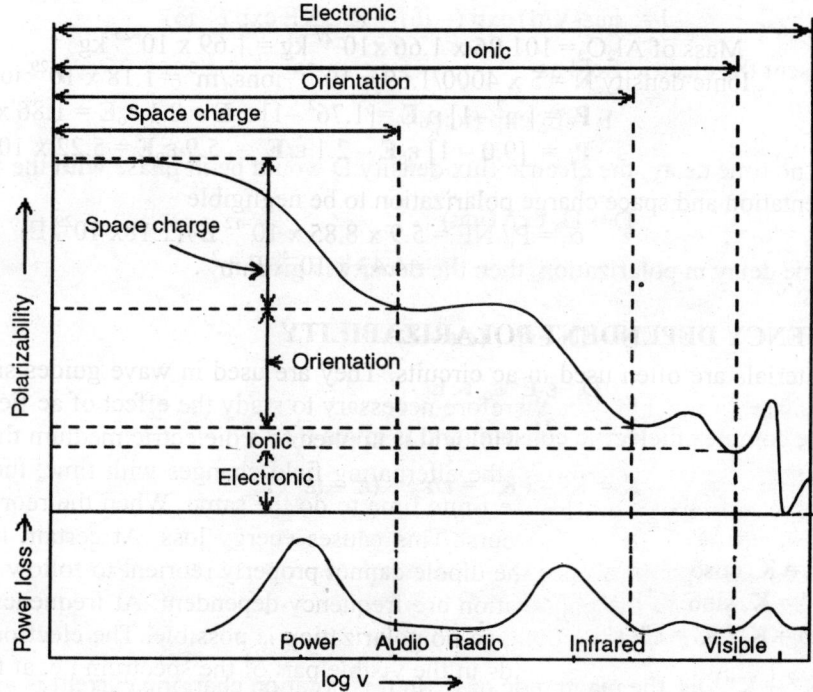

Fig. 20.3 Variation of polarizability with frequency

Sinusoidal voltage

When a sinusoidal voltage

$$V = V_o \exp(i\omega t) \tag{20.28}$$

is applied across the capacitor C with a plate separation l, a charging current I develops,

$$I = dQ/dt = d(CV)/dt = C\ dV/dt = i\omega CV = \omega CV_o \exp\ i(\omega t + \pi/2) \tag{20.29}$$

which is ahead in phase by $\pi/2$ in relation to the applied voltage. When the period of the applied voltage is much greater than the relaxation time of a polarization process, the polarization is complete during each cycle of alternating voltage so there is no loss of electrical energy during charging. If the period of the applied voltage is much less than the relaxation time, then no polarization will occur. When the period is of the same range as that of the relaxation time, resonance occurs. In this case, current leads the voltage by $(90-\delta)$, δ being the loss angle. Then the current I is expressed as

$$I = \omega CV_o \exp\ i(\omega t + \pi/2 - \delta) = i\omega CV \exp\ (-i\delta) \tag{20.30}$$

The energy loss due to resonance is proportional to $\tan \delta$. Now, the current I is not 90° phase ahead in relation to V. The component of the current parallel to voltage produces I^2R loss and component perpendicular to voltage is the charging current of the ideal capacitor. The current density J is expressed as

$$J = i\omega CV \exp\ (-i\delta)\ /A \tag{20.31}$$

In case of parallel plate capacitor with electric field **E** across the plate,

$$J = i\omega\varepsilon(V/d)\exp(-i\delta) = i\omega\varepsilon_o KE \exp(-i\delta) \qquad (20.32)$$

One can represent the electric field as

$$E = E_o \exp(i\omega t) \qquad (20.33)$$

When there is no time delay, the electric flux density D would be in phase with the field i.e.

$$D = D_o \exp(i\omega t) \qquad (20.34)$$

If there is a time delay in polarization, then the density is given by

$$D = D_o \exp i(\omega t - \delta) \qquad (20.35)$$

since $$D = \varepsilon E = \varepsilon_o KE$$

where K is the complex dielectric constant and is given by

$$K = K' - i K'' = \varepsilon/\varepsilon_o = (\varepsilon' - i\varepsilon'')/\varepsilon_o \qquad (20.36)$$

where

$$K' = K_s \cos\delta$$
$$K'' = K_s \sin\delta$$
$$\tan\delta = K''/K' = \varepsilon''/\varepsilon'$$

and $K_s = \sqrt{(K'^2 + K''^2)}$ is the magnitude of K. In this notation charging current is given by

$$J_c = i\omega \varepsilon' E \qquad (20.37)$$

and loss current is

$$J_e = \omega \varepsilon'' E = \sigma E \qquad (20.38)$$

where σ is the conductivity of the dielectric medium.

Effect of temperature on polarization

When a material possesses permanent dipole moment, the orientation polarization occurs in the presence of electric field. As the temperature of the material is increased, the thermal energy randomizes the orientations of dipoles and orientation polarization decreases. So the dielectric constant decreases with temperature. The other effects of rise in temperature are the following:

(i) Thermal energy accelerates the diffusion of ions in space charge polarization.
(ii) Large polar molecules e.g. nitrobenzene in solid material may not easily orient in the presence of electric field. If the material is softened or melted, the orientation polarization is observed.

Hence in such materials dielectric constant increases with temperature.

20.5 DIELECTRIC STRENGTH

We know that the dielectric materials are quite commonly used as insulators. Dielectric strength of material is the measure of the maximum field strength, which it can withstand without electrical breakdown. It is commonly defined as the breakdown voltage and is measured in volts per meter. Dielectric strength and dissipation factor for certain dielectric materials are given in Table 20.3.

Table 20.3 Dielectric strength and dissipation factor
for selected dielectric materials

Material	Dielectric strength in $V.m^{-1} \times 10$	$\tan\delta$ at 10^6 Hz
Al_2O_3	0.6	0.001
$BaTiO_3$	1.2	-
Epoxy	1.8	-
Fused silica	1.0	0.0004
Mica	4.0	-
Nylon(polyamide)	2.0	0.04
Paraffin wax	1.0	-
Phenolic	1.2	0.05
Polyethylene	2.0	0.0001
Polystyrene	2.0	0.0002
PVC	4.0	0.05
Rubber	2.4	-
Sodalime glass	1.0	0.009
Steatite	-	0.003
Teflon (PTFE)	2.0	0.00007
TiO_2	0.8	0.0002
Zircon	-	0.001

The charge carriers arising due to electronic or ionic imperfections contribute to the dc conductivity in the presence of low electric field. When the electric field strength increases, dc conductivity also increases. When the electric field strength is very high, field emissions from the electrodes result in large number of electrons to contribute to a burst of current which produces jagged holes, breakdown channel etc., thus rendering the dielectric useless. In this case, the valence electrons are not only excited to the conduction band but also acquire very high velocity due to high accelerating electric force. These electrons excite more electrons; thus an avalanche of conduction electrons starts flowing which causes physical breakdown. Impurities create energy levels in the energy gap and facilitate the excitation of electrons in the conduction band. Thermal breakdown can also occur in dielectric due to the excessive rise in temperature. The energy loss of dielectric is usually through dissipation in the form of heat. Sometimes heat may be generated in excessive quantity so that dielectric material either melts or disintegrates e.g. bakelite. The breakdown of dielectric may also occur due to defective surface, which contains cracks and pores. Presence of moisture in the atmosphere can also cause surface breakdown.

Dielectric loss

We have already discussed about the lag in polarization that results in the dielectric displacement vector **D** to lag behind the applied alternating field **E**. Thus if D is plotted against E, a hysteresis loop is obtained. The maximum E and maximum D are not coincident. A coercive field has to be applied to make the displacement vector i.e. polarization zero. The energy loss per cycle, W is the area of the hysteresis loop and is given by the expression

$$W = \int_0^{2\pi} D \, dE \qquad (20.39)$$

From Eqs. (20.30) and (20.35),

$$W = \int_0^{2\pi} D_o \, Re[\exp i(\omega t - \delta)] \, E_o \, Re[i \, \exp(i\omega t)] d(\omega t)$$

$$W = D_o E_o \int_0^{2\pi} Re[\exp i(\omega t - \delta)] \, Re[i \exp(i\omega t)] \, d(\omega t)$$

$$= - D_o E_o \int_0^{2\pi} \cos(\omega t - \delta) \sin(\omega t) \, d(\omega t)$$

$$= - \pi E_o D_o \sin\delta = - \pi E_o^2 \epsilon_o K \sin\delta \qquad (20.40)$$

Since T is the time period, the rate of energy loss i.e. power loss is given by

$$P = (\pi/T) E_o^2 \epsilon_o K \sin \delta \ = (\omega/2) E_o^2 \epsilon_o K \sin \delta \approx (\omega E_o^2 / 2) \, \epsilon_o K \tan\delta. \qquad (20.41)$$

If δ is small $\tan \delta$ is also called dissipation factor

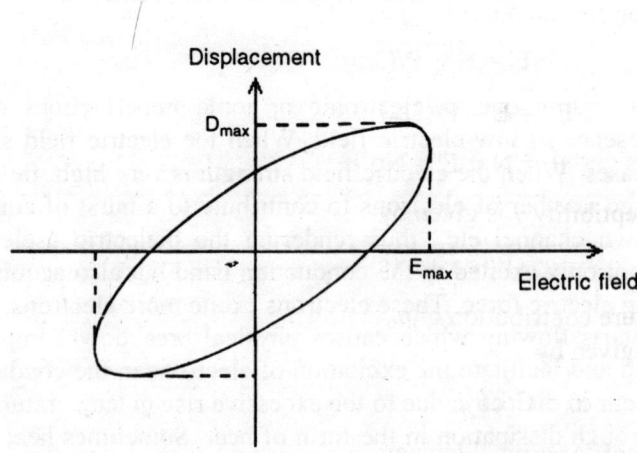

Fig. 20.4 E-D loop in the presence of dielectric loss.

Example 20.5

Calculate the power loss if a fused silica is placed in an electric field 50kV/m and frequency 10^6 Hz.

Solution:

$$\text{Power loss} = (\omega \, \epsilon_o E_o^2 \, K \tan \delta)/2$$

Using Tables 20.1 and 20.3 we get

$$\text{Power loss} = (10^6/s) \, (8.85 \times 10^{-12}) \times (50 \times 10^3)^2 \times 3.8 \times .0004/2 \ = 16.81 \ \text{W/m}^3.$$

Example 20.6

The value of loss factor is 0.01 for soda lime glass at 20°C in electric field of 2 kV/m. What should be the maximum frequency of the applied electric field if the power loss is not to exceed 10^3 W/m^3.

Solution:

$$10^3 = \omega (2 \times 10^3)^2 \times (8.854 \times 10^{-12})(6.9 \times 0.01)/2$$
$$\omega = 10^3/(4 \times 10^6 \times 8.854 \times 10^{-12} \times 0.0345) = 8.18 \times 10^8 \text{ rad/s}$$
$$\nu = 1.3 \times 10^8/\text{s}$$

Ferroelectricity

There are dielectric materials that exhibit spontaneous polarization in the absence of electric field. These materials are ferroelectrics. The ferroelectric materials have permanent dipoles. These electric dipoles are all aligned in the same direction even in the absence of electric field. The most common example of ferroelectric materials is $BaTiO_3$. The source of ferroelectricity is the local field E', which increases with the increase of polarization and the polarization further increases the local electric field; thus at a low temperature when thermal agitation is less, nearly all the electric dipoles line up in parallel arrays. As a result, the polarization becomes very large. The expression for polarization is given by

$$P = (\varepsilon_r - 1) \varepsilon_0 E = N\alpha E' \tag{20.42}$$

Using the expression for local field called as Lorentz field,

$$E' = E + P/(3\varepsilon_0)$$

$$E = E' - P/(3\varepsilon_0) = E' (1 - N\alpha/(3\varepsilon_0)) \tag{20.43}$$

But

$$P = N \alpha E' = N\alpha E/(1 - N\alpha/(3\varepsilon_0)) \tag{20.44}$$

The electrical susceptibility χ is given by

$$\chi = P/(\varepsilon_0 E) = (\varepsilon_r - 1) = N \alpha/ (\varepsilon_0 - N\alpha/3) \tag{20.45}$$

At low temperature contribution of α_0 is much greater than that of α_e and α_i so that $\alpha \approx \alpha_0$. The expression of α_0 is given by

$$\alpha_0 = C/kT , \tag{20.46}$$

where C is a constant. At critical temperature $T = T_c$, $N\alpha/(3\varepsilon_0) \to 1$, P as well as χ tend to infinity.

$$N\alpha/3\varepsilon_0 = N(C/kT_c)/(3\varepsilon_0) = 1$$

i.e.

$$T_c = N C/(3\varepsilon_0 k) = NT\alpha/(3\varepsilon_0)$$

Hence

$$T_c/T = N\alpha_0/(3\varepsilon_0) \tag{20.47}$$

So,

$$\chi = 3(T_c/T)/ (1 - (T_c/T)) = 3T_c/(T - T_c) \tag{20.48}$$

Thus, the susceptibility is inversely proportional to $(T - T_c)$ which is called the Curie-Weiss law. The law explains the experimental measurement above the critical temperature T_c, which is also called the Curie temperature. This law cannot explain the behaviour of a ferroelectric material near or below the Curie temperature.

Barium titanate has cubic symmetrical perovskite structure at a temperature above $T_c(120°C)$. Ba^{2+} ions occupy corner positions; O^{2-} ions face center and Ti^{4+} body center positions, Fig. 20.5. Although each unit cell has a dipole moment, this structure does not have any permanent dipole moment. At a temperature below T_c, there is less thermal agitation and greater density and the crystal structure changes from cubic to tetragonal. As a result there is distortion of structure, Fig. 20.5.

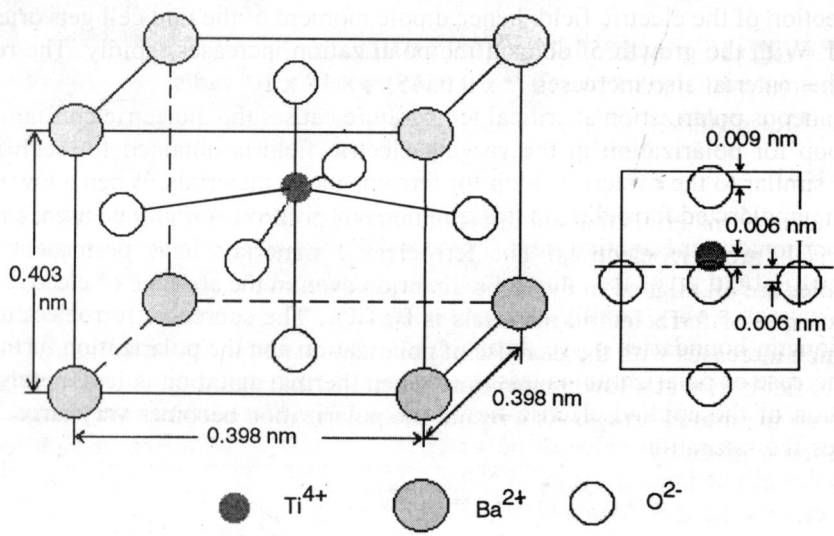

Fig. 20.5 Structure of BaTiO$_3$

Example 20.7

Calculate the dipole moments arising due to displacements of Ba^{2+}, Ti^{4+} and O^{2-} ions relative to the midplane of oxygen ions. Hence find out the total dipole moment of BaTiO$_3$ and polarization. Find the charge stored per unit area of the crystal.

Solution:

Referring to Fig. 20.5 we find that the dipole moments with respect to the plane at c/2:

$$p_{Ba2+} = 4(1/8)(2 \times 1.6 \times 10^{-19})(c/2) + 4(1/8)(2 \times 1.6 \times 10^{-19}) \times (-c/2) = 0$$

$$p_{Ti4+} = 4 \times 1.6 \times 10^{-19} \times 6 \times 10^{-12} = 3.84 \times 10^{-30} \text{ C.m/ion}$$

$$p_{O2-} \text{ (from 4 side face centers)} = (4/2) \times (-2 \times 1.6 \times 10^{-19}) \times (-6 \times 10^{-12}\text{m}) = 3.84 \times 10^{-30} \text{ C.m/ion}$$

$$\begin{aligned} p_{O2-} \text{ (from top and bottom face centers)} &= (1/2)(-2 \times 1.6 \times 10^{-19}) \times (c/2 - 9 \times 10^{-12}\text{m}) \\ &+ (1/2)(-2 \times 1.6 \times 10^{-19}) \times (-c/2 - 9 \times 10^{-12}\text{m}) \\ &= 2.88 \times 10^{-30} \text{ C.m/ion} \end{aligned}$$

Total dipole moment = $(3.84 + 3.84 + 2.88) \times 10^{-30}$ C.m /unit cell = 1.056×10^{-29} C.m /unit cell

Polarization = $(1.056 \times 10^{-29}\text{C.m /cell})/(3.98 \times 3.98 \times 4.03) \times 10^{-30}$ m^3)
= 0.165 C.m^{-2}

The charge per m^2 = 0.165 C. The corresponding electric field $\mathbf{E} = \mathbf{P}/\{\varepsilon_0(\varepsilon_r - 1)\} = 4.36 \times 10^6$ V/m where $\varepsilon_r = 4300$. This electric field is quite high. However, such high electric field is never experienced by touching BaTiO$_3$ crystal. This high electrostatic field gives rise to high electrostatic energy that align electric dipoles. The reason for not experiencing high electric field is that the whole crystal is decomposed into an array of grain-like domains. In each domain the electric dipoles are aligned in a particular direction and there is resultant dipole moment of each domain. Since the domains are randomly oriented, the net polarization of the material is zero.

If an external electric field is applied, the domains, which are already oriented parallel to the electric field, grow due to the fact that the other dipoles also try to orient parallel to the electric field. In barium titanate, titanium ions move towards negative terminal and oxygen ions, in the opposite direction of the electric field; hence dipole moment of the unit cell gets oriented along the applied field. With the growth of domain the polarization increases rapidly. The resultant dipole moment of the material also increases.

The spontaneous polarization at critical temperature causes the dielectric constant to increase. A hysteresis loop for polarization in the varying electric field is obtained for such materials, Fig. 20.6. This is similar to the hysteresis loop for ferromagnetic materials. When a low electric field is applied to an unpolarized ferroelectric material, the polarization is set up which is reversible and is directly proportional to the applied field. The slope of the polarization versus electric field curve at a specific electric field gives the value of dielectric constant. As the electric field is increased, the polarization increases very rapidly as a result of aligning of ferroelectric domains. During this alignment, domain boundaries move through the crystal. At a very high field, it is observed that the rate of increase of polarization is less and the polarization saturates when all the domains align in the direction of the applied electric field. The extrapolation of this curve back to the E = 0 ordinate gives the saturation value of polarization which is equal to spontaneous polarization, P_s with all dipoles aligned parallely.

When the electric field is decreased, the polarization is also reduced at a much slower rate than that of increase of polarization. As the electric field is reduced to zero, the polarization remains at a finite value called remnant polarization P_r. This happens due to the fact that the aligned domains cannot go back to their earlier random orientations, since this process also requires energy. The reversed electric field is applied to reduce the polarization. As the reversed electric field is increased, the polarization gradually reduces to zero. The electric field at which polarization becomes zero is called coercive field, E_c.

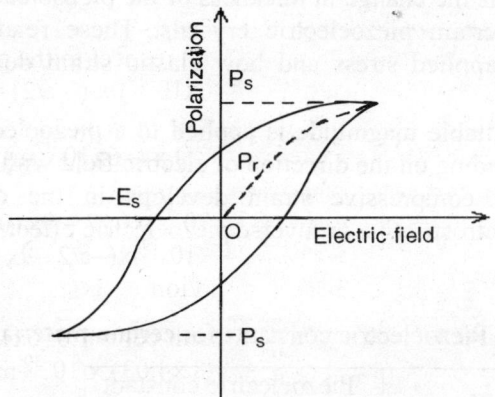

Fig. 20.6 Ferroelectric hysteresis

At low temperatures the hysteresis loops become fatter, and the coercive field becomes greater, Since larger amount of energy is required to reorient the domains. At higher temperatures, the thermal energy is utilized to randomize the domain orientations and hence the less coercive force is required to have zero polarization. At temperature greater than a specific value (T_c), no hysteresis remains and there is only a single value for the dielectric constant.

20.6 POLARIZATION RELAXATION

The residual polarization decays when the domains start growing in the direction of the applied reverse electric field. The rate of decay of polarization dP/dt is proportional to the remnant polarization P at a particular time.

$$- dP/dt \propto P$$

or
$$P = P_r \exp (- t/ \tau_p) \qquad (20.49)$$

where P_r is the amount of remnant polarization at time t = 0 and τ_p is the relaxation times i.e., the time during which the polarization decreases by a factor of 1/e. The more the relaxation time better will be the material to be used for computer memory.

20.7 PIEZOELECTRICITY

Piezoelectricity or literally pressure electricity is the generation of voltage in a material when it is subjected to a compressive stress. Such materials are called piezoelectric materials. Ferroelectric crystals also show piezoelectricity. When a compressional stress σ is applied, there is change in the state of electric polarization, ΔP. This causes the flow of electrons through an external circuit. The tensile stress causes flow of electrons in the opposite direction. Vibrational stress causes flip-flop of polarization.

The change in polarization generates an electric field ΔE, which in turn produces strain in the crystal. The piezoelectric equations, in one dimensional notation, are given by

$$\Delta P = \varepsilon_o (K-1) \Delta E \qquad (20.50)$$

$\Delta E = C_1 \sigma$ and strain $\varepsilon = C_2 \Delta E$, where C_1, C_2 are constants of medium. In case of tensile strain

$$\Delta P/P = \Delta l/l = \sigma/Y \qquad (20.51)$$

where l is the thickness and Δl is the change in thickness of the piezoelectric material. The Table 20.4 gives the constants of certain piezoelectric crystals. These relations explain how the polarization changes with the applied stress and how elastic strain depends on electric field generated due to ΔP.

When the electric field of suitable magnitude is applied to a piezoelectric crystal, the dipole spacing extends or shrinks depending on the direction of electric field with respect to the direction of polarization i.e. tensile or compressive strain develops in the crystal. Therefore, the piezoelectric crystal exhibits electrostrictive or inverse piezoelectric effect, Fig. 12.5 (c).

Table 20.4 Piezoelectric constants for certain materials

Material	Piezoelectric constant $C_2 \, mV^{-1} \, (\times 10^{-12})$
Quartz	2.3
$BaTiO_3$	100
$PbZrTiO_3$	150

Example 20.8

When a compressional stress is applied in the polarity direction of a slab of thickness 1 mm and cross-section 3 mm x 3 mm. If the polarization of the dielectric medium is 0.04 C/m^2 and strain produced is 0.008, find the number of electron flowing after the ends are connected.

Solution:

Change in polarization $\Delta P = (\Delta l/l) P = 0.008 \times 0.04$ $C/m^2 = 0.00032$ C/m^2

$$\text{Number of electrons} = \frac{3.2 \times 10^{-4} \times (3 \times 10^{-3})^2}{1.6 \times 10^{-19}} = 1.8 \times 10^{10}$$

Example 20.9

A piezoelectric capacitor of cross-sectional area 2mm x 2mm and thickness 1 mm is subjected to a compressional stress such that it produces an electric field of 5000 V/mm that will enable a spark to be generated in air. If the initial polarization P of dielectric medium is 40 C/m^2, K=75 and the Young's modulus Y = 74 GPa. Calculate the force to be applied on the surface.

Solution:

The change in polarization, P due to compressional stress
$$= 8.85 \times 10^{-12} \times (75-1) \times 5 \times 10^6 \ C/m^2 = 3.27 \times 10^{-3} \ Cm^{-2}$$
$$\text{Strain} = \Delta P/P = 3.27 \times 10^{-3}/40 = 8.18 \times 10^{-5},$$
$$\text{Stress} = Y \times 8.18 \times 10^{-5} = 6.05 \times 10^6 \ N.m^2$$
Force to be applied on the surface = $6.05 \times 10^6 \times 4 \times 10^{-6}$ N = 24.2N

Certain materials are piezoelectric but not ferroelectric e.g. quartz whereas $BaTiO_3$ is both ferroelectric and piezoelectric.

Piezoelectric materials are used in transducers and actuators. Transducers convert any mechanical stress, pressure, light, heat etc. into electrical signal. Actuators provide displacement of electromechanical devices for example motors. Phonograph cartridges and actuators translate specimens in the scanning tunneling microscope. In sonar detectors both direct and inverse piezoelectric effects are used. An ac electric power is feed into a piezoelectric transmitter, which generates a mechanical vibration that in turn causes the propagation of sound waves into water. When these sound waves hit a remote submarine and reflect back, they are detected by an array of hydrophones. The transducers in these hydrophones convert the pressure wave of echo into electrical signals. Another important application of piezoelectric effect is to use ultrasound to detect unwanted growth of fetus in the womb. Other applications of piezoelectric materials are in microphones, strain gauges, ultrasonic generators etc. The characterization of these materials having complex structure with a low degree of symmetry is difficult. The piezoelectric materials are sometimes heated above Curie temperature and cooled to room temperature in the presence of high electric field to improve their piezoelectric responses.

EXERCISES

20.1 Find the polarizability and dielectric constant of argon gas, as a result of electronic polarization at 760 mm Hg and 0°C. The estimated radius of argon atom is 1.09×10^{-10}m. Calculate the displacement of electrons with respect to nucleus when an electric field of 10^4 V/m is applied.
Ans: 1.44×10^{-40} F.m², 1.0004, 5×10^{-19} m

20.2 When the pressure is applied in the polarity direction of a 1 cm × 1 cm × 1 cm slab of a piezo-electric material of polarization 0.08 Cm^{-2}, a strain of 0.008 is produced. How many electrons will flow between the ends if they are joiuned?
Ans: 4×10^{11}

20.3 Two parallel plates with a material of dielectric constant 8 form a capacitor. The plates are separated by 6 x 10 ^{3}m and each has an area of 12.5 x 10 ^{4}m^2. A potential of 110V is applied across the plates. Find the following:
a) Capacitance
b) Magnitude of charge on each plate
c) Dielectric displacement D
d) Polarization
Ans: (a) 1.475 x 10 ^{11} F; (b) 1.62 x 10 ^{9} C; (c) 1.298 x 10 ^{6} C/m^2; (d) 1.136 x 10 ^{6} C/m

20.4 Qualitatively explain the reason for increase in charge storing capacity when a dielectric material is inserted within the plates of a parallel plate capacitor.

20.5 The polarization of a dielectric material within a parallel plate capacitor is 5 x 10 ^{6} C/m^2. Calculate the dielectric constant and dielectric displacement if an electric field of 5 × 10^{4}V/ m is applied.
Ans: 12.3, 5 x 10 ^{6} C/m^2

20.6 A parallel plate capacitor is to be made so as to store 10 ^{5} C at a potential of 2000 V. The separation between the plates is 0.5 mm. Calculate the area of the plates if there is
a) Vacuum ($\varepsilon_r = 1$)
b) Alumina ($\varepsilon_r = 9$)
as the dielectric between the plates?
Ans: 0.28 m^2, 0.0314 m^2

20.7 Xenon has a polarization of 10 ^{10} C/m^2 at 27°C in a voltage gradient of 10^4 volts/m. Calculate the pressure. Polarizability of xenon is 3.52 x 10 ^{40} F.m^2.
Ans: 1.16 atm

20.8 A piezoelectric crystal has a Young's modulus of 6 x 10^8 psi. What stress must be applied to change its polarization from 0.2105 to 0.2095 C/m^2? Given that 1 atm = 1.47 psi.
Ans: 2.85 x 10^6 psi

20.9 The electronic polarizability of He is 0.18 x 10 ^{40} F/m^2. Calculate its dielectric constant at 273°K and at 1 atm pressure.
Ans: 1.000055

20.10 The dielectric constant increases when ice melts into water whereas it decreases in the case of melting of HCl. Explain.

APPENDICES

Appendix I

PROPERTIES OF ELEMENTS

Element	Symbol	Atomic Number	Atomic Weight amu	Density $10^3 Kg.m^{-3}$	Molar Volume $10^{-6} m^3$	Thermal Expansion $10^{-6} k^{-1}$	Young's Modulus $GN\ m^{-2}$	Melting Point °C
Actinium	Ac	89	227	--	--	14.9	35	1050
Aluminium	Al	13	26.982	2.70	9.99	23.1	71	660.2
Americium	Am	95	243	11.7	20.8	--	--	--
Antimony	Sb	51	121.750	6.62	18.4	10.9	55	630
Argon	Ar	18	39.948	--	--	--	--	-189
Arsenic	As	33	74.922	5.72	13.09	4.28	39	817
Astatine	At	85	210	--	--	--	--	302
Barium	Ba	56	137.340	3.5	39.0	18.8	12.7	714
Berkelium	Bk	97	249	--	--	--	--	--
Beryllium	Be	4	9.012	1.85	4.90	11.5	289	1277
Bismuth	Bi	83	208.980	9.80	21.3	13.41	34	271
Boron	B	5	10.811	2.34	4.62	8.3	440	2030
Bromine	Br	35	79.909	3.12	25.6	--	--	-7
Cadmium	Cd	48	112.400	8.65	13.0	30.6	62	321
Calcium	Ca	20	40.080	1.55	25.86	22.4	19.5	838
Californium	Cf	98	251	--	--	--	--	--
Carbon (gr)	C	6	12.011	2.25	5.33	3.8	8.3	3550
Cerium	Ce	58	140.13	6.77	17.03	8.5	30	804
Cesium	Cs	55	132.905	1.90	69.84	97	1.75	28
Chlorine	Cl	17	35.453	--	--	--	--	-101
Chromium	Cr	24	51.996	7.19	7.23	8.4	243	1875
Cobalt	Co	27	58.993	8.85	6.66	12.4	206	1495
Copper	Cu	29	63.540	8.96	7.09	16.7	124	1083
Curium	Cm	96	247	--	--	--	--	--
Dysprosium	Dy	66	162.50	8.55	19.01	10.0	63	1407
Einsteinium	Es	99	254	--	--	--	--	--
Erbium	Er	68	167.3	9.15	18.28	12.3	73	1497
Europium	Eu	63	152	5.25	28.98	33.1	15	826
Fermium	Fm	100	255	--	--	--	--	--

Element	Symbol	Atomic Number	Atomic Weight amu	Density $10^3Kg.m^{-3}$	Molar Volume $10^{-6}m^3$	Thermal Expansion $10^{-6}k^{-1}$	Young's Modulus GN m^{-2}	Melting Point °C
Fluorine	F	9	18.998	--	--	--	--	-220
Francium	Fr	87	223	--	--	102	1.7	27
Gadolinium	Gd	64	157.25	7.86	20.01	8.28	56	1312
Gallium	Ga	31	69.720	5.91	11.8	18.1	92.5	30
Germanium	Ge	32	72.590	5.32	13.6	5.75	99	937
Gold	Au	79	196.967	19.32	10.20	14.1	78	1063
Hafnium	Hf	72	178.490	13.09	13.64	6.01	137	2222
Helium	He	2	4.003	--	--	--	--	-270
Holmium	Ho	67	164.93	6.79	24.3	10.7	67	1461
Hydrogen	H	1	1.008	--	--	--	--	-259
Indium	In	49	114.820	7.31	15.71	31.4	10.5	156
Iodine	I	53	126.904	4.94	25.7	--	--	114
Iridium	Ir	77	192.2	22.5	8.54	6.63	528	2454
Iron	Fe	26	55.847	7.87	7.1	11.7	210	1535
Krypton	Kr	36	83.800	--	--	--	--	-157
Lanthanum	La	57	138.92	6.19	22.44	10.4	38	920
Lawrencium	Lw	103	257	--	--	--	--	--
Lead	Pb	82	207.190	11.36	18.27	29.0	15.7	327
Lithium	Li	3	6.940	0.53	12.99	45	11.5	181
Lutetium	Lu	71	174.98	9.85	17.76	8.12	84	1652
Magnesium	Mg	12	24.321	1.74	14.0	25.7	44	650
Manganese	Mn	25	54.938	7.43	7.39	22.6	198	1245
Mendelevium	Md	101	256	--	--	--	--	--
Mercury	Hg	80	200.590	13.55	14.81	61	--	-38
Molybdenum	Mo	42	95.940	10.22	9.39	4.98	328	2610
Neodymium	Nd	60	144.240	7.00	20.61	9.98	38	1019
Neon	Ne	10	20.183	--	--	--	--	-249
Neptunium	Np	93	237	--	--	27.5	100	637
Nickel	Ni	28	58.710	8.90	6.59	12.7	193	1453
Niobium	Nb	41	92.906	8.57	10.8	7.07	105	2468
Nitrogen	N	7	14.007	--	--	--	--	-210
Nobelium	No	102	255	--	--	--	--	--
Osmium	Os	76	190.2	22.57	8.43	4.7	540	2700
Oxygen	O	8	15.999	--	--	--	--	-219
Palladium	Pd	46	106.400	12.02	8.88	11.5	124	1552
Phosphorus	P	15	30.974	1.83	16.92	124	4.6	44
Platinum	Pt	78	195.09	21.45	9.09	8.95	170	1769
Plutonium	Pu	94	242	19.5	12.3	55	96.5	640
Polonium	Po	84	210	--	--	23	25.5	254
Potassium	K	19	39.102	0.86	45.47	83	3.5	64

Element	Symbol	Atomic Number	Atomic weight Amu	Density $10^3 Kg.m^{-3}$	Molar Volume $10^{-6}m^3$	Thermal Expansion $10^{-6}k^{-1}$	Young's Modulus GN m^{-2}	Melting Point °C
Praseodymium	Pr	59	140.92	6.77	20.82	6.79	33	919
Promethium	Pm	61	145	--	--	9.0	42	1027
Protactinium	Pa	91	231	15.4	15.0	7.3	100	1230
Radium	Ra	88	226	5.0	45.00	20.2	16	700
Radon	Rn	86	222	--	--	--	--	-71
Rhenium	Re	75	186.20	21.04	8.85	6.63	460	3180
Rhodium	Rh	45	102.905	12.44	8.27	8.40	372	1966
Rubidium	Rb	37	85.470	1.53	55.87	88.1	2.7	39
Ruthenium	Ru	44	101.070	12.2	8.29	9.36	410	2500
Samarium	Sm	62	150.35	7.49	20.07	10.4	34	1072
Scandium	Sc	21	44.956	2.99	14.89	10.0	79	1539
Selenium	Se	34	78.960	4.79	16.48	36.9	58	217
Silicon	Si	14	28.086	2.33	12.06	3.07	103	1410
Silver	Ag	47	107.870	10.49	10.28	19.2	80.5	961
Sodium	Na	11	22.990	0.97	23.67	70.6	8.9	98
Strontium	Sr	38	87.620	2.60	34.00	20	13.5	768
Sulphur	S	16	32.064	2.07	15.5	64	19.5	119
Tantalum	Ta	73	180.948	16.6	10.9	6.55	181	2996
Technetium	Tc	43	99	--	--	8.06	370	2130
Tellurium	Te	52	127.600	6.24	20.45	16.77	41	450
Terbium	Tb	65	158.92	8.25	19.26	10.3	57.5	1356
Thallium	Tl	81	204.370	11.85	17.25	29.4	8	303
Thorium	Th	90	232.038	11.66	19.90	11.2	74	1750
Thulium	Tm	69	169	9.31	18.15	13.3	75	1545
Tin (gray)	Sn	50	118.690	7.30	16.26	5.3	52	232
Titanium	Ti	22	47.900	4.51	10.63	8.35	106	1668
Tungsten	W	74	183.85	19.3	9.53	4.59	396	3410
Uranium	U	92	238.030	19.07	12.48	12.6	186	1132
Vanadium	V	23	50.942	6.1	8.35	8.3	132	1900
Xenon	Xe	54	131.300	--	--	--	--	-112
Ytterbium	Yb	70	173.04	6.96	24.86	24.96	18	824
Yttrium	Y	39	88.905	4.47	19.89	12.0	65	1509
Zinc	Zn	30	65.376	7.13	9.17	29.7	92	420
Zirconium	Zr	40	91.220	6.49	14.06	5.78	92	1852

Appendix II
PHYSICAL CONSTANTS

Avogadro number	N_A	6.023×10^{23} mole^{-1}
Boltzmann constant	k	1.380×10^{-23} J °K^{-1} = 8.625×10^{-5} eV °K^{-1}
Gas constant	R	8.314 J mole^{-1} °K^{-1}
Planck constant	h	6.626×10^{-34} J s
Electronic charge	e	1.602×10^{-19} C
Electron rest mass	m	9.109×10^{-31} kg
Velocity of light	c	2.998×10^8 ms^{-1}
Bohr magneton (magnetic moment)	μ_B	9.273×10^{-24} A m^2
Permittivity of free space	ε_0	8.854×10^{-12} F m^{-1}
Permeability of free space	μ_0	$4\pi \times 10^{-7}$ H m^{-1} = 1.257×10^{-6} H m^{-1}
Faraday constant	F	96.49 kC mole^{-1} (of electrons)
Atomic mass unit	amu	1.660×10^{-27} kg
Acceleration due to gravity	G	9.81 m s^{-2}
Molar volume (0°C, 760 mm Hg)	V_m	22.4 litres

Appendix III
SELECTED CONVERSIONS

1 gauss/oersted = $4\text{л} \times 10^{-7}$ Henry/m

1 dyne = 10^{-5} N

1 erg = 10^{-7} J

1 J = 10^7 erg = 0.239 cal

1 cal = 4.18 J

1 eV = 1.602×10^{-19} J = 1.602×10^{-12} erg = 0.38×10^{-19} cal

1 eV/atom = 23,100 cal/mole

1 kcal = 4185 J = 4.185×10^{10} ergs

1 ft^3 H$_2$O = 62.4 lb H$_2$O = 8.33 gal H$_2$O

1 oersted = $(10^3/4\text{л})$ A/m = 79.6 A/m

1 gauss = 10^{-4} weber/m^2 = 10^{-4} volt s/m^2

1 psi = 6.89×10^4 dynes/cm^2 = 6.89 kN/m^2

1 lb/cu.in = 27680 kg/m^3

1 atm = 0.101325 MPa

1 bar = 0.1 MPa

1 poise = 0.1 Pa s

ln x = 2.3 log$_{10}$ x

Index